Lecture Notes in Mechanical Engineering

Series Editors

Francisco Cavas-Martínez, Departamento de Estructuras, Universidad Politécnica de Cartagena, Cartagena, Murcia, Spain

Fakher Chaari, National School of Engineers, University of Sfax, Sfax, Tunisia

Francesco Gherardini, Dipartimento di Ingegneria, Università di Modena e Reggio Emilia, Modena, Italy

Mohamed Haddar, National School of Engineers of Sfax (ENIS), Sfax, Tunisia

Vitalii Ivanov, Department of Manufacturing Engineering Machine and Tools, Sumy State University, Sumy, Ukraine

Young W. Kwon, Department of Manufacturing Engineering and Aerospace Engineering, Graduate School of Engineering and Applied Science, Monterey, CA, USA

Justyna Trojanowska, Poznan University of Technology, Poznan, Poland

AF803188

Lecture Notes in Mechanical Engineering (LNME) publishes the latest developments in Mechanical Engineering—quickly, informally and with high quality. Original research reported in proceedings and post-proceedings represents the core of LNME. Volumes published in LNME embrace all aspects, subfields and new challenges of mechanical engineering. Topics in the series include:

- Engineering Design
- Machinery and Machine Elements
- Mechanical Structures and Stress Analysis
- Automotive Engineering
- Engine Technology
- Aerospace Technology and Astronautics
- Nanotechnology and Microengineering
- Control, Robotics, Mechatronics
- MEMS
- Theoretical and Applied Mechanics
- Dynamical Systems, Control
- Fluid Mechanics
- Engineering Thermodynamics, Heat and Mass Transfer
- Manufacturing
- Precision Engineering, Instrumentation, Measurement
- Materials Engineering
- Tribology and Surface Technology

To submit a proposal or request further information, please contact the Springer Editor of your location:

China: Dr. Mengchu Huang at mengchu.huang@springer.com
India: Priya Vyas at priya.vyas@springer.com
Rest of Asia, Australia, New Zealand: Swati Meherishi at swati.meherishi@springer.com
All other countries: Dr. Leontina Di Cecco at Leontina.dicecco@springer.com

To submit a proposal for a monograph, please check our Springer Tracts in Mechanical Engineering at http://www.springer.com/series/11693 or contact Leontina.dicecco@springer.com

Indexed by SCOPUS. The books of the series are submitted for indexing to Web of Science.

More information about this series at http://www.springer.com/series/11236

Jarosław Stryczek · Urszula Warzyńska
Editors

Advances in Hydraulic and Pneumatic Drives and Control 2020

Springer

Editors
Jarosław Stryczek
Wrocław University of Science and Technology, Faculty of Mechanical Engineering
Wrocław, Poland

Urszula Warzyńska
Wrocław University of Science and Technology, Faculty of Mechanical Engineering
Wrocław, Poland

ISSN 2195-4356 ISSN 2195-4364 (electronic)
Lecture Notes in Mechanical Engineering
ISBN 978-3-030-59508-1 ISBN 978-3-030-59509-8 (eBook)
https://doi.org/10.1007/978-3-030-59509-8

© The Editor(s) (if applicable) and The Author(s), under exclusive license to Springer Nature Switzerland AG 2021
This work is subject to copyright. All rights are solely and exclusively licensed by the Publisher, whether the whole or part of the material is concerned, specifically the rights of translation, reprinting, reuse of illustrations, recitation, broadcasting, reproduction on microfilms or in any other physical way, and transmission or information storage and retrieval, electronic adaptation, computer software, or by similar or dissimilar methodology now known or hereafter developed.
The use of general descriptive names, registered names, trademarks, service marks, etc. in this publication does not imply, even in the absence of a specific statement, that such names are exempt from the relevant protective laws and regulations and therefore free for general use.
The publisher, the authors and the editors are safe to assume that the advice and information in this book are believed to be true and accurate at the date of publication. Neither the publisher nor the authors or the editors give a warranty, expressed or implied, with respect to the material contained herein or for any errors or omissions that may have been made. The publisher remains neutral with regard to jurisdictional claims in published maps and institutional affiliations.

This Springer imprint is published by the registered company Springer Nature Switzerland AG
The registered company address is: Gewerbestrasse 11, 6330 Cham, Switzerland

Preface

The purpose of the monograph "Advances in Hydraulic and Pneumatic Drives and Control 2020" was to present state-of-the-art research and development works in the fields of hydraulics and pneumatics carried out in various scientific and research centers. Scientific articles published in this monograph were selected from a large number of submitted papers for "The International Scientific and Technical Conference NSHP 2020" which is organized cyclically in Poland by the Polish Association of Engineers and Technicians SIMP Wrocław and the Faculty of Mechanical Engineering of Wrocław University of Science and Technology. Authors presented in the articles their work carried out in many European countries such as Austria, Czech Republic, Poland, Romania, Serbia, Ukraine, and also in the USA. The works reflect the current development trends prevailing in hydraulics and pneumatics. The academy's and industry's pursuit of scientific, yet technical, work is evident. There is also a tendency to widely use computer tools in both theoretical and experimental works. The monograph shows a number of technical solutions that may interest both specialists in the fields of hydraulics and pneumatics as well as a wider group of mechanical engineers. All articles were reviewed by an international group of reviewers who are well-known and respected specialists.

The editors of the monograph would like to thank the authors for their effort in writing the articles, the reviewers for insightful, substantive, and courteous reviews, and the Springer Publishing House for fruitful cooperation.

Jarosław Stryczek
Urszula Warzyńska

Contents

Hydraulic Components

Research on Flow Forces in the USAB10 Control Valve Using a CFD Method

Grzegorz Filo(✉) and Edward Lisowski

Faculty of Mechanical Engineering, Cracow University of Technology, Jana Pawla II 37, Cracow, Poland
{filo,lisowski}@mech.pk.edu.pl

Abstract. Proportional control valves are characterized by smooth opening of flow channels by means of the spool control. At the initial stage of opening, the flow gap is of a small cross-section which then gradually expands. This enables smoother start-up of a device as well as more precise control of the working motion compared to conventional control valves. Among the electromagnetically controlled proportional control valves, some complex solutions equipped with spool displacement transducers, compensating valves, advanced electronics, etc. can be distinguished. This allows compensation for the harmful effects of flow-related forces.

The article presents the results of flow tests through the USAB10 proportional control valve, designed and manufactured by PONAR Wadowice. The valve can be both used as a single unit or included in the sectional block which supplies multiple actuators. There are many points of fluid flow direction change as well as cross-sectional area change inside the valve block. This leads to the occurrence of forces associated with the flow which are difficult to determine using standard mathematical formulas. Hence, the research process included building of geometrical models, next preparation of discrete models and then carrying out simulation tests using CFD methods. The tests were conducted in Ansys/Fluent environment which allowed velocity distribution, pressure distribution and values of forces acting on the valve spool to be obtained.

Keywords: 3D modelling · Proportional directional control valve · Flow forces · CFD analysis

1 Introduction

Proportional directional spool valves with electromagnetic control allow the machine or device to be started softly and smoothly. This is possible due to the longitudinal slots in the spool which enable throttling of the flow. The subject of modeling and testing the characteristics of this type of valves is taken up in numerous scientific research centers. The aim of the studies is usually to improve valve dynamic characteristics or increase control capabilities. Zhang [1] proposed a novel control valve design that employs two independent spools instead of the traditional single-spool solution. Due to the new structure,

© The Editor(s) (if applicable) and The Author(s), under exclusive license to Springer Nature Switzerland AG 2021
J. Stryczek and U. Warzyńska (Eds.): NSHP 2020, LNME, pp. 3–14, 2021.
https://doi.org/10.1007/978-3-030-59509-8_1

the influence of dead zone and damping on the valve performance was lightened, while the dynamic characteristics and control accuracy was significantly improved. Results of a directional 4/3 control valve analysis carried out by means of a Direct Numerical Simulation (DNS) method instead of a commonly used RANS technique was proposed by Posa [2]. The used method provides important details on the instantaneous structures of the flow, affecting the valve performance. Posa studied the discharge coefficient and flow forces for different valve openings and pressure drops in the 2D domain. The results of simulations showed that the axial force acting on the spool, which has the greatest impact on valve characteristics, is roughly proportional to the pressure drop and to the valve opening. In addition, Posa showed that for small openings and pressure drops the flow is steady and attached to the wall of the outflow chamber, while in the opposite case the flow is strongly unsteady and organized like a free jet. A series of studies on forces acting on the spool of a control valve was also conducted by Amirante. The research concerned, among others, the effect of flow forces and covered various types of valves, including a 4/3 hydraulic open center directional control valve [3] and a direct (single stage) proportional valve [4]. The conducted numerical analysis provided important fluid dynamic indications about the efflux angle values and the flow rate distribution inside the valve, in particular during the first opening phase. Amirante proposed geometrical modification of the valve spool in the form of a cylindrical hole created on the top of the hemi-spherical notch to improve fluid metering at small valve openings. Then he provided experimental validation of the proposed methodology for the design of the spool surfaces [5]. The influence of shape and number of spool notches on the discharge characteristics of a hydraulic distributor metering edge, including discharge coefficient, velocity coefficient and flow angle by means of theoretical and experimental studies was studied by Borghi [6]. Similarly, Simic [7] proposed optimization of both spool and housing geometry in a small hydraulic seat valve to enable the reduction of the axial flow forces to a minimum value, while Ye [8] carried out research aimed at clarification of the groove shape effect on flow characteristics by means of CFD method.

The issue addressed in this article is pressure compensation in the proportional spool control valve without the use of any additional correction valves, sensors nor feedback loops. This work is a continuation of previous research that concerned pressure compensation in a multi-section proportional directional control valve [9]. It is also based on the results of 3D CFD analysis of flow forces acting on the spool of a solenoid-operated directional control valve [10], results of flow force analysis in the initial phase of throttle gap opening of a proportional control valve [11] as well as determination of flow coefficient of a proportional directional valve with integrated spool position controller [12]. In general, compensation of the pressure effect requires modification of the system of forces acting on the valve spool. The sum of forces includes electromagnetic force, spring force, viscous friction force and flow forces. The main purpose of the work was to modify the valve parameters in such a way that the flow characteristics were maximally independent of pressure. This allows the operator to use an excessive pump capacity to supply other actuators that have different pressure. The assumed goal was achieved by modification of the spool geometry. The study was carried out using the CFD method on a three-dimensional model of the valve fluid. During the analyzes, values of flow coefficients, jet angles in the gaps and axial component of flow force acting on the spool were determined. The object of study was USAB10 proportional control valve designed by PONAR Wadowice.

2 Working Principle of the Studied Control Valve

The object of the analysis is USAB10 control valve designed for maximum flow rate $Q = 80\ \mathrm{dm^3/min}$ and pressure $p = 35$ MPa. The valve is controlled by two electromagnets and return springs. Its characteristic feature is the ability to connect in sections if there is a need to supply multiple actuators from a single pump. Geometrical model of the valve is presented in Fig. 1. There is one supply connector P and two return connectors T_1, T_2 on the side wall of the valve block. In the neutral position, the flow through the valve is cut off. Switching on the power supply of one of the solenoids results in opening the flow in the $P-A$ and $B-T$ or $P-B$ and $A-T$ directions.

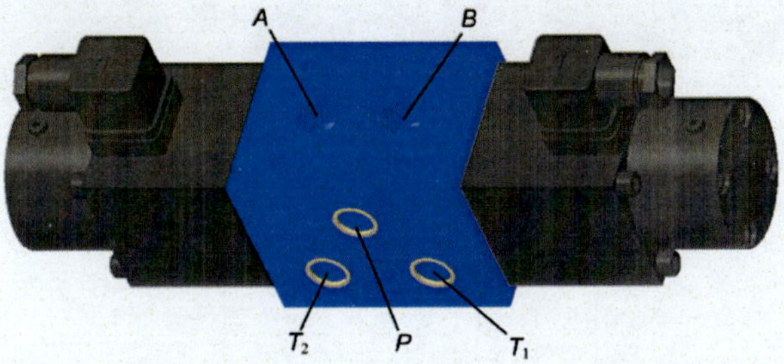

Fig. 1. Control valve geometrical model, P – supply port, A, B – flow channel ports, T_1, T_2 – return ports.

Figure 2 shows longitudinal section of the valve model. The main component is valve body (1) with a spool chamber and transversely hollowed P, A, B, T_1, T_2 flow channels. There is a spool (2), two thrust sleeves (3) and two springs (4) inside the chamber. Two control blocks which are attached on both sides contain solenoids (5) and power connectors (6).

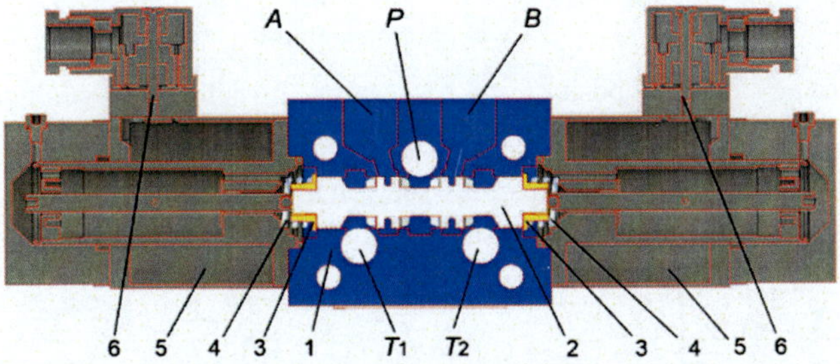

Fig. 2. Longitudinal section of the valve; P, A, B, T_1, T_2 – flow channels, 1 – valve block, 2 – spool, 3 – thrust sleeve, 4 – spring, 5 – electromagnet, 6 – power connector.

The proposed design solution of the spool is shown in Fig. 3. The spool has cutouts in which a flow gap is created by removing material in a direction perpendicular to the longitudinal axis. The cutouts are relatively deep in order to create flow conditions as close as possible to the normal direction of the spool axis. There are four cutouts arranged symmetrically on the circumference of each control edge. The overlap of the spool is $x_{p0} = 0.4$ mm, while the cutout length is 3.6 mm. The cutouts on two control edges have been additionally extended by small arc-shaped undercuts on a 0.2 mm stroke, which provide a small flow before the main gap is opened, while the overlap is reduced to $x_p =$ 0.2 mm. Function of the gap area against the spool position (Fig. 4) can be approximated using the second-degree polynomial (Eq. 1).

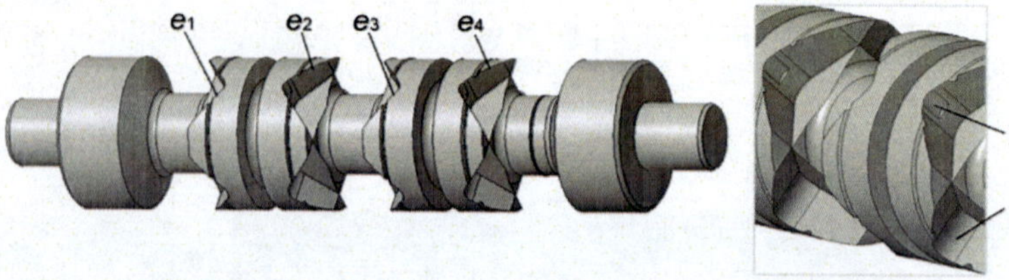

Fig. 3. Valve spool model; e_1, ..., e_4 – control edges, 1 – base cutout, 2 – cutout with additional undercut.

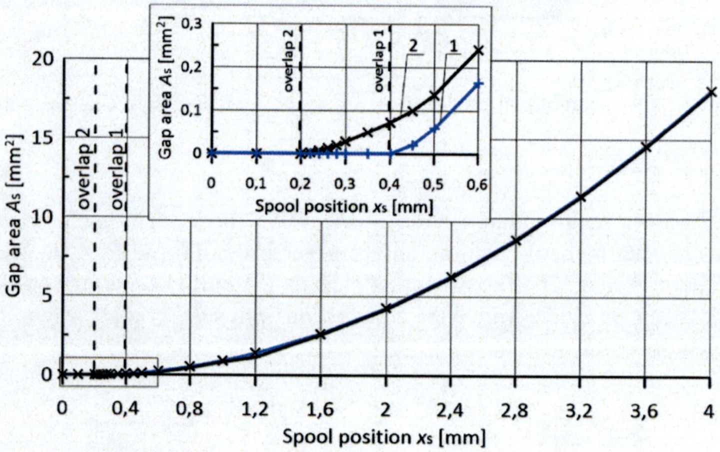

Fig. 4. Gap area against spool position; 1 – base cutout, 2 – cutout with additional undercut.

The approximated function of the cross-sectional gap area against the spool position for the cutout with additional undercut has the following form:

$$A_s(x_s) = \begin{cases} 0, & for\, x_s \le x_p \\ a_1 \cdot (x_s - x_p) + a_2 \cdot (x_s - x_p)^2, & for\, x_s > x_p \end{cases} \tag{1}$$

$$where: a_1 = 0.6868, a_2 = 1.2015, x_p = 0.2.$$

3 Flow Channel Models and CFD Analysis

For the purpose of CFD analysis, a number of geometric models of the fluid flow channels were built to match different spool positions. This was done on the basis of 3D geometric models of the valve body and the spool.

3.1 Models of Flow Channels

Geometric models of the fluid flow paths were generated in the CREO Parametric program for the each considered spool position using the *Boolean* operations. Due to the fact that the *P–A* and *B–T* flow channels have the identical geometry as *P–B* and *A–T*, only *P–A* and *B–T* were selected for analysis (Fig. 5). Since the individual channels are geometrically separable, the analysis of each channel can be carried out independently of the other one.

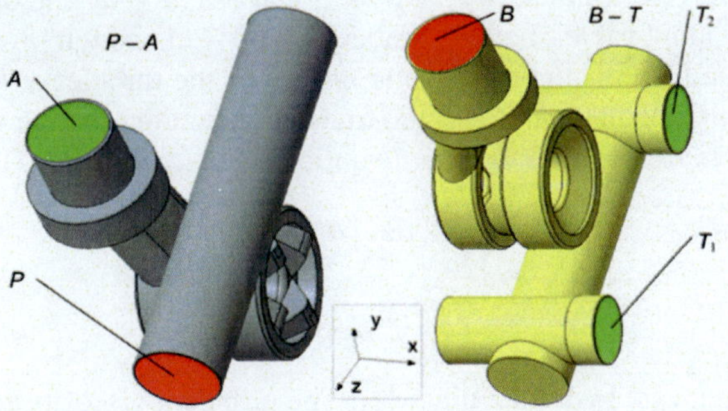

Fig. 5. Geometrical models of *P – A* and *B – T* flow channels.

The FEM mesh and flow models were made using the ANSYS/Fluent program. Figure 6 shows discrete models of flow channels generated for the following spool gap width: $x_g = x_s - x_p = 0.8$ mm.

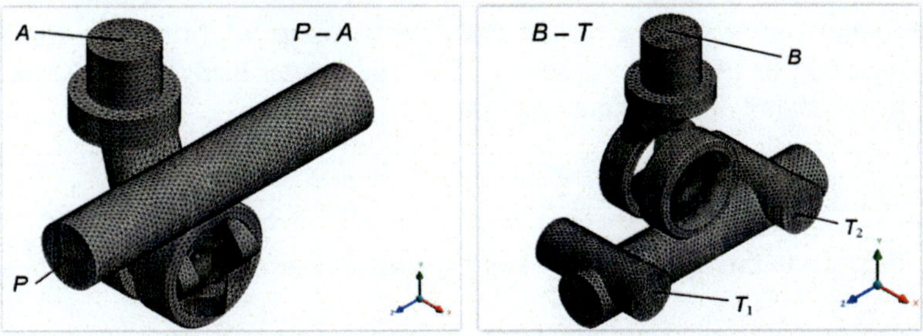

Fig. 6. Discrete models of *P–A* and *B–T* flow channels.

Some advanced mesh-generating functions, as a local mesh refinement, were used to make the models. The presented *P–A* and *B–T* models contain over 350 000 cells and over 120 000 nodes each. In the same way, FEM models for the following gap widths were made: $x_g = 0.4$ mm, 0.8 mm, 1.2 mm, 1.6 mm, 2.0 mm, 2.4 mm and 3.5 mm. The pressure-velocity coupling was performed using the *Pressure based* solver. The convergence criterion of the absolute value of residuals of mass and momentum equal to 10^{-3} was adopted. The grid was created according to general Fluent recommendations and the following quality criteria were achieved: orthogonal quality with minimal value 0.74, skewness 0.51 and maximum aspect ratio 5.97.

3.2 CFD Model Parameters

In the first step of the CFD model creation the flow kind was determined. Based on the geometry of flow channels, fluid viscosity and the flow rate range, the Reynolds number varies from $Re = 10^3$ to $Re = 6 \cdot 10^3$. Hence, the turbulent pattern was assumed to simulate the flow. ANSYS/Fluent provides multiple turbulence models [13], including *Standard*, *RNG* and *Realizable* $k - \varepsilon$, *Standard* and *SST* $k - \omega$, and others. In the case of flow through a spool valve, the $k - \varepsilon$ model works well enough [8–11], hence it was adopted for the simulation studies. Kinetic energy of the turbulence k and dissipation factor ε were computed on the basis of the transport equations, which can be found e.g. in [9], using the *Intensity* and *Length scale* options (Eq. 2) and (Eq. 3), respectively.

$$I = 0.16 \cdot (Re)^{-1/8}, \tag{2}$$

$$\ell = 0.07 \cdot D_H, \tag{3}$$

where D_H is the relevant hydraulic diameter. The turbulent viscosity μ_t was calculated using the k and ε values (Eq. 4), while the other model parameters were assigned default values: $C_{1\varepsilon} = 1.44$, $C_{2\varepsilon} = 1.92$, $C_\mu = 0.09$, $s_k = 1.0$, $s_\varepsilon = 1.3$.

$$\mu_t = \rho \cdot C_\mu \cdot k^2 \cdot \varepsilon^{-1}. \tag{4}$$

The following conditions were defined: fluid velocity at the inlet and pressure at the outlet. The velocity magnitude was defined in the *Boundary Conditions/Velocity Specification Method* option as normal to the boundary. The outlet pressure was defined using the *Outlet condition* as a *Gauge Pressure* of value 0.1 MPa. The force acting on the spool along its axis was introduced as F_x parameter using the *Calculators and Expressions* command of the *Results* module.

3.3 Flow Equations

Volumetric flow rate through the valve gap Q_g can be determined from the equation:

$$Q_g = \mu \cdot A_g(x_g) \cdot \sqrt{2 \cdot \Delta p \cdot \rho^{-1}}. \tag{5}$$

Hence the average fluid velocity:

$$v_g = \frac{Q_g}{A_g} = \mu \cdot \sqrt{2 \cdot \Delta p \cdot \rho^{-1}}. \tag{6}$$

Figure 7 and Fig. 8 show flow characteristics determined from the CFD analysis of the *P–A* and *B–T* channels, respectively. As it arises from the drawings, similar values of pressure loss were obtained, which results from identical cross-sectional area of the throttling gaps. Slight divergence is due to the differences in geometry of the inflow and outflow channels. The average flow coefficient for both channels $\mu = 0.75$ was determined on the basis of the obtained characteristics.

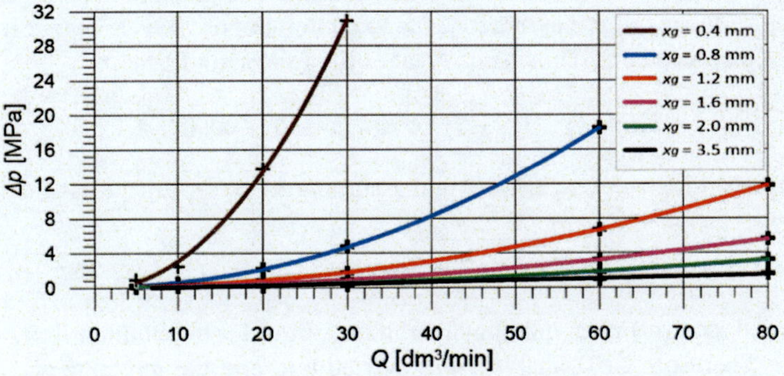

Fig. 7. Flow characteristics of the *P–A* channel based on CFD results.

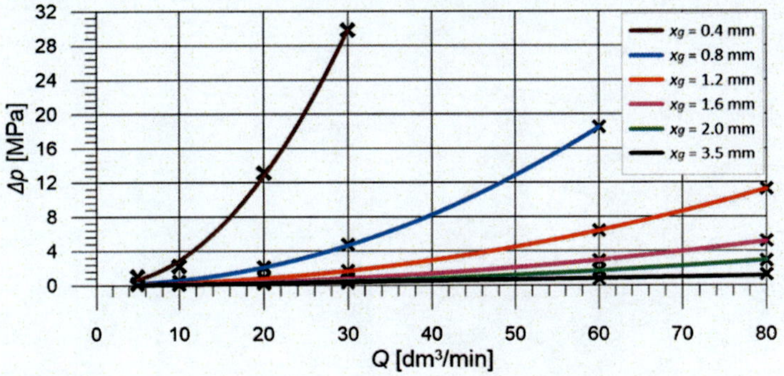

Fig. 8. Flow characteristics of the *B–T* channel based on CFD results.

The valve spool equation of motion has the following form:

$$m_s \frac{d^2 x_s}{dt^2} + F_{ts} + F_{ss} + F_{ds} = F_{el}. \tag{7}$$

Viscous friction force F_{ts} and spring force F_{ss} can be defined as:

$$F_{ts} = \varphi_s \cdot \frac{dx_s}{dt},\ F_{ss} = c_s \cdot x_s + F_{ss0}, \tag{8}$$

where φ_s – viscous friction coefficient, c_s – spring stiffness and F_{ss0} – spring tension. The electromagnet force F_{el} is proportional to the control signal I_{el}:

$$F_{el} = k_{el} \cdot I_{el}, \tag{9}$$

Vector equation of flow forces $\boldsymbol{F}_{ds}$ can be formulated basing on the momentum conservation equation [13]:

$$\boldsymbol{F}_{ds} = \frac{d(m\boldsymbol{v})}{dt} = \frac{dm}{dt} \cdot (\boldsymbol{v}_2 - \boldsymbol{v}_1) = \rho \cdot Q_g \cdot (\boldsymbol{v}_2 - \boldsymbol{v}_1), \tag{10}$$

where ρ – fluid density, $\boldsymbol{v}_1, \boldsymbol{v}_2$ – fluid velocity vectors at input and output. Since the radial component of the flow force acts mainly on the channel walls and is compensated to a large extent by means of the symmetry, the axial component has the greatest influence on the spool displacement. Thus, (Eq. 9) takes the following form:

$$F_{ds} = \rho \cdot Q_g \cdot (v_2 \cdot \cos(\alpha_2) - v_1 \cdot \cos(\alpha_1)). \tag{11}$$

After substituting v_2 by v_g (Eq. 6) and assuming $\alpha_1 = \frac{\pi}{2}$, Eq. 11 has the form:

$$F_{ds} = \rho \cdot Q_g \cdot \mu \cdot \sqrt{2 \cdot \Delta p \cdot \rho^{-1}} \cdot \cos(\alpha_2) = \lambda \cdot Q_g \cdot \sqrt{2 \cdot \Delta p \cdot \rho}, \tag{12}$$

where $\lambda = \mu \cdot \cos(\alpha_2)$ is an unknown parameter. In order to obtain λ, first α_2 must be determined. Therefore CFD analysis was carried out, and the α_2 value was estimated as an average jet angle at the valve gap output. Figure 9 shows velocity vectors for the *P–A* flow channel in the *X–Y* plane for the spool gap width $x_g = 0.4$ mm and $x_g = 2.0$ mm, respectively. The analogous results obtained for the *B–T* channel are presented in Fig. 10.

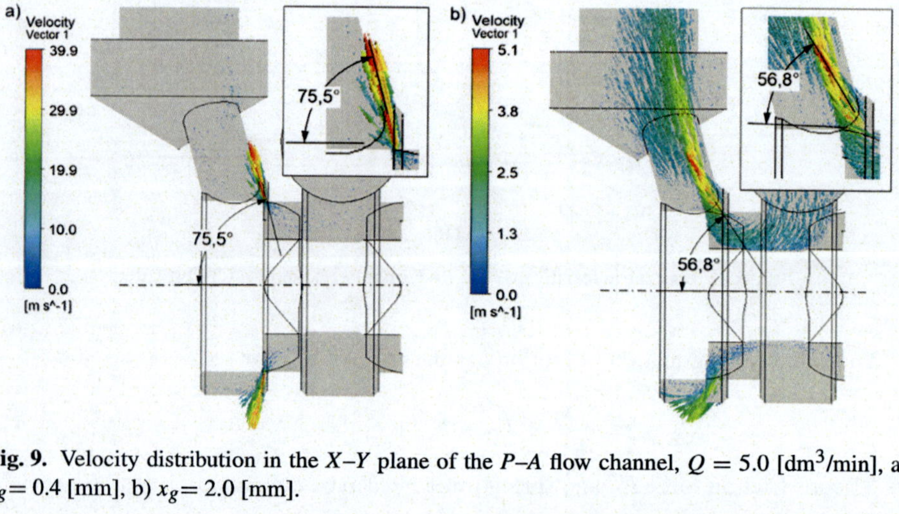

Fig. 9. Velocity distribution in the *X–Y* plane of the *P–A* flow channel, $Q = 5.0$ [dm^3/min], a) $x_g = 0.4$ [mm], b) $x_g = 2.0$ [mm].

As it arises from Fig. 9 and Fig. 10, the α_2 value is inversely proportional to the gap width x_g in both considered flow channels. Hence, increase in x_g decreases α_2 and thus increases λ which is proportional to $\cos(\alpha_2)$. As a result, axial component of flow force rises. In order to take into account the dependency between λ and x_g in the

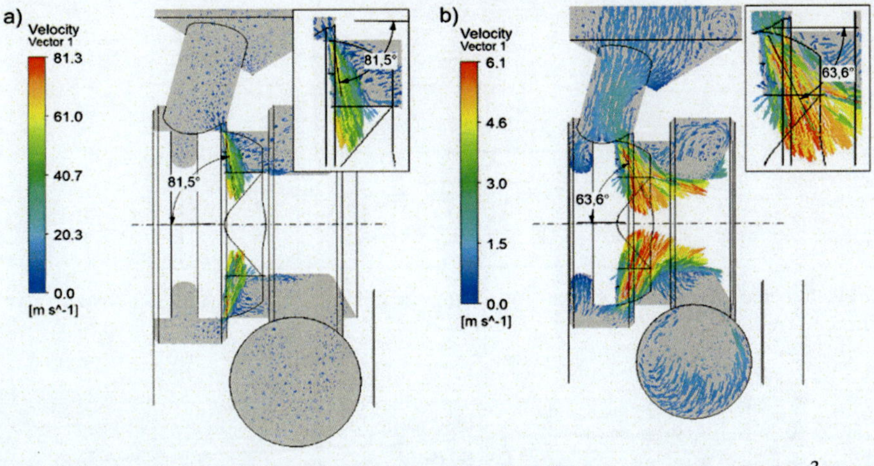

Fig. 10. Velocity distribution in the X–Y plane of the B–T flow channel, $Q = 5.0$ [dm^3/min], a) $x_g = 0.4$ [mm], b) $x_g = 2.0$ [mm]

control characteristics of the valve, linearization based on the CFD analysis results was performed.

The linearized functions of $\lambda(x_g)$ obtained for P–A and B–T flow channels are shown in Eqs. 13 and 14, respectively:

$$\lambda_{P-A}(x_g) = 0.1320848 + 0.1392931 \cdot x_g, \quad (13)$$

$$\lambda_{B-T}(x_g) = 0.0552060 + 0.1391350 \cdot x_g. \quad (14)$$

Both obtained lines have similar inclination angle, while the total value is significantly higher in case of the P–A channel, due to the larger free term value. Thus, in the next step control characteristics of the P–A channel were determined.

3.4 Control Characteristics

Control characteristics of the valve were determined individually for each flow path. Figure 11 shows a scheme used for testing the P–A channel. The fixed capacity pump (2) supplies the tested valve (1). Pressure at the valve inlet is set by an adjustable relief valve (3), while the electromagnet force is adjusted by means of the I_{el} signal.

Control characteristics can be calculated by solving system of equations including flow equations and a spool motion equation. The P–A flow channel characteristics obtained for the with fixed values of the control signal in the range from 40% to 100% are shown in Fig. 12. As it arises from the figure, after exceeding the initial threshold pressure value, the volumetric flow rate as a function of pressure drop is close to constant in the considered range of the control signal.

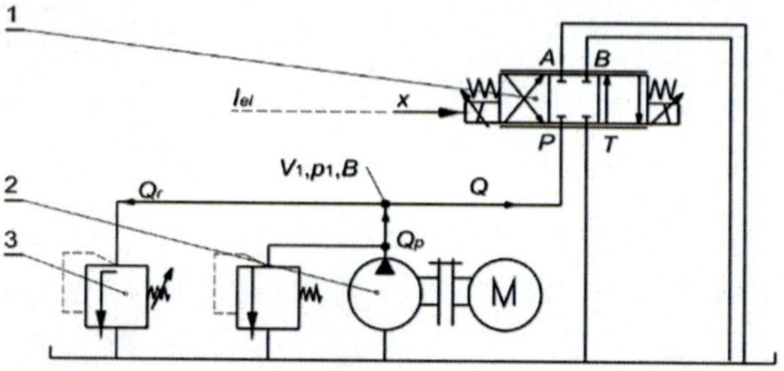

Fig. 11. Scheme of hydraulic system for testing the control characteristics; 1 – tested valve, 2 – pump, 3 – relief valve.

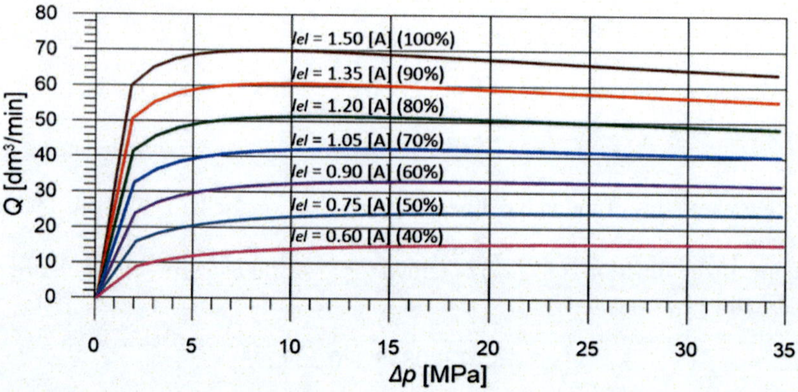

Fig. 12. Flow rate against pressure drop (*P–A* channel) for fixed values of control signal.

4 Test Bench Experiments

The test bench used to conduct experiments is shown in Fig. 13. It was equipped with pressure, flow rate, position, temperature and current sensors. The test cycle was carried out in the following way: first a maximum control signal value of $I_{el} = 1.5$ A was set while the relief valve was opened ($p = 2.0$ MPa). During the measurement, the pressure was gradually increased to 35 MPa and then similarly decreased to 2.0 MPa. Then the control signal was reduced by 10% and the cycle was repeated.

Comparison of theoretical and experimental characteristics are shown in Fig. 14. The presented results were obtained for the control signal value $I_{el} = 1.50$ A (100%), $I_{el} = 1.05$ A (70%) and $I_{el} = 0.60$ A (40%), respectively.

The obtained results of laboratory experiments confirmed that the valve kept the set flow rate relatively stable in the entire considered pressure range. The flow rate obtained during the laboratory experiments varied between 12 dm^3/min and 15 dm^3/min for I_{el} = 0.60 A and between 40 dm^3/min and 45 dm^3/min for $I_{el} = 1.05$ A. In these cases the maximum relative difference between simulation and laboratory results did not exceed 7%.

Fig. 13. Test bench; 1 – tested valve, 2 – pressure transducers, 3 – pressure gauge, 4 – spool position transducer.

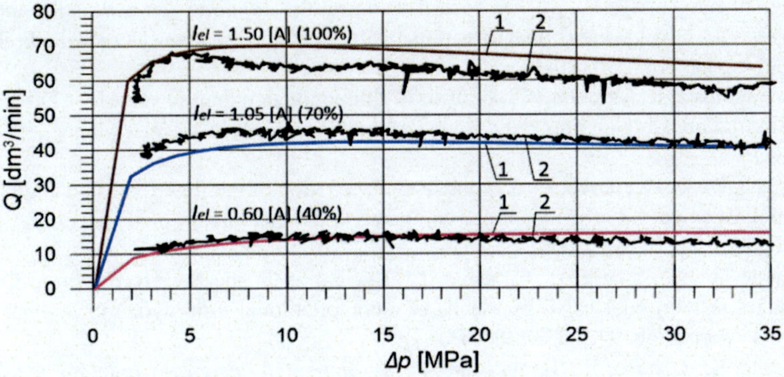

Fig. 14. Comparison of control characteristics obtained for fixed values of control signal; 1 – theoretical curve, 2 – experimental curve.

At the maximum control signal $I_{el} = 1.50$ A, simulation and experimental curves are well matched in the low pressure range up to 5 MPa. Next, a significant flow rate decrease from 68 dm^3/min to 63 dm^3/min is observed between $\Delta p = 5$ MPa and $\Delta p = 10$ MPa, which differs from the simulation curve. As a result, the obtained flow rate values on most of the experimental curve are about 10% lower than those resulting from the simulation. The revealed discrepancy constitute a starting point for further research, including electromagnet tests and model refining.

5 Summary

The article provides an analysis of the possibility to compensate impact of pressure on flow characteristics in a proportional directional spool valve. The compensation was made by means of flow forces, modifying the spool geometry accordingly. First, CFD analysis was performed on a 3D model for the fixed spool positions and different flow rates. It was noticed from the results of the CFD analysis that the jet angle in the throttling gap depends mainly on the gap width (spool position) and, to a significantly limited extent, on the pressure. This allowed an analytical function of the jet angle influence on the flow force value to be determined. Then, control characteristics of the valve were calculated and compared to results of test bench experiments. Detailed conclusions:

- the use of flow forces to compensate pressure requires the appropriate jet angle value, which can be achieved by adjusting the spool gap geometry,
- this solution can be applied in the practice in case, when precise control of the speed of actuator is not required,
- the proposed method does not require any additional valves nor control systems.

References

1. Zhang, J., Lu, Z., Xu, B., Su, Q.: Investigation on the dynamic characteristics and control accuracy of a novel proportional directional valve with independently controlled pilot stage. ISA Trans. **93**, 218–230 (2019)
2. Posa, A., Oresta, P., Lippolis, A.: Analysis of a directional hydraulic valve by a Direct Numerical Simulation using an immersed-boundary method. Energy Convers. Manag. **65**, 497–506 (2013)
3. Amirante, R., Del Vescovo, G., Lippolis, A.: Evaluation of the flow forces on an open centre directional control valve by means of a computational fluid dynamic analysis. Energy Convers. Manag. **47**, 1748–1760 (2006)
4. Amirante, R., Moscatelli, P., Catalano, L.: Evaluation of the flow forces on a direct (single stage) proportional valve by means of a computational fluid dynamic analysis. Energy Convers. Manag. **48**, 942–953 (2007)
5. Amirante, R., Distaso, E., Tamburrano, P.: Sliding spool design for reducing the actuation forces in direct operated proportional directional valves: experimental validation. Energy Convers. Manag. **119**, 399–410 (2016)
6. Borghi, M., Milani, M., Paoluzzi, R.: Influence of notch shape and number of notches on the metering characteristics of hydraulic spool valves. Int. J. Fluid Power **6**, 5–18 (2005)
7. Simic, M., Herakovic, N.: Reduction of the flow forces in a small hydraulic seat valve as alternative approach to improve the valve characteristics. Energy Convers. Manag. **89**, 708–718 (2015)
8. Ye, Y., Yin, C., Li, X., Zhou, W., Yuan, F.: Effects of groove shape of notch on the flow characteristics of spool valve. Energy Convers. Manag. **86**, 1091–1101 (2014)
9. Lisowski, E., Filo, G., Rajda, J.: Pressure compensation using flow forces in a multi-section proportional directional control valve. Energy Convers. Manag. **103**, 1052–1064 (2015)
10. Lisowski, E., Czyżycki, W., Rajda, J.: Three dimensional CFD analysis and experimental test of flow force acting on the spool of solenoid operated directional control valve. Energy Convers. Manag. **70**, 220–229 (2013)
11. Lisowski, E., Filo, G., Rajda, J.: Analysis of Flow forces in the initial phase of throttle gap opening in a proportional control valve. Flow Meas. Instrum. **59**, 157–167 (2018)
12. Lisowski, E., Filo, G.: Analysis of a proportional control valve flow coefficient with the usage of a CFD method. Flow Meas. Instrum. **53**, 269–278 (2017)
13. Ansys-Fluent: Users Guide, 13.0 ed. (2011)

Design, Modeling and Simulation of Gearing for Improving Gerotor Pump Performance

Lozica Ivanović(✉)

Faculty of Engineering, University of Kragujevac, Sestre Janjić 6, 34000 Kragujevac, Serbia
lozica@kg.ac.rs

Abstract. Gerotor pumps are a special type of gear pumps with internal trochoidal gearing. Due to their compact design and simple structure, they are characterized by high reliability and low cost. They are represented in many applications, mostly in the automotive industry. Increasingly stringent environmental regulations require automotive manufacturers to improve vehicle design by increasing the efficiency of their components, including the gerotor pump. This paper presents a procedure for determining the optimal combination of trochoidal gearing design parameters, in order to improve the performance of a gerotor pump. Gerotor performance is affected by sliding velocity, volume efficiency and contact stress. An appropriate mathematical model was developed in order to analyze variation effect of gear design parameters related to value of the pump functional characteristics. In order to reduce the maximum contact stresses, forces and moments' analysis was performed by analytical and numerical method. Through consideration of the gear tooth profile with technological gaps, their impact on the volume losses and gear ratio of the pump was identified. The verification of the developed models was carried out through the realization of physical gear pairs and performing of a laboratory experiments with simulation of pump operating conditions. These results establish valuable guidelines for constructors to design more efficient construction solutions of the gerotor pumps.

Keywords: Trochoidal gearing · Gerotor pump design · Volumetric flow rate · Contact stress

1 Introduction

The gerotor pumps are a special kind of positive displacement pumps and used widely in industry, mostly in a lubrication system for internal combustion engines. The pump impeller is the gear pair with internal trochoidal gearing, has one tooth difference. There are several methods for the geometric profiles design of the gerotor gear pair set [1, 2]. One group of researchers proposed generating a profile by equidistant modification of the epitrochoid and its conjugated envelope [3–5]. The second one, used cycloidal gear tooth design method [6] and third one using the circular pin gear generating technique [7]. Also, there are numerous approaches to improve the pump performance and efficiency. Significant contributions to the investigation of the benefits of the of plastics components

© The Editor(s) (if applicable) and The Author(s), under exclusive license
to Springer Nature Switzerland AG 2021
J. Stryczek and U. Warzyńska (Eds.): NSHP 2020, LNME, pp. 15–27, 2021.
https://doi.org/10.1007/978-3-030-59509-8_2

application in trochoidal machines have been made in the [6, 8–10]. The higher area efficiency, the outlet pressure and the outlet flow, with a larger span angle of the gerotor pump tooth profile, is shown in [11]. The impact of real-life technological clearances on the performance of the pump is studied by numerical simulation in [12, 13].

This paper is designed to provide researchers in the field of trochoid machines with a reference that provides opportunities to improve the performance of gerotor pumps. It is not intended to be a comprehensive review but rather provides guidance and information to keep researchers aware of the potential performance improvements.

2 Gear Profiles Design

Gear profile design approach proposed in [3–5] is adopted. The internal gear profile is equidistance of the epitrochoid generated by rolling the inside of a larger circle r_a on a smaller fixed circle r_t. The external gear profile is the equidistance of the conjugate external envelope, described by circular arc with r_c radius (see Fig. 1).

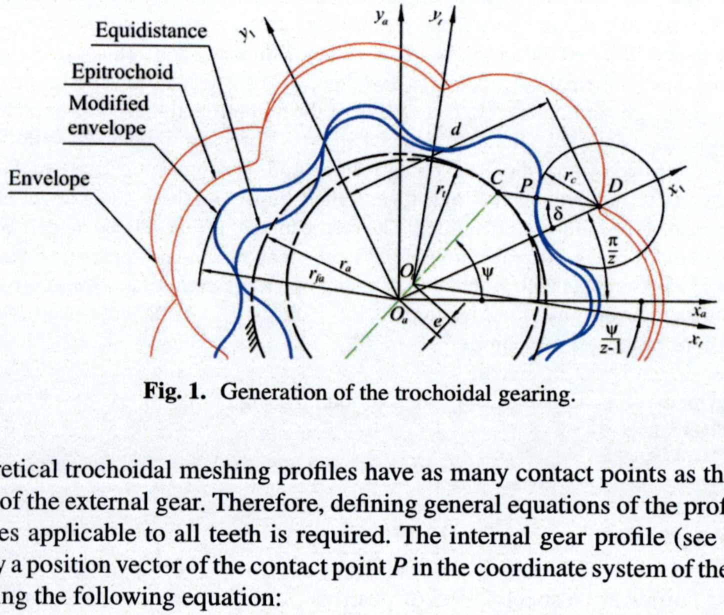

Fig. 1. Generation of the trochoidal gearing.

Theoretical trochoidal meshing profiles have as many contact points as the number of teeth z of the external gear. Therefore, defining general equations of the profile points coordinates applicable to all teeth is required. The internal gear profile (see Fig. 1) is defined by a position vector of the contact point P in the coordinate system of the trochoid $O_t x_t y_t$ using the following equation:

$$x_{P_i}^{(t)} = e\left\{z\lambda \cos\left[\tau_i + \frac{\psi}{z-1}\right] - \cos\frac{z\psi}{z-1} - c\cos\left[\tau_i + \frac{\psi}{z-1} + \delta_i\right]\right\}$$
$$y_{P_i}^{(t)} = \left\{z\lambda \sin\left[\tau_i + \frac{\psi}{z-1}\right] - \sin\frac{z\psi}{z-1} - c\sin\left[\tau_i + \frac{\psi}{z-1} + \delta_i\right]\right\} \quad (1)$$

where i is the ordinal number of the tooth, λ is the trochoid coefficient, $\lambda = d/ez$, c is the equidistant coefficient, $c = r_c/e$, $\tau_i = \pi(2i-1)/z$ and δ is the leaning angle [5].

The derived equations can be applied to different kinematic schemes of trochoidal machines. In pumps with fixed shaft axes where the drive shaft is fixed to the internal gear, the following expressions define the relations between the referent angle ψ and the shaft rotation angles $\varphi_a = -\psi$ and $d\,\varphi_t = (z/z - 1)d\varphi_a$.

3 Procedure for Selection of the Improved Teeth Profile

3.1 Defining the Problem

In order to achieve high level functional characteristics of the pump it is necessary to provide the best combination of gear geometrical parameters. In the pump process design usual parameters, such as: the gear width *b*, eccentricity *e*, and the root radius of the external gear r_{fa} are given (see Fig. 1). It is necessary to determine the external gear teeth number *z* and the values of the coefficients λ and *c*, which define the optimum gear teeth profile. Before analyzing the influence of the parameter variation on the gear pair characteristics, the primary aspect of the selection of the optimum gear profile should be defined, and that would be the minimum curvature at the critical contact point, and hence the higher tooth flank resistance to destruction.

3.2 Dimensional Limits Parameters

In order to define the practical application domain for the geometric parameters of the trochoidal pump gear pair, it was necessary to determine the geometric and kinematic conditions that should be fulfilled by the profiles so that the proper meshing could be achieved. In addition to the limits that prevent several types of interference, the condition for uniform gear teeth wear is set, as well as the condition for achieving a maximum of the minimum equivalent curvature radius profile curve. Their geometric and kinematic conditions for trochoidal gearing design are given in the [14].

3.3 Geometric Parameters Selection

By varying the geometric parameters using the defined limits, their best combinations are selected. The mathematical model given in [14] was used for this analysis, with the purpose of correction of the existing lubrication pump gear set. Design parameters of commercial gear set, Model 1, are $\lambda = 1.575$, $c = 3.95$, while the Model 2 was designed based on calculations with better characteristics, $\lambda = 1.375$, $c = 2.75$. For both gear sets the parameters are: $z = 6$, $e = 3.56$ mm, $r_s = 26.94$ mm, $\omega_t = 2\pi n_t = 50\pi\ \mathrm{s}^{-1}$ and $\Delta p = 0.6$ MPa. The selected parameter values are further checked against the criteria set to evaluate the gear set solution.

3.4 Gear Set Design Evaluation

Kinematic Characteristics. During the meshing the trochoidal gearing profiles are rolling in the same time and sliding one to each other. The sliding of the profiles in the contact point is the consequence of the difference of the relative velocities intensity of the points on the profiles of the internal respectively the external gear. Except the sliding velocity, for the analyzing of the phenomenon of wear, the summary rolling velocity is also significant [5]. Figure 2 illustrates diagrams of the leaning angle, sliding velocity and summary rolling velocity for both considered gear sets.

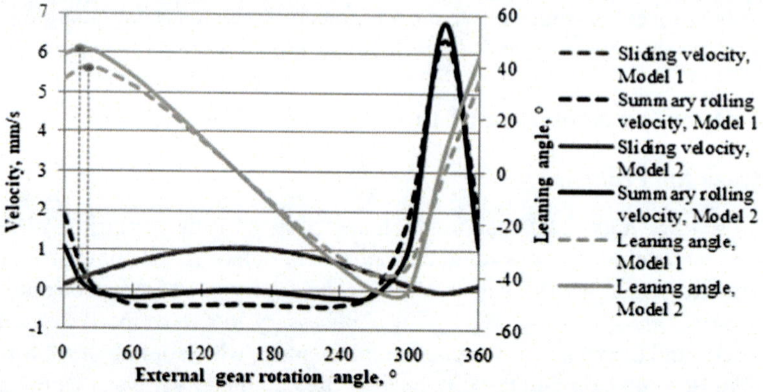

Fig. 2. Leaning angle, sliding velocity and summary rolling velocity.

Based on the graphical interpretation given, it can be concluded that the curves of the sliding velocity and the summary rolling velocities intersect at the point when the leaning angle reaches the maximum value. In that moment the rolling velocity of the external gear tooth profile reaches zero. The change in sign of the sliding velocity and the summary rolling velocity influences the moment direction change of the sliding and rolling friction forces, and thus the mechanical power losses. Fatigue pitting development is expected on the profile section where the sliding velocity is less than the summary rolling velocity. On the rest of the profile decrease of the lubricant layer thickness, increase of the contact temperature and the risk of scoring are expected.

Volumetric Characteristics. Modern gerotor pump design methods require the use of appropriate mathematical models of processes that occur in actual pump structures. For that reason, the instantaneous pump chamber volume was first considered [15] and then, using the certain trigonometric transformations, the following formula to determine the theoretical volume variation in the active chamber is obtained:

$$\frac{\mathrm{d}V_\mathrm{i}}{dt} = \omega_t b e^2 z \left\{ 2\lambda \, \sin\frac{\pi}{z} \, \sin\left(\frac{2\pi i}{z} - \psi\right) - \frac{c}{z}\left[1 + \lambda^2 - 2\lambda \, \cos(\tau - \psi)\right]^{\frac{1}{2}} \Bigg|_{\tau_\mathrm{i}}^{\tau_\mathrm{i+1}} \right\} \quad (2)$$

as well as that of theoretical gerotor pump flow rate which is determined by the formula:

$$Q = \sum_{i=m}^{n} \frac{dV_i}{dt} \quad (3)$$

where m and n are the ordinal numbers of starting and the final chambers which can be found at the same time in the thrust phase. Figure 3 shows diagrams of the volume characteristics of a gerotor pump with gear set Model 2 as a function of external gear rotation angle, based on which their size and variation can be observed. The differences of pump flow variation for both considered gear sets are negligible. This is confirmed by the calculated values of the flow rate irregularity, which for both models are is equivalent to $\delta_q \approx 10\%$.

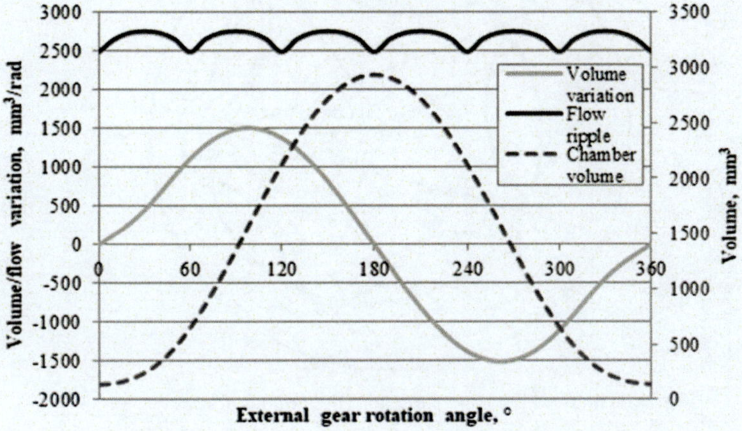

Fig. 3. Volumetric characteristics for Model 2.

Dynamic Characteristics. In order to create conditions that reduce contact forces and thus reduce wear, the forces and moments affecting the gear pair of the gerotor pump are analyzed. Conventional calculation methods of the pump load start from the hypothesis that the pressure in all the chambers of the same zone (inlet and delivery) is constant. In this case, the force of fluid pressure, which separates the inlet zone from the delivery zone, is a continual force that can be presented as an equivalent concentrated pressure force, as shown in Fig. 4 [15, 16]. Starting with the specific conditions in which the load is transmitted simultaneously at a number of contact points, for the calculation of contact force at the current contact point, the expression can be written:

$$F_{ni} = \frac{M_{F_n(O_t)} \sin \alpha_{ni}}{e(z-1) \sum\limits_{j=p}^{q} \sin^2 \alpha_{nj}} \tag{4}$$

where p and q are the ordinal numbers of the initial and the final tooth of the external gear that transfer the load, α_{ni} is the angle between the axis x_f and the normal CP_i, and $M_{F_n(O_t)}$ is the total moment of normal forces with respect to the centre of the gear that is equal to the sum of the drive moment and the pressure force moment with respect to the center of the internal gear [15].

During the pump working process in every chamber, due to fluid flow, the pressure p_i is changing. In a gerotor pump the fluid distribution is done through the holes with variable fluid flow area, because it is assumed that fluid flow area is equal to instantaneous

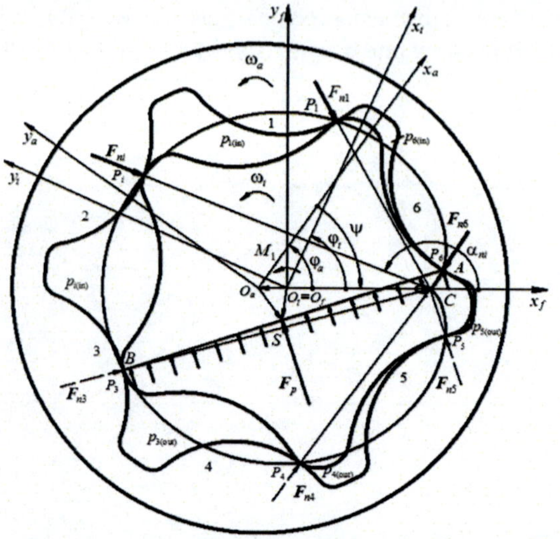

Fig. 4. The fluid pressure force and the contact forces ($\varphi_a = 55°$).

chamber area. This assumption led to expression for calculation of the pressure variation Δp_i during the fluid flow in the active chamber [15].

In order to obtain more accurate results, the pressure variation due to fluid flow should be taken into account while calculating forces and moments. However, in that case, it is difficult to obtain an analytical solution, so it is advised to use a numerical method. The CATIA® module for structural analysis was applied for finite element analysis (FEA) of the gear loads in the gerotor pump. The boundary conditions were such that the support surface of the drive shaft was completely fixed. To obtain results for the moments and forces, the sensor was defined in the gear cross-section plane.

The analytical-numerical calculation consisted of two phases. In the first phase, only relative fluid pressure on the intake and discharge side was given as load, and the output result was the resultant fluid pressure and its moment in relation to the center O_a. This moment value is equivalent to the contact forces moment $M_{F_n(O_t)}$ [15] and it was used for analytical calculation of the forces using the Eq. (4). In the second phase of the numerical calculations, the fluid pressure and the contact forces were given as load. The output results were the support reaction force and driving moment.

The calculation procedure involved two cases. In the first case, the pressure variation Δp_i due to the fluid flow was neglected ($p_i = \text{const}$), while in the second case it was taken into account ($p_i \neq \text{const}$). Then, the fluid pressure value was calculated for each chamber and it was given as load of the active chamber surface (between the two adjacent profile contact lines). Figure 5 shows the FEA model with the motion restrain, loads and results at the angular position of $\varphi_\text{a} = 5°$.

For maximum contact stress calculation, when the gears are made of the same material, $E_t = E_a = E$ and $\nu_t = \nu_a = \nu$, the following expression is evaluated through the classic Hertz theory [16]:

$$\sigma_{ni} = \sqrt{\frac{1.57 F_{ni} E}{\pi^2 b \left(1 - \nu^2\right) \rho_{ekv}}} \tag{5}$$

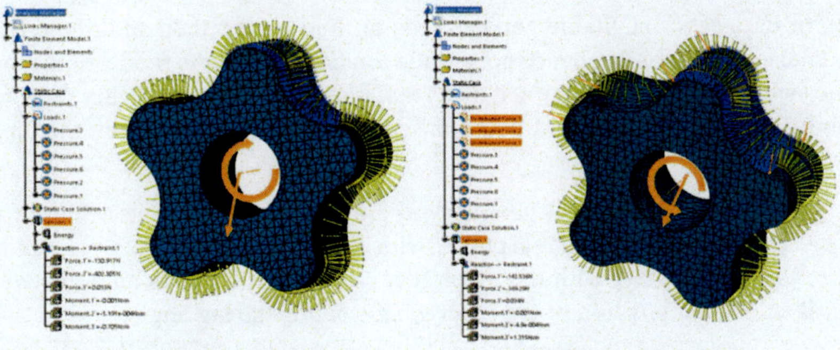

a) I phase, Case II b) II phase, Case II

Fig. 5. Visualisation of the discrete gear load and results ($\varphi_a = 5°$).

Based on the analysis of the forces and the developed methodology, a program was made in the software package Mathematica® for the identification of teeth in contact and calculation of contact forces and stresses in the trochoidal pump gear set.

Figure 6 shows values of the driving torque and the maximum contact stresses of the different angular positions during one working process phase, obtained by the analytical-numerical calculation for both considered cases. Simplified calculation in case I give lower values for the driving torque. Diagrams show that in both cases the newly generated Model 2 yields lower values of the maximum driving torque, as well as of the maximum contact stresses than the value obtained with the commercial pump model, up to 16.8%, which confirms the choice of the better solution.

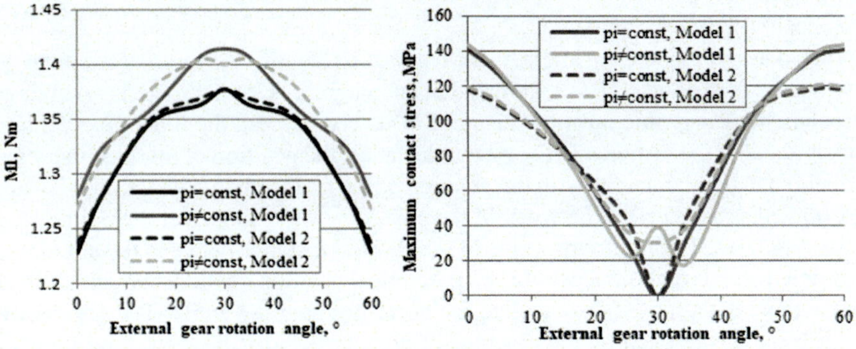

a) b)

Fig. 6. Comparison of the driving torque (a) and maximum trochoidal gear contact stress (b) for two cases.

Torque Pulsations in Gerotor Pump with Technological Clearances. The previously described research considered a theoretical trochoidal profile with assumptions about ideally accurate geometry. However, in real constructions, the tooth profiles are made with technological clearances, which are necessary to allow proper mounting and meshing of the gears. Inevitable gap can lead to fluid losses, then additional dynamic forces, reduce stability and enhance noise and vibration. In order to model of the meshing of profiles with clearance it is adopted that a real profile of the internal gear was obtained by a modification of the trochoid with constant difference increased by the clearance size, ε. [17, 18].

For gearing pairs with ideal geometry, the transfer of motion is achieved with a constant gear ratio. However, for real pumps with gaps, the inner gear rotates an additional angle $\Delta\theta_i$ to make contact with the outer gear (see Fig. 7, a). This angle between the theoretical and actual position of the inner gear is called the lag angle.

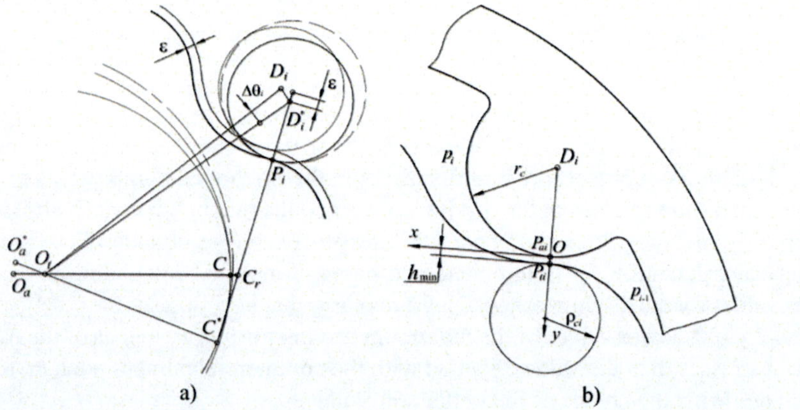

Fig. 7. Geometric relations used to determine the lag angle (a) and minimum gap (b).

When determining the lag angle, it is necessary to identify the tooth of the outer gear that will first get in contact. It is assumed that an imaginary circular tooth is revolving on the trochoidal gear profile until the distance of its center from the inner gear center O_t becomes equal to the distance D_iO_t. At that instance, the position of the imaginary tooth and trochoidal gear coincides with the position of the real teeth in contact. The lag angle is determined as a difference between of the position vectors angles of the points D_i and D_i* (see Fig. 7, a). The minimum value of the shortest distance between the circular toot profile and the real trochoidal profile at a theoretical position identifies which tooth gets in contact and is then used to calculate the corresponding lag angle. The gap between other gear pairs is defined by the distance of points P_{ti} and P_{ai}, when those points are located at the common normal of the profile. By varying the value of the angle ψ_i for the observed tooth of the outer gear, the value angle that defines the position of the common normal and point P_{ti} on the internal gear profile is obtained and then used to determine the normal distance between points D_i and P_{ti}.

Finally, the minimum clearance between gear profiles can be calculated as:

$$(h_{\text{min}})_i = \overline{D_iP_{ti}} - r_c, i = 1, 2, \ldots, z. \tag{6}$$

Variation of the gap height is analyzed for a period of one outer gear revolution. Diagram shows the results for two adjacent gear teeth by which can be concluded that the same teeth pair stays in contact during one phase (see Fig. 8, a). Consequently, in short angular interval, there is simultaneous meshing of two adjacent gear teeth. Diagrams show peaks, the first and the most expressive of them taking place at the moment when contact between adjacent couple of gear teeth ends, while the second of them takes place after a period that corresponds to one phase angle. Further decrease in gap has nearly linear character with noticeable deviation at the moment when new couple of gear teeth comes into contact. It could be also concluded that the width of the gear teeth profile active part, as well as the length of the interval of double gearing, is larger for the Model 2.

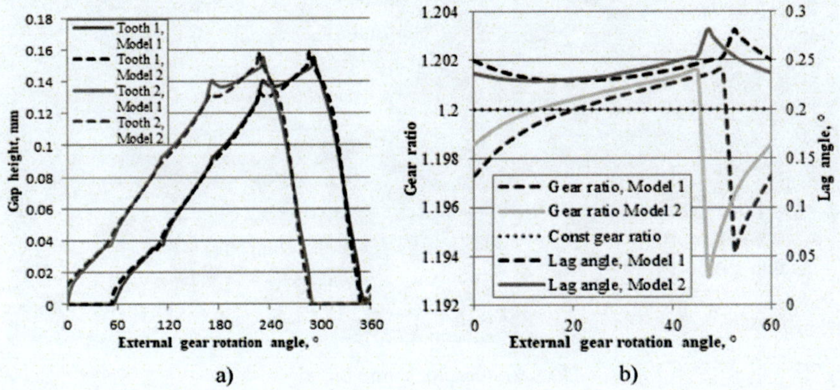

Fig. 8. Comparison of the gap height (a) and lag angle and gear ratio (b) for two gerotor gear sets with technological clearance, $\varepsilon = 0.07$ mm.

After identifying the tooth in contact, as well as position of the current pitch point C_r of the inner gear (see Fig. 7, a), the real pump gear ratio can be calculated [17]. For the presented example, values of a lag angle and the current gear ratio are determined and obtained results as a function of external gear rotation angle are given in Fig. 8, b.

Diagrams of a lag angle and the gear ratio show the peaks whose position concurs with peaks on the clearance height diagram. The fluctuations in the gear ratio directly affect the variation of the output torque, and consequently the appearance of noise and vibration in the machine. For this reason, the goal is to achieve as little gear ratio deviations as possible from the ideal, constant value ($u = 1.2$).

Based on the presented results the conclusion was made that more beneficial indicators are derived from the Model 2 proposed as a solution with improved characteristics compared to the existing commercial ones Model 1.

4 Experimental Validation

The experimental testing of gerotor pump with two considered gear sets was done in laboratory with the simulation of real operating conditions (see Fig. 9). The layout of the test rig was in accordance with the hydraulic circuit shown in Fig. 10. The tested gerotor pump was driven by electric motor up to 2.2 kW. The working medium of the pump was SAE 15W30 mineral oil, and the oil temperature was maintained within 40 ± 2 °C. The speed is adjusted by a speed controller and measured by a contactless tachometer. In doing so, three values of the pump rotational speed were varied (500 rpm, 1000 rpm and 2000 rpm).

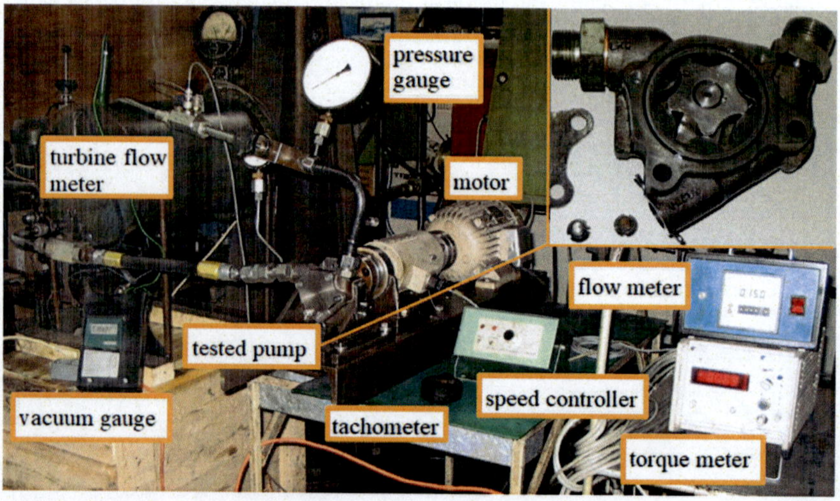

Fig. 9. Gerotor pump test rig.

A vacuum gauge was used to measure the intake pressure of the pump working fluid, while a pressure gauge was used to measure the discharge pressure of the working fluid. The test rig enabled measurements of the outlet flow rate by a flow meter. Measuring the pump flow rate was performed at the discharge pressure of the pump from 0 up to 1 MPa in increase steps of 0.1 MPa.

The results of measurements of the pump flow rate and volumetric efficiency, as a function of discharge pressure, are shown in the Fig. 11. Based on the shown diagram, it can be observed that there is a linear relationship between the flow rate and the discharge pressure, and the output flow was approximately 13 l/min at 1000 rpm, which is about 7% less than the calculated value pump capacity (14 cc/rev). In addition, it was observed that the value of the volumetric efficiency is higher for the Model 2 and at higher rotational speed.

The deviation between the calculated and experimentally measured flow was found during the experiment, but as the pumps tested on the same device from the regular manufacturing process showed the same difference, the newly made model was considered to satisfy the experimental validation and as such could be to use in practice.

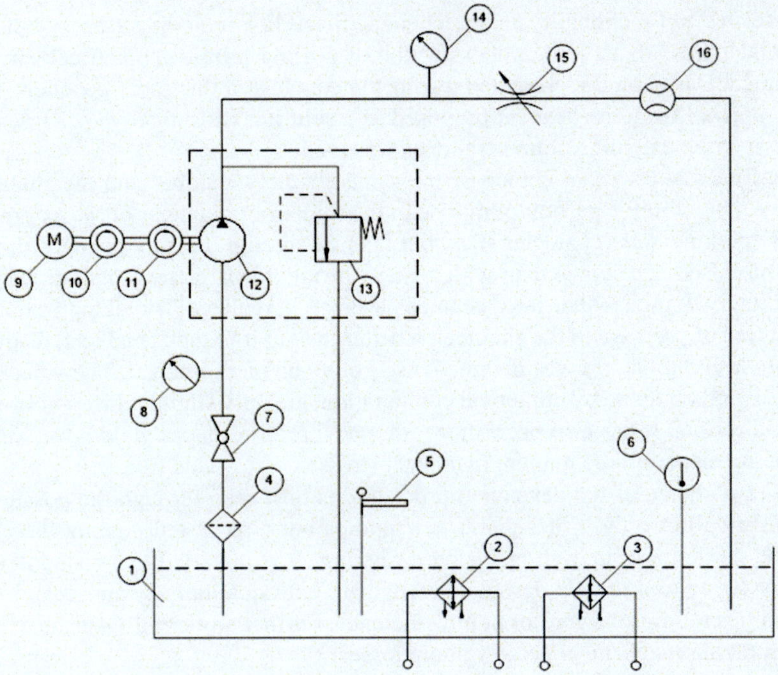

Fig. 10. ISO schematic of test rig: 1-tank, 2-cooler, 3-heater, 4-filter, 5-oil level indicator, 6-temperature gauges, 7-ball valve, 8-vacuum gauge, 9-motor, 10-tachometer, 11-torque meter, 12-tested pump, 13-safety valve, 14-pressure gauge, 15-valve, 16- flow meter.

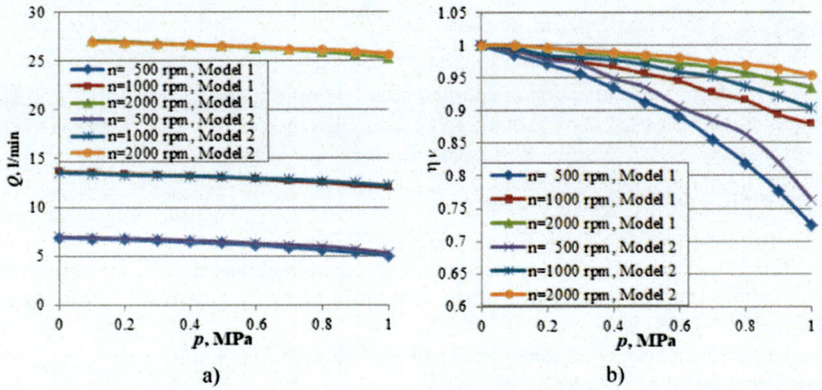

Fig. 11. Flow rate (a) and volumetric efficiency diagrams (b).

5 Conclusion

Based on the trochoidal gearing generation mathematical model the procedure for the selection of the improving gerotor gear set is defined. The most significant constraining factors related to functionality and practical applicability, including aspect of the maximal contact stress reduction, in the selection of gearing profile geometrical parameters are adopted. Based on the presented results the conclusion that more beneficial indicators are derived from the gear set proposed as a solution with improved characteristics compared to the existing commercial ones was made.

It has been shown that choice of the same chamber number and the smaller trochoid coefficient in the gerotor pumps with the same root diameter of the external gear changes the kinematic characteristics, but it does not significantly changes the pump flow. It has also been shown that, with the new profile form, larger length of the double gearing interval and smaller inter chamber leakage is realized. By taking into account the variation of pressure in the chamber, a better insight into the actual load distribution and more accurate values was obtained, used finite element method. The values of the drive moment, obtained by numerical computation, are only slightly different from their analytical results. It has also been shown that new profile shapes yield lower values of the drive moment and the maximum contact stresses.

The importance of the developed procedure is best seen through the possibility of its application not only in designing new pumps but also in redesigning the existing ones, and particularly in high precision designing. It is intended to assisted develop a computer-based tool to help designers recognize, both qualitatively and quantitatively, opportunities to improve gerotor pump efficiency. Further work will focus on exploring and minimizing mechanical gerotor pump losses.

Acknowledgment. Research presented in this paper was supported by Ministry of Education, Science and Technological Development of Republic of Serbia, Grant TR 35033.

References

1. Nang, A., Maiti, R.: Unification of epitrochoid origin profile design approaches for external lobed star member used in hydrostatic and gear units. J. Mech. Sci. **227**(2), 299–310 (2012)
2. Gamez-Montero, P.J., Castilla, R., Codina, E.: A review of gerotor technology in hydraulic machines. Energies **12**, 2423 (2019)
3. Maiti, R., Sinha, G.L.: Kinematics of active contact in modified epitrochoid generated rotary piston machines. Mech. Mach. Theory **23**, 39–45 (1988)
4. Gamez-Montero, P.J., Castilla, R., Codina, E.: Methodology based on best practice rules to design a new-born trochoidal gear pump. Proc. Inst. Mech. Eng. Part C J. Mech. Eng. Sci. **232**(6), 1057–1068 (2018)
5. Ivanovic, L., Josifovic, D.: Specific sliding of trochoidal gearing profile in the gerotor pumps. FME Trans. **34**, 121–127 (2006)
6. Stryczek, J., Bednarczyk, S., Biernacki, K.: Gerotor pump with POM gears: design, production technology, research. Arch. Civil Mech. Eng. **14**(3), 391–397 (2014)
7. Beard, J.E., Yannitell, W., Pennock, G.R.: The effects of the generating pin size and placement on the curvature and displacement of Epitrochoidal gerotors. Mech. Mach. Theory **27**, 373–389 (1992)
8. Stryczek, J., Banaś, M., Krawczyk, J., Marciniak, L., Stryczek, P.: The fluid power elements and systems made of plastics. Procedia Eng. **176**, 600–609 (2017). Elsevier

9. Gamez-Montero, P.J., Antoniak, P., Castilla, R., Freire, J., Krawczyk, J., Stryczek, J., Codina, E.: Magnet-sleeve-sealed mini trochoidal-gear pump prototype with polymer composite gear. Energies (10), 1458 (2017)
10. Gamez-Montero, P.J., Garcia-Vilchez, M., Raush, G., Freire, J., Codina, E.: Teeth clearance and relief grooves effects in a trochoidal-gear pump using new modules of GeroLAB. J. Mech. Des. Trans. ASME **134** (5) (2012)
11. Hsieh, C.F.: Flow characteristics of gerotor pumps with novel variable clearance designs. J. Fluids Eng.-Trans. ASME **137**(4), 041107 (2015)
12. Pellegri, M., Vacca, A.: Numerical simulation of Gerotor pumps considering rotor micro-motions. Meccanica **52**(8), 1851–1870 (2017)
13. De Martin, A., Jacazio, G., Sorli, M.: Optimization of Gerotor Pumps with asymmetric profiles through an evolutionary strategy algorithm. Machines **7**(1), 17 (2019)
14. Ivanovic, L.: Reduction of the maximum contact stresses by changing geometric parameters of the trochoidal gearing teeth profile. Meccanica **51**(9), 2243–2257 (2016)
15. Ivanović, L., Devedžić G., Mirić N., Ćukovic S.: Analysis of forces and moments in the gerotor pumps. Proc. Inst. Mech. Eng. Part C J. Mech. Eng. Sci. **1**(1), 1–13 (2010)
16. Maiti, R.: Active contact stresses at epitrochoid generated rotor-stator set of fixed axis or equivalent system ROPIMA type hydrostatic units. J. Eng. Ind. **113**, 465–473 (1991)
17. Ivanovic, L., Devedzic, G., Cukovic, S., Miric, N.: Modeling of the meshing of trochoidal profiles with clearances. J. Mech. Design **134**(4), 041003–1/041003-9 (2012)
18. Rundo, M.: Models for flow rate simulation in gear pumps: a review. Energies **10**, 1261 (2017)

Volumetric and Torque Efficiency of Pumps During Start-up in Low Ambient Temperatures

Ryszard Jasiński(✉)

Faculty of Mechanical Engineering, Department of Mechanics and Mechatronics, Gdańsk University of Technology, ul. G. Narutowicza 11/12, 80-233 Gdańsk, Poland
ryszard.jasinski@pg.edu.pl

Abstract. The hydraulic system start-up in low ambient temperature is accompanied by numerous phenomena, which include changes in physical properties of hydraulic fluid, dimensions of cooperating working elements and clearances between these elements, friction and flow resistance. This changes the system efficiency, increases wearing of components and changes other operational features.

The volumetric, torque and total efficiency of hydraulic pumps change during start-up in low ambient temperatures, as well as in thermal shock conditions, until they reach the set conditions. The time to reach steady-state conditions depends on many parameters: ambient temperature, flow rate, fluid temperature, design of hydraulic units.

The author carried out tests of hydraulic pumps in the Hydraulics Laboratory of the Faculty of Mechanical Engineering of the Gdańsk University of Technology. In the research four types of pumps were investigated: several gear pumps, two axial piston pumps of different designs and one radial piston pump. It was proved, that in thermal shock conditions some malfunctions may occur, especially with large temperature difference between oil and pump.

Experimental research of various hydraulic pumps operating in low ambient temperatures, undertaken by the author, allows to answer the question how the volumetric and torque efficiency changes in such conditions and what parameters does it depend on.

Keywords: Hydraulic drive · Hydraulic pump · Thermal shock

1 Introduction

In winter, hydraulic machines and equipment fail more than in summer. The most common cause of failure is damage of parts (elements) of the hydraulic system components during improper operation. In some cases of starting the hydraulic system in low ambient temperatures, the efficiency of pump is reduced. This is due to increased flow resistance, lack of sufficient lubrication of mating parts of components, icing of components, insufficient clearance between the mating parts, high viscosity of the working fluid (mineral oil).

© The Editor(s) (if applicable) and The Author(s), under exclusive license to Springer Nature Switzerland AG 2021
J. Stryczek and U. Warzyńska (Eds.): NSHP 2020, LNME, pp. 28–39, 2021.
https://doi.org/10.1007/978-3-030-59509-8_3

Information related to the operation of hydraulic components and systems in low ambient temperatures can generally be derived from several sources [2, 21, 22, 27, 28].

Hydraulic pumps, hydraulic motors and valves [33–35] are the basic hydraulic system components, in which energy losses occur. The losses are classified into volumetric, pressure and mechanical ones and determined for a given hydraulic device in steady conditions, with neglecting impact of heat exchange in such device considering as insignificant in total balance of energy losses [4, 19, 26, 30–32]. In process of starting hydraulic machine the heat exchange on the path of working fluid - elements of hydraulic components - the environment has great impact on the efficiency of energy conversion in these components of devices, especially during start-up of cold components supplied with hot working fluid, which was proved in [6–18].

Volumetric and torque efficiency of pump during start-up in low ambient temperature is affected by the use of appropriate oil in the system. One of the disadvantages of hydraulic mineral oils is large change in viscosity as the temperature changes. In low temperature, the kinematic viscosity of these working fluids can be as high as 10000 mm^2/s or even higher. Manufacturers of hydraulic components recommend that the oil viscosity during start-up (depending on the design) should not exceed 300 to 2000 mm^2/s [27].

The author has been conducting research on hydraulic components (pumps, valves, motors) in low ambient temperatures for many years. Analyzing the starting conditions of hydraulic systems (pumps) in low ambient temperature, two cases of supplying a cooled pump with cold or warm working fluid can be distinguished (Fig. 1).

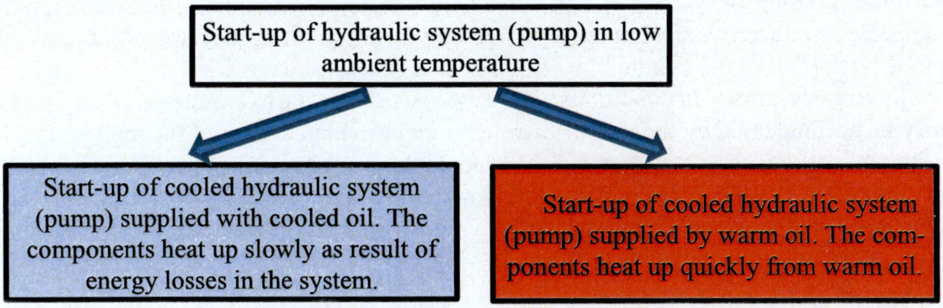

Fig. 1. Cases of hydraulic systems during start-up

In the first case, at start-up the entire hydraulic system, including oil, has a temperature equal to the ambient temperature. However, in the second case, just before the start of the system, the oil is heated in the tank to temperature much higher than the low ambient temperature and components (thermal shock conditions).

The author decided to perform experimental tests of four types of pumps (several gear pumps, two axial piston pumps of different designs and one radial piston pump) started in low ambient temperatures in order to assess the operation of hydraulic components and systems in these conditions. These tests were intended to enable measurements of changes in volumetric, torque and total efficiency of hydraulic pumps during start-up in low ambient temperatures for two cases of supplying cold pumps with cold or warm working fluid.

2 Start-up of Hydraulic System in Low Ambient Temperature

2.1 Start-up and Heating of the Cooled Hydraulic System

As a result of energy losses [1, 2, 5, 24, 27–29], during start-up of the cooled hydraulic system (Fig. 2), the process of components heating up occurs.

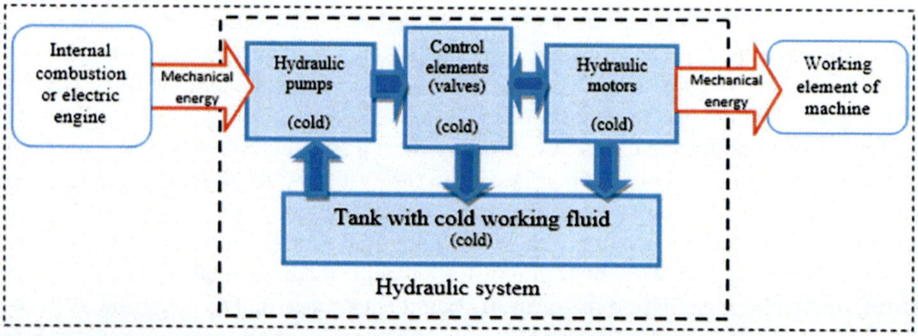

Fig. 2. Diagram of a machine with cold hydraulic system supplied with cold working fluid from tank [6]

In hydraulic systems operating in low ambient temperatures, the working fluid should have the right viscosity. This can be achieved by using a hydraulic oil that is adapted to operate at subzero temperatures or by heating the oil to lower its viscosity, to ensure proper operation of the system.

Starting the pumps in conditions where the viscosity of the hydraulic oil is very high may be accompanied by the risk of extremely undesirable cavitation. Incomplete fluid filling of working chambers (cavitation) exposes the pump to damage, accelerated wear of elements and reduction of volume efficiency of pump.

2.2 Start-up and Heating Process of Cold Hydraulic Components Supplied with Warm Oil

Starting the pump in low ambient temperature conditions can be achieved with warm working fluid (Fig. 3). It is a commonly used method in practice as an alternative to starting with the use of cooled special oils, which viscosity is sufficient. The pump is then exposed to work in thermal shock conditions, which is accompanied by uneven heating of the elements of this component. This causes different thermal expansion of cooperating elements, affecting the value of clearances in the transient process.

The hydraulic system, in which the pump is exposed to work in thermal shock conditions is system in which the oil tank is equipped with heating system and the pump is located at distance from it. There may be instances of pump start-up in which, as result of break in the operation of the hydraulic system, the pump has cooled significantly in relation to the heated hydraulic oil in the tank. However, in practice, before the warm working medium is delivered to the pump, the cold oil must be replaced by the warm in the pump and in the suction line.

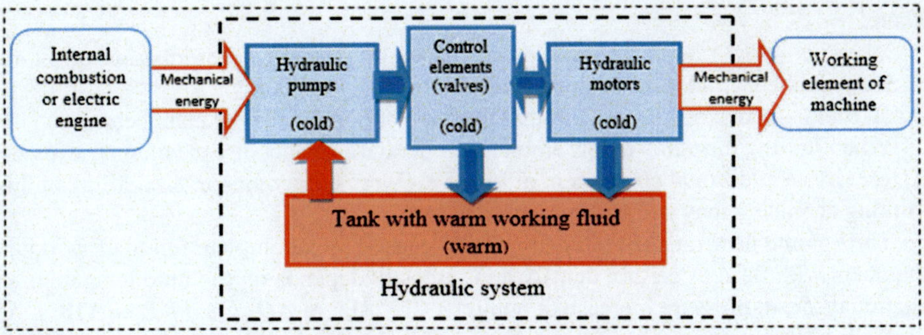

Fig. 3. Diagram of a machine with cold hydraulic system supplied with warm working fluid from tank [6]

The time needed for the supply of warm working fluid depends on the length of the cold oil filled pipe and the average flow velocity depending on the pipe diameter and flow rate. The longer the hose connecting the tank to the pump is, the longer it takes to start up when supplied with cold oil.

The longer the time of supplying cold working fluid to the pump is, the more likely the decrease in volume efficiency occurs. These effects are described in Sect. 2.1.

The shorter the time of supplying cold working fluid to the pump is, the faster the thermal shock conditions will occur, causing the possibility of disappearance of clearance between the cooperating elements of pump. This can affect change volumetric, torque and total efficiency.

3 Efficiency of Hydraulic Pumps Operated in Low Ambient Temperatures

When starting the hydraulic system in thermal shock condition, dynamic process of heating the pump elements occurs. It has large impact on changes in clearances between cooperating elements and the properties of the working fluid. As a result, the pump efficiency changes as function of time.

During start-up in thermal shock conditions, the pump efficiency changes as a function of time to steady state.

The change in pump volumetric efficiency $\eta_{Pv}(\tau)$ over time τ is calculated from the ratio of actual flow rate $Q_P(\tau)$ to the theoretical flow rate Q_{Pth}:

$$\eta_{Pv}(\tau) = \frac{Q_P(\tau)}{Q_{Pth}} \tag{1}$$

or:

$$\eta_{Pv}(\tau) = \frac{Q_{Pth} - \Delta Q_{Strat}(\tau)}{Q_{Pth}} \tag{2}$$

The pump's volume losses are determined by the sum of its components:

$$\Delta Q_{Strat}(\tau) = \Delta Q_S(\tau) + \Delta Q_V(\tau) \tag{3}$$

where:

$\Delta Q_S(\tau)$ – flow rate of volume losses caused by periodical deformation of pump working chambers and fluid compressibility,

$\Delta Q_V(\tau)$ – leakage flow rate through the gaps of pump working chambers.

When starting the pump in low ambient temperature, an increase in volumetric losses $\Delta Q_S(\tau)$ by an additional component of losses $\Delta Q_N(\tau)$ of incomplete fluid filling of the working chambers may occur.

The volume flow rate of volumetric losses caused by incomplete filling of working chambers with fluid of certain density and viscosity depends on the fluid temperature, rotational speed, pressure increase in pump [2, 27, 28], i.e. $\Delta Q_N(\tau) = f(T, n_P, \Delta p_P)$.

Such phenomena in operating conditions of hydraulic machines and devices in low ambient temperatures should not be allowed.

4 Stand for Testing Hydraulic Pumps in Low Ambient Temperatures

In order to detect phenomena occurring during start-up of hydraulic pumps (axial piston, radial piston and gear pumps) in low ambient temperatures experimental tests were carried out [3, 6–18, 20].

The Hydraulics Laboratory at the Faculty of Mechanical Engineering of the Gdańsk University of Technology is equipped with, among others, multi-pump supply devices with oil temperature stabilization, devices for testing components and hydraulic systems as well as a system for measuring and recording mechanical, hydraulic and therma quantities.

Fig. 4. Stand for testing gear pumps in low temperature chamber

The tested pumps were cooled to −25 °C in low temperature chamber. The tests were carried out without forced air flow in the chamber. The temperature of oil supplying the pump during tests in thermal shock conditions was set in the range of 20 to 60 °C (usually

50 °C) by means of oil temperature stabilization system containing heater, cooler and temperature controller.

During start-up of hydraulic pump, the following parameters were measured: pressure at the inlet and outlet of pump, flow rate, rotational speed of pump shaft, temperature in low temperature chamber, oil temperature, temperature of pump elements, torque of pump.

Hydraulic systems (pumps) were tested in low ambient temperatures in cases of:

1. cold pump start, which sucks cold working medium from tank (Figs. 4 and 5),

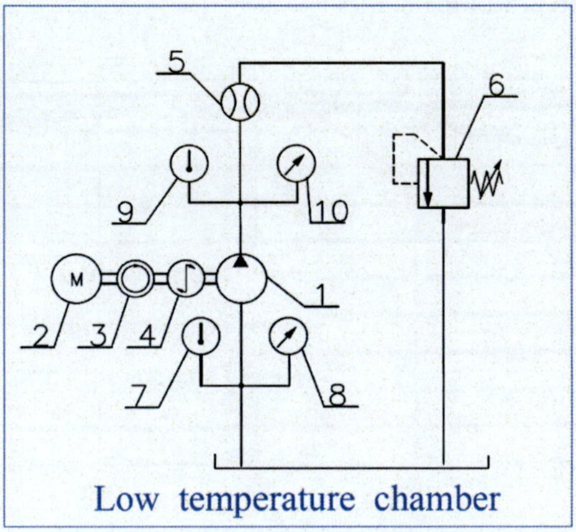

Fig. 5. Diagram of stand for testing pumps during start-up in low ambient temperature (cold pump and cold working fluid in tank): 1 - tested pump, 2 - electric motor, 3 - tachometer, 4 - torque meter, 5 - flow meter, 6 - overflow valve, 7, 9 - temperature sensors, 8, 10 - pressure sensors

2. cold pump start under thermal shock condition [6].

5 Tests of Hydraulic Gear Pumps in Low Ambient Temperatures

The author carried out tests of several gear pumps in low-temperature chamber. The cold pumps (theoretical displacements 12.2 cm^3/rev, 19.5 cm^3/rev, 20 cm^3/rev) were tested during start-up with cold oil (Fig. 5), as well as in thermal shock conditions (theoretical displacements of 5 cm^3/rev, 7 cm^3 rev, 12.2 cm^3/rev, 19.5 cm^3/rev, 20 cm^3/rev). In tested gear pumps the bodies were made of aluminum alloy, while the moving parts were made of steel. Aluminum alloy has almost twice the value of the coefficient of linear expansion in relation to steel, as a result there was no disappearance of clearance between gear vertices and the body bore.

One of the tests was carried out when the cold pump (theoretical displacement q_{Pth} = 19.5 cm^3/rev) sucked in cold oil from tank (Bartran HV 22 BP [36]) and pumped through gear flow meter and overflow valve to tank [8]. The pump was driven by an electric motor whose shaft rotated depending on the pump load at rotational speed 1500 rpm. Based on the determined volumetric efficiency, it was found that undesirable effects occur when the pump is started in low temperature of −20 °C [8].

Other tests were carried out for cold gear pump (theoretical displacement of 19.5 cm^3/rev) sucking in oil (Total Azolla ZS 46) at same temperature of 0 °C, 5 °C, 10 °C. After starting the pump (low ambient temperatures: 0 °C) for 160 s, the gas releases from the fluid and the pump working chambers are not completely filled with oil, and the fluid is pumped with noise. The volumetric efficiency is increased slowly from 0.63 to about 0.95 (Fig. 6). The period of incomplete filling of a cooled pump fluid during start-up depends on the ambient conditions, pump operating parameters and properties as well as the temperature of the fluid.

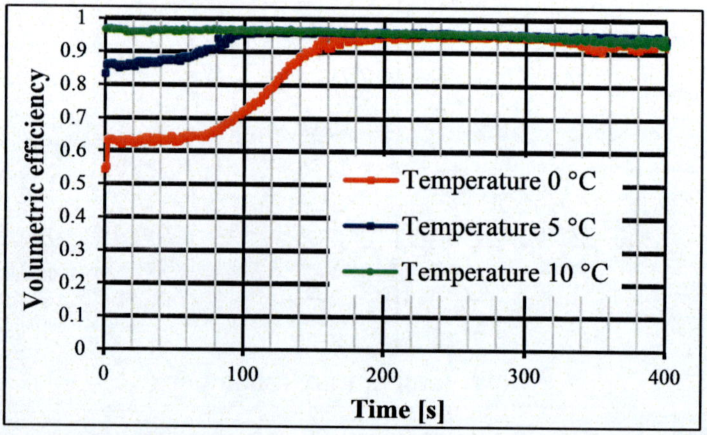

Fig. 6. Volumetric efficiency of gear pump (theoretical displacement 19.5 cm^3/rev) as function of time during start-up in low ambient temperatures: 0 °C, 5 °C, 10 °C (oil Total Azolla ZS 46) [6]

After starting the pump in ambient temperature of 5 °C for about 100 s, the volumetric efficiency of pump was reduced (Fig. 6).

Based on the test (Figs. 7 and 8) of the cold gear pump (theoretical displacement of 20 cm^3/rev), which sucks cold working fluid from tank, changing volumetric efficiency values (Fig. 7) can be seen. In the initial period of start-up process, the pump chambers are not completely filled with liquid. Only after 150 s, there is an increase in suction pressure and volume efficiency. The increase in volumetric efficiency indicates reduction of gas extraction from oil. It can be stated that the suction pressure value of the pump started in low ambient temperature is of great importance for its operation and affects volumetric efficiency.

The volumetric efficiency of the pump, in which the working chambers were not completely filled with fluid, changed as a function of time. The period of incomplete fluid filling of working chambers depends on the load, type of oil and change in fluid temperature (viscosity).

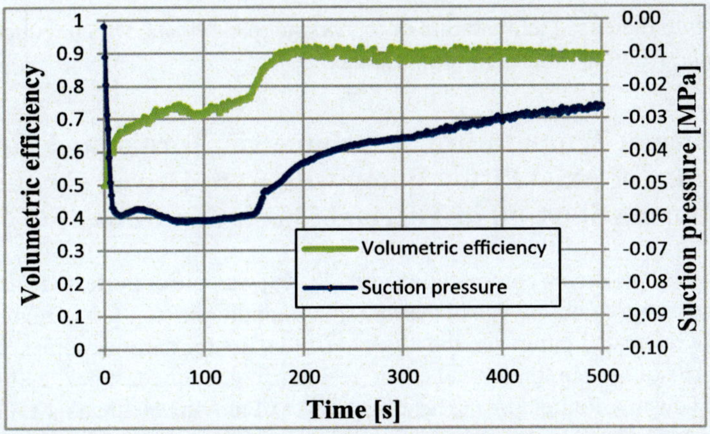

Fig. 7. Volumetric efficiency of gear pump (theoretical displacement 20 cm^3/rev) and suction pressure as function of time during start-up in low ambient temperature (temperature of pump before start-up: 0 °C, oil Total Azolla ZS 46)

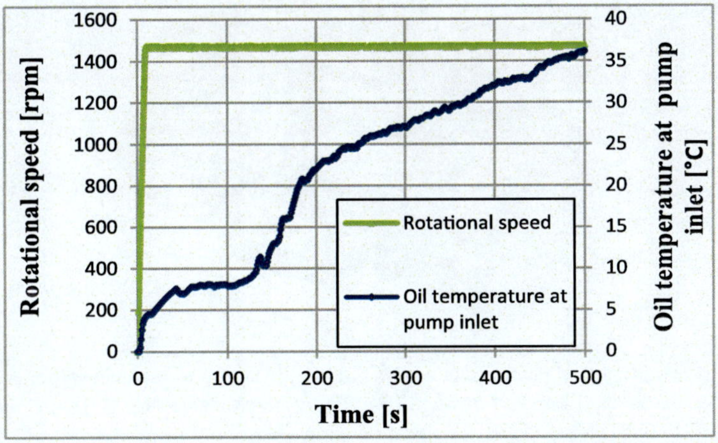

Fig. 8. Rotational speed of gear pump (theoretical displacement 20 cm^3/rev) and oil temperature at the pump inlet as function of time during start-up in low ambient temperature (temperature of pump before start-up: 0 °C, oil Total Azolla ZS 46)

Test results of many gear pumps operated in thermal shock conditions were presented in [6–8, 11]. One of tests of the pump (theoretical displacement $q_{Pth} = 19.5$ cm^3/rev) in thermal shock conditions was performed for parameters: oil temperature $T_{ol} = 46$ °C, ambient temperature $T_{ot} = -12$ °C, flow rate $Q = 28$ dm^3/min and torque $M = 15$ Nm [11]. Volumetric efficiency in start-up is usually close to 100%. This is because of the small radial clearance, resulting from low ambient temperature and low oil temperature. Volumetric efficiency decreases when the temperature of pump elements rise. In the initial phase of start-up, torque efficiency is low and then gradually increases. This is associated with increased temperature of oil and pump elements, which reduces oil flow resistance and friction, and changing clearances.

6 Influence of Temperature on Volumetric, Torque and Total Efficiency of Axial Piston Pump with Cam-Driven Commutation Unit During Start-up in Thermal Shock Conditions

The pump with cam-driven commutation unit [23–25] was one of the tested piston pumps in low ambient temperatures, also in thermal shock conditions [6, 13, 15]. Figure 9 shows the design of the tested pump and the holes in fixed parts for thermocouples [6, 13, 15]. The fixed parts of the pump include: cover, real housing, cylinder block, cylinder block cover, front housing. For testing the heating process of moving elements, i.e. pistons and slipper, they were made in special version with holes for mounting thermocouples.

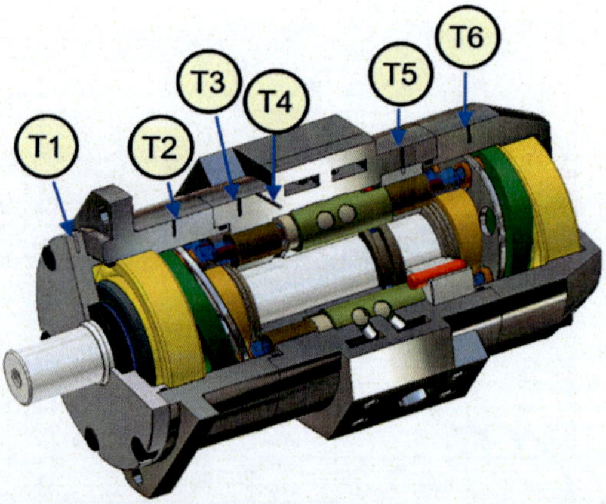

Fig. 9. Pump design, placed the thermocouples (T1–T6) in the holes made in stationary elements of pump [13]: T1-cover, T2-front housing, T3-cylinder block - area No. 1, T4-cylinder block - area No. 2, T5- cylinder block cover, T6 - rear housing

Continuous measurement: oil pressure, flow rate, leakage flow, torque, temperature of oil and elements made it possible to determine volumetric, torque and total efficiency of axial piston pump during start-up in thermal shock conditions. The pump was tested in thermal shock conditions using oil Total Azolla ZS 46 or Bartran HV 22 [36].

One of the tests of pump during start-up was made for the parameters: oil pressure on pump outlet 9.15 MPa, rotational speed 1100 rpm, oil temperature 49 °C, ambient temperature −12 °C.

Oil temperature on the inlet and outlet of the pump after 100 s increased from −12 °C to 38 °C (Fig. 9). Then oil temperature slowly rises to 49 °C.

In initial period of operation of pump, the temperature of oil leaks increases much more slowly than the temperature of the main working fluid stream at the inlet and outlet (Fig. 10). However, after some time, the leakage stream reaches temperature higher than the supply oil temperature. The fastest temperature rise in the cylinder block (Fig. 10) occurs near the pump internal channels through which the main stream of warm oil flows and in the vicinity of the pump timing (measurements with thermocouple No. 4). The pump cover (thermocouple No. 1) heats up slowly.

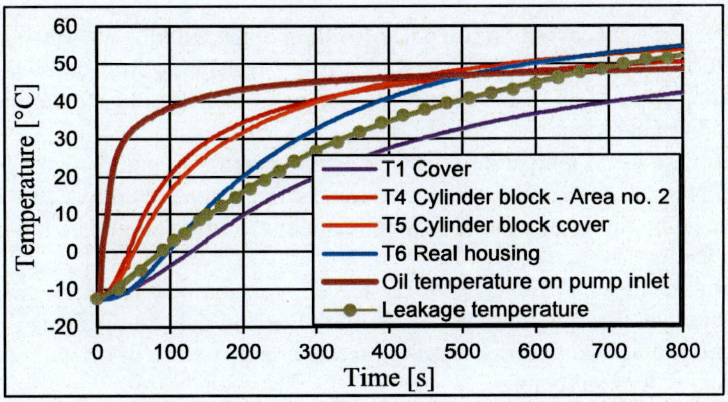

Fig. 10. Oil temperature on pump inlet, leakage temperature, elements temperature of axial piston pump during start-up in thermal shock conditions (oil Total Azolla ZS 46) [13]

The axial piston pump with cam-driven commutation unit is characterized by very high volumetric efficiency 0.98 (Fig. 11). In 800 s, the volumetric efficiency decreases to almost 0.96. 100 s after starting of the system, the torque efficiency of the pump remains at 0.8. The efficiency depends to a large extent on the pump load, which in this case was 9.15 MPa. The higher is the working pressure of the pump, the higher its torque efficiency, and in the same time the lower its volumetric efficiency.

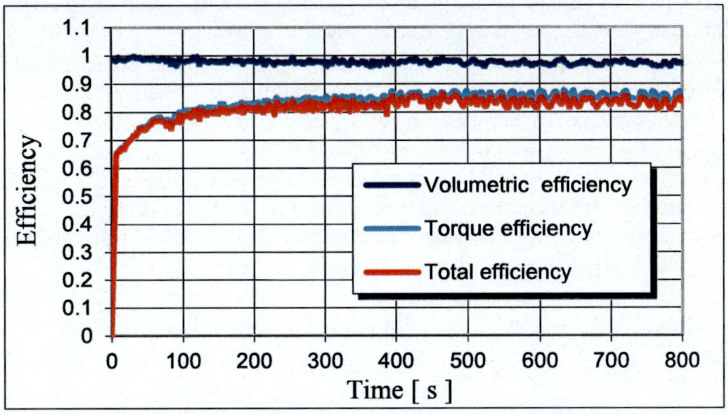

Fig. 11. Total, volumetric and torque efficiency of axial piston pump with cam-driven commutation unit for parameters: pumping pressure 9.15 MPa, flow rate 30 dm^3/min, ambient temperature −12 °C, oil temperature 49 °C [13]

7 Summary

Based on the results of experimental tests of pumps, it was found that in self-sucking hydraulic pumps (gear, piston) operated in low ambient temperatures, the working chambers may not be completely filled with fluid. This reduces the volumetric efficiency of the pumps.

In order to prevent cavitation (incomplete filling of the working chambers) in pumps during start-up in low ambient temperatures, it is advisable to heat the working fluid before starting the system or pre-supply the pumps to increase the suction pressure by using an additional pump, e.g. screw pump.

The decrease in oil temperature as well as the increase in pump shaft speed cause increase in pressure loss in the suction section of the pump. Therefore, the rotational speed of the pump during start-up in low ambient temperatures should be lower than the nominal value.

The torque and total efficiency of the tested gear and piston pumps, which were started in thermal shock conditions, in the initial period depend on many parameters among others: rotational speed, pumping pressure, temperature of oil and elements of pump, design of hydraulic units.

References

1. Balawender, A.: Energy Analysis and Methodology for Research Low-Speed Hydraulic Motors. Scientific Notebooks of the Gdańsk University of Technology, Gdańsk (1988). (in Polish)
2. Biały, J.: Research of gear pumps at reduced temperature. Ph.D. thesis, Military Technical Academy, Warszawa (1981). (in Polish)
3. Bocian, M.: Tests of gear pump at low ambient temperatures. Thesis, supervisor: Jasiński, R. Gdańsk University of Technology, Gdańsk (2013). (in Polish)
4. Deptuła, A., Osiński, P., Partyka, M.: Identification of influence of part tolerances of 3PWR-SE pump on its total efficiency taking into consideration multi-valued logic trees. Pol. Marit. Res. **24**(1) (2017). https://doi.org/10.1515/pomr-2017-0006
5. Dindorf, R.: Modeling and Simulation of Non-Linear Elements and Control Systems of Fluid Power Drives. Politechnika Świętokrzyska Publishing House, Kielce (2004). (in Polish)
6. Jasiński, R.: Operation of components of hydraulic drive of machines in low ambient temperatures. Monograph **166**, Gdańsk (2018). (in Polish)
7. Jasiński, R.: Research and method for assessment of operation of hydraulic drive components started in low ambient temperature and supplied with warm working medium. Report on execution of the research project financed by Polish Ministry of Science and Higher Education (in Polish), No. 4 T07C042 30, Gdańsk (2010). (in Polish)
8. Jasiński, R.: Research of hydraulic pumps in low ambient temperatures. Chapter in the monograph, Research, design, production and operation of hydraulic systems". Klich, A., Kozieł, A., Palczak, E. (eds.) CYLINDER 2010. Instytut Techniki Górniczej KOMAG, Gliwice (2010). (in Polish)
9. Jasiński, R.: Operation of hydraulic drives in low ambient temperatures. Part I. Hydraulika i Pneumatyka **1** (2011). (in Polish)
10. Jasiński, R.: Operation of hydraulic drives in low ambient temperatures. Part II. Hydraulika i Pneumatyka **2** (2011). (in Polish)

11. Jasiński, R.: Influence of type of material on performance of hydraulic components in thermal shock conditions. Solid State Phenom. **183** (2012)
12. Jasiński, R.: Problems of the starting and operating of hydraulic components and systems in low ambient temperature (Part I). Pol. Marit. Res. **4** (2008)
13. Jasiński, R.: Problems of the starting and operating of hydraulic components and systems in low ambient temperature (Part II). Pol. Marit. Res. **1** (2009)
14. Jasiński, R.: Problems of the starting and operating of hydraulic components and systems in low ambient temperature (Part III). Pol. Marit. Res. **4** (2009)
15. Jasiński, R.: Problems of the starting and operating of hydraulic components and systems in low ambient temperature (Part IV). Pol. Marit. Res. **3** (2017)
16. Jasiński, R.: Methods of determination of correct operation area for hydraulic component in low ambient temperatures. Dev. Mech. Eng., Gdańsk (2008)
17. Jasiński, R.: Experimental tests of axial piston pump PWK 27 from HYDROTOR in low ambient temperatures. Napędy i Sterowanie **4** (2008). (in Polish)
18. Jasiński, R., Lewandowski P.: Modeling of heating process of piston pump 27 PWK company Hydrotor during start-up in low ambient temperatures. Napędy i Sterowanie, No. 6 (2008). (in Polish)
19. Kollek, W., Osiński, P., Stosiak, M., Wilczyński, A., Cichoń, P.: Problems relating to high-pressure gear micropumps. Arch. Civil Mech. Eng. **14**(1) (2014)
20. Lewandowski, P.: Experimental tests of operation of piston pump PWK 27 during start-up in low ambient temperature conditions. Thesis, supervisor. Jasiński, R. Gdańsk (2006). (in Polish)
21. Minav, T., Heikkinen, J., Schimmel, T. Pietola, M.: Direct driven hydraulic drive: effect of oil on efficiency in sub-zero conditions. Energies **12**(2) (2019)
22. Murrenhoff, H., Gels, S., Zahn, A., Paßolat, O.: Tieftemperaturuntersuchung von Zahnrad-pumpen. O + P **4** (2012)
23. Osiecki, L.: Piston pump with the cam commutation. Seminar papers: "Napędy i Sterowanie 2003", Gdańsk (2003). (in Polish)
24. Osiecki, A.: Hydrostatic Drive of Machines. WNT, Warszawa (1998). (in Polish)
25. Osiecki, L., Patrosz, P., Landvogt, B., Piechna, J., Zawistowski, T., Żyliński, B.: Simulation of fluid structure interaction in a novel design of high pressure axial piston hydraulic pump. Arch. Mech. Eng. **60**(4), 509–529, 2 (2013)
26. Osiński, P., Patrosz, P.: Research on dynamic pressure courses in displacement pumps using piezoelectric sensors, Napędy i Sterowanie, no. 11 (2016). (in Polish)
27. Przychodzień, T., et al.: Operation of Heavy Work Machines in Winter Conditions. WNT, Warszawa (1990). (in Polish)
28. Szewczyk, K.: Effect of Reduced Temperature on Characteristics of Hydraulic Systems of Heavy Work Machines. Politechnika Krakowska, Kraków (1991). (in Polish)
29. Stryczek, S.: Hydrostatic Drive. WNT, Warszawa (1992). (in Polish)
30. Śliwiński, P.: The influence of water and mineral oil on mechanical losses in the displacement pump for offshore and marine applications. Pol. Marit. Res. **25**(S1 (97)), 178–188 (2018). https://doi.org/10.2478/pomr-2018-0040
31. Śliwiński, P.: New satellite pumps. Fundam. Mach. Des. Key Eng. Mater. **490**. https://doi.org/10.4028/www.scientific.net/KEM.490.195
32. Śliwiński, P.: The methodology of design of axial clearances compensation unit in hydraulic satellite displacement machine and their experimental verification. Arch. Civil Mech. Eng. **19**(4), August 2019. https://doi.org/10.1016/j.acme.2019.04.003
33. Catalogues from Hydrotor company
34. Catalogues from Bosch-Rexroth company
35. Catalogues from Parker company
36. Data sheet of oil Bartran HV22 company BP

Flow Analysis of a 2URED6C Cartridge Valve

Edward Lisowski[1], Janusz Rajda[2], Grzegorz Filo[1(✉)], and Paweł Lempa[1]

[1] Faculty of Mechanical Engineering, Cracow University of Technology, Jana Pawla II 37, Cracow, Poland
{lisowski,filo}@mech.pk.edu.pl, plempa@pk.edu.pl
[2] PONAR S.A, ul. Wojska Polskiego 29, Wadowice, Poland
janusz.rajda@ponar-wadowice.pl

Abstract. Hydraulic systems are widely used in drive and control systems of multiple devices. Various types of control valves are used to determine the appropriate direction of the working fluid movement. This article provides flow analysis of a two-way hydraulic control valve in a cartridge housing. The flow rate through the valve is controlled by a spring and an electromagnet. Design of a poppet ensures tightness in both directions in the cut-off position. Accordingly, there is no need to use additional shut-off valves. However, the fluid stream entering the valve is divided into several jets, due to the cartridge design. This makes the flow complex and difficult to be modelled. Therefore, the advanced ANSYS/Fluent software was used to carry out the analysis. The conducted tests allowed the influence of selected geometrical and physical parameters on the valve characteristics to be examined.

Keywords: Cartridge valve · Seat valve · 3D modelling · Flow analysis · CFD

1 Introduction

Electromagnetically controlled two-way valves of 2URED6 type belong to the popular group of cartridge valves designed for hydraulic drives. Cartridge valves are used in both stationary and mobile hydraulics, enabling the implementation of various functions required in the hydraulic system, such as pressure, flow rate and direction control of the working fluid. Among the valves for controlling flow direction, the most popular are simple two-way valves, in particular electromagnetically controlled. These types of valves perform functions of cutting off or opening the flow, therefore are often known as the *on-off* valves. They allow virtually any configuration of connections between hydraulic lines with actuators to be made. At the same time, they can implement a modern control logic using modules based on microprocessor technology. This type of directional control valves is particularly suitable for small and medium flow rates. The valves are characterized by low weight and relatively simple construction, which makes them reliable in operation.

Among the electromagnetically controlled directional valves, two types of constructions can be distinguished: spool control valves, which cut off flow paths without ensuring

© The Editor(s) (if applicable) and The Author(s), under exclusive license to Springer Nature Switzerland AG 2021
J. Stryczek and U. Warzyńska (Eds.): NSHP 2020, LNME, pp. 40–49, 2021.
https://doi.org/10.1007/978-3-030-59509-8_4

the complete tightness, and directional seat valves (also called poppet valves), which ensure the tightness themselves. Hence, spool valves require additional hydraulic elements, such as check valves or controlled check valves, which is not necessary in the case of poppet valves. The above mentioned features of construction and functional advantages make directional seat valves often and willingly used for the design of hydraulic systems, especially in mobile applications. To ensure the complete tightness, closing of the flow most often takes place at the circular edge of the valve seat and at the surface of the mating element which usually is conical. Sealing rings are applied to balance the forces from the static pressure acting on the cylindrical surface of the valve poppet. Despite ensuring the static balance, directional seat valves may be characterized by relatively higher values of flow forces acting on the closing element (poppet), compared to directional spool valves. This results in a significant limitation of the valve operating range, assuming the use of an electromagnet of a similar power as in a spool valve.

The issue of research on flow forces affecting valve components by means of CFD method is known in the literature. Simic [1] conducted optimization of spool and housing geometry in a small hydraulic seat valve to enable the reduction of the axial flow forces. Sapra [2] presented simulation and experimental results of flow around conical element, Huovinen [3] provided results of experimental and numerical study of a choke valve in a turbulent flow, while Liu [4] showed realization of direct proportional flow control with load-pressure compensation feature on a load control valve. Similarly, modelling of flow through a three-stage large-flow directional valve which includes two cartridge poppet valves was a subject undertaken by Xu [5]. This work is a continuation of previous research that concerned improvement of a control valve geometry by means of CFD methods related to a differential switching valve installed inside a throttle-check valve block [6, 7]. The main aim of the article is to study the possibility of reducing flow forces acting on a poppet head, thereby increasing the operational range of a directional seat control valve.

2 Object of Research

A 2-way directional seat valve for cartridge installation (2URED6C5) designed and manufactured by PONAR Wadowice is the object of research. The work concerns modernization of the valve in order to increase maximum operating pressure from 25 MPa to 35 MPa, extend the flow rate range to at least 30 dm^3/min and improve the boundary curve of the operating range. Increase in maximum operating pressure requires the use of a new type of electromagnet, which must meet the strength requirements due to the increased pressure at the *P* connection. A 3D model of the valve is shown in Fig. 1.

The longitudinal section through the hydraulic working part of the valve is shown in Fig. 2. The valve opens after supplying voltage to the coil of solenoid (4). The poppet (2) moves inside the valve body (1) in the x direction after the electromagnet overcomes initial tension of a return spring (3). Turning off the supply voltage results in return of the poppet to its initial position due to the force generated by the return spring. As a result, the *P*–*A* channel is closed and flow through the valve is cut off.

Depending on the pressure difference at ports *A* and *P*, the fluid flow can take place from port *P* to *A* or vice versa. The leak-tightness isolation of the flow is maintained in

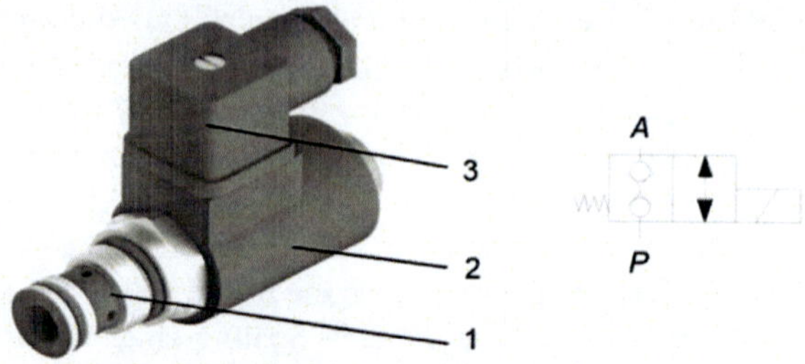

Fig. 1. General view and graphic symbol of the 2URED6C5 cartridge seat valve; 1 – body of the hydraulic working part, 2 – electromagnet, 3 – power supply connector, A, P – connection ports.

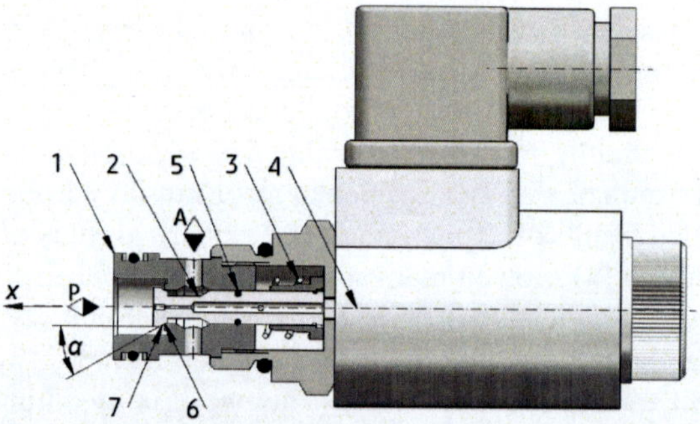

Fig. 2. Longitudinal cross section through the valve: 1 – body, 2 – poppet, 3 – return spring, 4 – electromagnet, 5 – sealing of the cylindrical part of poppet, 6 – poppet seat, 7 – conical surface of the poppet, α – inclination angle of the cone.

both directions in case when the solenoid is switched off. The tightness is ensured by sealing (5) placed on the cylindrical part of the poppet as well as the way of closing the valve by pressing the conical part of the poppet against the sharp edge of the seat (6) using the return spring (4). Thus, requirements for the accuracy of circularity of both elements determine the level of tightness.

The catalogue flow resistance characteristic of the 2URED6C5 valve is shown in Fig. 3, while an operating range is presented in Fig. 4. As it arises from Fig. 4, the operating range for high pressure and flow rate values is significantly narrowed by the boundary curve, particularly in the *A-P* direction.

3 Numerical Model and Results of CFD Analysis

Numerical analysis included the choice of turbulence model, mesh creation, conducting simulations and presentation of the obtained results.

3.1 Turbulence Model

In order to carry out the analysis, first kind of the flow must be determined. Due to the fact that there are no suitable condition for laminar flow formation in the directional control

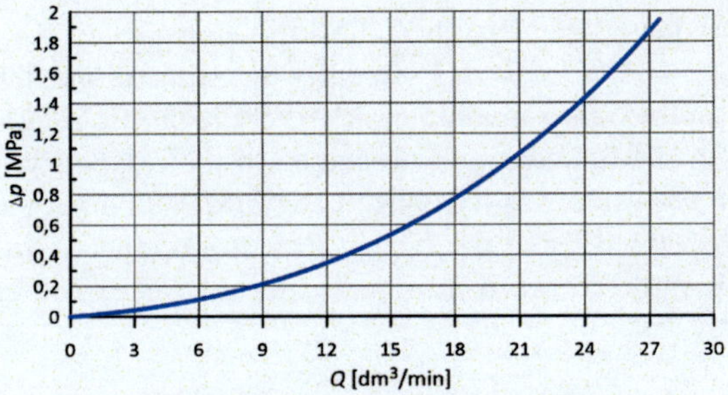

Fig. 3. Flow resistance characteristic of the 2URED6C5 valve.

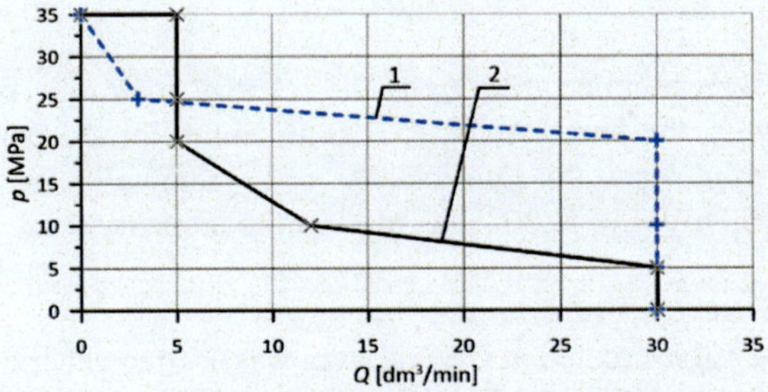

Fig. 4. Limit curve of the 2URED6C5 valve operating range with a standard poppet design (cone angle $\alpha = 30°$); 1 – flow in the *P–A* direction, 2 – flow in the *A–P* direction.

valve channels, the turbulent flow pattern has been applied. The ANSYS/Fluent system provides several different models of the turbulence [8], including $k-\varepsilon$, $k-\omega$, $k-kl-\omega$, v^2-f and others. In the case of flow through a directional control valve the $k-\varepsilon$ model works well enough, which was stated e.g. by Ye [9], Pan [10] and Domagala [11]. There are several variants of $k-\varepsilon$ model, including *Standard*, *RNG* and *Realizable*. In this work the *Realizable* $k-\varepsilon$ model was chosen for simulation. According to the model, the following equations (Eq. 1) and (Eq. 2) are used to calculate kinetic energy of the turbulence k and dissipation factor ε, respectively:

$$\frac{\partial(\rho k)}{\partial t} + \frac{\partial(\rho k u_i)}{\partial x_i} = \frac{\partial}{\partial x_j}\left[\left(\mu + \frac{\mu_t}{\sigma_k}\right)\frac{\partial k}{\partial x_j}\right] + G_k + G_b - \rho\varepsilon - Y_M + s_k, \quad (1)$$

$$\frac{\partial(\rho \varepsilon)}{\partial t} + \frac{\partial(\rho \varepsilon u_i)}{\partial x_i} = \frac{\partial}{\partial x_j}\left[\left(\mu + \frac{\mu_t}{\sigma_k}\right)\frac{\partial \varepsilon}{\partial x_j}\right] + C_{1\varepsilon}\frac{\varepsilon}{k}(G_k + C_{3\varepsilon}G_b) - C_{2\varepsilon}\rho\frac{\varepsilon^2}{k} + s_\varepsilon. \quad (2)$$

The G_k parameter defines increase in the kinetic energy of turbulence caused by gradient of average velocities, while G_b and Y_M represent energy generated by the buoyancy phenomenon and the compressibility of fluid, respectively. The $C_{1\varepsilon}$, $C_{2\varepsilon}$ and $C_{3\varepsilon}$ parameters are model constants, s_k, s_ε are Prandtl numbers. The μ_t turbulent viscosity is calculated according to Eq. 3:

$$\mu_t = \rho \cdot C_\mu \frac{k^2}{\varepsilon}. \quad (3)$$

Model constants have been assigned the following values: $C_{1\varepsilon} = 1.44$, $C_{2\varepsilon} = 1.92$, $C_\mu = 0.09$, $s_k = 1.0$, $s_\varepsilon = 1.3$. To carry out simulations, the boundary conditions were defined in the *Setup* module of ANSYS/Fluent program. The inlet velocity magnitude normal to the boundary was assumed in the *Boundary Conditions* option. The *Velocity magnitude* value was introduced as *Workbench* input parameter, while the outlet pressure was defined using the *Gauge Pressure* option of the *Outlet condition.* Values of k and ε parameters were specified using the *Intensity I* and *Length scale* ℓ options:

$$I = 0.16 \cdot (Re)^{-1/8}, \tag{4}$$

$$\ell = 0.07 \cdot D_H, \tag{5}$$

where Re is Reynold's number and D_H is the relevant hydraulic diameter, which is $D_H = 8$ mm. Hence, the length scale $\ell = 0.56$ mm. For the considered flow rate $Q = 5 - 30$ dm^3 min^{-1} and fluid viscosity $\nu = 35$ mm^2 s^{-1}, the Reynolds number varies from $Re = 900$ to $Re = 5200$. This implies the intensity value $I = 5.5 - 6.9$ %.

Geometrical and Discrete Fluid Model

The 3D model of the valve created in CAD system was used to generate a model of fluid. The fluid model is shown in Fig. 5a. In addition, to visualize the internal details, Fig. 5b presents a longitudinal cross-section through the fluid model with a plane crossing the axis of symmetry of the valve.

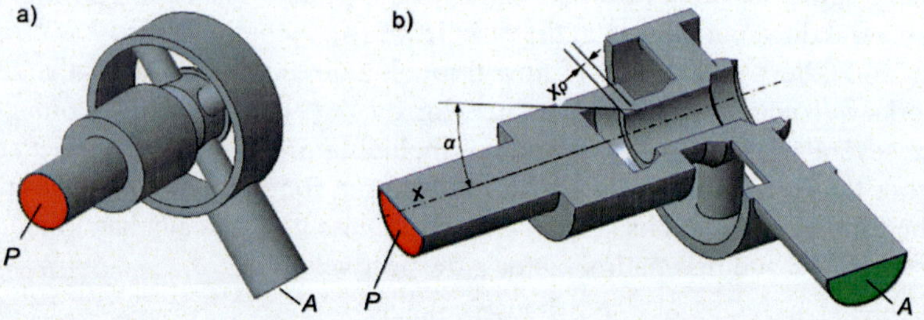

Fig. 5. Fluid model; P, A – connection ports, x_p – poppet gap width, α – poppet head angle.

The generated geometrical model of fluid was next processed in ANSYS/Fluent program in order to obtain a discrete model (finite element mesh). The created mesh is shown in Fig. 6. It includes about 350 000 cells and 150 000 nodes.

The boundary conditions and physical parameter values were set as follows:

- average fluid velocity at the inlet defined by the input flow rate,
- static pressure at the outlet, value $p = 0.1$ MPa,
- considering the flow rate range of the valve, the following flow rate values were selected for the simulation: $Q = 5, 10, 20, 30$ dm^3/min,
- channel walls are smooth with no fluid slip effect on the surfaces,

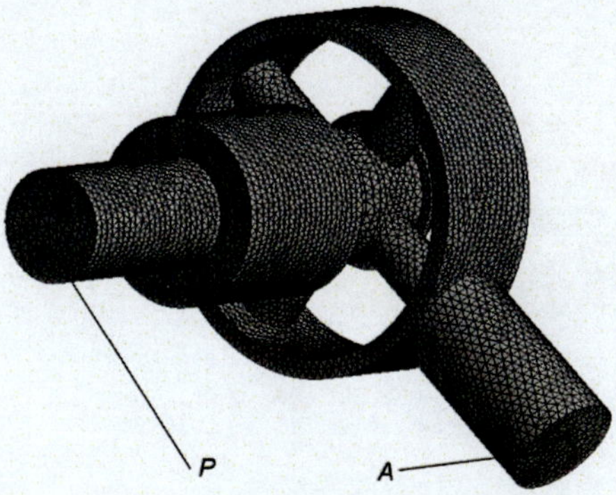

Fig. 6. Mesh of the fluid domain.

- working fluid is hydraulic oil with the parameters: density $\rho = 870\ \text{kg/m}^3$, dynamic viscosity $\mu = 0.013\,\text{Pa}\cdot\text{s}$.

The *Pressure based* solver was used to carry out the pressure-velocity coupling. The absolute value of residuals of mass and momentum equal to 10^{-4} was adopted as a convergence criterion. The model was meshed in accordance with the general Fluent recommendations, including skewness below 0.5 and aspect ratio less than 50. The following quality criteria of the grid were achieved: skewness 0.42–0.45 and maximum aspect ratio around 5.

3.2 Results of Simulations

The series of calculations were carried out in ANSYS/Workbench using the previously created geometry and the assumed boundary conditions for all selected flow rate values in both directions: *P–A* and *A–P*. The analysis revealed that the most adverse flow-related forces occur at the flow from *P* to *A*. Figure 7a shows pressure distribution on the valve walls in case of the *P–A* flow at a flow rate $Q = 20\ \text{dm}^3/\text{min}$ and the standard value of a poppet head inclination angle $\alpha = 30°$. The pressure drop for this case was $\Delta p = 0.65$ MPa.

Fluid velocity distribution along the streamlines in the vertical plane of symmetry of the valve and in the horizontal plane of symmetry of the A port is presented in Fig. 7b. The stream of inflowing fluid is divided into four smaller streams and then reconnected at the entrance to the valve seat chamber. The fluid then flows around the conical head of the poppet in the seat chamber. This generates an axial force associated with the flow which acts in the opposite direction of the liquid flow.

In order to reduce the axial force acting on the poppet, it was decided to modify the cone angle of the poppet head. Due to the fact that axial component of flow force is reversely proportional to the value of the cone angle, simulations were carried out with an increased value of the cone angle. Figure 8 shows pressure distribution on the valve walls and velocity distribution in selected planes for the *A–P* direction, the flow rate value $Q = 20\ \text{dm}^3/\text{min}$ and the cone angle $\alpha = 60°$. The obtained pressure drop was

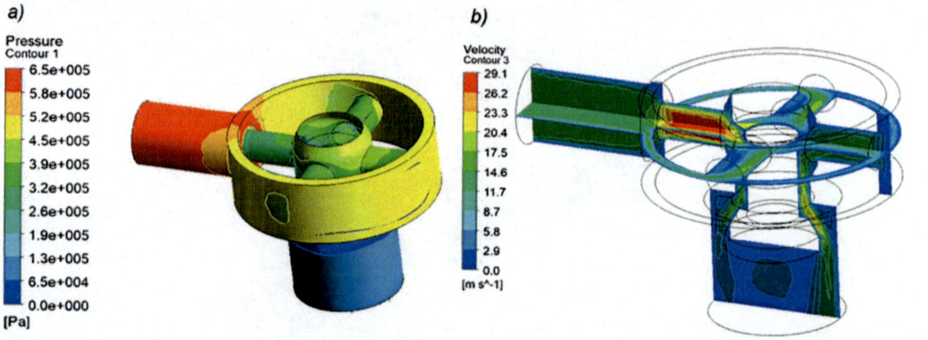

Fig. 7. CFD simulation results for $\alpha = 30°$ and $Q = 20$ dm^3/min; a) pressure distribution, b) velocity distribution in the plane of symmetry of the valve and in the plane of symmetry of the *A* port.

$\Delta p = 0.69$ MPa, which means that increase in the cone angle from 30° to 60° caused increase in flow resistance by about 6%.

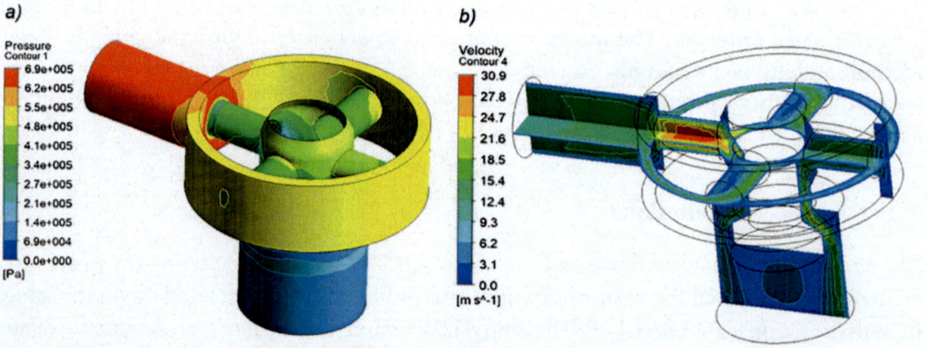

Fig. 8. CFD simulation results for $\alpha = 60°$ and $Q = 20$ dm^3/min; a) pressure distribution, b) velocity distribution in the plane of symmetry of the valve and in the plane of symmetry of the *A* port.

3.3 Estimation of Axial Flow Force

Based on the results of CFD analysis, the flow force was determined for different values of the poppet head cone angle, $\alpha = 30°$, 45°, 50° and 60°, respectively. The obtained results in the form of chart of the axial component of flow force against the cone angle are shown in Fig. 9. It arises from the results, that increase in the value of the cone angle results in a significant reduction in the axial component of flow force. However, this is also associated with an increase in flow resistance by several percent and a slight deterioration of the sealing conditions of the poppet.

When considering the cone angle $\alpha = 60°$, decrease in the F_x axial component of the flow force is over 50% compared to the value of the base solution $\alpha = 30°$. This result accordingly contributes to a significant increase in the 2URED6C5 valve operating range.

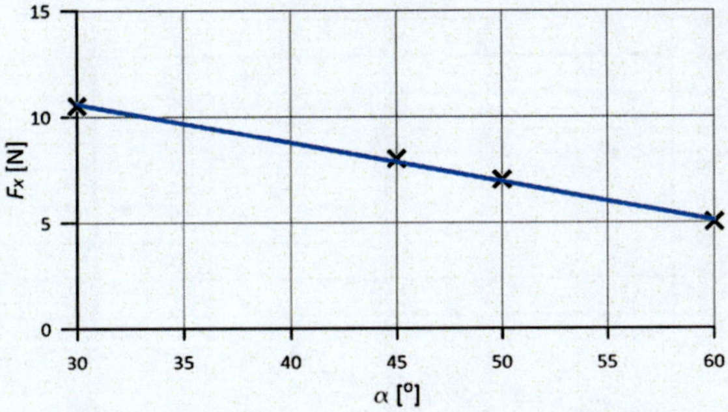

Fig. 9. Axial component of flow force F_x against cone angle of the poppet head α

4 Test Bench Experiments with the Modified Poppet

In the next step, a prototype valve poppet with the cone angle $\alpha = 60°$ was designed and manufactured. Then, experimental research was carried out on the test bench shown in Fig. 10.

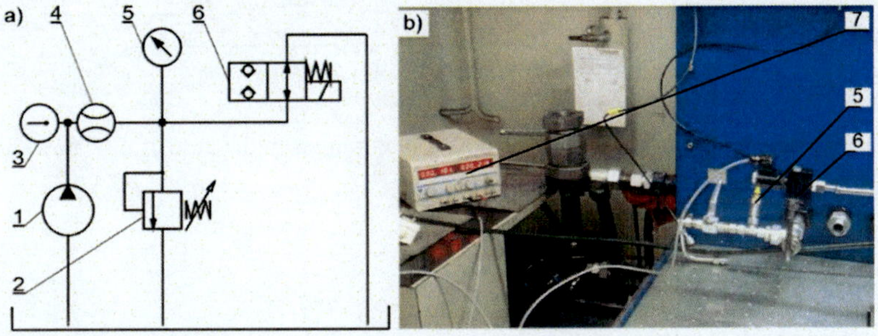

Fig. 10. Test bench: a) scheme, b) view; 1 – pump, 2 – relief valve, 3 – temperature meter, 4 – flowmeter, 5 – pressure transducer, 6 – tested valve, 7 – power supply.

The experiments were aimed at determining the boundary curves of flow characteristics of the tested 2URED6C5 valve. Results of the conducted test bench experiments are presented in Figs. 11 and 12. Comparison of the characteristics obtained for both flow directions (*P–A* and *A–P*) revealed a significant increase in the operating range of the valve, particularly in the case of flow in the *A–P* direction.

As a criterion of the operating range improvement, comparison of the surface area contained under the limit curves was adopted. Guided by this criterion, it can be assessed that the improvement in the operating range is as follows:

- 50% – for the *P–A* flow direction,
- 250% – for the *A–P* flow direction.

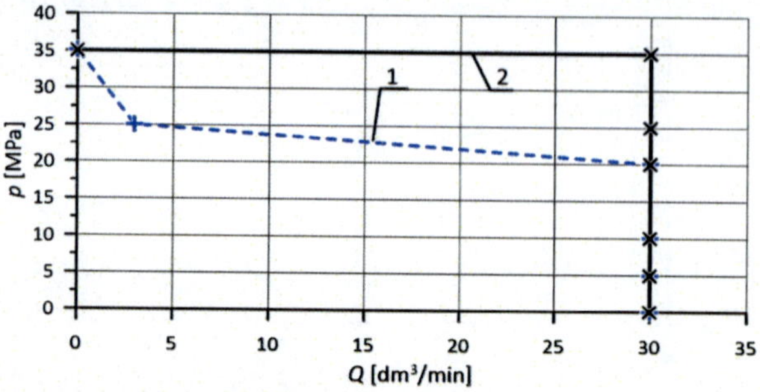

Fig. 11. Comparison of limit curves of valve operating ranges in the case of *P–A* flow direction; 1 – cone angle $\alpha = 30°$, 2 – cone angle $\alpha = 60°$.

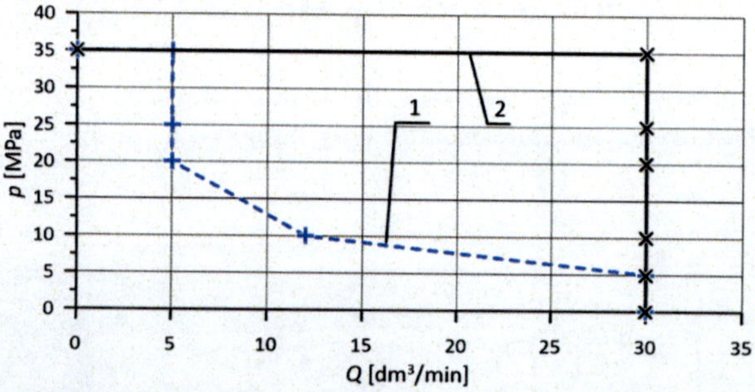

Fig. 12. Comparison of limit curves of valve operating ranges in the case of *A–P* flow direction; 1 – cone angle $\alpha = 30°$, 2 – cone angle $\alpha = 60°$.

5 Summary

During the experimental testing of the poppet prototype with a cone angle $\alpha = 60°$, a slightly worse sealing tightness was found. Therefore, further research is needed in order to clarify the causes. In addition, the obtained performance characteristics of the valve indicate that its operation may be expected at the flow rate range up to 35 dm^3/min, or even up to 40 dm^3/min, at the operating pressure close to the maximum value $p =$ 35 MPa.

Within the future work, a compromise solution, consisting in making and testing a valve prototype with the poppet head angle $\alpha \sim 45°$ is considered. With this angle value, it is expected that the operational range would be still significantly increased (which results from the chart presented in Fig. 9) with a slight increase in flow resistance and maintained sufficient conditions for tight closing of the valve.

References

1. Simic, M., Herakovic, N.: Reduction of the flow forces in a small hydraulic seat valve as alternative approach to improve the valve characteristics. Energy Convers. Manage. **89**, 708–718 (2015)

2. Sapra, M.K., Bajaj, M., Kundu, S.N., Sharma, B.S.V.G.: Experimental and CFD investigation of 100 mm size cone flow elements. Flow Meas. Instrum. **22**, 469–474 (2011)
3. Huovinen, M., Kolehmainen, J., Koponen, P., Nissilä, T., Saarenrinn, P.: Experimental and numerical study of a choke valve in a turbulent flow. Flow Meas. Instrum. **45**, 151–161 (2015)
4. Liu, J., Xie, H., Hu, L., Yang, H., Fu, X.: Realization of direct flow control with load pressure compensation on a load control valve applied in overrunning load hydraulic systems. Flow Meas. Instrum. **53**, 261–268 (2017)
5. Xu, B., Ding, R., Zhang, J., Su, S.: Modeling and dynamic characteristics analysis on a three-stage fast-response and large-flow directional valve. Energy Convers. Manage. **79**, 187–199 (2014)
6. Lisowski, E., Filo, G., Rajda, J.: Flow analysis of a switching valve with innovative poppet head geometry by means of CFD method. Flow Measur. Instrum. **70**, 101643 (2019)
7. Filo, G., Lisowski, E., Kwiatkowski, D., Rajda, J.: Numerical and experimental study of a novel valve using the return stream energy to adjust the speed of a hydraulic actuator. J. Mech. Eng. **65**, 103–112 (2019)
8. Ansys-Fluent: Users Guide, 13.0 ed (2011)
9. Ye, Y., Yin, C., Li, X., Zhou, W., Yuan, F.: Effects of groove shape of notch on the flow characteristics of spool valve. Energy Convers. Manage. **86**, 1091–1101 (2014)
10. Pan, X., Wang, G., Lu, Z.: Flow field simulation and a flow model of servo-valve spool valve orifice. Energy Convers. Manage. **52**, 3249–3256 (2011)
11. Domagala, M.: CFD analysis of a flow control valve. In: Proceedings of 5th FPNI PhD Symposium. Kraków, pp. 445–450 (2008)

Strength Calculation Methodology for Circumferential Backlash Compensation with Integrated Lips

Piotr Osiński(✉)

Wroclaw University of Science and Technology, 27 Wybrzeże Wyspiańskiego St., 50-370 Wrocław, Poland
piotr.osinski@pwr.edu.pl

Abstract. Contemporary gear pumps evolve towards greater internal tightness and consequently, higher working pressures. This trend necessitates the development of novel pump designs with the so-called circumferential compensation. In order to develop a new method of circumferential backlash compensation one needs a computational model for determining stresses and displacements in selected critical point. Thus the adoption of a mathematical model of the compensating element is the starting point for analytical strength calculations. In the considerations presented in this paper the semi-circular shape of the compensating lip is replaced with a model of a curved bar resting in several places on the tooth tips of the gears.

The presented material is the result of research work conducted under the project entitled: Development of high-pressure gear pump designs. The project received funding from Applied Research Programme path A, contract no. PBS3/A6/22/2015. The research was carried out in the Hydraulic Drives and Machine Vibroacoustics Laboratory at Wrocław University of Science and Technology in collaboration with the Hydrotor SA company.

Keywords: Gear pump · Gear pump break-in · High pressure gear pump

1 Introduction

The first pump design with circumferential backlash compensation was based on patent application P.397539 (exclusive right no. PL221099 [16]). Initially, constant lip thickness g_w of 2 mm was assumed in this design (Fig. 1).

This design solution was characterized by high internal tightness, starting from rotational speed n ≈ 500 rpm. Prior to assembly the lips had been bent towards the axis of the gears to ensure the elastic pressing of the compensating element against the points of the teeth. Durability tests showed that the structure would reach maximum pressure ptmax ≈ 5 MPa. Operation above the maximum pressure would result in lip breaking (Fig. 2 left side). A material flow effect, indicating stresses equal to or exceeding the yield point, can be seen in the lip. Figure 2 (right side) shows a larger part of the

© The Editor(s) (if applicable) and The Author(s), under exclusive license to Springer Nature Switzerland AG 2021
J. Stryczek and U. Warzyńska (Eds.): NSHP 2020, LNME, pp. 50–64, 2021.
https://doi.org/10.1007/978-3-030-59509-8_5

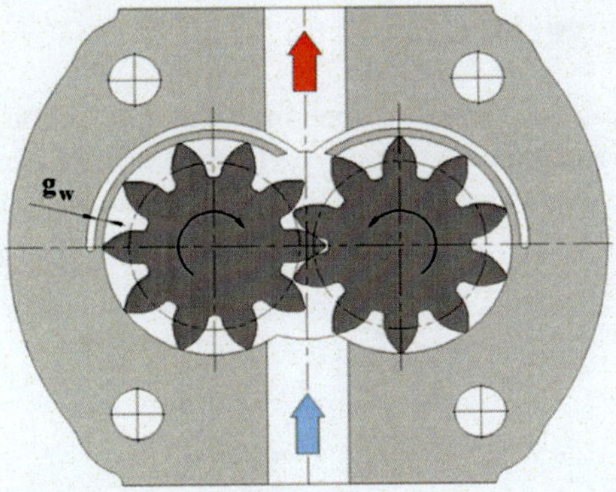

Fig. 1. First version of circumferential compensation with constant lip thickness, acc. to patent PL221099 (patent application no. P.397539 [16]).

compensating element with a local narrowing, resembling the "neck" occurring in an elastoplastic material subjected to the tension test. The breaking stress was reached on the lip's left side. The character of the lip breakage indicated a separating fracture, while the elongation of the lip indicates the occurrence of considerable tensile forces due to friction between the tooth points and the compensating lip.

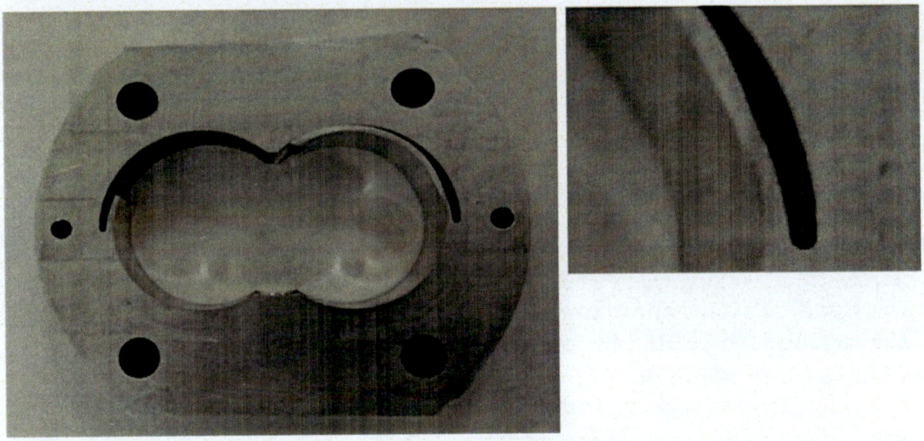

Fig. 2. Inner casing in first version of circumferential compensation: left side – after durability tests, right side – view of narrowing in compensating lip cross section.

This means that because of the relatively low strength of the lip this design solution can be used exclusively in high-efficiency and low-pressure pumps. Considering that gear pumps are marginally used in low-pressure systems and the tendency to increase the efficiency [1, 15, 17, 18] and operating pressure of gear pumps [2–5, 14], further research was focused on increasing the strength of the structure [8]. Designs with lips of variable thickness [8] with integrated [6, 7, 12] and separable [6, 8, 11, 16] lips were considered.

2 Designs with Integrated Lip

Designs with the integrated lip were claimed in patent application no. P.404801 (patent PL223649 [7]) and in patent application no. P.418261 (patent PL230845 [12]). The idea of this design solution, covered by patent no. PL223649 (patent application no. P.404801 [7]), is shown in Fig. 3.

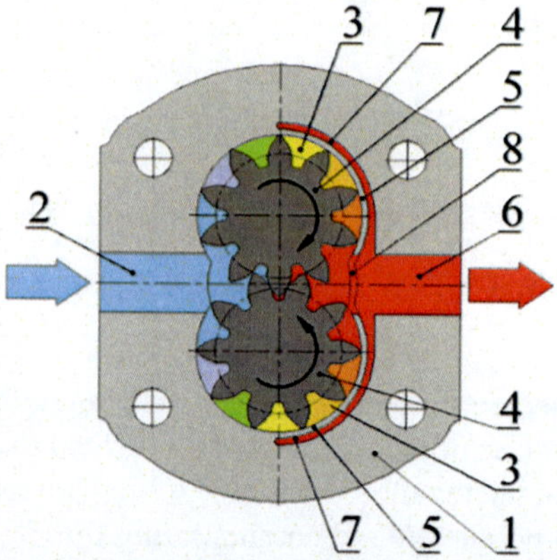

Fig. 3. Cross section through internal casing of pump with circumferential backlash compensation, acc. to patent no. PL223649 (patent application no. P.404801 [7]): 1 – pump's inner casing, 2 – suction channel, 3 – intertooth space, 4 – gears, 5 – circumferential backlash compensating lip, 6 – pressure channel, 7 – constant-pressure recess, 8 – channel connecting working chamber with pressure channel.

Circumferential backlash compensating partition 5 extends symmetrically along the whole width of gears 4, whereby additional channel 8, located symmetrically in the flexible partition, had to be introduced. Thanks to this channel the working medium can be forced from working chamber 3 to forcing channel 6. The forcing pressure acts on the outer surface of circumferential backlash compensating partition 5 through channels 7. Circumferential backlash compensating partition 5 is simultaneously pressed against two meshing gear wheels 4.

An exemplary inner casing with circumferential backlash compensation, made acc. to patent PL223649 (patent application no. P.404801 [7]), is shown in Fig. 4. Lips with variable thickness were used in this design. Moreover, thanks to the lip's dovetail shape the compensation can deform only in the assumed upper range. The displacement limiter is claimed in patent application no. P.427264 [10].

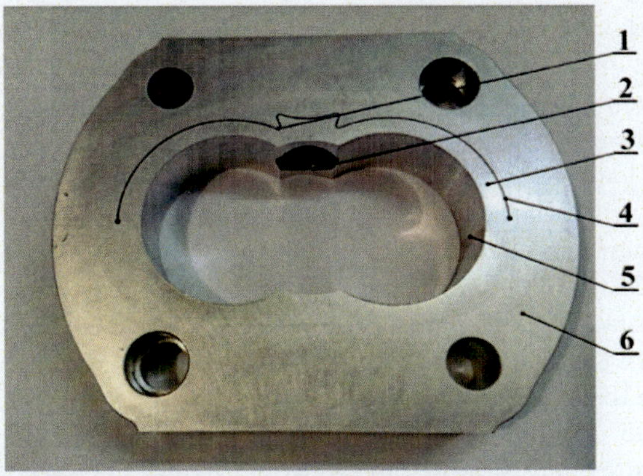

Fig. 4. Pump casing with integrated circumferential backlash compensation with variable lip thickness, acc. to patent PL223649 (patent application P.404801 [7]), and with displacement limiter acc. to patent application no. P.427264 [10]: 1 – channel connecting constant-pressure recess with pressure space, 2 – channel connecting working chamber with pressure channel, 3 – circumferential backlash compensating lip, 4 – constant-pressure recess, 5 – working chamber, 6 – outer casing.

3 Strength Calculations of Circumferential Backlash Compensation with Integrated Lip

In order to develop a new method of circumferential backlash compensation a computational model for determining stresses or displacements in selected critical points needs to be adopted. Thus a mathematical model of the compensating element load is the starting point for analytical strength calculations. In further considerations the semi-circular shape of the compensating lip is replaced with a model of a curved bar in several places

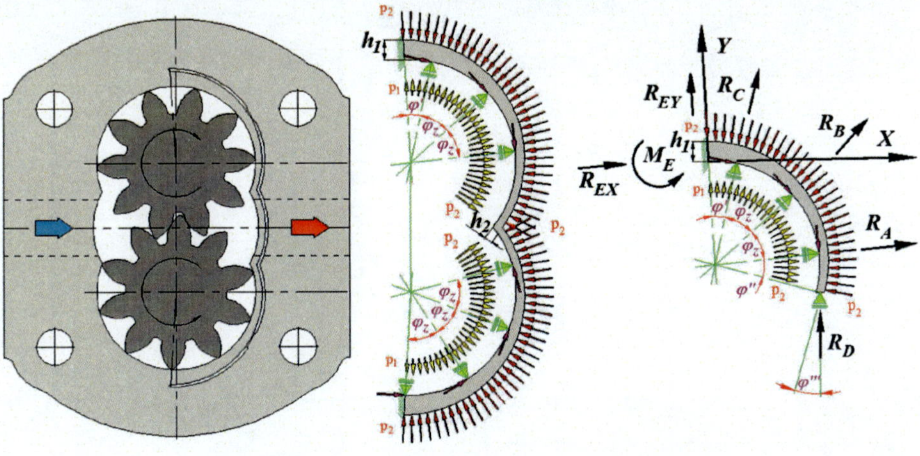

Fig. 5. Model of integrated compensating lip load.

resting on the points of gear wheels. This model also reflects the character of the interaction between the lip and the gear wheel. A four times statically indeterminate model with no support in the place of restraint of the compensating element (a wheel with 10 teeth cut in it) is presented here.

The system of loads in the model is internally balanced with reactions R_A, R_B, R_C, R_D in the supports, with friction forces T_A, T_B, T_C, pressure p_2 on the lip's outer side, the pressure linearly increasing from p_1 to p2 on the lip's inner side, reaction forces R_E and moment M_E in the restraint point. The lip's width in the direction perpendicular to the load surface is constant and amounts to b. Reaction R_D was introduced as an equivalent force resulting from the action of the symmetric integrated lip. Moreover, reaction R_D is perpendicular to the axis of symmetry of the casing, and so to the axis of symmetry of the integrated lip. Applying the above assumptions to the adopted loading scheme (Fig. 5) one can write the following three equations describing the stable equilibrium:

$$\sum P_x = R_{Dx} + R_{Ex} + R_{Cx} + R_{Bx} + R_{Ax} + T_{Cx} + T_{Bx} + T_{Ax} - P^b_{kx} = 0 \tag{1}$$

$$\sum P_y = R_{Dy} + R_{Ey} + R_{Cy} + R_{By} + R_{Ay} - T_{Cy} - T_{By} - T_{Ay} - P^b_{ky} = 0 \tag{2}$$

$$\begin{aligned} \sum M_E = M_E + R_D \cdot \cos\varphi''' \cdot r_w \cdot \sin(\varphi' + 2\cdot\varphi_z + \varphi'') \\ + R_C \cdot r_w \cdot \sin\varphi' + R_B \cdot r_w \cdot \sin(\varphi' + \varphi_z) + R_A \cdot r_w \cdot \sin(\varphi' + 2\cdot\varphi_z) \\ - T_C \cdot (r_w - r_w \cdot \cos\varphi') - T_B \cdot (r_w - r_w \cdot \cos(\varphi' + \varphi_z)) \\ - T_A \cdot (r_w - r_w \cdot \cos(\varphi' + 2\cdot\varphi_z)) - P^b_k \cdot r_w \cdot \sin\left(\frac{\varphi' + 2\cdot\varphi_z + \varphi''}{3}\right) = 0 \end{aligned} \tag{3}$$

Consistently with the Cartesian coordinate system, in formulas (1), (2) and (3) the following components of the reaction forces and the friction forces were used:

$$R_{Ax} = R_A \cdot \sin(\varphi' + 2\cdot\varphi_z) \tag{4}$$

$$R_{Ay} = R_A \cdot \cos(\varphi' + 2\cdot\varphi_z) \tag{5}$$

$$R_{Bx} = R_B \cdot \sin(\varphi' + \varphi_z) \tag{6}$$

$$R_{By} = R_B \cdot \cos(\varphi' + \varphi_z) \tag{7}$$

$$R_{Cx} = R_C \cdot \sin\varphi' \tag{8}$$

$$R_{Cy} = R_C \cdot \cos\varphi' \tag{9}$$

$$R_{Dx} = R_D \cdot \frac{\sin(\varphi' + 2\cdot\varphi_z + \varphi'')}{\sin\varphi'''} \tag{10}$$

$$R_{Dy} = R_D \cdot \frac{\cos(\varphi' + 2\cdot\varphi_z + \varphi'')}{\cos\varphi'''} \tag{11}$$

$$T_{Ax} = T_A \cdot \cos(\varphi' + 2 \cdot \varphi_z) \tag{12}$$

$$T_{Ay} = T_A \cdot \sin(\varphi' + 2 \cdot \varphi_z) \tag{13}$$

$$T_{Bx} = T_B \cdot \cos(\varphi' + \varphi_z) \tag{14}$$

$$T_{By} = T_B \cdot \sin(\varphi' + \varphi_z) \tag{15}$$

$$T_{Cx} = T_C \cdot \cos \varphi' \tag{16}$$

$$T_{Cy} = T_C \cdot \sin \varphi' \tag{17}$$

$$P_{kx} = P_k \cdot \sin \frac{\varphi_c}{3} \tag{18}$$

$$P_{ky} = P_k \cdot \cos \frac{\varphi_c}{3} \tag{19}$$

$$\varphi_c = \varphi' + 2 \cdot \varphi_z + \varphi'' \tag{20}$$

The value of equivalent force P_k^b for the constant load exerted by the hydrostatic pressure of the liquid on the compensating element amounts to

$$P_k^b = \frac{1}{2} b \cdot r_w \cdot (p_2 - p_1) \cdot \varphi_c \tag{21}$$

The above concentrated force is applied in the point corresponding to angle $\frac{\varphi_c}{3}$ measured relative to the point of restraint. For the adopted coordinate system XY the force can be resolved into the following components

$$P_{kx}^b = P_k^b \cdot \sin \frac{\varphi_c}{3} \tag{22}$$

$$P_{ky}^b = P_k^b \cdot \cos \frac{\varphi_c}{3} \tag{23}$$

Friction forces T_A, T_B and T_C resulting from the interaction between the tooth point and the compensating lip can be calculated, assuming uniform pressure, from relation (24) for friction loss moment ΔM in intertooth gaps versus theoretical moment M_t and hydromechanical efficiency η_{hm}.

$$\Delta M = M_t \cdot \left(\frac{1}{\eta_{hm}} - 1 \right) = r_w \cdot z_T \cdot T \tag{24}$$

where average friction force T following from the above assumption is

$$T_A = T_B = T_C = T_i = T \tag{25}$$

The average friction force T is applied to the radius r_w of the tooth tip. Variable z_T specifies the number of tips of the teeth interacting with the casing and the compensating element in the gear pump. Relations (24) and (25) can be used when the hydromechanical losses in the considered pump are known and if a significant predominance of mechanical losses over hydraulic losses is assumed. In addition, the distribution of pressures on the teeth interacting with the compensation was assumed. If the above criteria are not satisfied, forces T_A, T_B and T_C should be calculated from relations (26), (27) and (28) describing the kinetic (dynamic) friction forces occurring between the two elements moving relative to each other. Then the friction forces are linearly dependent on friction coefficient μ_{kin} being a function of rotational speed $\mu_{\text{kin}} = \mu_{\text{kin}}(n)$. During motion, friction coefficient μ_{kin} is lower than static friction coefficient μ_{sta} and increases with rotational speed n.

$$T_A = R_A \cdot \mu_{kin} \tag{26}$$

$$T_B = R_B \cdot \mu_{kin} \tag{27}$$

$$T_C = R_C \cdot \mu_{kin} \tag{28}$$

The value of the friction coefficient μ_{kin} is calculated from the friction loss moment ΔM (24), which can be determined after measuring the static characteristics of hydromechanical efficiency η_{hm}. [5]. The value μ_{kin} determined in this way is overestimated, because the other friction sources are not taken into account. From the point of view of strength, this is the most demanding criterion. Acceptance of a higher value for the coefficient provides overload reserve.

The considered beam is a four times statically indeterminate system. Reactions R_A, R_B, R_C and R_D in the beam's supports were assumed as statically indeterminate quantities. Hence in order to calculate them one should additionally formulate five equations based on Menabrea's theorem, assuming that the deflection in supports A, B, C and D is equal to zero.

$$f_A = \frac{\partial V}{\partial R_A} = \frac{1}{EI}\int_0^{\varphi_c} M_g \cdot \frac{\partial M_g}{\partial R_A} d\varphi = 0 \tag{29}$$

$$f_B = \frac{\partial V}{\partial R_B} = \frac{1}{EI}\int_0^{\varphi_c} M_g \cdot \frac{\partial M_g}{\partial R_B} d\varphi = 0 \tag{30}$$

$$f_C = \frac{\partial V}{\partial R_C} = \frac{1}{EI}\int_0^{\varphi_c} M_g \cdot \frac{\partial M_g}{\partial R_C} d\varphi = 0 \tag{31}$$

$$f_D = \frac{\partial V}{\partial R_D} = \frac{1}{EI}\int_0^{\varphi_c} M_g \cdot \frac{\partial M_g}{\partial R_D} d\varphi = 0 \tag{32}$$

Therefore the equations of bending moments in the particular intervals of the beam are as follows:

– for interval I $(0 < \varphi \leq \varphi'')$

$$M_g^I(\varphi) = -\frac{b \cdot (p_2 - p_1) \cdot r_w^2 \cdot \varphi^2}{2 \cdot \varphi_c} \cdot \sin\frac{\varphi}{3} - R_D \cdot r_w \cdot \sin\varphi''' \cdot \sin\varphi + \\ + R_D \cdot \cos\varphi''' \cdot (r_w - r_w \cdot \cos\varphi) \tag{33}$$

$$M_g^I(\varphi) = -\frac{b \cdot (p_2 - p_1) \cdot r_w^2 \cdot \varphi^2}{2 \cdot \varphi_c} \cdot \sin\frac{\varphi}{3} - 2 \cdot R_D \cdot r_w \cdot \sin\left(\varphi''' - \frac{\varphi}{2}\right) \cdot \sin\left(\frac{\varphi}{2}\right) \tag{34}$$

$$\frac{\partial M_g^I}{\partial R_A} = 0 \tag{35}$$

$$\frac{\partial M_g^I}{\partial R_B} = 0 \tag{36}$$

$$\frac{\partial M_g^I}{\partial R_C} = 0 \tag{37}$$

$$\frac{\partial M_g^I}{\partial R_D} = -2 \cdot r_w \cdot \sin\left(\varphi''' - \frac{\varphi}{2}\right) \cdot \sin\left(\frac{\varphi}{2}\right) \tag{38}$$

– for interval II $\varphi'' < \varphi \leq \varphi'' + \varphi_z)$

$$M_g^{II}(\varphi) = M_g^I(\varphi) + R_A \cdot r_w \cdot \sin(\varphi - \varphi'') - T_A \cdot \left[r_w - r_w \cdot \cos(\varphi - \varphi'')\right] \tag{39}$$

$$\frac{\partial M_g^{II}}{\partial R_A} = r_w \cdot \sin(\varphi - \varphi'') \tag{40}$$

$$\frac{\partial M_g^{II}}{\partial R_B} = 0 \tag{41}$$

$$\frac{\partial M_g^{II}}{\partial R_C} = 0 \tag{42}$$

$$\frac{\partial M_g^{II}}{\partial R_D} = -2 \cdot r_w \cdot \sin\left(\varphi''' - \frac{\varphi}{2}\right) \cdot \sin\left(\frac{\varphi}{2}\right) \tag{43}$$

– for interval III ($\varphi'' + \varphi\_z < \varphi \leq \varphi'' + 2\varphi_z$)

$$M_g^{III}(\varphi) = M_g^{II}(\varphi) + R_B \cdot r_w \cdot \sin(\varphi - (\varphi'' + \varphi_z)) \\ - T_B \cdot \left[r_w - r_w \cdot \cos(\varphi - (\varphi'' + \varphi_z))\right] \tag{44}$$

$$\frac{\partial M_g^{III}}{\partial R_A} = r_w \cdot \sin(\varphi - \varphi'') \tag{45}$$

$$\frac{\partial M_g^{III}}{\partial R_B} = r_w \cdot \sin(\varphi - (\varphi'' + \varphi_z)) \tag{46}$$

$$\frac{\partial M_g^{III}}{\partial R_C} = 0 \tag{47}$$

$$\frac{\partial M_g^{III}}{\partial R_D} = -2 \cdot r_w \cdot \sin\left(\varphi''' - \frac{\varphi}{2}\right) \cdot \sin\left(\frac{\varphi}{2}\right) \tag{48}$$

– for interval IV ($\varphi'' + 2\varphi\_z < \varphi \leq \varphi'' + 2\varphi_z + \varphi''$)

$$M_g^{IV}(\varphi) = M_g^{III}(\varphi) + R_C \cdot r_w \cdot \sin(\varphi - (\varphi'' + 2 \cdot \varphi_z)) - T_C \cdot \left[r_w - r_w \cdot \cos(\varphi - (\varphi'' + 2 \cdot \varphi_z))\right] \tag{49}$$

$$\frac{\partial M_g^{IV}}{\partial R_A} = r_w \cdot \sin(\varphi - \varphi'') \tag{50}$$

$$\frac{\partial M_g^{IV}}{\partial R_B} = r_w \cdot \sin(\varphi - (\varphi'' + \varphi_z)) \tag{51}$$

$$\frac{\partial M_g^{IV}}{\partial R_C} = r_w \cdot \sin(\varphi - (\varphi'' + 2 \cdot \varphi_z)) \tag{52}$$

$$\frac{\partial M_g^{IV}}{\partial R_D} = -2 \cdot r_w \cdot \sin\left(\varphi''' - \frac{\varphi}{2}\right) \cdot \sin\left(\frac{\varphi}{2}\right) \tag{53}$$

Taking into account the equations of bending moments in the particular beam intervals, expressions 29–32 assume the following form:

$$\int_0^{\varphi''} \frac{M_g^I}{EI} \cdot \frac{\partial M_g^I}{\partial R_A} d\varphi + \int_{\varphi''}^{\varphi''+\varphi_z} \frac{M_g^{II}}{EI} \cdot \frac{\partial M_g^{II}}{\partial R_A} d\varphi + \int_{\varphi''+\varphi_z}^{\varphi''+2\cdot\varphi_z} \frac{M_g^{III}}{EI} \cdot \frac{\partial M_g^{III}}{\partial R_A} d\varphi + \int_{\varphi''+2\cdot\varphi_z}^{\varphi_c} \frac{M_g^{IV}}{EI} \cdot \frac{\partial M_g^{IV}}{\partial R_A} d\varphi = 0 \tag{54}$$

$$\int_0^{\varphi''} \frac{M_g^I}{EI} \cdot \frac{\partial M_g^I}{\partial R_B} d\varphi + \int_{\varphi''}^{\varphi''+\varphi_z} \frac{M_g^{II}}{EI} \cdot \frac{\partial M_g^{II}}{\partial R_B} d\varphi + \int_{\varphi''+\varphi_z}^{\varphi''+2\cdot\varphi_z} \frac{M_g^{III}}{EI} \cdot \frac{\partial M_g^{III}}{\partial R_B} d\varphi + \int_{\varphi''+2\cdot\varphi_z}^{\varphi_c} \frac{M_g^{IV}}{EI} \cdot \frac{\partial M_g^{IV}}{\partial R_B} d\varphi = 0 \tag{55}$$

$$\int_0^{\varphi''} \frac{M_g^I}{EI} \cdot \frac{\partial M_g^I}{\partial R_C} d\varphi + \int_{\varphi''}^{\varphi''+\varphi_z} \frac{M_g^{II}}{EI} \cdot \frac{\partial M_g^{II}}{\partial R_C} d\varphi + \int_{\varphi''+\varphi_z}^{\varphi''+2\cdot\varphi_z} \frac{M_g^{III}}{EI} \cdot \frac{\partial M_g^{III}}{\partial R_C} d\varphi + \int_{\varphi''+2\cdot\varphi_z}^{\varphi_c} \frac{M_g^{IV}}{EI} \cdot \frac{\partial M_g^{IV}}{\partial R_C} d\varphi = 0 \tag{56}$$

$$\int_{0}^{\varphi''}\frac{M_g^I}{EI}\cdot\frac{\partial M_g^I}{\partial R_D}d\varphi+\int_{\varphi''}^{\varphi''+\varphi_z}\frac{M_g^{II}}{EI}\cdot\frac{\partial M_g^{II}}{\partial R_D}d\varphi+\int_{\varphi''+\varphi_z}^{\varphi''+2\cdot\varphi_z}\frac{M_g^{III}}{EI}\cdot\frac{\partial M_g^{III}}{\partial R_D}d\varphi+\int_{\varphi''+2\cdot\varphi_z}^{\varphi_c}\frac{M_g^{IV}}{EI}\cdot\frac{\partial M_g^{IV}}{\partial R_D}d\varphi=0 \tag{57}$$

The above model of the pump with circumferential compensation via the integrated lip (Fig. 5) can be solved using numerical methods, e.g. by means of the Mathematica software. After carrying out differentiation and integration and taking into account the static equilibrium conditions one gets the values of statically indeterminate force R_E and statically indeterminate moment M_E in the point of restraint, i.e. in the point most critical as regards the stress intensity of the analysed beam. The above loads produce a complex stress state in the cross section:

- the shear stress generated by force R_{Ey} reaches the highest value in the beam's middle fibres,
- the flexural stress generated by moment M_E reaches its maximum in the beam's outer fibres,
- the tensile stress generated by normal force R_{Ex} has a constant value in the beam's whole cross section.

Considering the above, the beam with width b and thickness h should satisfy two strength conditions. The first condition applies to the maximum shear stress k_t:

$$\tau_{max}=\frac{3}{2}\cdot\frac{|R_{Ey}|}{b\cdot h}\le k_t \tag{58}$$

$$\text{hence } h_{\min(\tau)}=\frac{3}{2}\cdot\frac{|R_{Ey}|}{b\cdot k_t} \tag{59}$$

The second strength condition specifying the allowable tensile σ_r and compressive stress σ_g in the outer fibres, induced by the normal force R_{Ex} and the bending moment M_E. The second condition applies to the maximum tensile stress k_r:

$$\sigma=|\sigma_r|+|\sigma_g|\le k_r \tag{60}$$

$$\text{where } \sigma_r=\frac{R_{Ex}}{bh} \tag{61}$$

$$\sigma_g=\frac{6\cdot M_E}{bh^2} \tag{62}$$

After substituting Eqs. (61) and (62) into (60) one gets the following equation for the beam's minimal thickness $h_{\min(\sigma)}$:

$$b\cdot k_r\cdot h_{\min(\sigma)}^2-|R_{Ex}|\cdot h_{\min(\sigma)}-6\cdot|M_E|=0 \tag{63}$$

$$h_{\min(\sigma)}=\frac{|R_{Ex}|+\sqrt{R_{Ex}^2+24\cdot|M_E|\cdot b\cdot k_r}}{2\cdot b\cdot k_r} \tag{64}$$

The highest value obtained from formulas (59) and (64) should be assumed as the minimal beam width $h_{\min}$, i.e.

$$h_{min} = max\left(h_{\min(\tau)}, h_{\min(\sigma)}\right) \quad (65)$$

The above is the complete methodology for strength calculations of the circumferential backlash compensating element. The derived relations are valid for designs with an integrated lip with wrapping angle φ_c of 108–144° and number of teeth $z = 10$. In the case of a gear pump with another compensating lip geometry and for a solution in which a different number of teeth interact with the lip, the above models should be extended or reduced by adding/removing appropriate equations. Study [6] presents a model for:

- designs with wrapping angle φ_c of 144–180°and number of teeth $z = 10$ (a four times statically indeterminate model with a separable lip),
- designs with wrapping angle φ_c of 120–150° and number of teeth $z = 12$ (a five times statically indeterminate model with an integrated lip).

A three times statically indeterminate model with a separable circumferential backlash compensating lip of variable thickness with wrapping angle φ_c of 108–144° and number of teeth $z = 10$ is described in monograph [8]. Publication [11] presents a model with wrapping angle $\varphi_c = 102°$, 132° and 169°, number of teeth $z = 10$ and with separable lip. Additionally, the model in paper [11] allows calculations for gear in a strictly defined position (the gear tip is located in the place where the lip is attached).

The model for calculating stress in the point of restraint was also validated by FEM calculations presented in [13]. The calculated maximum stress in the point of restraint of the integrated compensation did not exceed 300 MPa (Fig. 6) and it is 28.6% lower than the yield point for alloy PA9 ($R_{0.2} = 420$ MPa[1]). The analytically calculated (using the above model) shear stress amounted to $\tau = 187$ MPa, while the tensile and compressive stress amounted to $\sigma = 95$ MPa. According to the HMH hypothesis, reduced stress σ_{zr} for normal stress σ and shear stress τ amounts to:

$$\sigma_{zr} = \sqrt{\sigma^2 + 3 \cdot \tau^2} = 337\,\text{MPa} \quad (66)$$

Moreover, it was shown in [13] that the highest stress values in the considered structures occur in the compensating lip in the region of the opening connecting the working chamber with the pressure channel (Fig. 6). The high stresses in this place are due to the considerable narrowing of the integrated lip by the opening connecting the displacement chamber with the delivery channel. A more detailed analysis of the results of analytical calculations and validated by FEM will be presented in subsequent publications.

[1] The value of yield stress $R_{0.2}$ for PA9 was assumed acc. to standard PN-EN 755-2: 2001.

Fig. 6. Distribution of reduced stress, acc. to HMH hypothesis, in casing with integrated lips for point of support located in point of restraint of compensating lip [13].

4 Conclusions

The methodology of strength calculation of circumferential compensation presented in this paper assumes the occurrence of static loads. These loads are the result of hydrostatic pressure of the working fluid as well as forces and reaction on the contact of the gear with the body. In the presented analytical model, it was assumed that all hydraulic and mechanical losses result from friction between the gear vertices and the body. Such simplification means that the loads assumed in the model are greater than in reality. In the context of the strength of circumferential compensation, this is an extremely unfavorable load case, which indirectly increases the safety factor. The prototype designed based on the adopted model confirmed the correct strength of the element compensating for peripheral clearance. The verification of strength was carried out experimentally during the tests. The experimental tests presented in [6] confirmed the high volumetric efficiency of the design with the integrated lip and possibilities that the rated pressures can be increased to 36 MPa. Traces of interaction between gears and casing indicate cooperation on both suction and pressure sides. In conventional constructions, cooperation on the pressure side does not occur. Since the area without interaction is small (area 2 in Fig. 7) a slight increase in the wrapping angle should be considered in subsequent constructions with an integrated lip.

The presented calculation methodology is the starting point for further analyzes. In the next step, the FEM model should be expanded and compared with analytical calculations. A more detailed model would allow an analysis of the size of the pump's internal gaps. The issue of fatigue strength was also not recognized. Destructive tests have shown that the lip is damaged at the fixation site (Fig. 8) or a few millimeters

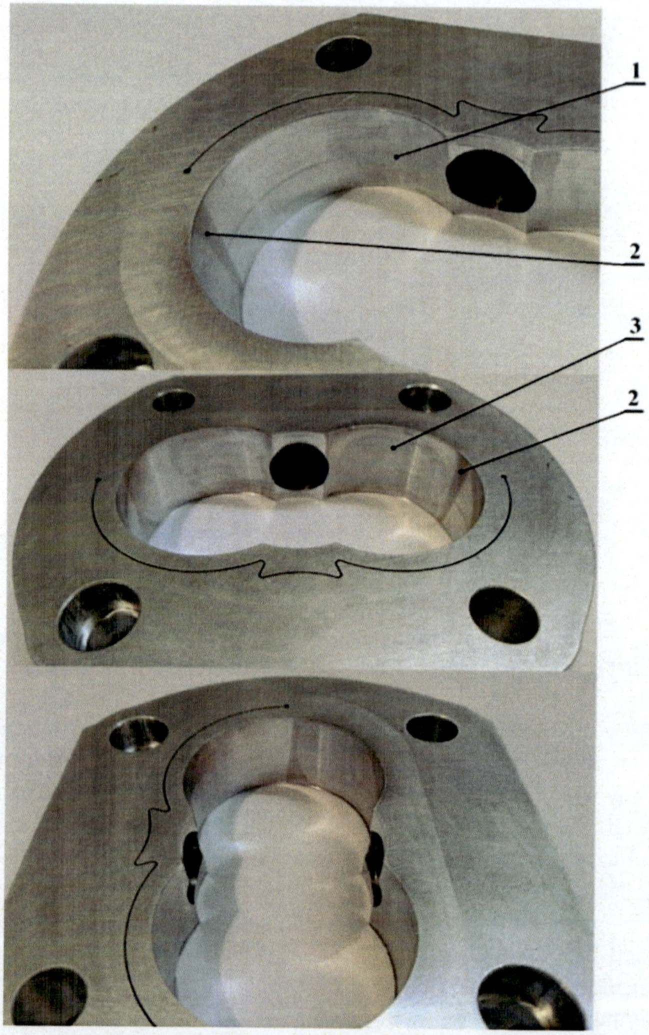

Fig. 7. Traces of interaction between gears and casing in integrated lips acc. to patent PL223649 (patent application P.404801 [7]), and with displacement limiter acc. to patent application no. P.427264 [10]: 1 – trace of interaction on delivery side, 2 – area without interaction, 3 – traces of interaction on suction side.

behind it (Fig. 2). Explanation of the cause of the crack point shift requires microscopic examination of the material structure.

The analysis of the static characteristics of the prototype unit carried out in [6] indicates that further research aimed at developing a pump design with a pre-loaded circumferential compensating element is needed. Such a solution would be characterized by greater compliance at lower operating pressure values and would ensure greater

Fig. 8. Inner casing in first version of circumferential compensation after durability tests.

internal tightness during starting. An example here can be the solution contained in patent application no. P.427263 [9].

References

1. Borghi, M., Zardin, B., Specchia, E.: External gear pump volumetric efficiency: numerical and experimental analysis. SAE Technical Paper No. 2009-01-2844; SAE International: Chicago, IL, USA (2009)
2. Casoli, P., Vacca, A., Berta, G.L.: Potentials of a numerical tool for the simulation of flow in external gear machines. In: The Tenth Scandinavian International Conference on Fluid Power, May 21–23, Tampere, Finland (2007)
3. Frosina, E., Senatore, A., Rigosi, M.: Study of a high-pressure external gear pump with a computational fluid dynamic modelling approach. Energies **10**, 1113 (2017)
4. Schiffer, J., Benigni, H., Jaberg, H.: Development of a novel miniature high-pressure fuel pump with a low specific speed. J. Autom. Eng. (2013). https://doi.org/10.1177/0954407013476820
5. Kollek, W.: Gear pumps – Design and Operation (in Polish). Ossolineum, Wrocław (1996)
6. Osiński, P.: High-Performance Gear Pumps (in Polish). Wrocław University of Science and Technology Publishing House, Wrocław (2019)
7. Osiński, P.: Patent. Poland, no. 223649. Gear pump: Int. Cl. F04C 2/08, F04C 29/00, F04C 28/20. Patent application no. 404801 z 22.07.2013. Pub. 31.10.2016 (in Polish)/Applicant name: Wrocław University of Science and Technology, Wrocław, PL
8. Osiński, P.: High-pressure and low-pulsation external gear pumps (in Polish)/Osiński, P., Wrocław University of Science and Technology Publishing House (2013). ISBN 978-83-7493-815-0, Habilitation dissertation
9. Osiński, P., Bury, P.J., Cieślicki, R.A.: External gear pump (in Polish). Patent application no. P.427263, 01.10.2018/Osiński, P., P. Bury, R. Cieślicki. Applicant name: WUS&T
10. Osiński, P., Bury, P.J., Cieślicki, R.A.: External gear pump (in Polish). Patent application no. P.427264, 01.10.2018/Applicant name: Wrocław University of Science and Technology, Wrocław, PL
11. Osiński, P., Chruścielski, G.: Strength calculations of an element compensating circumferential backlash in the external gear pump. J. Theoret. Appl. Mech. **54**(1), 251–262 (2016)

12. Osiński, P., Cieślicki, R.A.: Patent. Poland, no. 230845. External gear pump (in Polish): Int. Cl. F04C 2/24, F04C 2/08, F04C 2/14. Patent application no. P 418261 z 10.08.2016. Pub. 31.12.2018/Applicant name: Wrocław University of Science and Technology, Wrocław, PL; HYDROTOR PLC, Tuchola, PL
13. Osiński, P., Warzyńska, U.: Strength calculations of casings with a circumferential backlash compensating lip, and gear wheels of high-pressure pumps (in Polish)/P. Osiński, U. Radziwanowska. Reports of Mechanical Engineering Faculty of Wrocław University of Science and Technology. Series SPR no. 92 (2018)
14. Stryczek, S.: State-of-the-art and trends in the development of the hydraulic drive (in Polish). Przegląd Mechaniczny. 1988 R. 47, [Part 1] no. 9, pp. 5–9, [Part 2] no. 10, pp. 20, 29–33, [Part 3] no. 12, pp. 13–14, 23–24, [Part 4] no. 13, pp. 11–15, [Part 5] no. 14, pp. 10–15, [Part, 6] no. 15, pp. 26–29
15. Śliwiński, P.: The influence of water and mineral oil on mechanical losses in the displacement pump for offshore and marine applications. Polish Maritime Res. **25**(97), 178–188 (2018). https://doi.org/10.2478/pmr-2018-0040
16. Wiczkowski, E., Osiński, P., Kollek, W., Kania, A.: Patent. Poland, no. 221099. External gear pump (in Polish): Int. Cl. F04C 2/24. Application no. 397539, 22.12.2011. Pub. 29.02.2016/Wrocław University of Science and Technology, Wrocław, PL
17. Zardin, B., Natali, E., Borghi, M.: Evaluation of the Hydro-Mechanical Efficiency of External Gear Pumps, MDPI (2019)
18. Zhou, J., Vacca, A., Casoli, P.: A novel approach for predicting the operation of external gear pumps under cavitating conditions. Simul. Modelling Practice Theory **45**, 35–49 (2014). www.elsevier.com/locate/simpat

Optimizing the Break-in Process of High-Pressure Gear Pumps

Piotr Osiński(✉), Paweł Bury, and Rafał Cieślicki

Wroclaw University of Science and Technology, 27 Wybrzeże Wyspiańskiego St., 50370 Wrocław, Poland
{piotr.osinski,pawel.bury,rafal.cieslicki}@pwr.edu.pl

Abstract. The subject of the study is the analysis of the brake-in process of high-pressure external gear pumps. Pumps operating under discharge pressure p_t above 30 MPa can be included in the group of highly stressed devices. The durability of such devices is not only influenced by the choice of materials or surface quality after machining, but also by first commissioning and initial working conditions. Hydrostatic drive components, mainly pumps and motors due to their design require break-in but this process is often overlooked, which can have a negative effect on the efficiency and durability. The mathematical model shows that the displacement of gear shafts within the clearance in the bearing depends on the discharge pressure. It was concluded that the brake-in process should be conducted with a variable pressure step. On this basis, an attempt was made to prepare the break-in cycle, which is characterized by a constant displacement step of the gear shaft in the bearings. This newly prepared brake-in process will therefore ensure an optimal, constant volume of material removal from the housing.

The material presented in the article is the result of research done under the project "Development of high-pressure external gear pump".

Keywords: Gear pump · Gear pump break-in · High pressure gear pump

1 Introduction

Reliability and durability of gear pumps [1, 2] and their energy efficiency [3–9] depends not only on the tooth profile [10–12], technology of manufacture [13–17], used materials [18] or the conditions in which the pump is operated, but also from the break-in process [19, 20]. During the process of break-in tooth tips of gears are pushed against the body of the pump with high force. This cause "milling" of body surface and mechanically remove excess material. The milling depth depends on the clearance in the bearings and is associated with the occurrence of hydrostatic pressure on the gears [21]. This pressure forces indicates the displacement of gears towards the low-pressure suction port [22, 23]. In a properly break-in unit, the width of the break-in trace accurately corresponds to the width of the gears, which are always made within the tolerances in the production process. For this reason, it is recommended to pair the active and passive wheels so

© The Editor(s) (if applicable) and The Author(s), under exclusive license to Springer Nature Switzerland AG 2021
J. Stryczek and U. Warzyńska (Eds.): NSHP 2020, LNME, pp. 65–75, 2021.
https://doi.org/10.1007/978-3-030-59509-8_6

that they are dimensionally as close as possible to each other. This pairing can take place under the so-called selection group. The correctness of the break-in process can be pre-determined visually (Fig. 1). The cooperating surfaces are smooth, without visible scuffing and scratches as well as thermal discoloration [19].

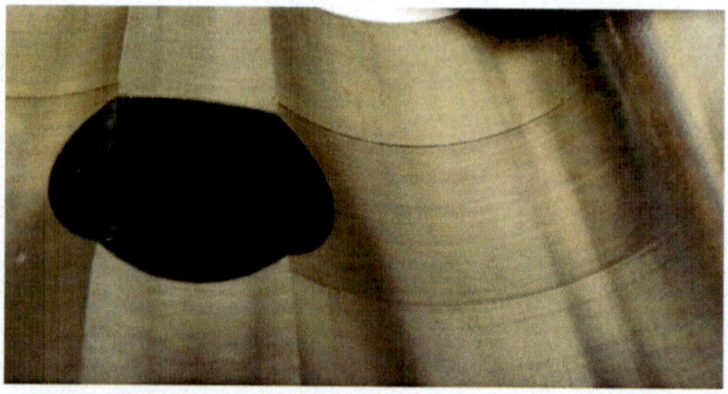

Fig. 1. An example of a correctly break-in body. Experimental pump type KPF1, $p_{tmax} = 40$ MPa [19].

Fig. 2. Seal defective due to excessive temperature rise [19].

The introduction of a new generation of high pressure pumps was possible due to a number of design changes. The new geometry of the gears improved the lubrication conditions of the tooth tip [24, 25]. Also used alloys with increased abrasion resistance known for use in slide bearings. Such changes, together with an improperly break-in process, have very adverse effects as shown in Fig. 3. Thus, the tooth geometry, which improves lubrication conditions during normal operation, causes the temperature to rise during the break-in process, and then plasticizes the material, followed by damage to the associated body components and seals Fig. 2.

Gears with modified tips require a two-step break-in process. In the first phase, the gears rub in with a sharp, unmodified edge, thus removing excess body material. In

Fig. 3. Damage to the body and tip of the tooth due to improper break-in process (40 MPa) [19].

the second stage, the wheels work as intended, so the edge improving the lubrication conditions rubs onto the body [22].

Correctly conducted break-in process does not guarantee proper wear of all elements. Due to the worse thermal conductivity of aluminium bronze, if the critical temperature is exceeded, thermal discoloration may appear on the surface of the body (Fig. 4).

Fig. 4. The body surface is smooth, despite thermal discoloration (40 MPa) and the mating wheel without visible discoloration [19].

The break-in process also reveals errors in: mapping the involute's outline, manufacturing tolerances, material defects or allows to detect irregularities in the thermochemical treatment process. An example would be wheels with too much residual austenite or too little side clearance (Fig. 5).

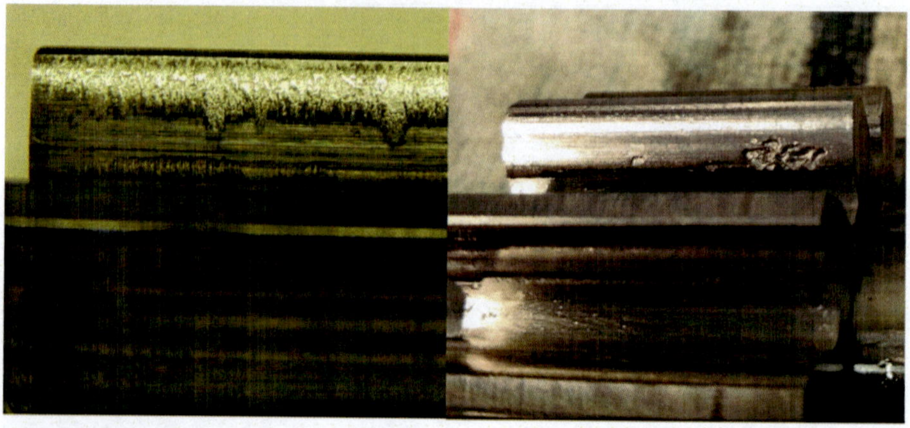

Fig. 5. Gear damage as a result of: incorrect thermo-chemical treatment (left), too little side clearance (right) [19].

2 Gear Position in the Pump Load Process

Pump gears are usually cut directly on the shafts, which are mounted by means of slide bearings (formerly rolling bearings) in the body. In order for a plain bearing to be able to carry a given load, it is necessary to apply appropriate clearance. As the load on the bearing increases, the shaft journal moves in the bush to form an eccentric ε. Such conditions favour the emergence of a lubricant wedge, which generates a load balancing lifting force W in the bearing in accordance with the equation [26]:

$$W = \frac{U\mu\varepsilon L^3}{c^2\left(1-\varepsilon^2\right)^2}\frac{\pi}{4}\sqrt{\left(\frac{16}{\pi}-1\right)\varepsilon^2+1} \tag{1}$$

where:

W — lifting force,
U — slip speed in bearing,
μ — dynamic viscosity coefficient,
ε — eccentricity,
L — bearing length,
c — radial clearance in the bearing,

The smallest clearance height h_0 between the shaft and the bearing is determined by the formula [26]:

$$h_0 = c(1-\varepsilon) \tag{2}$$

From dependencies (1) and (2) it follows that the minimum gap height is a non-linear function of the load W, which depends linearly on the discharge pressure p_t. The following diagram (Fig. 6) shows how the discharge pressure p_t affects the smallest gap height in the bearing for various clearances c, resulting from the permissible tolerances of bearing design (3PW-BPF-24 pumps).

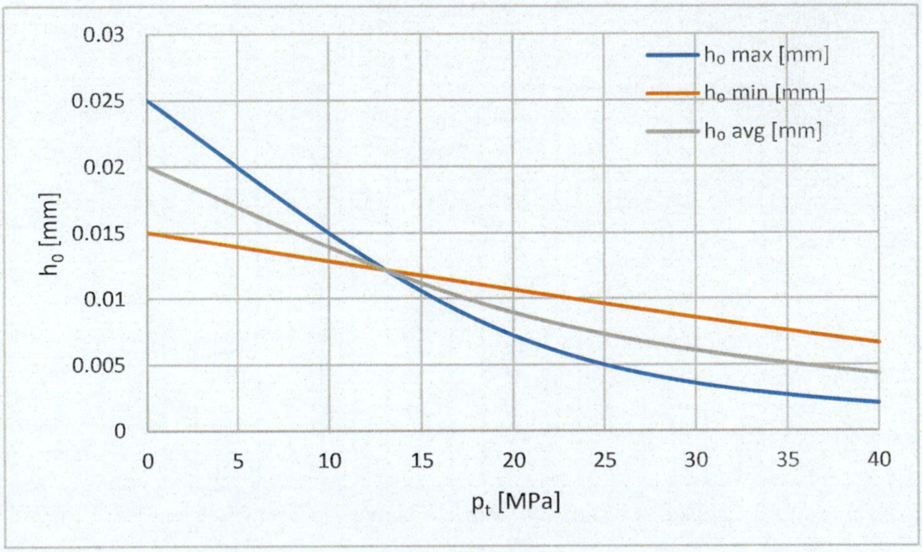

Fig. 6. The minimum clearance in the bearing depending on the pumping pressure p_t, where *max* is the largest value of the clearance in the bearing, *min* - the smallest, while the *avg.* - the average [19].

3 Break-in Process

The standard process of break-in gear units consists of periodically loading and unloading the pump, each time increasing the pressure by a certain constant value - usually by 2, 2.5 or 4 MPa [19]. The operating time of the pump under load as well as the idling time (no load) between successive pressure increases are usually assumed constant throughout the process (Fig. 7). Also, the pressure increase gradient is assumed to be constant at 1 MPa/s. It is assumed that the working time under load should not be less than 15 s. Similarly, the idling time between successive load cycles should be a minimum of 15 s. This phase of the break-in cycle allows the pump surface to cool down.

In further analysis, due to the non-linear characteristics of the minimum clearance height under pressure, the absolute and relative journal offset values in the bearing were introduced:

$$\Delta h_0 = h_0(p_{ti+1}) - h_0(p_{ti}) \tag{3}$$

$$\delta h_0 = \frac{\Delta h_0}{c} \cdot 100\% \tag{4}$$

Using the above relationships, the values of the smallest clearance height h_0 and the absolute and relative value of the shaft offset in the bearing (Δh_0 and δh_0) were determined for a standard break-in cycle with a stroke of 2 MPa. The obtained data are presented in the table below (Table 1). Analysing the relative displacement of the shaft, it can be observed that the largest displacement occurs at low pressures (depending on

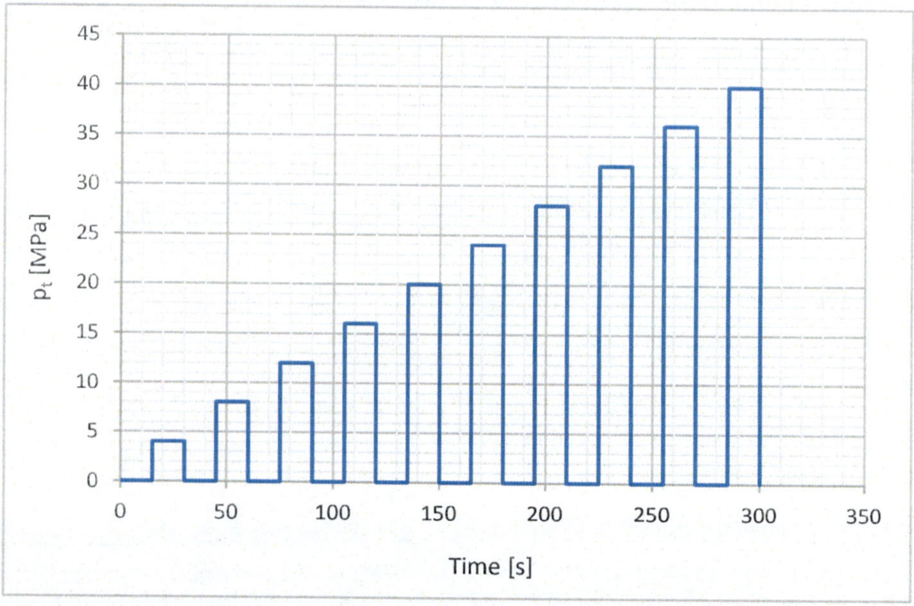

Fig. 7. The standard process of break-in gear units [19].

the bearing design 3–8.1%), while at maximum pressures the values fall even below one percent. It can be concluded that the break-in process should be conducted with a variable pressure step, lower at low pressures, higher at higher.

The above statement implies that break-in process should be conducted with variable pressure step, so that the value of the displacement between the break-in cycles were constant. This break-in process will result in a roughly constant volume of body material being removed [27]. In accordance with this principle, a break-in cycle was developed (Table 2 and Fig. 8), which is particularly recommended for high-pressure pumps, in which the body is made of materials with increased resistance to abrasion. It can be seen a lower increase in pressure amplitude for lower pressures and higher for higher pressures. The developed cycle can be regarded as universal for units with similar spigot and bearing geometry

Table 1. The standard process of break-in gear units [19].

p_t [MPa]	h_0 [mm]		Δh_0 [mm]		δh_0 [%]		
	max	min	max	min	max	min	avg.
0	0.02500	0.01500	–	-	–	–	–
2	0.02298	0.01456	0.00202	0.00044	8.0	3.0	5.5
4	0.02096	0.01413	0.00202	0.00044	8.1	2.9	5.5
6	0.01895	0.01369	0.00201	0.00044	8.0	2.9	5.5
8	0.01696	0.01326	0.00199	0.00044	8.0	2.9	5.4
10	0.01501	0.01282	0.00195	0.00044	7.8	2.9	5.4
12	0.01314	0.01238	0.00187	0.00044	7.5	2.9	5.2
14	0.01139	0.01195	0.00175	0.00044	7.0	2.9	4.9
16	0.00980	0.01151	0.00158	0.00043	6.3	2.9	4.6
18	0.00841	0.01108	0.00139	0.00043	5.6	2.9	4.2
20	0.00722	0.01065	0.00119	0.00043	4.8	2.9	3.8
22	0.00621	0.01022	0.00101	0.00043	4.0	2.9	3.4
24	0.00537	0.00980	0.00084	0.00043	3.4	2.8	3.1
26	0.00467	0.00937	0.00070	0.00042	2.8	2.8	2.8
28	0.00409	0.00896	0.00058	0.00042	2.3	2.8	2.6
30	0.00360	0.00855	0.00049	0.00041	1.9	2.7	2.3
32	0.00319	0.00814	0.00041	0.00040	1.6	2.7	2.2
34	0.00285	0.00775	0.00035	0.00039	1.4	2.6	2.0
36	0.00255	0.00737	0.00029	0.00038	1.2	2.6	1.9
38	0.00230	0.00699	0.00025	0.00037	1.0	2.5	1.7
40	0.00208	0.00663	0.00022	0.00036	0.9	2.4	1.6

In addition to the pressure values in subsequent cycles, it is important to choose appropriate working times at the set discharge pressure and idle operation. In order for the break-in process to be fully effective, it is recommended to experimentally establish these working times. When break-in the first units, the torque behaviour on the pump shaft must be carefully monitored. After applying pressure, the moment drops initially and then stabilizes. Time counted from applying load to stabilization of torque for each step of the cycle is important. After testing several test units, the appropriate working time under load can be determined as the longest time increased by 20–30%. It is recommended that the working time under the applied pressure is not less than 15 s. The idling time between successive loads should also be not less than 15 s due to the time needed to cool the lapped surfaces of the pump. The break-in schedule outlined above (Fig. 8) has been arranged for the shortest break-in periods - 15s/15s.

Table 2. Break-in cycle according to the principle of a fixed shaft offset step [19].

p_t [MPa]	h_0 [mm]		Δh_0 [mm]		δh_0 [%]		
	max	min	max	min	max	min	avg.
0	0.02500	0.01500	-	–	–	–	–
1	0.02399	0.01478	0.00101	0.00022	4.0	1.5	2.7
2	0.02298	0.01456	0.00101	0.00022	4.0	1.5	2.7
3	0.02197	0.01435	0.00101	0.00022	4.0	1.5	2.7
4	0.02096	0.01413	0.00101	0.00022	4.0	1.5	2.7
5	0.01996	0.01391	0.00101	0.00022	4.0	1.5	2.7
6	0.01895	0.01369	0.00100	0.00022	4.0	1.5	2.7
7	0.01795	0.01347	0.00100	0.00022	4.0	1.5	2.7
8	0.01696	0.01326	0.00099	0.00022	4.0	1.5	2.7
9	0.01597	0.01304	0.00098	0.00022	3.9	1.5	2.7
10	0.01501	0.01282	0.00097	0.00022	3.9	1.5	2.7
11	0.01406	0.01260	0.00095	0.00022	3.8	1.5	2.6
12	0.01314	0.01238	0.00092	0.00022	3.7	1.5	2.6
13	0.01224	0.01217	0.00089	0.00022	3.6	1.5	2.5
14	0.01139	0.01195	0.00086	0.00022	3.4	1.4	2.4
15	0.01057	0.01173	0.00081	0.00022	3.3	1.4	2.4
16	0.00980	0.01151	0.00077	0.00022	3.1	1.4	2.3
17	0.00908	0.01130	0.00072	0.00022	2.9	1.4	2.2
18	0.00841	0.01108	0.00067	0.00022	2.7	1.4	2.1
20	0.00722	0.01065	0.00119	0.00043	4.8	2.9	3.8
22	0.00621	0.01022	0.00101	0.00043	4.0	2.9	3.4
24	0.00537	0.00980	0.00084	0.00043	3.4	2.8	3.1
26	0.00467	0.00937	0.00070	0.00042	2.8	2.8	2.8
28	0.00409	0.00896	0.00058	0.00042	2.3	2.8	2.6
30	0.00360	0.00855	0.00049	0.00041	1.9	2.7	2.3
32	0.00319	0.00814	0.00041	0.00040	1.6	2.7	2.2
34	0.00285	0.00775	0.00035	0.00039	1.4	2.6	2.0
36	0.00255	0.00737	0.00029	0.00038	1.2	2.6	1.9
38	0.00230	0.00699	0.00025	0.00037	1.0	2.5	1.7
40	0.00208	0.00663	0.00022	0.00036	0.9	2.4	1.6

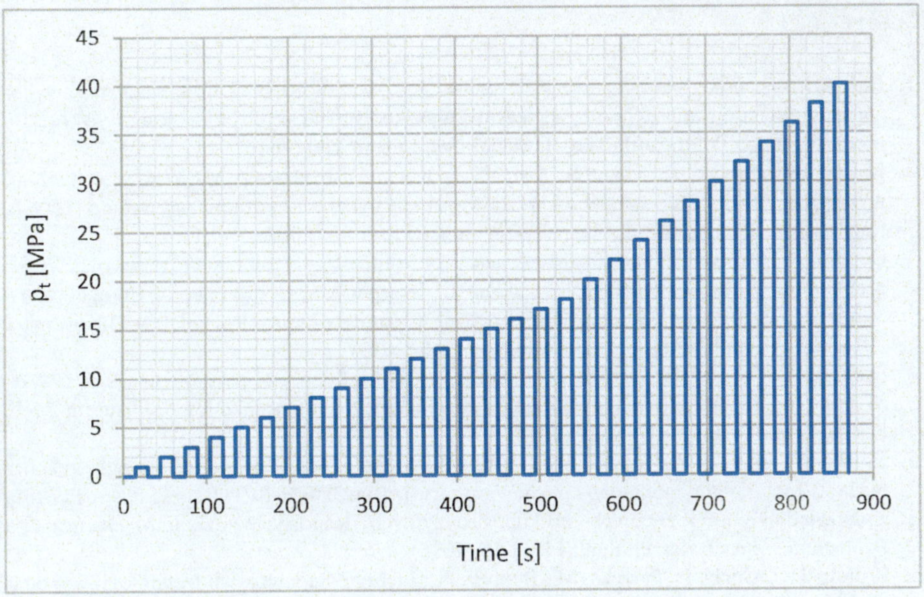

Fig. 8. Break-in cycle according to the principle of a fixed journal offset step [19].

4 Conclusions

The break-in process in high-performance units has a significant impact on pump durability and efficiency. As mentioned at the beginning, this process is often underestimated, which can result in damage shown in the pictures (Figs. 2, 3, 4 and 5). Often, due to the inability to conduct a controlled break-in process for components in the finished device (hydraulic power supply, drive systems of working machines), elements such as pumps or engines should be given to the break-in by the manufacturer at automatic stations designed for this task. Stations of this type should ensure collection and storage of data from the entire break-in process. In particular, all hydraulic, mechanical and thermal quantities should be recorded. Hydraulic parameters will allow primarily to assess the energy efficiency of the pump and the degree of oil contamination. Mechanical parameters are mainly the moment measured on the drive shaft. The magnitude of the torque allows the assessment of mechanical losses of the tested unit. Thermal parameters indicate whether heat energy is properly discharged from the fluid stream energy generator, which is also one of the main sources responsible for thermal gains [28] generated in a hydrostatic drive system.

The break-in cycle presented in this article, despite being designated for a specific unit (3PW-BPF-24), can be treated as approximate for geometrically similar units. In the case of units belonging to another size group, the cyclogram should be developed in a manner analogous to the presented example.

References

1. Bury, P.J., Cieślicki, R.A., Osiński, P., Różański, R., Zakrzewski, B.: Zaprojektowanie i wykonanie oprzyrządowania technologicznego dla wysokociśnieniowych pomp. Raporty Wydziału Mechanicznego Politechniki Wrocławskiej. Ser. SPR nr 180 (2016)
2. Kollek, W., Kudźma, Z., Maga, K., Osiński, P., Rutański, J., Stosiak, M.: Wpływ technologii wykonania powierzchni zębów na właściwości akustyczne i hydrauliczne pomp zębatych. Raporty Inst. Konstr. Ekspl. Masz. PWroc. Ser. SPR nr 85 (2007)
3. Deptuła, A., Osiński, P., Partyka, M.: Analysis of a dimensional tolerances' series of 2PW-SE gear pumps using multi valued logical. In: III International Conference of Computational Methods in Engineering Science (CMES 2018), Kazimierz Dolny, Poland, 22–24 November 2018. [Les Ulis : EDP Sciences], 06002, pp. 1–6 (2019)
4. Deptuła, A., Osiński, P., Partyka, M.: Identification of influence of part tolerances of 1PWR-SE pump on its total efficiency taking into consideration multi-valued logic. J. Autom. Mob. Robot. Intell. Syst. **12**(4), 28–41 (2018)
5. Kollek, W., Osiński, P., Partyka, M., Deptuła, A.: Identyfikacja wpływu technologii wykonania konstrukcji pomp modelowych o nowym zarysie (typ 3PWR-SE) na sprawność całkowitą z uwzględnieniem logicznych struktur decyzyjnych. Raporty Wydziału Mechanicznego Politechniki Wrocławskiej. Ser. SPR nr 46 (2015)
6. Kollek, W., Osiński, P., Partyka, M., Deptuła, A.: Identyfikacja wpływu technologii wykonania konstrukcji pomp modelowych o nowym zarysie (typ 2PWR-SE) na prawność całkowitą z uwzględnieniem logicznych struktur decyzyjnych. Raporty Wydziału Mechanicznego Politechniki Wrocławskiej. Ser. SPR nr 45 (2015)
7. Kollek, W., Osiński, P., Partyka, M., Deptuła, A.: Identyfikacja wpływu technologii wykonania konstrukcji pomp modelowych o nowym zarysie (typ 1PWR-SE) na sprawność całkowitą z uwzględnieniem logicznych struktur decyzyjnych. Raporty Wydziału Mechanicznego Politechniki Wrocławskiej. Ser. SPR nr 44 (2015)
8. Kollek, W., Osiński, P., Partyka, M., Deptuła, A.: Identyfikacja wpływu technologii wykonania konstrukcji pomp modelowych o nowym zarysie (typ 2PW-SEW) na sprawność całkowitą z uwzględnieniem logicznych struktur decyzyjnych. Raporty Wydziału Mechanicznego Politechniki Wrocławskiej. Ser. SPR nr 114 (2015)
9. Zardin, B., Natali, E., Borghi, M.: Evaluation of the Hydro—Mechanical Efficiency of External Gear Pumps, MDPI (2019)
10. Deptuła, A., Osiński, P.: The optimization of three-involute tooth outline with taking into consideration multi-valued logic trees. In: Proceedings of the 13th International Scientific Conference: Computer Aided Engineering, pp. 99–107. Springer, Heidelberg (2017)
11. Kollek, W., Osiński, P.: Badania hydrauliczne i akustyczne prototypu pompy z podciętą stopą zęba. Raporty Inst. Konstr. Ekspl. Masz. PWroc. Ser. PRE nr 6 (2002)
12. Kollek, W., Osiński, P.: Badania modelowe i wdrożenie typoszeregu pomp zębatych V generacji o zmodyfikowanym zarysie ewolwenty: projekt celowy Nr 6-T07-2003-C/06266: zestawienie sprawozdań. Raporty Inst. Konstr. Ekspl. Masz. PWroc. Ser. SPR nr 86 (2007)
13. Dudley, W.: Handbook of Practical Gear Design. Technomic Publishing (1994)
14. Dudziński, W.: Ekspertyza metalograficzna kół zębatych. Dokument wewnętrzny, Politechnika Wrocławska (2016)
15. Kollek, W., Maga, K., Osiński, P., Rutański, J.: Badania akustyczne i hydrauliczne nowo projektowanych pomp modelowych typu 1PW-SES o zębach skośnych szlifowanych. Raporty Wydziału Mechanicznego Politechniki Wrocławskiej. Ser. SPR nr 38 (2016)
16. Osiński, P., Bury, P.J.: Opracowanie zarysu i technologii wykonania dla kół zębatych stosowanych w wysokociśnieniowej pompie zębatej. Raporty Wydziału Mechanicznego Politechniki Wrocławskiej. Ser. SPR nr 95 (2015)

17. Osiński, P., Różański, R., Zakrzewski, B.: Ocena jakości wytwórczej technologii wykonania wysokociśnieniowych pomp zębatych. Raporty Wydziału Mechanicznego Politechniki Wrocławskiej. Ser. SPR nr 181 (2016)
18. Cieślicki, R.A., Osiński, P.: Ocena konstrukcji i analiza uszkodzeń w prototypowych wysokociśnieniowych pompach zębatych poddanych testom niszczącym. In: Innowacyjne techniki i technologie dla górnictwa: bezpieczeństwo - efektywność - niezawodność: monografia, [Dokument elektroniczny: praca zbiorowa/red nauk. A. Klich, A. Kozieł]. Gliwice: Instytut Techniki Górniczej KOMAG, s. 213–225 (2017)
19. Osiński, P., Bury, P.J.: Opracowanie procedury docierania wysokociśnieniowych pomp zębatych na zmodernizowanym stanowisku. Raporty Wydziału Mechanicznego Politechniki Wrocławskiej. Ser. SPR nr 167 (2016)
20. Osiński, P., Różański, R., Zakrzewski, B.: Opracowanie dokumentacji techniczno-ruchowej prototypowych pomp niskopulsacyjnych ze szczególnym uwzględnieniem procesu docierania. Raporty Wydziału Mechanicznego Politechniki Wrocławskiej. Ser. SPR nr 184 (2015)
21. Osiński, P.: Pompy zębate o obniżonym poziomie emisji hałasu. Oficyna Wydawnicza Politechniki Wrocławskiej, Wrocław (2017)
22. Osiński, P.: Wysokosprawnościowe pompy zębate. Oficyna Wydawnicza Politechniki Wrocławskiej, Wrocław (2019)
23. Osiński, P.: Wysokociśnieniowe i niskopulsacyjne pompy zębate o zazębieniu zewnętrznym. Oficyna Wydawnicza Politechniki Wrocławskiej, Wrocław (2013)
24. Bury, P., Osiński, P.: Analiza wpływu kształtu wierzchołka zęba na rozkład ciśnienia hydrodynamicznego w szczelinie korpus-wierzchołek zęba. Raporty Wydziału Mechanicznego Politechniki Wrocławskiej. Ser. SPR nr 212 (2015)
25. Osiński, P., Bury, P.J.: Polish Patent, nr 230846. Pompa zębata o zazębieniu zewnętrznym. Politechnika Wrocławska, Wrocław
26. Stachowiak, G., Batchelor, A.: Engineering Tribology, 4th edn. Elsevier/Butterworth-Heinemann, Oxford (2014). ISBN 978-0-12-397047-3
27. Dalpiaz, G., D'elia, G., Mucchi, E., Fernández Del Rincón, A.: Modeling run in process in external gear pumps. In: ASME 8th Biennial Conference on Engineering Systems Design and Analysis, pp. 927–936 (2006)
28. Osiński, P., Huss, W.M., Bury, P.J., Kiec, K.: Badania mocy cieplnej w pompie zębatej 3PZ4. Napędy i sterowanie **20**(3), 110–114 (2018). ISSN 1507-7764

Check for updates

Influence of Gaps' Geometry Change on Leakage Flow in Axial Piston Pumps

Piotr Patrosz(✉)

Gdansk University of Technology, ul. Narutowicza 11/12, 80-233 Gdansk, Poland
piotr.patrosz@pg.edu.pl

Abstract. Axial piston pumps are the core elements of most medium and high-pressure hydraulic systems. High volumetric efficiency and high working pressures are the main advantages of this type of pumps. Therefore it is necessary to accurately calculate the leakage flow through the gaps to design efficient high-pressure pump. The main difficulty of such calculations is the fact, that the gaps change their geometry during pumping process. The change of gaps' geometry can be caused by the motion of pump's elements or can be pressure-induced. Both of these factors are included in the article. The paper describes the leakage in: gap between piston and cylinder block, gap in commutation window, gaps in hydrostatic bearings, gap between commutation plate and cylinder block. The article includes methodology and results of FEM and CFD simulation of flow through deformed and undeformed gaps. Basing on the CFD results the mathematical models of such gaps were prepared and added to the article.

Keywords: Hydraulics · Pumps · CFD

1 Introduction

Axial piston pumps are the core elements of most medium and high-pressure hydraulic systems. Thanks to their high volumetric efficiency, high working pressure and potential to work as variable displacement machines they are considered one of the most versatile hydraulic pumps. They are the only type of pumps that are used in closed loop circuits and dominated the market of variable displacement pumps used in medium and high-pressure open circuits. Thanks to this features the engineers are facing a hard challenge to develop new, even more efficient pumps, that can beat the competition. To increase the efficiency of the piston pumps it is necessary to lower the leakage flow rate through the gaps. The problems of gap flow is well described in the literature [1]. However most of the mathematical models are stationary and does not include the change of gap geometry caused by the motion of pump elements or by deformation of these elements.

1.1 Design of Axial Piston Pumps

To fully understand how these gaps are changing it is necessary to shortly describe two types of axial piston pumps that are the subject of the article. First is the axial piston

© The Editor(s) (if applicable) and The Author(s), under exclusive license
to Springer Nature Switzerland AG 2021
J. Stryczek and U. Warzyńska (Eds.): NSHP 2020, LNME, pp. 76–89, 2021.
https://doi.org/10.1007/978-3-030-59509-8_7

pump with plate commutation (Fig. 1) and the second is axial piston pump with cam driven commutation (Fig. 2), which is designed and developed at Gdansk University of Technology [2]. The first type is the most common. It consists of a shaft 1 that is combined with rotary cylinder block 3. Pistons 4 are rotating with cylinder block and are cooperating with stationary, slanted swash plate 7 using slippers 6. The reciprocating movement of pistons cause the pumping and sucking of the fluid. The commutation plate 5 (Fig. 1b) is responsible for connecting right channels during pumping and sucking process, in order to ensure, that the fluid is pumped out of the high-pressure outlet P and sucked through the low-pressure inlet S (Fig. 1c). During the switching process working chamber may be disconnected from both P and S channels, what will depending how this channels overlay or not the channel in the cylinder block during the switching process.

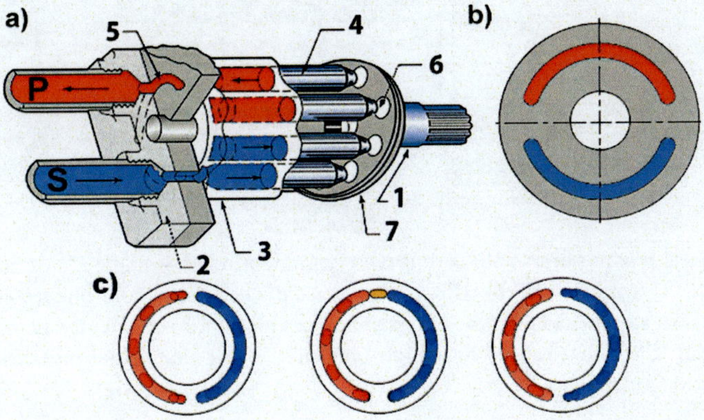

Fig. 1. Axial piston pump with plate commutation [3]: a) cross section of the pump; b) commutation plate c) basics of operation of commutation unit (1 – shaft; 2 – casing; 3 – cylinder block; 4 – piston; 5 – commutation plate; 6 – slipper; 7 – swash plate; P – outlet; S – inlet; red color – high pressure; blue color – low pressure, yellow color – disconnected channel to the working chamber).

The second type of the axial piston pump that is taken into consideration is the PWK axial piston pump with cam driven commutation (Fig. 2). The main difference of this pump is that there are two groups of pistons 4 placed inside a stationary cylinder block that is a part of a casing 2. The pistons movement is generated by the rotating swash plates 7 connected with shaft 1. Such configuration makes it impossible to use plate commutation. Therefore cam driven commutation mechanism was used. This mechanism consists of commutation tubes 9 cooperating with cam 11 through plastic slippers 10. The cam 11 is rotating with the shaft 1. Its movement pushes the commutation tubes 9, so that the commutation window can connect a working chamber with P channel during pumping process and S channel during suction. The pump enables the displacement change by changing cam 11 position relative to the shaft 1. The displacement change is obtained thanks to the mechanism consisting of a stepper motor 14, planetary gear 13 and connecting shaft 12. In case of the pump with cam driven commutation the displacement change does not directly affect the value of the leakage flow rate, because the angle of the swash plate remains constant. The displacement impacts indirectly the leakage flowrate only during short time, when the working chamber is cut off by the commutation. Although this phenomenon is very important when calculating pressure peaks [4], due to its short time of occurrence it can be treated as negligible in overall leakage calculations. In case of pumps with plate commutation the displacement change affects the leakage flowrate, because the angle of the swash plate changes and this affects values of forces working on the pistons what is included further in the article.

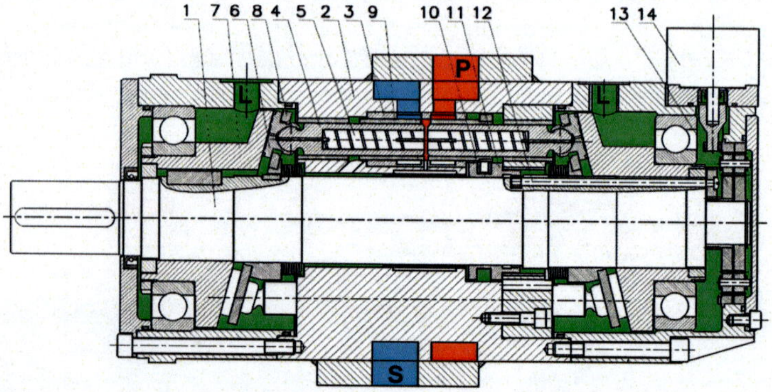

Fig. 2. PWK axial piston pump with cam driven commutation (1 – shaft; 2 – casing; 3 – bronze sleeve; 4 – piston; 5 – spiral choke; 6 – slipper; 7 – swash plate; 8 – separator plate; 9 – commutation sleeve; 10 – plastic slipper; 11 – cam; 12 – connecting shaft; 13 – planetary gear; 14 – stepper motor; P – outlet; S – inlet; red color – high pressure; blue color – low pressure, green color – casing pressure) [3].

1.2 Main Leakage Sources in Axial Piston Pumps

In both described types of axial piston pumps the sources of leakage (Fig. 3 and 4) are similar:

- leakage through an annular gap between the piston and cylinder block (Fig. 3 and 4 purple color),
- leakage through a hydrostatic bearing under a slipper (Fig. 3 and 4, red color),
- leakage through a gap in commutation (Fig. 3 and 4, green color).

Only in case of a pump with plate commutation, there will be an additional leakage source through a hydrostatic bearing under a cylinder block (Fig. 4, orange color).

Almost every mentioned above gap change its shape when the pump is working. This change can be a result of a movement of elements or may be caused by pressure growth. Additionally thermal deformations are also possible, however they will not be taken into consideration in this article, but the interested reader will be able to find lots of information about them in the literature [6, 7].

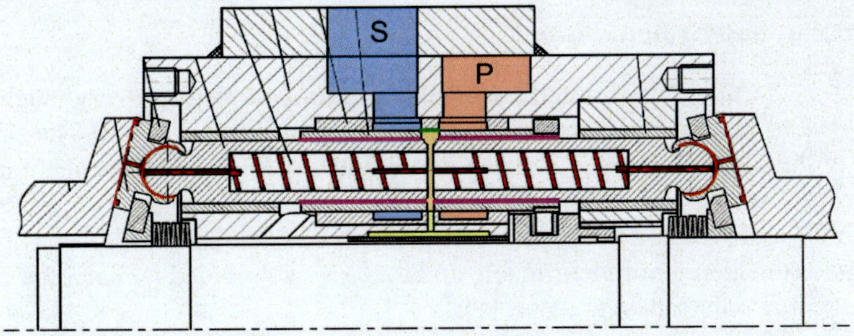

Fig. 3. Leakage sources in PWK axial piston pump with cam driven commutation (red – leakage through a hydrostatic bearing under a slipper; green – leakage through a commutation window; purple – leakage through an annular gap around piston; light blue – low pressure, pink – high pressure) [4].

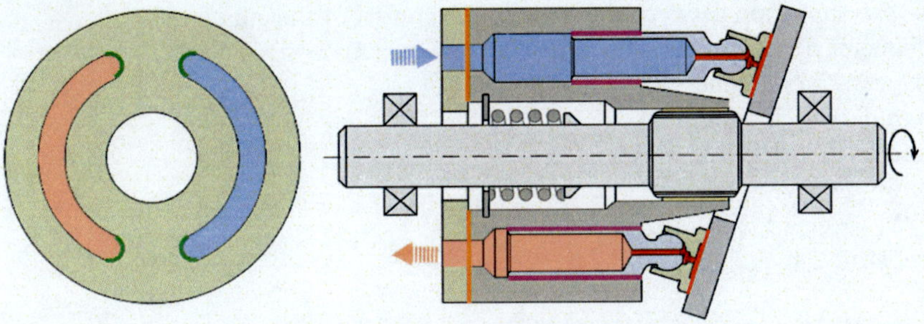

Fig. 4. Leakage sources in axial piston pump with plate commutation (red – leakage through a hydrostatic bearing under a slipper; green – leakage through a commutation window; purple – leakage through an annular gap around piston; orange – leakage through a hydrostatic bearing under cylinder block; light blue – low pressure, pink – high pressure) [5].

2 Models of Leakage Flow Rate in Gaps

Since the internal leakage of the pump is hard to measure, and distinguishing its source is even harder, CFD and mathematical models were prepared. Their main task is to predict the value of leakage flow rate through each gap.

2.1 Annular Gap Around Piston

Definition of the flow rate through an annular gap around piston is very much similar in both types of axial piston pumps. It has been described in [1, 12–15] The simplified sketches of the possible gap geometries are presented in Fig. 5. The annular gap can be coaxial, eccentric or skewed, however fast moving piston has tendency to position itself parallel to the cylinder. Therefore eccentric gap will be taken into further consideration. If the gap is considered non-deformable and the flow is assumed laminar, the equation defining the flow rate can be noted as [1]:

$$Q_A = \frac{\pi \cdot \Delta p_A \cdot d_t \cdot h_A^3}{12 \cdot \nu \cdot \rho \cdot L_t}\left(1 + 1.5(\varepsilon)^2\right) \tag{1}$$

where:

Δp_A – pressure drop between the inlet and the outlet of the annular gap,
d_t – diameter of the annular gap,
h_A – annular gap height,
L_t – annular gap length,
ν – fluid's kinematic viscosity,
ρ – fluid's density,
$\varepsilon = e/h_A$ – eccentricity of the piston,
e – eccentric displacement.

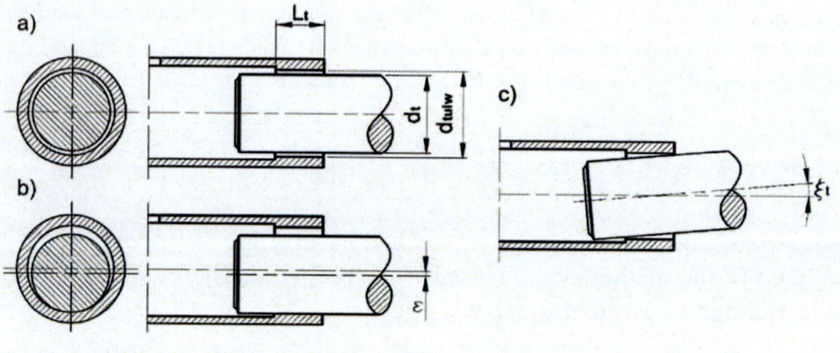

Fig. 5. Annular gap: a) coaxial; b) eccentric; c) skewed [4]

The Eq. (1) can be considered as correct for pumps with plate commutation, since the pistons are fitted directly into virtually thick and rigid cylinder block. Pistons in pumps with cam driven commutation are fitted in relatively elastic commutation sleeves which can be deformed by pressure. Using Lamé's Eq. (2) [8] to determine radial deformation

Δr, of a sleeve and Eq. (1), Eq. (3) was derived [4].

$$\Delta r = \Delta p_A \left\{ \frac{d_{tulw}^2}{2E\left(d_{tulz}^2 - d_{tulw}^2\right)} \left[(1 - \upsilon) d_{tulw} + (1 + \upsilon) \frac{d_{tulz}^2}{d_{tulw}} \right] \right\} = \Delta p_A \cdot A \quad (2)$$

$$Q_{Adef} = \frac{\pi \cdot d_t}{12 \cdot \nu \cdot \rho \cdot L_t} \left(\frac{\Delta p_A^4 \cdot A^3 + 4h_A \cdot \Delta p_A^3 \cdot A^2 + 6h_A^2 \cdot \Delta p_A^2 \cdot A + 4h_A^3 \cdot \Delta p_A}{4} + 1.5\varepsilon^2 \cdot h_A \cdot \Delta p_A + 0.75A \cdot \Delta p_A^2 \varepsilon^2 \right) \quad (3)$$

where:

E – Young modulus,
d_{tulz} – external diameter of a sleeve,
d_{tulw} – internal diameter of a sleeve,
υ – Poisson's ration.

Using the Eq. (3) it is possible to calculate the flow rate through the deformable annular eccentric gap. It includes the fact that the pressure at the inlet of the gap is much higher than in the outlet. Therefore the gap height will be greater on the inlet side then on the outlet.

2.2 Slipper-Swash Plate Gap

One of the most important gaps in every piston pump is the gap between slipper and swash plate. The leakage through this gap is necessary for supporting hydrostatic bearing between cooperating elements. The created gap is a radial gap limited by diameters d_{phsw} and d_{phsz} (Fig. 6).

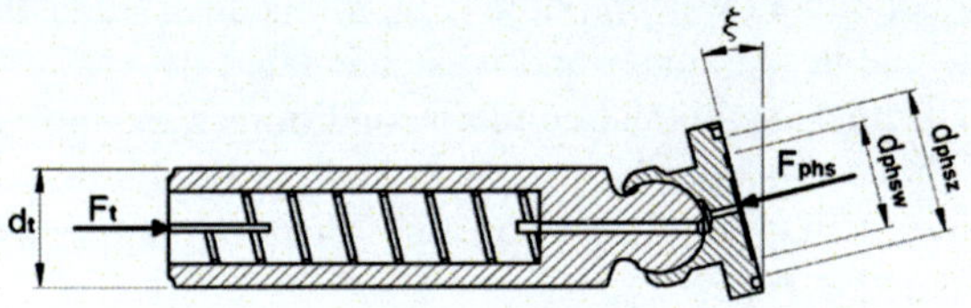

Fig. 6. Sketch of the piston and slipper [4]

To calculate the flow through it, the pressure p_{phs} in the hydrostatic bearing has to be determined. Using the force equilibrium:

$$F_t = F_{phs} \cos\xi, \quad (4)$$

Equation (5), is derived:

$$p_{phs} = (p_p - p_c) \cdot \frac{2 \cdot d_t^2 \cdot \ln\left(\frac{d_{phsz}}{d_{phsw}}\right)}{d_{phsz}^2 - d_{phsw}^2} \cos\xi \quad (5)$$

where p_p is pumping pressure and p_c is the casing pressure.

Flow rate through the radial gap, if this gap is nondeformable might be calculated using the equation:

$$Q_{phs} = \frac{\pi \cdot p_{phs} \cdot h_{phs}^3}{6 \cdot \nu \cdot \rho \cdot ln\left(\frac{d_{phsz}}{d_{phsw}}\right)} \tag{6}$$

Unfortunately gap height h_{phs} is hard to determine. Therefore the Eq. (6) is difficult to use without numerical methods. To overcome that problem the flow can be indirectly calculated using Eq. (7) [9] defining the flow in spiral choke (Fig. 7):

$$Q_{phs} = k_{dl} \frac{(p_p - p_{ph}) \cdot h_{dl}^4}{\nu \cdot \rho \cdot L_{dl}} \tag{7}$$

where:

k_{dl} – geometry coefficient,
L_{dl} – length of a spiral.

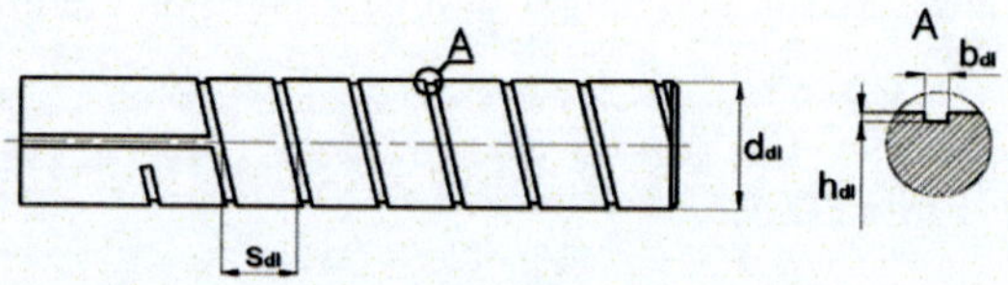

Fig. 7. Sketch of a spiral choke [4]

k_{dl} coefficient is obtained from CFD simulation equals 0,087. Finite element model used to compute k_{dl} coefficient was prepared (Fig. 8). It consisted of 400 000 hexahedral elements, didn't include heat transfer and was calculated in ANSYS CFX using SSQ turbulence model. The inlet and outlet boundary conditions were applied on the opposing ends of the spiral choke. To the rest of the model walls, the no-slip wall boundary condition was applied. CFD model proved to be accurate comparing to the results of measurements presented in [9].

Since the method of calculating the flow in slipper-swash plate gap is indirect, it is not highly sensitive to the gaps height variation. To assess how the deformation of the gap, caused by pressure, can impact gap height, resulting in change of pressure distribution, the FEM calculations were conducted. Slipper FEM model (Fig. 9) consisting of 38945 mainly hexahedral finite elements and loaded with pressure gradient with maximal value of 25 MPa applied to the slipper pocket. This pressure will appear in hydrostatic bearing when the pumping pressure is $p_p = 30$ MPa. The slipper is supported using frictionless support marked as blue region in Fig. 9a, and by elastic support marked as yellow region. This support was very elastic and didn't affect the solution, but stabilized the solver. Considering these boundary conditions the height of radial gap may increase locally only by 0,6 μm. This will not influence the pressure distribution significantly.

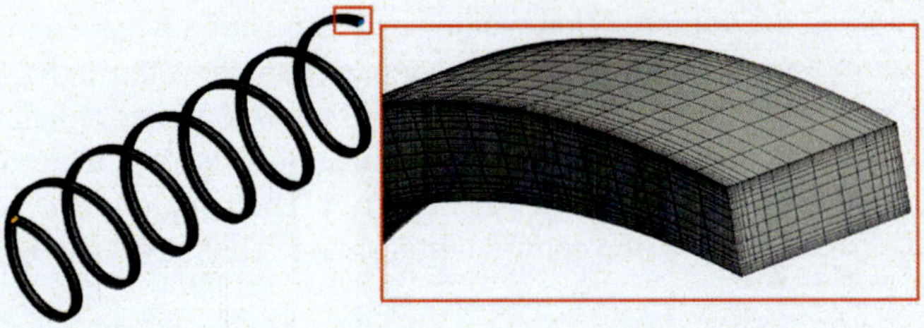

Fig. 8. Model of a spiral choke used in CFD calculations.

Therefore the influence of the deformation, caused by the pressure can be treated as negligible when using Eq. (7).

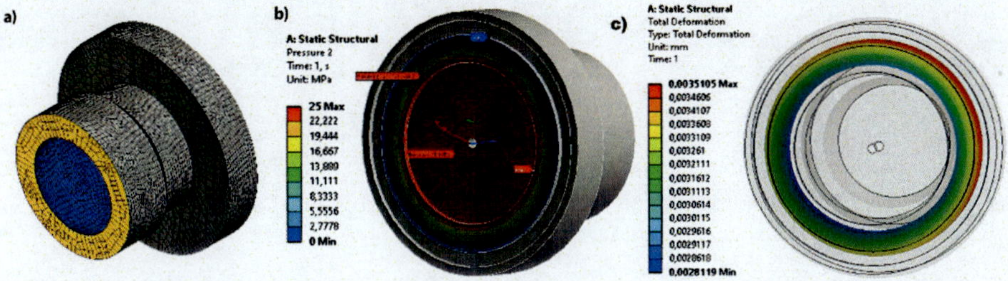

Fig. 9. Slipper deformation FEM model and results: a) mesh and supports, b) load pressure gradient, c) deformation of the gap.

The elastohydrodynamic lubrication effects also should be taken into consideration, because the motion of the fluid film, caused by pistons or swashplate rotation, can change the pressure gradient under the slipper, positioning the slipper under slight angle to the swash plate. This phenomenon would be very important when calculating the flow rate directly through a slipper using Eq. (6). However assuming that the hydrostatic pressure in a pocket under the slipper is relatively high and its changes, even considering the EHL effects, are not significant, the indirect calculations of the flow through a choke in a piston, would give good results with reasonably low error, maintaining relatively simple mathematical notation (7), in comparison with direct calculations which will have to include FSI numerical solution to obtain accurate results.

For pumps with different chokes inside the pistons, Eqs. (4) and (5) are still applicable, however Eq. (7) has to be replaced with the other equation corresponding with the right choke type.

2.3 Commutation Gap

Commutation gap is a gap that is created between the commutation window and inlet or outlet channels. The flow through this gap is very specific for every pump. It strongly depends on the shape of commutation window and on the position of cylinder block or commutation sleeve. Additionally this gap rapidly changes its length and is only opened for a short time when the commutation window is in a position between inlet and outlet channels. Commutation gap is a dynamic and complex gap. Flow through this gap cannot be accurately calculated without doing experimental measurements or using CFD tools. Therefore in this article the commutation gap leakage flow rate calculation is shown on an example of a pump with cam driven commutation. The simulations were conducted for the commutation window in the commutation sleeve. The model used for simulation is presented in Fig. 10.

The CFD model included: commutation window inside the sleeve, low pressure channel and annular gap between them. In the model the window is moving away from the channel so that the flow between them occurs only through the annular gap. The results of simulation are shown in Fig. 11.

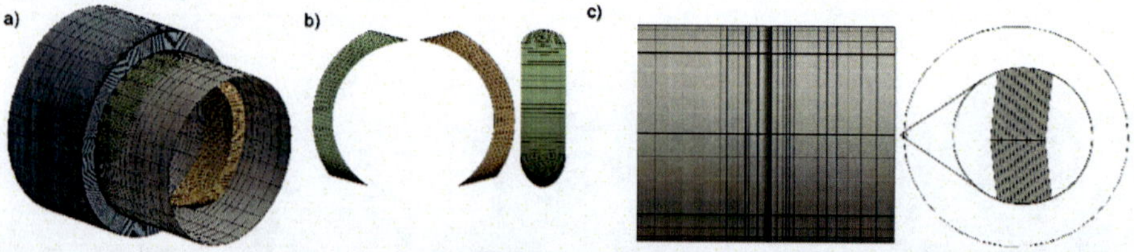

Fig. 10. Model of gap in commutation window: a) full model, b) model of the window, c) annular gap.

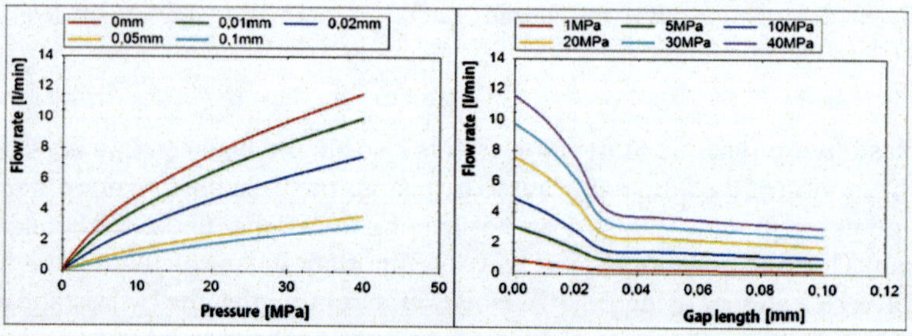

Fig. 11. Flow rate through the commutation gap in function of pressure and gap length calculated using the CFD model presented in Fig. 10 [4].

Sketch of the gap used for further analytical calculations is presented in Fig. 12.

Mathematical model of the commutation gap (Fig. 12) is defined with the equation [4]:

$$Q_{ok} = Q_{lam} \cdot f(b_{ok}) + Q_{tur} \cdot (1 - f(b_{ok})) \tag{8}$$

where:

$f(b_{ok})$ is the transition function given by empirically specified logistic function [4]:

$$f(b_{ok}) = \frac{1}{1 + e^{-18\left(\frac{b_{ok}}{5[mm]}+1\right)}} \tag{9}$$

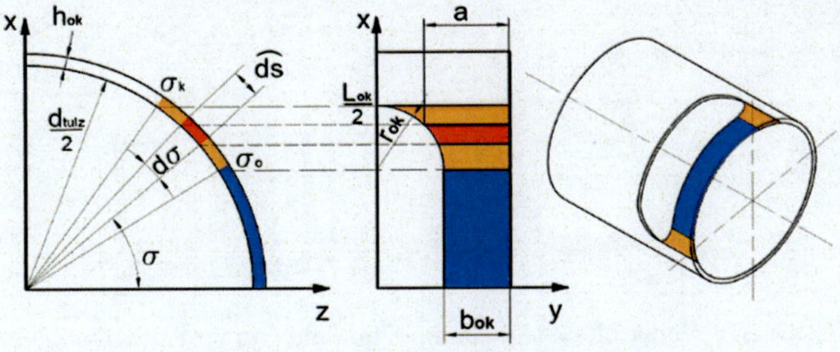

Fig. 12. Sketch of the commutation gap.

Q_{lam} is the laminar component of the flow and is given by the equation [4]:

$$Q_{lam} = \frac{(p_P - p_C) \cdot d_{tulz} \cdot h_{ok}^3}{12 \cdot \nu \cdot \rho} \left[\frac{\sigma_o}{b_{ok}} + \int_{\sigma_o}^{\sigma_k} \frac{d\sigma}{b_{ok} + r_{ok} - \sqrt{r_{ok}^2 - \left(\frac{d_{tulz}}{2} \sin \sigma - \frac{L_{ok}}{2} + r_{ok} \right)^2}} \right] \tag{10}$$

Q_{tur} is the laminar component of the flow and is given by the equation [4]:

$$Q_{tur} = h_{ok} \cdot L_{ok} \sqrt{\frac{2(p_P - p_C)}{\zeta_{ok} \cdot \rho}} \tag{11}$$

where ζ_{ok} is the empirical flow resistance coefficient.

2.4 Cylinder Block Commutation Plate Gap

The last and the most complex gap is the gap between cylinder block and the commutation plate. The fluid film in this gap provides a hydrostatic bearing which supports rotating cylinder block, which is pushed towards the stationary commutation plate by the forces from pistons and pressures inside the working chambers. The distribution of the pressure in the gap is presented in Fig. 13a [11]. This pressure gradient constantly changes, depending on the position of the cylinder block. This gradient should cancel the forces mentioned earlier, but because of the fact that number of pressurized working chamber is constantly changing (Fig. 1c), and because of a geometric features of the cylinder block and the commutation plate, value and position of the gradient isn't ideal, what causes the gap to be skewed, as its shown in Fig. 13b.

Additionally the flow in the commutation gap doesn't occur only to the casing of the pump but it also occurs between the channels in the commutation plate, therefore the flow rate through gap depends on an angular position of the cylinder block (Fig. 1c) and the overlap of commutation window, which will determine for how long the working chamber may remain cut off from both channels in the commutation plate. This may lead to a sudden change of pressure in the working chamber and cause an appearance of unbalanced forces in the hydrostatic bearing which may lead to a temporary leakage flowrate change.

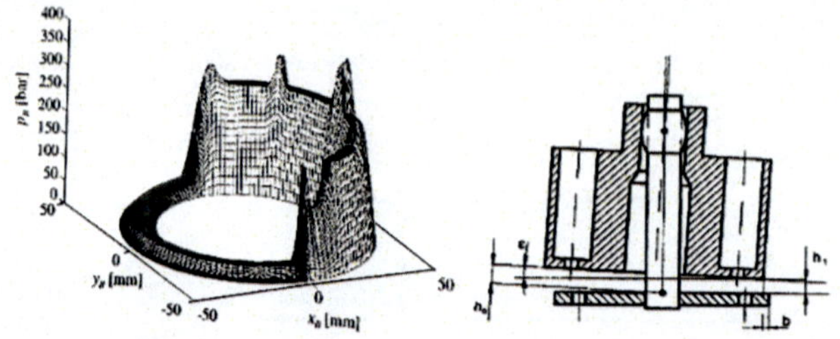

Fig. 13. Gap between cylinder block and commutation plate a) pressure distribution in the gap [11] b) gap shape [2].

The only accurate method of determining the flow rate through this gap is the CFD and structural FEM analysis. Unfortunately, description of this method exceeds the capacity of that article. Therefore it is recommended for the interested reader to check the articles [10, 11] and especially a book [2] from, which some of the results are presented in next chapter.

3 Results and Discussion

Using the equations from the previous chapter Fig. 14, 15, 16, 17 were created. Figure 14 presents the comparison of the flow through a deformable annular eccentric gap between. As it is presented the influence of gap deformation is considerable, therefore omitting this phenomenon is not recommended especially in pumps with cam driven commutation. Such big difference is caused by the fact, that in Eq. (3) gap height is in third power. Therefore, if height of the undeformed gap is about 10 μm and the radial deformation can exceed 5 μm, the flow rate through the gap can increase over 3.3 times.

Flow rate through the gap between a single slipper and swash is presented in Fig. 16. The chart presents difference between the results of CFD simulations (red) and results of calculations using Eq. (7). As it is visible the chart obtained from simulations is a little nonlinear, which indicates that the flow is not fully laminar. However the analytical model (7) assumes the laminar flow. The difference between the simulation and calculations is small and for most needs can be treated as negligible.

The flow rate through the gap in commutation window is presented in a Fig. 17. On the chart the analytical and CFD models are compared. Compatibility of the results is between 80–90%, what in most cases can be assumed as good convergence, especially taking into account the fact, that the gap is opened only for a short period, when the commutation window is between the inlet and outlet channel. For the rest of a time, when the length of the gap exceeds 1 mm the flow rate through it is negligible.

The flow rate Q through the gap between the cylinder block and commutation plate is supposed to be responsible for 30–50% [10] of volumetric losses in piston pumps. Additionally it is highly dependable on angular position φ of the cylinder block. Unfortunately all the simulation in the literature [2, 10] does not separate the flow through this gap to the casing and between the channels in the commutation plate. The example of this results are shown in Fig. 17.

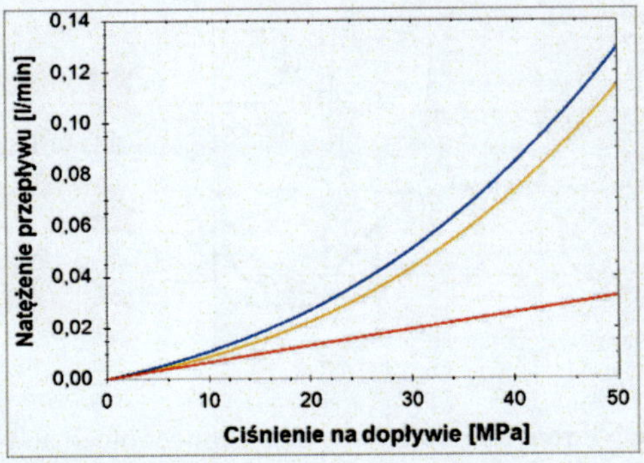

Fig. 14. Flow rate through the annular gap between the piston and cylinder [4]: red - nondeformable central gap; yellow – deformable central gap, blue – deformable eccentric gap.

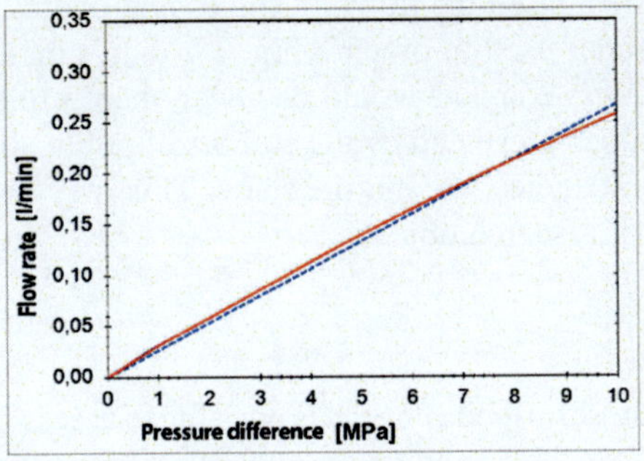

Fig. 15. Flow rate through the spiral choke and slipper-swashplate radial gap: red – simulation result, blue – result of calculations using Eq. 7 [4].

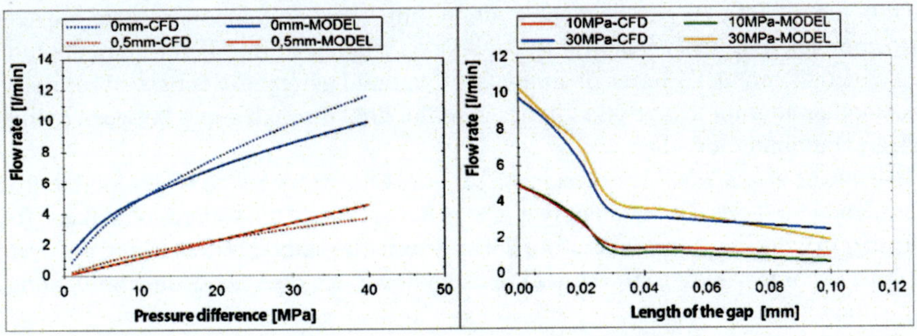

Fig. 16. Flow rate through the commutation gap

The presented in Fig. 17 results of CFD simulation show also the influence of the overlap and usage of unloading groves on the flow rate. Figure 17 presents that when zero or positive overlap is used, the significant flowrate spikes can be observed. As it was written before the unbalanced forces working on the cylinder block cause the deformation

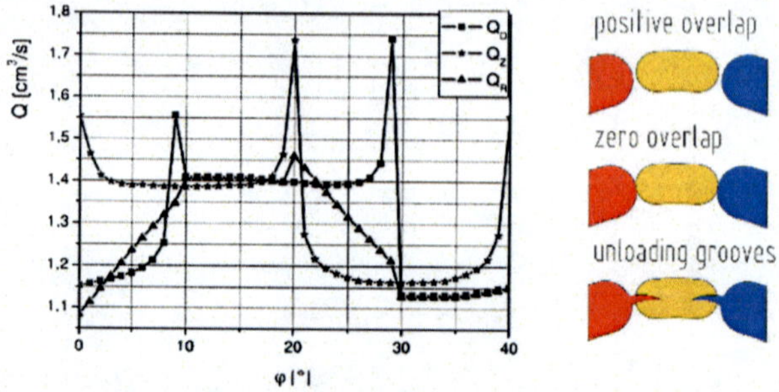

Fig. 17. Leakage flow rate through a gap between the cylinder block and the commutation plate [2]: Q_D – positive overlap; Q_Z – zero overlap; Q_R – unloading groves

of the gap and as a consequence sudden flowrate change. Usage of the unloading groves in the presented model decreased the flowrate spikes and overall leakage compering to positive overlap. The model used to prepare Fig. 17, didn't include compressibility of the fluid therefore the flowrate spikes would be lower, thanks to fluid compression.

Unfortunately the values presented by the authors of publications [2, 10, 12] despite presenting the similar tendencies, are not the same. This may be the result of different assumptions and different research objects.

4 Conclusion

Using the methods shortly described above, it is possible to assess the leakage of the pump without building prototypes and conducting expensive laboratory tests. In most cases analytical models were prepared and verified using CFD simulations. The verification confirmed that the maximal error of the presented models does not exceed 20% and in most cases is even significantly lower.

Only the model of a gap between cylinder block and commutation plate proved to be too complex to reliably simplify it to mathematical equation. Therefore in the future it is planned to conduct a series of simulations, to build a bigger data base of results and make another attempt to analytically describe the flow through a gap between cylinder block and commutation plate.

It would be also advisable to conduct a series of laboratory test to verify models using experimental methods. Unfortunately it is very hard separate the sources of leakage from each other in working pump. Therefore the only reliable method is to measure the overall leakage and comparing it to the sum of leakage flow rates obtained from simulations or calculations.

Acknowledgments. The article includes the research introduced in PhD thesis [5] and continued in project funded by The National Centre for Research and Development within the framework of program LIDER:

Project no: LIDER/22/0130/L-8/16/NCBR/2017.

Project title: Hydro-mechanical automatic gearbox for agricultural vehicles and heavy machinery.

Funding value: 1 197 500,00 PLN.

References

1. Osiecki, A.: Hydrostatyczny napęd maszyn. WNT, Warszawa (1998)
2. Złoto, T.: Modelowanie odciążenia hydrostatycznego i analiza zjawisk przepływowych w szczelinie rozrządu tarczowego pompy wielotłoczkowej osiowej. Politechnika Częstochowska, Częstochowa (2007)
3. Osiecki, A. Osiecki, L.: Hydrostatic axial piston machine. European patent 0742870 (1999)
4. Patrosz, P.: Compensation of pressure peaks in variable displacement piston pump with cam driven commutation. Ph.D. thesis, Gdansk (2017)
5. Chao, Q., Zhang, J., Xu, B., et al.: Centrifugal effects on cavitation in the cylinder chambers for high-speed axial piston pumps. Meccanica **54**(6), 815–829 (2019)
6. Jasiński, R.: Influence of type of material on performance of hydraulic components in thermal shock conditions. Solid State Phenom. **183**, 95–100 (2012)
7. Jasiński, R.: Determination of ability of hydrotronic systems to start in low ambient temperatures. Solid State Phenom. **164**, 31–36 (2010)
8. Walczyk, Z.: Wytrzymałość materiałów - Teoria i przykłady Tom 2. Politechnika Gdańska (1999)
9. Zastempowski, B.: Badanie przepływu oleju przez dławik śrubowy. Ph.D. thesis, Gdansk University of Technology, Gdańsk (1981)
10. Ivantysynova, M., Baker, J.: Power loss in the lubricating gap between cylinder block and valve plate of swash plate type axial piston machines. Int. J. Fluid Power **10**(2), 29–43 (2009)
11. Cho, J., Zhang, X., Manring, N.D., Nair, S.S.: Dynamic modelling and parametric studies of an indexing valve plate pump. Int. J. Fluid Power **3**(3), 37–48 (2002)
12. Wieczorek, U., Ivantysynova, M.: Computer aided optimization of bearing and sealing gaps in hydrostatic machines—the simulation tool caspar. Int. J. Fluid Power **3**(1), 7–20 (2002). https://doi.org/10.1080/14399776.2002.10781124
13. Pelosi, M., Ivantysynova, M.: A novel fluid-structure interaction model for lubricating gaps of piston machines. WIT Trans. Built Environ. **105**, 13–24 (2009)
14. Ivantysynova, M., Huang, C.: Investigation of the gap flow in displacement machines considering elastohydrodynamic effect. In: Proceedings of the JFPS International Symposium on Fluid Power, vol. 2002, no. 5–1, pp. 219–229 (2002)
15. Hamrock, B.J.: Fundamentals of Fluid Film Lubrication, vol. 1255. Nasa Publication (1991)

Check for updates

Design of Asymmetric Gerotor Pumps

Andrew J. Robison(✉) and Andrea Vacca

Purdue University, West Lafayette, IN 47906, USA
{arobiso,avacca}@purdue.edu
https://engineering.purdue.edu/Maha/

Abstract. A method to generate an asymmetric gerotor gearset based on an elliptical outer gear tooth rotated about its center is presented. While this profile type has been known since the mid 20th century, it has yet to be discussed in scientific literature. A case study is performed to quantify the effects of asymmetry on the size, flow ripple, wear rate, and contact stress of two pumps with the same geometric displacement. The results showed that applying asymmetry can have either a positive or a negative effect on each of the performance goals depending on the values of the other input parameters.

Keywords: Gerotor · Optimization · Gear geometry · Hydraulic pump

1 Introduction

Gerotor pumps are among the most common positive displacement pumps because of their low cost, manufacturability, durability, low noise, and low outlet flow ripple. They use a special internal gearset with one tooth difference, where each tooth of the outer gear contacts the inner gear throughout their rotation. Discrete control volumes are present between the teeth of the gears that are sealed by the contact points. These volumes increase and decrease in size with rotation, which is used to generate a pumping action. The most common applications for gerotor pumps are in low-pressure applications such as pre-charge, lubrication, automatic transmission, and fuel pumps.

Recent trends in Fluid Power, such as electrification, are imposing new challenges for gerotor pumps to be more compact; energy efficient; quiet; and to operate at higher speeds, pressures, and with lower viscosity working fluids than in the past [5,8]. Much of the recent research in gerotor design has been focused on these challenges by developing fluid dynamic and lubrication models (Castilla and Gamez-Montero et al. [2], Altare and Rundo [1], and Pellegri and Vacca [9] or on using advanced plastics for the gear material Stryczek et al. [11]). However, the shape of the gears is another area of research that can offer an increase in performance.

A significant challenge in designing a gerotor gearset is that an infinite number of curve types can be used to generate a suitable gerotor gearset, and an

© The Editor(s) (if applicable) and The Author(s), under exclusive license
to Springer Nature Switzerland AG 2021
J. Stryczek and U. Warzyńska (Eds.): NSHP 2020, LNME, pp. 90–101, 2021.
https://doi.org/10.1007/978-3-030-59509-8_8

infinite number of curves within each profile type exist. This is in contrast to spur gears for power transmission applications, where the involute is widely accepted as the best type of tooth shape. Several different profile types for gerotors have been described in literature and refer to the shape of the outer gear tooth. Examples include circular [4], cycloidal [3], elliptical [4], hypotrochoidal [7], and sinusoidal [4]. Other novel profile types almost assuredly exist that can offer improvement over these standard profile types.

An example of a novel profile type that could offer improvement is an asymmetric profile, an example of which could be using an ellipse that is rotated about its center. This profile type was first described in a patent by Hill [6], who claimed that the asymmetry can lead to a better pressure angle between the gears, which can lead to reduced wear and contact stress. However, a description of this profile type and evaluation of these claims are absent from scientific literature.

The goal of this work is to give a description of how to generate an asymmetric gerotor profile using an outer gear tooth formed by an ellipse rotated about its center and to show how the asymmetry can affect the dimensions of a pump for a given displacement and the kinematic flow ripple. This lays the foundation for a more detailed study of the potential advantages of an asymmetric gearset in comparison to conventional profile shapes.

2 Profile Description

Although an asymmetric gerotor may appear to be more complicated than a conventional gerotor, it can be generated in the exact same way. Nearly any smooth curve can be selected as the outer gear tooth, and then a matching inner gear is generated by applying the Law of Gearing. The Law of Gearing states that for two curves to transmit rotational motion at a constant ratio, the common normal to the curves must pass through the pitch point. The basic tooth form for an asymmetric gerotor is in shown in Fig. 1. Parametric equations to describe the tooth are given by Eqs. (1) and (2), where x and y are the abscissa and ordinate of the outer gear respectively parameterized on the interval $-\pi \leq \phi \leq \pi$. The gearset has m outer gear teeth and $m-1$ inner gear teeth. In this case, the outer gear root is formed by an arc of a circle with radius r_r concentric with outer gear. Any shape can be used for the root provided it does not cause interference, and fillets can be added to connect the elliptical tooth to the circular root to reduce stress concentrations, to reduce bending stress, and to ease manufacture. The shape of the outer gear root does not affect the displacement and flow ripple of the pump, but it does influence its packaging size and inlet flow velocity.

$$x = \rho - d\cos(\phi)\cos(\alpha) + k\sin(\phi)\sin(\alpha) \tag{1}$$

$$y = d\cos(\phi)\sin(\alpha) + k\sin(\phi)\cos(\alpha) \tag{2}$$

After the outer gear tooth has been defined, the inner gear can be generated using the approach given by Colbourne [4]. In this approach, a kinematic inversion is performed so that the outer gear is fixed, and the inner gear orbits about

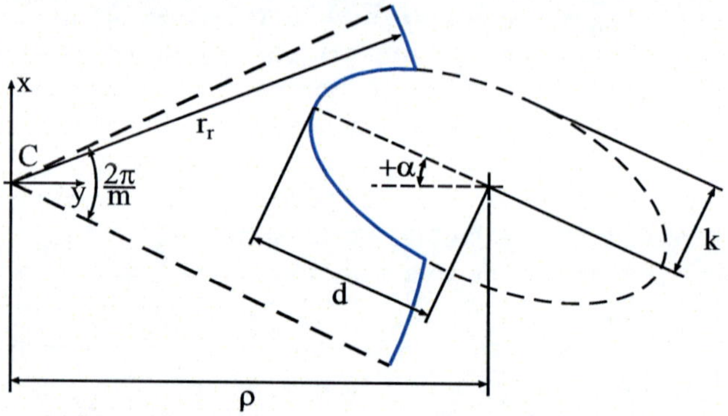

Fig. 1. Outer gear tooth formed by an ellipse rotated about its center for $m = 7$, $\rho = 25\,\text{mm}$, $d = 10\,\text{mm}$, $k = 5\,\text{mm}$, $\alpha = 25°$, $r_r = 21\,\text{mm}$

the outer gear center. This is illustrated in Fig. 2. As the inner gear orbits, its axes ξ, η will rotate clockwise by the angle ψ relative to x, y. The line of centers will then rotate by $(m-1)\psi$ in the counterclockwise direction. Note that two solutions for ψ exist for every point of the outer gear tooth. They correspond to the two times each point of the outer gear tooth contacts the inner gear. This is represented in Fig. 2, where the blue and red curves show the position of the inner gear as it orbits and makes contact with the same point on the outer gear. The common normal to the curves is shown in black and intersects the outer gear pitch circle twice, and a magnified view of the coordinate axes is also shown in Fig. 2.

The two relative rotation angles are found for each point on the outer gear, and then a coordinate transformation is performed to give the coordinates of the inner gear relative to its center. The relative rotation angles are given by Eq. (3), where derivatives are taken with respect to ϕ, and the four quadrant arctangent function is used. The angle ψ is related to the outer gear rotation angle in the standard reference frame with the line of centers fixed by Eq. (4), where θ_o is the outer gear rotation angle. A zero value of θ_o corresponds to the point shown in Fig. 4. In this way, the outer gear rotation angle at which each point of the outer gear tooth contacts the inner gear can be found explicitly. After angle ψ is determined, a coordinate transformation is performed to give the coordinates of the contact point with respect to the inner gear center in Eqs. (5) and (6), where ξ and η are the abscissa and ordinate of the inner gear respectively, and e is the center distance (eccentricity) of the gearset. This gives the inner gear tooth in three sections shown in Fig. 3. The primary contact region in Fig. 3 is shown in black, corresponds to ψ_1, and has a positive pressure angle. The secondary contact region is split in two parts shown in red and blue in Fig. 3. The red portion corresponds to positive values of ψ_2, while the black portion corresponds to negative values of ψ_2. After the inner gear tooth is defined, both the inner and outer gear can be formed by rotating the tooth about its respective center as shown in Fig. 4.

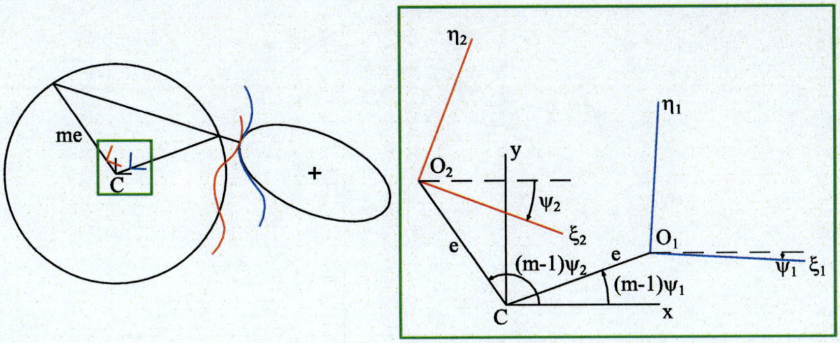

Fig. 2. Illustration of kinematic inversion for a gearset of $m = 7$, $e = 2\,\text{mm}$, $\rho = 25\,\text{mm}$, $d = 10\,\text{mm}$, $k = 5\,\text{mm}$, $\alpha = 25°$

$$\psi_{1,2} = \frac{1}{m-1}\text{atan}\left(\frac{y'(xx'+yy') \mp x'\sqrt{m^2e^2(x'^2+y'^2)-(xx'+yy')^2}}{x'(xx'+yy') \pm y'\sqrt{m^2e^2(x'^2+y'^2)-(xx'+yy')^2}}\right) \quad (3)$$

$$\theta_o = -(m-1)\psi \quad (4)$$

$$\xi_{1,2} = -e\cos(m\psi_{1,2}) + x\cos(\psi_{1,2}) - y\sin(\psi_{1,2}) \quad (5)$$

$$\eta_{1,2} = -e\sin(m\psi_{1,2}) + x\sin(\psi_{1,2}) + y\cos(\psi_{1,2}) \quad (6)$$

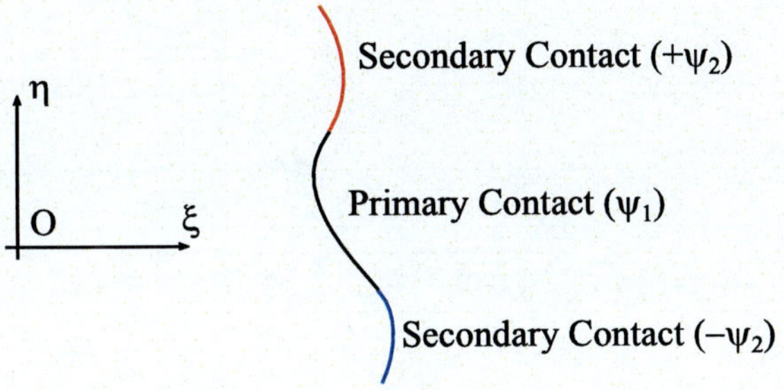

Fig. 3. Inner gear tooth formed by an ellipse rotated about its center for $m = 7$, $e = 2\,\text{mm}$, $\rho = 25\,\text{mm}$, $d = 10\,\text{mm}$, $k = 5\,\text{mm}$, $\alpha = 25°$

Only a portion of the outer gear tooth will contact the inner gear, as only the points on the outer gear tooth whose normal line intersect the pitch circle can satisfy the Law of Gearing. The minimum and maximum values of the allowable gear parameters are given by $\phi_{m_{1,2}}$ and are found by finding the two solutions to Eq. (7) on the interval $-\pi \leq \phi \leq \pi$. The points with parameters $\phi_{m_{1,2}}$ have

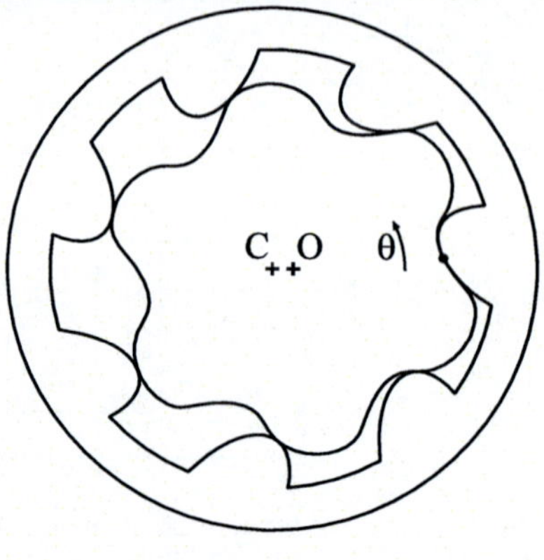

Fig. 4. Gerotor gearset using an asymmetric elliptical outer gear tooth for $m = 7$, $e = 2\,\text{mm}$, $\rho = 25\,\text{mm}$, $d = 10\,\text{mm}$, $k = 5\,\text{mm}$, $\alpha = 25°$, $r_r = 21\,\text{mm}$

corresponding relative rotation angles $\psi_{m_{1,2}}$. These two points are illustrated in Fig. 5, where the normal lines to the tooth curve at those points are tangent to the outer gear pitch circle. For contact at these two points, the pressure angle is zero and $\psi_1 = \psi_2 = \psi_m$.

$$m^2e^2(x'^2 + y'^2) - (xx' + yy')^2 = 0 \tag{7}$$

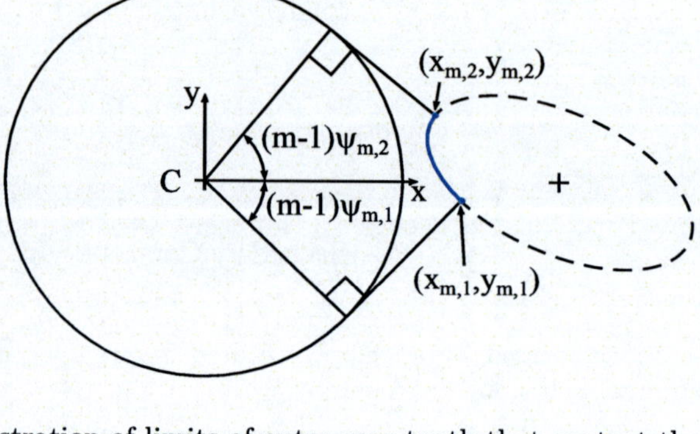

Fig. 5. Illustration of limits of outer gear tooth that contact the inner gear

The maximum allowable center distance for the asymmetric gerotor is found by first finding the two points on the outer gear tooth whose tangent line has a slope of zero denoted as $(x_{\text{max}_{1,2}}, y_{\text{max}_{1,2}})$, which correspond to points with gear parameters $\phi_{\text{max}_{1,2}}$ given by the two solutions to Eq. (8). The maximum allowable center distance is given by $e_{\max}$ in Eq. (9). If $e > e_{\max}$, no point on the outer gear tooth has a normal line that is tangent to the pitch circle, so continuous action throughout the whole rotation does not exist. However, usually cusps will form in the inner gear with increasing center distance before reaching this condition.

$$\phi_{\max_{1,2}} = \operatorname{atan}\left(\frac{k}{d}\cot\alpha\right) \tag{8}$$

$$e_{\max} = \frac{1}{m}\min\{x_{\max_1}, x_{\max_2}\} \tag{9}$$

The number of input variables required to define the gerotor gearset can be reduced by using three non-dimensional parameters in addition to the number of teeth and the tooth angle α to describe the aspect ratios of the gearset. The non-dimensional parameters used are defined in Eqs. (10)–(12), where λ_p is the ratio of the pitch circle radius relative to the tooth position, λ_d is the tooth size ratio, and λ_t is the tooth aspect ratio. A pump with a certain aspect ratio can then be scaled to achieve a desired displacement for a given face width.

$$\lambda_p = \frac{me}{\rho} \tag{10}$$

$$\lambda_d = \frac{d}{\rho} \tag{11}$$

$$\lambda_t = \frac{k}{d} \tag{12}$$

Not every combination of non-dimensional input parameters can give a geometrically feasible pump without cusps or self intersections. Each of the ways an infeasible geometry can be generated are illustrated in Fig. 6. In (a), the tooth extends past the gear center due to the combination of a high value of both λ_t and α. The center distance is too great for continuous contact in (b), as no normal line to the bottom half of the tooth can be tangent to the pitch circle. In (c), the minimum allowable root radius does not intersect the outer gear tooth. Then in (d), the outer gear teeth overlap before they intersect the root. Lastly in (e), the inner gear contains cusps that form when using too great of a center distance.

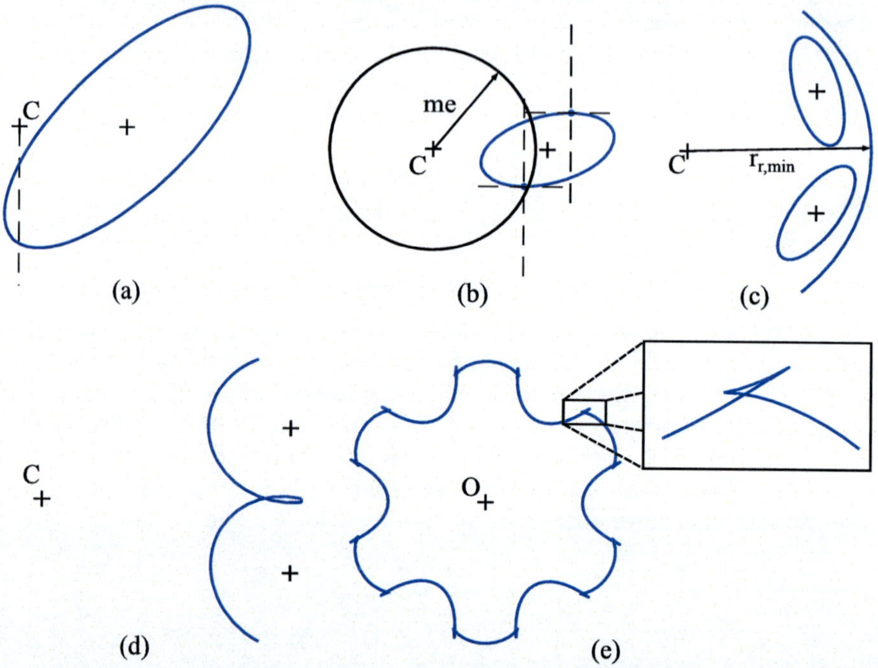

Fig. 6. Illustration of infeasible pump geometries

3 Effect of Asymmetry

A case study was performed on two reference pumps using different tooth rotation angles to see how asymmetry affects the size, flow, wear, and contact stress of a pump. The pumps used for the study are shown in Fig. 7 at different tooth rotation angles. A value of $\alpha = 0$ corresponds to a symmetric gerotor, so a comparison between symmetric and asymmetric gerotors can be made. The pumps were all scaled to have a geometric displacement of $10\,\text{cm}^3/\text{rev}$ for a $1\,\text{cm}$ face width. The minimum allowable root radius for pumps as a function of tooth rotation angle for is shown in Fig. 8. For the pump with $\lambda_t < 1$, the minimum allowable radius increased by applying asymmetry. However for the pump with $\lambda_t > 1$, the minimum allowable radius was decreased by applying asymmetry.

The kinematic flow ripple per unit face width for gerotor is defined in Eq. (13), where q_j is the contribution from the jth displacement chamber defined in Eq. (14). Applying asymmetry to the two reference pumps resulted in a change of both phase and magnitude in the kinematic flow ripples as shown in Fig. 9, where the flow ripples are plotted for $\alpha = -45°$, $\alpha = 0°$, and $\alpha = 45°$.

$$Q = \sum_{j=1}^{m} q_j \tag{13}$$

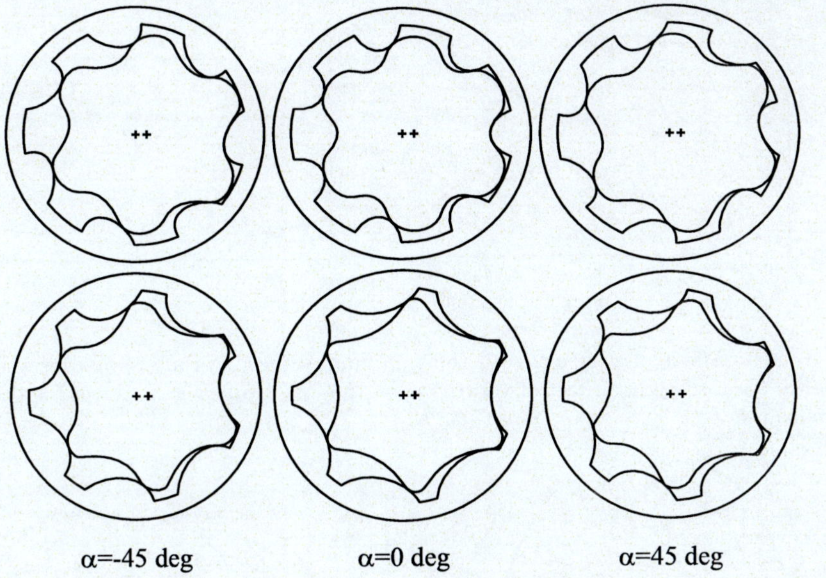

Fig. 7. Illustration of pumps used in case study to observe the effect of changing α. Top: $m = 7$, $\lambda_p = 0.65$, $\lambda_d = 0.3$, $\lambda_t = 0.7$. Bottom: $m = 7$, $\lambda_p = 0.65$, $\lambda_d = 0.3$, $\lambda_t = 1.3$

$$q_j = \begin{cases} -\frac{dA_j}{d\theta_i} & \frac{dA_j}{d\theta_i} \leq 0 \\ 0 & \frac{dA_j}{d\theta_i} > 0 \end{cases} \tag{14}$$

The difference between the flow ripples can be quantified by using the signal power of the ripple defined in Eq. (15). The signal power accounts for the energy in the ripple across all the frequencies, and it is plotted for both the reference pumps over the full range of α in Fig. 8. Applying asymmetry increased the flow ripple for the pump with $\lambda_t < 1$, but decreased it slightly for $\lambda_t > 1$.

$$\bar{E} = \frac{m-1}{2\pi} \int_0^{\frac{2\pi}{m-1}} \left[Q(\theta_i) - \bar{Q}\right]^2 \, d\theta_i \tag{15}$$

The shape of the gear profiles has a significant effect on the wear the profiles in operation, and the amount of wear can be quantified by the classic Archard wear model in Eq. (16), which predicts that the volume of worn material W is generally proportional to the work done by friction. In Eq. (16), F is the contact force, S is the sliding distance, and H is the material hardness. If small clearances on the order of 10–100 μm are introduced between the gears for manufacturing tolerances and the bearings are assumed to be stiff, the number of points in contact would be reduced to one. If only one point makes contact, the contact force can be determined directly from the required torque, position of the contact point, and pressure angle. The Archard model can then be normalized to the operating conditions, face width, and material properties to give a normalized

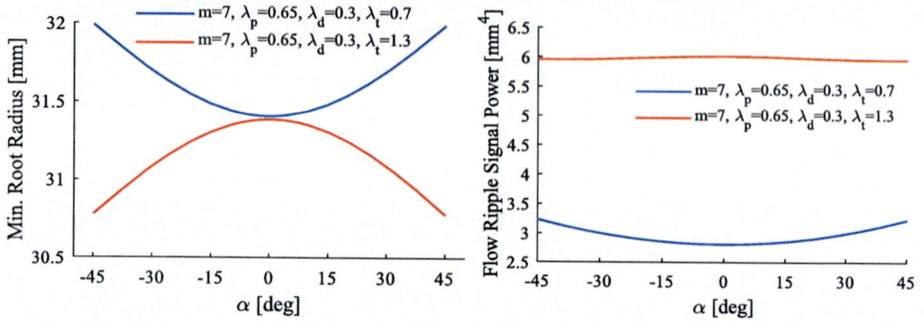

Fig. 8. Left: Effect of varying α on the minimum allowable root radius for the two reference pumps. Right: Effect of varying α on the flow ripple signal power for the two reference pumps

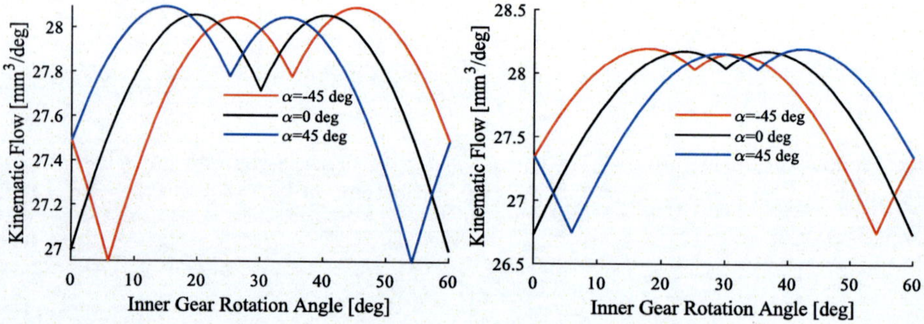

Fig. 9. Left: Effect of varying α on the flow ripple for a pump with $m = 7$, $\lambda_p = 0.65$, $\lambda_d = 0.3$, $\lambda_t = 0.7$. Right: Effect of varying α on the flow ripple for a pump with $m = 7$, $\lambda_p = 0.65$, $\lambda_d = 0.3$, $\lambda_t = 1.3$.

wear rate $\overline{W}$ in Eq. (17) that is dependent only on the gear geometry, where $\overline{F}$ is the normalized contact force, $\overline{S}$ is the normalized sliding speed, Δp is the operating pressure, and ω_i is the inner gear rotation speed. The kinematic flow is used to predict the normalized contact force defined in Eq. (18), where $\mathbf{r_i}$ is the position of the contact point with respect to the inner gear center and $\hat{\mathbf{n}}$ is the unit normal vector of the gear profiles at the contact point. The normalized sliding speed is then defined in Eq. (19), where $\mathbf{v_i}$ and $\mathbf{v_o}$ are the velocities of the contact point on the inner and outer gear respectively, and $\hat{\mathbf{t}}$ is the unit tangent vector of the gear profiles at the contact point. Note that a more detailed description of the normalized wear rate is given in previous work [10].

$$W = \frac{KFS}{H} \tag{16}$$

$$\overline{W} = \frac{WH}{Kh\Delta p\omega_i} = \overline{FS} \tag{17}$$

$$\overline{F} = \frac{F}{h\Delta p} = \frac{Q}{|\mathbf{r_i} \times \hat{\mathbf{n}}|} \tag{18}$$

$$\overline{S} = \frac{|\mathbf{v_i} \cdot \hat{\mathbf{t}} - \mathbf{v_o} \cdot \hat{\mathbf{t}}|}{\omega_i} \tag{19}$$

A plot comparing the peak values of the normalized wear rate as a function of α for the two reference pumps is shown in Fig. 10. For both pumps, the wear could be reduced by applying asymmetry. However, a greater improvement was achieved for the pump with $\lambda_t > 1$.

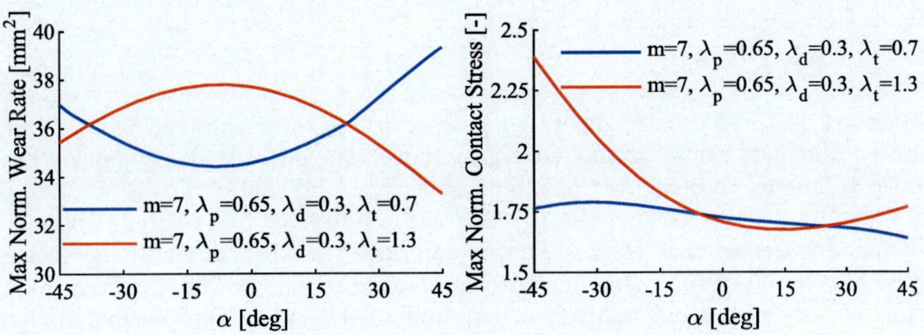

Fig. 10. Left: Effect of varying α on the wear rate for the two reference pumps. Right: Effect of varying α on the contact stress for the two reference pumps

The amount of contact stress in the gears is also affected by the profile geometry. The two teeth in contact can be modeled according to Hertzian contact theory as cylinders with radii equivalent to the radii of curvature of the gear surfaces. The peak contact pressure p_0 can be normalized to material properties and operating pressure in Eq. (20) to give a normalized contact pressure that is dependent only on the gear geometry. In Eq. (20), E^* is the combined modulus of elasticity given in Eq. (21), and κ_i and κ_o are the curvatures of the inner and outer gear surfaces at the contact point respectively. In Eq. (21), ν_i and ν_o are the Poisson's ratio for the inner and outer gear respectively, and E_i and E_o are the moduli of elasticity of the inner and outer gear respectively. Note that a more detailed description of the normalized contact pressure is also given in previous work [10].

$$\overline{p_0} = p_0\sqrt{\frac{\pi}{E^* \Delta p}} = \sqrt{\overline{F}\,|\kappa_i + \kappa_o|} \tag{20}$$

$$\frac{1}{E^*} = \frac{1-\nu_i^2}{E_i} + \frac{1-\nu_o^2}{E_o} \tag{21}$$

The peak values of the maximum contact pressure as a function of α for the two reference pumps are also plotted in Fig. 10. Both pumps can have a reduction in contact stress by applying positive asymmetry.

The overall effect of applying asymmetry to the gears depends largely on the other input parameters. For the reference pump with $\lambda_t < 1$, the contact stress or wear rate can be reduced, but it comes at the cost of an increase in the other performance goals. For the reference pump with $\lambda_t > 1$, all of the performance goals could be improved by applying positive asymmetry. Adding asymmetry increases the design freedom in comparison to a symmetric profile, which can open new areas of the design space that are improved with respect to the symmetric profile. However, more work is necessary to identify the optimal design space and to quantify its advantage over the symmetric profile in a more general sense.

4 Conclusions

A method was presented to generate a gearset for a gerotor pump using an ellipse rotated about its center as the outer gear tooth. The effect that asymmetry has on the minimum radius flow ripple, adhesive wear, and contact stress of a pump was investigated by varying the tooth rotation angle for two reference pumps. The results showed that the asymmetry can have the effect of either increasing or decreasing all of the performance goals depending on the values of the other input parameters. More analysis is required to determine the optimal design space and how the performance of asymmetric profile compares to conventional gerotor profiles.

Acknowledgements. This material is based upon work partially supported by the National Science Foundation Graduate Research Fellowship Program under Grant No. DGE-1333468. Any opinions, findings, and conclusions or recommendations expressed in this material are those of the authors and do not necessarily reflect the views of the National Science Foundation.

References

1. Altare, G., Rundo, M.: Advances in simulation of gerotor pumps: an integrated approach. Proc. Inst. Mech. Eng. Part C: J. Mech. Eng. Sci. **231**(7), 1221–1236 (2017)
2. Castilla, R., Gamez-Montero, P.J., Raush, G., Codina, E.: Method for fluid flow simulation of a gerotor pump using OpenFOAM. J. Fluids Eng. **139**(11), 111101 (2017). https://doi.org/10.1115/1.4037060
3. Choi, T.H., Kim, M.S., Lee, G.S., Jung, S.Y., Bae, J.H., Kim, C.: Design of rotor for internal gear pump using cycloid and circular-arc curves. J. Mech. Des. **134**(1), 011005 (2012)
4. Colbourne, J.R.: Gear shape and theoretical flow rate in internal gear pumps. Trans. Can. Soc. Mech. Eng. **3**(4), 215–223 (1975)
5. Gamez-Montero, P.J., Castilla, R., Codina, E., Freire, J., Morató, J., Sanchez-Casas, E., Flotats, I.: GeroMAG: in-house prototype of an innovative sealed, compact and non-shaft-driven gerotor pump with magnetically-driving outer rotor. Energies **10**(4), 1–14 (2017). https://doi.org/10.3390/en10040435

6. Hill, M.F.: Elliptical contours for rotor teeth. United States Patent 2,389,728 (1945)
7. Kwon, S.M., Kang, H.S., Shin, J.H.: Rotor profile design in a hypogerotor pump. J. Mech. Sci. Technol. **23**(12), 3459–3470 (2009). https://doi.org/10.1007/s12206-009-1007-y
8. Mancò, S., Nervegna, N., Rundo, M., Margaria, M.: Miniature gerotor pump prototype for automotive applications. In: 3rd International Fluid Power Conference (3rd IFK), pp. 153–167 (2002)
9. Pellegri, M., Vacca, A.: Numerical simulation of gerotor pumps considering rotor micro-motions. Meccanica 1–20 (2016). https://doi.org/10.1007/s11012-016-0536-6
10. Robison, A., Vacca, A.: Multi-objective optimization of circular-toothed gerotors for kinematics and wear by genetic algorithm. Mech. Mach. Theory **128**, 150–168 (2018). https://doi.org/10.1016/j.mechmachtheory.2018.05.011
11. Stryczek, J., Bednarczyk, S., Biernacki, K.: Gerotor pump with POM gears: design, production technology, research. Arch. Civ. Mech. Eng. **14**(3), 391–397 (2014). https://doi.org/10.1016/j.acme.2013.12.008

Check for updates

An Approximate, Closed Form Solution of Sealing Gap Induced Lateral Forces for Imperfect Sealing Land Geometries

Rudolf Scheidl(✉), Markus Resch, Matthias Scherrer, and Philipp Zagar

Johannes Kepler University Linz, Altenbergerstraße 69, 4040 Linz, Austria
rudolf.scheidl@jku.at

Abstract. It is a well-known fact that even small deviations from the perfect cylindrical geometry of either the bore or the sealing land of spool valves may cause substantial lateral forces and, in turn, a sticking of the valve. The reason is the pressure distribution in the sealing gap which loses its axisymmetry in such cases. This effect is counteracted by a number of circumferential grooves which even the pressure distribution. That valve sticking effect can be studied by solving the Reynolds equation for such gap geometries and computing the resulting radial force. From a design viewpoint it is valuable to have a compact understanding of the role of design parameters; in the case of non-uniform sealing gaps, of the role of the imperfection parameters. Gaining compact understanding of cause effect relations with numerical solutions is difficult if more than two parameters are involved since a graphical representation by diagrams is impossible. In this paper an approximate but closed form solution is derived for a class of imperfect sealing gap geometries. They provide that immediate insight by a compact formula for the resulting lateral force. The study is based on a Ritz method for a direct solution of the variational formulation of the Reynolds equation and asymptotic expansions concerning the role of the imperfection parameters.

Keywords: Sealing gap · Rayleigh-Ritz method · Reynolds equation · Valve sticking

1 Introduction

Sealing of hydraulic valves and pumps is mostly done by small gaps. Ideally, these gaps are very thin, perfectly cylindrical annular volumes. In reality, manufacturing imperfections and potentially also pressure induced elastic deformations cause a deviation from this ideal gap geometry. The resulting pressure fields of such imperfect geometries cause lateral forces. In best case, these forces stabilize the centric position of the spool or piston in the bore, in worst case, they exhibit a destabilizing action and high sticking forces [1]. This well known phenomenon is used, for instance, in zero friction hydraulic cylinders to realize a hydrostatic bearing by conical pistons in order to avoid solid body

© The Editor(s) (if applicable) and The Author(s), under exclusive license
to Springer Nature Switzerland AG 2021
J. Stryczek and U. Warzyńska (Eds.): NSHP 2020, LNME, pp. 102–111, 2021.
https://doi.org/10.1007/978-3-030-59509-8_9

contact with the cylinder wall [2]. Pressure balancing grooves in the sealing lands of valve spools are a standard means to avoid excessive sticking forces.

Such forces can be computed by solving the Reynolds equation [3] for the fluid film in the gap. This is trivial for the perfect cylindrical case but is usually done by numerical methods for non-cylindrical gap geometries because of the non-uniform conductivity term in the equation which is given by the third power of the gap height.

It is not the numerical computation of the Reynolds equation for a given gap geometry but the study of the influence of gap geometry parameters on the lateral forces which is challenging. Even simple versions of gap geometry families might have three or more parameters. A parameter study done by a series of numerical calculations for a representative set of parameter combinations not only requires computation time to solve the multitude of problem versions numerically but, even more important, is difficult to be comprehended by graphical representations of the results due to problem dimensionality.

In this paper a Rayleigh-Ritz method is used to obtain approximate closed form solutions of the fluid film force for a class of gap geometries [4]. This result makes the influence of the problem parameters obvious without any graphical representation and is in good agreement with numerical solutions obtained with a finite volume model.

2 Model of Gap Pressure Field

2.1 Gap Geometry

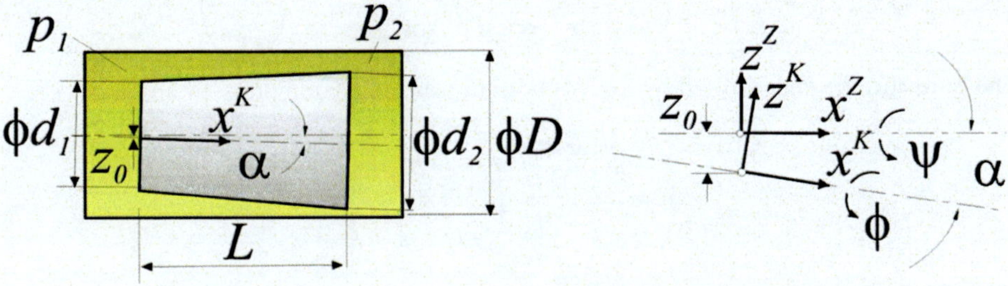

Fig. 1. Geometry of the considered gap conical gap geometries.

The problem is formulated for conical gap geometries. For the derivation of the gap function h a truncated cone with conicity k, length L, and diameters d_1, d_2 is placed in a perfect cylindrical bore with diameter D. The cone is shifted out of bore center by a value z_0 and rotated by an angle α. Obviously, it does not make a difference if the bore has a conical shape instead, since only the gap geometry is relevant. It is assumed that the gap is a usually small sealing gap, i.e., $h = O(D.10^{-3})$. That fact will be used later on, to simplify the results. Two Cartesian coordinate systems, one centered to the cone axis (indicated by a superscript K), the other to the bore axis (indicated by a superscript

Z), are used to describe the geometry. The truncated cone surface is described by Eq. (1) and spanned by the two surface coordinates x and ϕ (Fig. 1).

$$\begin{pmatrix} x_k \\ y_k \\ z_k \end{pmatrix}^{(K)} = \begin{pmatrix} x \\ \frac{1}{2}(d_1 + k\,x)\sin(\phi) \\ \frac{1}{2}(d_1 + k\,x)\cos(\phi) \end{pmatrix}; \; k = \frac{d_2 - d_1}{L} \tag{1}$$

It is transferred to the bore centered coordinate system by a shift operation and a rotation according to Eq. (2).

$$\begin{pmatrix} x_k \\ y_k \\ z_k \end{pmatrix}^{(Z)} = \begin{pmatrix} \cos\alpha & & \sin\alpha \\ & 1 & \\ -\sin\alpha & & \cos\alpha \end{pmatrix} \begin{pmatrix} x_k \\ y_k \\ z_k \end{pmatrix}^{(K)} + \begin{pmatrix} 0 \\ 0 \\ -z_0 \end{pmatrix} \tag{2}$$

The cone surface normal **n** in both coordinate systems is given the following equations; t is the line parameter and measures the distance of the normal line point from the cone surface.

$$\begin{pmatrix} n_x \\ n_y \\ n_z \end{pmatrix}^{(K)} = \begin{pmatrix} x - t^*k \\ \left(\frac{1}{2}(d_1 + k\,x) + t^*\right)\sin\phi \\ \left(\frac{1}{2}(d_1 + k\,x) + t^*\right)\cos\phi \end{pmatrix}; \quad t^* = \frac{t}{\sqrt{1+k^2}} \tag{3}$$

$$\begin{pmatrix} \mathrm{n_x} \\ \mathrm{n_y} \\ \mathrm{n_z} \end{pmatrix}^{(Z)} = \begin{pmatrix} \cos\alpha & & \sin\alpha \\ & 1 & \\ -\sin\alpha & & \cos\alpha \end{pmatrix} \begin{pmatrix} \mathrm{n_x} \\ \mathrm{n_y} \\ \mathrm{n_z} \end{pmatrix}^{(K)} + \begin{pmatrix} 0 \\ 0 \\ -z_0 \end{pmatrix}$$

The bore surface is described by Eq. (4) with the surface parameters x_Z and ψ.

$$\begin{pmatrix} x_{Zyl} \\ y_{Zyl} \\ z_{Zyl} \end{pmatrix}^{(Z)} = \begin{pmatrix} x_Z \\ \frac{1}{2}D\sin\psi \\ \frac{1}{2}D\cos\psi \end{pmatrix} \tag{4}$$

The gap height h is the distance between bore and cone surface measured along the cone surface normal **n**. It is identical to the value of t which equates the coordinates according to (4) and the cone coordinates according to (3). For the computation of t the vector identity $\mathbf{n} = \mathbf{x}_{Zyl}$ is first linearized with respect to the very small quantities α, k, $h_0/D = (1 - d_1/D)/2$, $\zeta_0 = z_0/h_0$, and $\Delta\phi = \psi - \phi$. The second and third row of these linearized equation are then multiplied by $\cos\phi$ and $\sin\phi$, respectively, and added up. This leads to the following approximate equation for the gap height h.

$$h = t = cos(\phi)(\alpha x + h_0\varsigma_0) - \frac{1}{2}kx + h_0 \tag{5}$$

2.2 Variational Form of the Reynolds Equation and Ritz Approximation

For the application of the Rayleigh-Ritz method for the approximate solution the variational form of the Reynolds equation is used.

$$\begin{gathered} \delta\Pi = 0; \quad \Pi = \int\limits_0^L \int\limits_0^{2\pi} \tfrac{1}{2}\tfrac{h^3}{\eta}(\nabla p)^2 d\phi dx \\ \tfrac{1}{2}(\nabla p)^2 = \tfrac{1}{2}\nabla p \cdot \nabla p = \tfrac{1}{2}\left(\left(\tfrac{\partial p}{\partial x}\right)^2 + \tfrac{1}{R^2}\left(\tfrac{\partial p}{\partial \phi}\right)^2\right) \\ R = \tfrac{D}{2} \end{gathered} \tag{6}$$

The existence of a variational formulation guarantees that the Rayleigh-Ritz method gives the best approximation of the real solution for a given Ritz ansatz in terms of a distance norm, which in this case corresponds to the continuity relation.The pressure field is approximated by the following Ritz ansatz.

$$p(x,\phi) = a_{00} + a_{01}\frac{x}{L} + a_{02}\left(\frac{x}{L}\right)^2 + \left(a_{10} + a_{11}\frac{x}{L} + a_{12}\left(\frac{x}{L}\right)^2\right)\cos(\phi) \tag{7}$$

The boundary conditions

$$p(0,\phi) = p_1; p(L,\psi) = p_2 \tag{8}$$

pose additional conditions on the coefficients a_{00}, a_{01}, a_{02}, a_{10}, a_{11}, a_{12}. Considering these leads to the final version of (6, 7)

$$\begin{aligned} p(x,\psi) &= p_1 + (p_2 - p_1 - a_{02})\frac{x}{L} + a_{02}\left(\frac{x}{L}\right)^2 \\ &+ \left(-a_{12}\frac{x}{L} + a_{12}\left(\frac{x}{L}\right)^2\right)\cos(\phi) \end{aligned} \tag{9}$$

with only two generalized coordinates a_{02}, a_{12}. They are computed by a direct solution of the variational equation.

$$\begin{gathered} \delta\Pi = \int\limits_0^L \int\limits_0^{2\pi} \frac{h^3}{\eta}\left(\left(\frac{\partial p}{\partial x}\right)\left(\frac{\partial \delta p}{\partial x}\right) + \frac{1}{R^2}\left(\frac{\partial p}{\partial \phi}\right)\left(\frac{\partial \delta p}{\partial \phi}\right)\right) d\phi dx = 0 \\ \delta p(x,\phi) = \delta a_{02}\left(-\frac{x}{L} + \left(\frac{x}{L}\right)^2\right) + \delta a_{12}\left(-\frac{x}{L} + \left(\frac{x}{L}\right)^2\right)\cos(\phi) \end{gathered} \tag{10}$$

Evaluating the expression for $\delta\Pi$ and separating for the independent variations δa_{02}, δa_{12} yields two linear equations for a_{02}, a_{12}, given by (11).

$$\begin{aligned} &\mathbf{A}.\begin{pmatrix} a_{02} \\ a_{12} \end{pmatrix} = \mathbf{b}(p_1 - p_2)L \\ &A_{1,1} = \left(240\varsigma_0^2 + 160\right)h_0^3 - 60L\left(\varsigma_0^2 k - 4\varsigma_0\alpha + 2k\right)h_0^2 + \\ &\quad 48L^2\left(-2\varsigma_0\alpha k + 2\alpha^2 + k^2\right)h_0 - 7L^3 k\left(6\alpha^2 + k^2\right) \\ &A_{1,2} = \left(60\varsigma_0^3 + 240\varsigma_0\right)h_0^3 - \left(120\left(-\left(\frac{3}{4}\right)\alpha\varsigma_0^2 + \varsigma_0 k - \alpha\right)\right)Lh_0^2 + \end{aligned}$$

$$24L^2\left(3\varsigma_0\alpha^2+\varsigma_0 k^2-4\alpha k\right)h_0+21\alpha\left(\alpha^2+k^2\right)L^3$$

$$A_{2,1}=\left(\frac{1}{4}\varsigma_0^3+\varsigma_0\right)h_0^3-\frac{1}{2}\left(-\frac{3}{4}\alpha\varsigma_0^2+\varsigma_0 k-\alpha\right)Lh_0^2$$

$$+L^2\left(3\varsigma_0\alpha^2+\varsigma_0 k^2-4\alpha k\right)\frac{h_0}{10}+\frac{7\alpha\left(\alpha^2+k^2\right)L^3}{80}$$

$$A_{2,2}=-\frac{L^5\alpha^2 k}{448R^2}-\frac{L^5k^3}{1344R^2}-L^4\alpha h_0 k\frac{\varsigma_0}{140R^2}-\frac{21L^3\alpha^2 k}{160}-7\frac{L^3k^3}{480}$$

$$-L^3h_0^2k\frac{\varsigma_0^2}{160R^2}-\frac{3L^2\alpha h_0 k\varsigma_0}{10}-\frac{3Lh_0^2k\varsigma_0^2}{16}+\frac{L^4\alpha^2 h_0}{140R^2}+\frac{L^4h_0k^2}{140R^2}+\frac{L^3\alpha h_0^2\varsigma_0}{40R^2}$$

$$+\frac{3L^2\alpha^2+L^2k^2}{10}h_0+\frac{h_0^3\varsigma_0^2L^2}{40R^2}+\frac{3L\alpha h_0^2\varsigma_0+3h_0^3\varsigma_0^2}{4}$$

$$-\frac{L^3h_0^2k}{40R^2}-\frac{Lh_0^2k}{4}+\frac{h_0^3L^2}{30R^2}+\frac{h_0^3}{3}$$

$$b_1=9\left(\left(\frac{20k\varsigma_0^2}{3}-\frac{80\alpha\varsigma_0}{3}+\frac{40k}{3}\right)h_0^2-20L\left(-2\varsigma_0\alpha k+2\alpha^2+k^2\right)\left(\frac{h_0}{3}\right)+L^2k\left(6\alpha^2+k^2\right)\right)$$

$$b_2=-\frac{1}{80}\left(9\left(\alpha^3+k^2\alpha\right)L^2+10L\left(3\varsigma_0\alpha^2+\varsigma_0k^2-4\alpha k\right)h_0-40h_0^2\left(-\frac{3}{4}\alpha\varsigma_0^2+\varsigma_0k-\alpha\right)\right) \quad (11)$$

2.3 Approximations Exploiting the Smallness of the Gap

The solutions for a_{02}, a_{12} are lengthy and are not given here. The smallness of ς_0, α, k suggests to take truncated Taylor series of the solutions with respect to these variables up to the first non-vanishing order. This yields the compact expressions (12).

$$a_{0,2}=\frac{-3\lambda(p_1-p_2)k}{8\mu};\quad a_{1,2}=\frac{15\lambda(p_1-p_2)(2\varsigma_0 k\mu+\alpha\lambda k+4\alpha\mu)}{8\mu^2(\lambda^2+10)}$$

$$\lambda=\frac{L}{R};\quad \mu=\frac{h_0}{D} \quad (12)$$

The corresponding approximate pressure field equation is given by (13).

$$p(x,\phi)=p_1+(p_2-p_1)\left(\tfrac{x}{L}-\tfrac{3\lambda k}{8\mu}\left(\tfrac{x}{L}-\left(\tfrac{x}{L}\right)^2\right)\right)+$$
$$(p_2-p_1)\left(\tfrac{15\lambda(4\mu\alpha+2k\mu\zeta_0+k\alpha\lambda)}{8\mu^2(10+\lambda^2)}\left(\tfrac{x}{L}-\left(\tfrac{x}{L}\right)^2\right)\right)\cos(\phi) \quad (13)$$

The pressure field is symmetrical to the vertical mid-plane (x-z). Thus, the resulting force has no y-component. The smallness of α and k makes the axial component of the force also a very small value and is not considered here. The approximate resulting z-component F_{res}, computed by the rule

$$F_{res}=-\int_0^L\int_0^{2\pi}p(x,\psi)\mathrm{R}\cos(\phi)d\phi dx. \quad (14)$$

becomes finally

$$F_{res}=-\frac{5\pi L^2\,(4\mu\alpha+2k\mu\zeta_0+k\alpha\lambda)}{16\mu^2(10+\lambda^2)}(p_2-p_1). \quad (15)$$

2.4 Interpretations of the Result

- A non-axisymmetric pressure field and a non-vanishing resultant vertical force F_{res} require that either α or ς_0 are nonzero. ς_0 has only an effect if also the conicity is nonzero ($k \neq 0$). A parallel displacement (ς_0) of a perfect spool does not create a force.
- A conicity ($k \neq 0$) causes a bulbous pressure field.
- The central position of the conical spool is stable w.r.t. a parallel displacement ($\alpha = 0$) if $k\,(p_1 - p_2) > 0$. Then F_{res} is positive if the cone is shifted downwards and negative if shifted upwards. In both cases the force tends to move the cone back to the center.
- A clockwise rotation ($\alpha > 0$) of a single cylindrical piston ($k = 0$) presses the spool downwards if $(p_2 - p_1) > 0$ and upwards else. That means that such a single cylinder is always unstable w.r.t. a rotation α if no other effects counteract this instability mechanism.
- However, a cylindrical spool with two sealing lands in a perfectly cylindrical bore and a pressure situation as indicated in Fig. 2 is stable w.r.t. a rigid body rotation, if if $p_2 > p_1$ and unstable if $p_1 > p_2$.

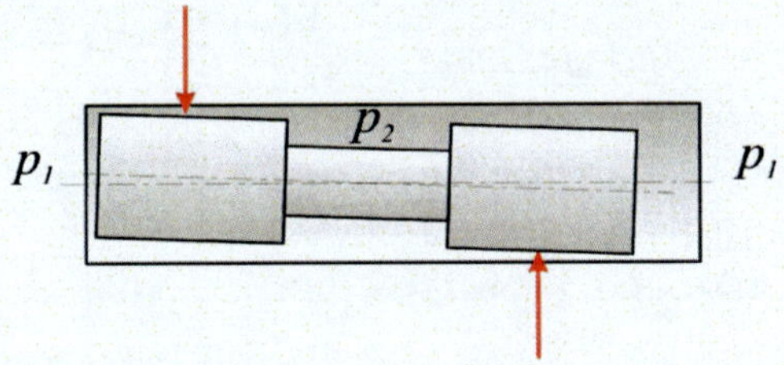

Fig. 2. A tilted perfectly cylindrical spool with two sealing lands is stable with respect to an inclination if $p_2 > p_1$ and unstable if $p_1 > p_2$.

3 Verification with a Numerical Model

3.1 Finite Volume Model

In order to assess the analytical model a finite volume model for the fluid film in the sealing gap is established. A rectangular grid of finite volumes (cells) represents the fluid film in the sealing gap, see Fig. 3. Pressure and gap height fields are discretized by values at the cell center points ($p_{i,j}$, $h_{i,j}$). The continuity law for the incompressible fluid requires the sum of the flows into the cell to be zero.

$$q_{i;j+1,j} + q_{i+1,i;j} + q_{i;j-1,j} + q_{i-1,i;j} = 0 \tag{16}$$

Since the spool is not moving in the bore the cell flows in (16) stem from diffusive terms only. They read

$$q_{i-1,i;j} = \frac{\left(p_{i-1,j} - p_{i,j}\right)}{\left(\Delta s_{\phi,\,i-1,j} + \Delta s_{\phi,\,i,j}\right)/2} \frac{\frac{\Delta s_{x,i-1,j}+\Delta s_{x,i,j}}{2}\left(\frac{h_{i-1,j}+h_{i,j}}{2}\right)^3}{12\eta}$$

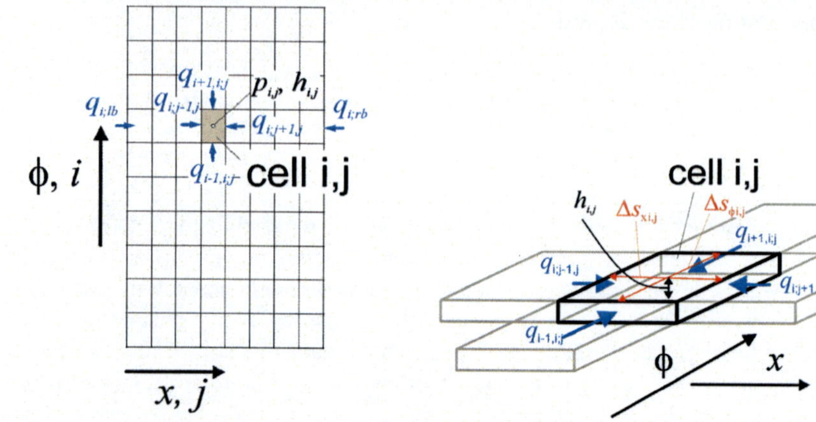

Fig. 3. Unwrapped grid of the finite volume model. The line index i refers to the circumferential direction ϕ, the index j to the lateral direction x. The pressure and gap height fields are discretized by the cell midpoint pressures $p_{i,j}$, and gaps $h_{i,j}$, respectively.

$$\begin{aligned}
q_{i+1,i;j} &= \frac{(p_{i+1,j}-p_{i,j})}{(\Delta s_{\phi,\,i+1,j}+\Delta s_{\phi,\,i,j})/2}\,\frac{\frac{\Delta s_{x,i+1,j}+\Delta s_{x,i,j}}{2}\left(\frac{h_{i+1,j}+h_{i,j}}{2}\right)^3}{12\eta}\\
q_{i,j-1;j} &= \frac{(p_{i,j-1}-p_{i,j})}{(\Delta s_{x,\,i,j-1}+\Delta s_{x,\,i,j})/2}\,\frac{\frac{\Delta s_{\phi,i,j-1}+\Delta s_{\phi,i,j}}{2}\left(\frac{h_{i,j-1}+h_{i,j}}{2}\right)^3}{12\eta}\\
q_{i,j+1;} &= \frac{(p_{i,j+1}-p_{i,j})}{(\Delta s_{x,\,i,j+1}+\Delta s_{x,\,i,j})/2}\,\frac{\frac{\Delta s_{\phi,i,j+1}+\Delta s_{\phi,i,j}}{2}\left(\frac{h_{i,j+1}+h_{i,j}}{2}\right)^3}{12\eta}
\end{aligned} \tag{17}$$

Flows into the cells at the left and right boundaries $q_{i,lb}$, $q_{i,rb}$ are

$$\begin{aligned}
q_{i,lb} &= \frac{(p_1-p_{i,1})}{\Delta s_{x,i,1}/2}\,\frac{\Delta s_{\phi,i,1}h_{i,1}^3}{12\eta}\\
q_{i,rb} &= \frac{(p_2-p_{i,N_x})}{\Delta s_{x,i,N_x}/2}\,\frac{\Delta s_{\phi,i,N_x}h_{i,N_x}^3}{12\eta}
\end{aligned} \tag{18}$$

$\Delta s_{\phi,i,j}$ and $\Delta s_{x,i,j}$ are the cell lengths, η is the dynamic viscosity of the fluid.Inserting the diffusive flow terms (17, 18) into the continuity Eq. (16) gives a linear equation system for the cell pressures $p_{i,j}$. The resulting vertical force is given by Eq. (19).

$$F_{res,num} = -\sum_{i=1}^{N_\phi}\sum_{j=1}^{N_x} p_{i,j}\Delta s_{\phi,i,j}\Delta s_{x,i,j}cos(\phi_{i,j}) \tag{19}$$

3.2 Comparison of Analytical and Numerical Model

In the following results concerning the pressure field and the resultant lateral force are shown for three exemplary cases according to Table 1. Figures show the geometric configuration (gap is 100 times magnified), the pressure field with the resulting vertical forces for both models, the gap field, and the relative pressure field error (Figs. 4, 5 and 6).

Table 1. Exemplary cases for comparing analytical and numerical models.

Case	*D [mm]*	*L [mm]*	h_0 *[mm]*	ζ_0	α *[rad]*	*k*
1	10	15	0.01	0.5	0.0	−0.000267
2	10	15	0.01	0.0	0.00033	+0.000267
3	10	15	0.01	0.2	0.00033	0.0

$p_1 = 1$ bar, $p_2 = 100$ bar in all cases

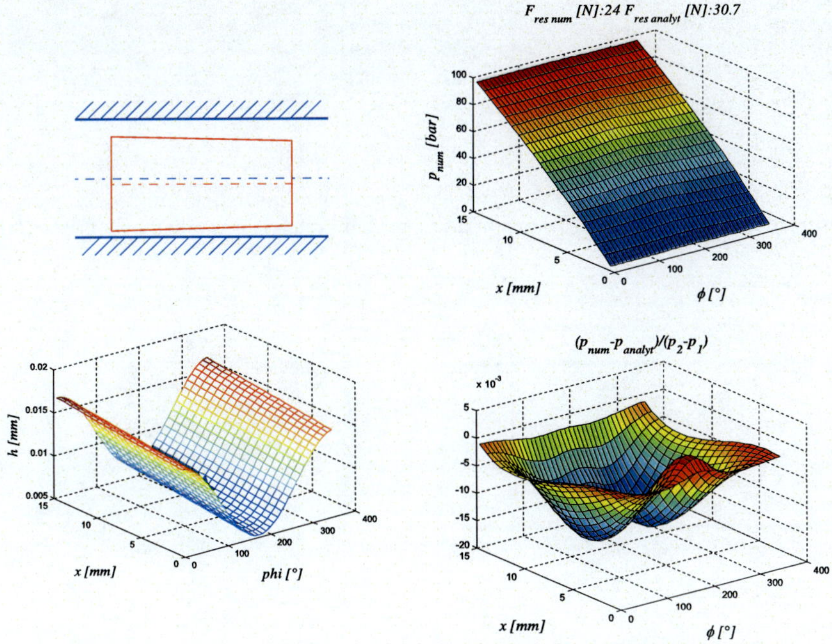

Fig. 4. Situation and results of numerical and approximate analytical model for parameters according to Case 1 of Table 1; (the gap in the upper left sketch is magnified by a factor 100).

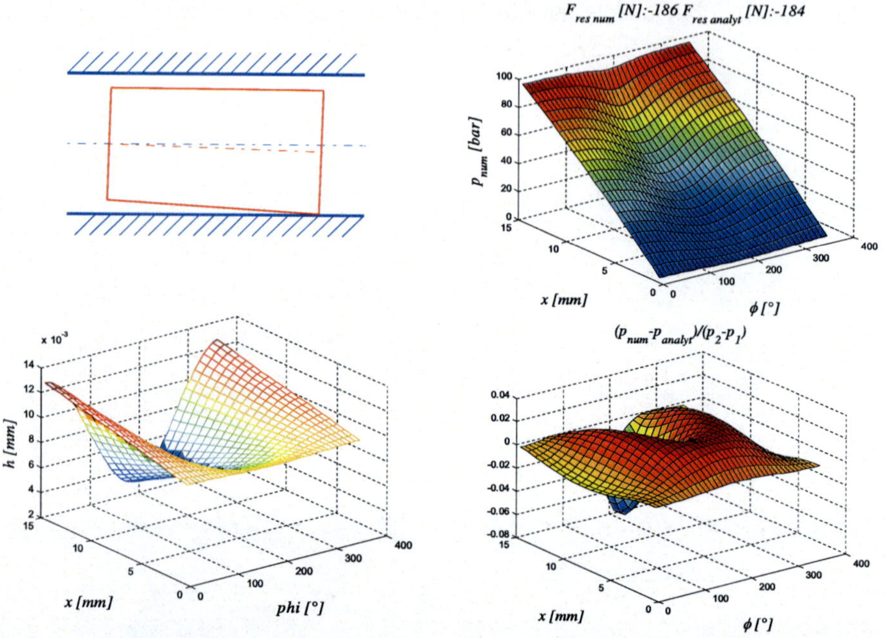

Fig. 5. Situation and results of numerical and approximate analytical model for parameters according to Case 2 of Table 1; (the gap in the upper left sketch is magnified by a factor 100).

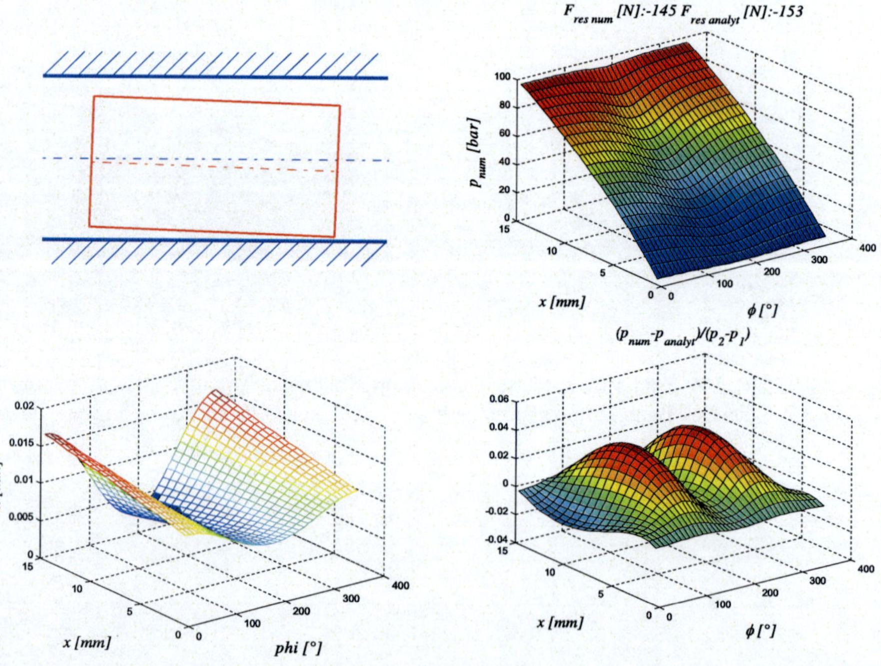

Fig. 6. Situation and results of numerical and approximate analytical model for parameters according to Case 3 of Table 1; (the gap in the upper left sketch is magnified by a factor 100).

4 Summary and Conclusions

An approximate analytical model for pressure and lateral force generated by the pressure field of an eccentric and tilted conical sealing gap is derived. To this end, a Rayleigh-Ritz method, a variational formulation of the Reynolds equation and an ansatz function for the pressure field, and series expansions to account for the smallness of several gap parameters are combined. The pressure field ansatz uses a first order Fourier expansion in the circumferential and quadratic polynomials in the axial gap coordinate. The results are compared with solutions obtained by a finite volume method. The pressure errors are a few percent of the pressure difference over the gap, the force error is up to 20%.

The errors are outweighed by the compactness of the formula for the force. It allows an immediate interpretation of the influences of the six geometric parameters and of the two pressures which cannot be achieved adequately with a numerical parameter study because of the dimensionality of the parameter space.

Acknowledgement. This works was done in the framework of the COMET K2 Center on Symbiotic Mechatronics, which is funded by the Austrian Federal Government, the State Upper Austria and by its Scientific and Industrial Partners.

References

1. Winkler, B., Mikota, G., Scheidl, R., Manhartsgruber, B.: Modelling and simulation of the elasto-hydrodynamic behavior of sealing gaps. Aust. J. Mech. Eng. (Formerly Trans. Mech. Eng.) **2**(1), 65–72 (2005)
2. Findeisen, D., Findeisen, F.: Ölhydraulik: Handbuch für die hydrostatische Leistungsübertragung in der Fluidtechnik, 5th edn. Springer, Heidelberg (2006)
3. Szeri, A.Z.: Fluid Film Lubrication. Cambridge University Press, Cambridge (2010)
4. Zheng, T., et al.: A Ritz model of unsteady oil-film forces for nonlinear dynamic rotor-bearing system. J. Appl. Mech. **71**(2), 219–224 (2004)

The Influence of Water and Mineral Oil on Pressure Losses in Hydraulic Motor

Pawel Sliwinski(✉) and Piotr Patrosz

Gdansk University of Technology, Narutowicza 11/12, 80-233 Gdansk, Poland
pawel.sliwinski@pg.edu.pl

Abstract. In this paper, pressure losses in hydraulic motor supplied with water and mineral oil (two liquids that have significantly different viscosity and lubricating properties) are described and compared. The experimental tests were conducted using an hydraulic satellite motor that is dedicated to working with different liquids, including water. The methodology of measuring the pressure drop in internal channels of this motor is also described. The known mathematical formulas describing the pressure losses in the internal channels are presented also. Experimental studies have shown that a high flow rate in the motor causes the pressure losses in the motor supplied with water to be a several percent greater than the pressure losses in the motor supplied with oil. However, low flow rates cause the pressure losses in the motor supplied with water to be about ten percent lower than the pressure losses in the motor supplied with oil. Furthermore, when the volume of the working chamber nears the minimum, additional pressure increase caused by compression of liquid occurs in this chamber.

Keywords: Pressure losses · Satellite motor · Water · Oil · Internal channels in motor

1 Introduction

The hydraulic motor in the hydraulic system is the actuator. Its purpose is to convert hydraulic energy into mechanical energy. The energy carriers in hydraulic systems are liquids. The type of liquid is determined by the requirements for the system. The liquid commonly used in hydraulic systems is mineral oil. However, there is a growing trend towards researching and developing components and hydraulic systems that are supplied with water [1–4]. Studies of hydraulics systems, where water is the working medium, are now important in many industries, such as the food industry, the mining industry and marine technology [4–6].

Pressure losses in hydraulic positive displacement machines are the pressure drop Δp_{ich} in internal channels in these machines. Channels inside the motor or pump body for supplying (inflow channels) working chambers and removing (outflow channels) the liquid from working chambers are defined as the internal channels. The geometry of these internal channels is important and have an influence on the pressure drop in

© The Editor(s) (if applicable) and The Author(s), under exclusive license to Springer Nature Switzerland AG 2021
J. Stryczek and U. Warzyńska (Eds.): NSHP 2020, LNME, pp. 112–122, 2021.
https://doi.org/10.1007/978-3-030-59509-8_10

these channels [7–9]. This pressure drop has a significant impact on the efficiency of energy conversion in a hydraulic motor and has an impact on the hydraulic system's work parameters [7, 10–15].

Thus far, there are no publications about the influence of water and mineral oil on the pressure losses in hydraulic motors. Of course, publications can be found, but they relate to pressure drop in basic elements such as, for example, a pipe [16]. There is also no specific information about the designs of motors that can be supplied with both water and mineral oil. Furthermore, there are no prominent studies in literature that compare the influence of the type of liquid on the flow characteristics in the internal channels. Consequently, the results of experimental tests of pressure losses in motor supplied with mineral oil and water and their comparison have been defined for this article.

2 Pressure Drop in the Internal Channel of the Hydraulic Motor

The pressure drop Δp_{ich} increases the value of the pressure drop Δp that is measured in motor ports according to the formula:

$$\Delta p = \Delta p_i + \Delta p_{ich} \tag{1}$$

where Δp_i is the pressure drop in the working chambers of the motor and is a function of motor load M and mechanical losses M_l in the motor.

Pressure drop Δp_{ich} in internal channels is the following sum:

$$\Delta p_{ich} = \Delta p_{ic1} + \Delta p_{ic2} \tag{2}$$

where Δp_{ic1} and Δp_{ic2} is the pressure drop in inflow and outflow internal channel (respectively).

In the literature following mathematical formulas, describing the pressure losses, are known [12–15, 17–20]:

$$\Delta p_{ich} = \left(\frac{C_1}{Re} + C_2\right) \cdot \rho \cdot \frac{c^2}{2} \tag{3}$$

$$\Delta p_{ich} = C_{ich} \cdot \rho \cdot \omega^2 \cdot \left(\frac{q_t}{2 \cdot \pi}\right)^{2/3} \tag{4}$$

$$\Delta p_{ich} = k_8 \cdot p_n \cdot \left(\frac{Q}{Q_t}\right)^{a_{Qs}} \cdot \left(\frac{\nu}{\nu_n}\right)^{a_{\nu s}} \tag{5}$$

$$\Delta p_{ich} = C_t \cdot \rho \cdot Q^2 + C_l \cdot \nu \cdot \rho \cdot Q \tag{6}$$

where:

- q_t – the theoretical displacement,
- ω – the angular speed of the motor shaft,
- Q_t – the theoretical delivery of the pump in hydraulic circuit;
- p_n – the nominal pressure of the pump;

- c – average speed of flow in motors internal channels,
- ρ – the density of liquid,
- ν – the kinematic viscosity,
- ν_n – the reference kinematic viscosity,
- a_{Qs} and $a_{\nu s}$ – exponents,
- C_1, C_2, C_{ich}, C_t, C_l, k_8, – coefficients.

For describing the pressure losses in a motor the formula (6) is recommended because [18–20]:

a) does not take into account the geometry of the motor's internal channels;
b) the pressure losses are expressed as a function of parameters having a direct impact on these losses (specifically the density and viscosity of the liquid and the liquid flow rate);
c) the values of constants C_t and C_l can be calculated quickly based on the equation of the trend line of the experimental characteristic $\Delta p_{ich} = f(Q)$.

3 Tested Motor

A prototype of a hydraulics satellite motor was selected for experimental tests. The design of this motor is presented in Fig. 1. The working mechanism of the satellite motor is a specific gear mechanism in which the rotor revolves around the shaft axis and the revolving motion is done by satellites which are in gear with the stator and the rotor (Fig. 2).

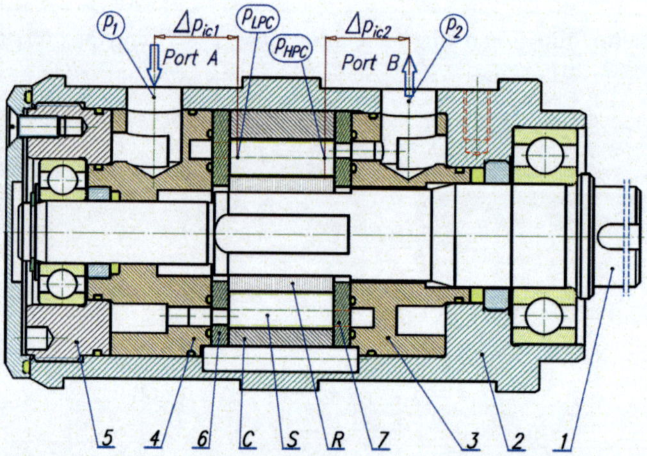

Fig. 1. Hydraulic satellite motor with marked pressure drop in internal channels: C – curvature (stator), S – satellite, R – rotor, 1 – shaft, 2 – casing, 3 and 4 – inflow and outflow manifolds, 5 – rear plate, 6 and 7 – distribution (commutation) plates [6, 15, 19, 20].

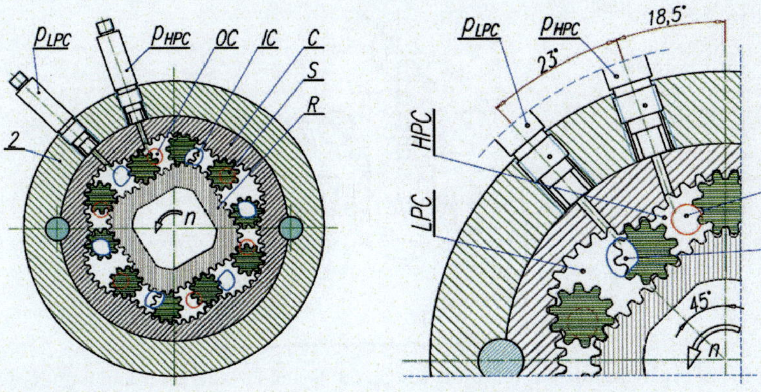

Fig. 2. Working mechanism of satellite motor with location of pressure sensors p_{LPC} and p_{HPC}: C – curvature (stator), R – rotor, S – satellite, LPC – low pressure working chamber, HPC – high pressure working chamber, IC – inflow hole, OC – outflow hole [18–20].

The toothed unit, shown in Fig. 2, is the satellite working mechanism of the motor. It consists of a toothed rotor R (4 humps), toothed stator C (6 humps) and ten wheels S (satellite). The working chambers in the satellite mechanism are closed by the compensation plates (Fig. 1 – elements 6 and 7, and Fig. 3), which also play the role of distribution plates.

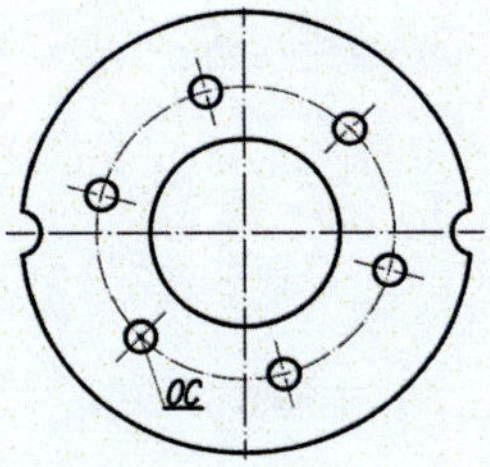

Fig. 3. Distribution (commutation) plate: OC – inflow/outflow hole [6, 19, 20].

The satellite motor used for the test had the theoretical displacement q_t = 32,94 cm^3/rev. and the diameter of inflow/outflow holes is 4,1 mm. The geometrical overlap in the commutation unit of working mechanism is zero. In the tested motor (Fig. 1), the internal channels on the supply side have the same dimensions as the internal channels on the outflow side. The diameter of port A and port B is 10 mm.

4 Test Stand and Parameters of Working Liquids

The measurement of pressure drop in an internal channel of a hydraulic motor can be carried out according to the following experimental method (Fig. 4). The hydraulic motor 1 works as a pump and is driven by an electric motor 2 with adjustable speed *n*

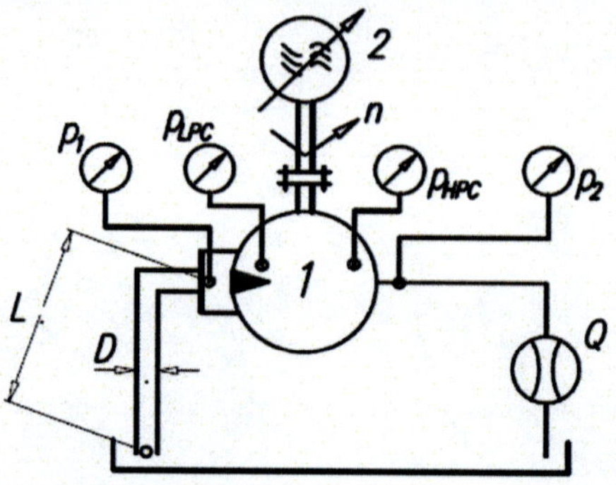

Fig. 4. Scheme of the hydraulic circuit of the test stand: D and L – diameter and length of suction tube, 1 – tested motor, 2 – electric motor controlled by frequency converter, Q – flow meter, n – rotational speed, p_{HPC} – pressure in high pressure working chamber, p_{HPC} – pressure in low pressure working chamber, p_1 – pressure in inflow motor port, p_2 – pressure in outflow motor port [18–20].

via a frequency converter. All measured data are recorded at the same time by the data acquisition system.

According to Fig. 4:

$$\Delta p_{ic1} = p_1 - p_{LPC} \tag{7}$$

$$\Delta p_{ic2} = p_{HPC} - p_2 \tag{8}$$

In the satellite motor the internal channels on the supply side have the same dimensions as the internal channels on the outflow side. Therefore, it is enough to measure only Δp_{ic1} and then:

$$\Delta p_{ich} = 2 \cdot \Delta p_{ic1} \tag{9}$$

During the test a measurement data acquisition system at the same time records:

a) the pressure p_1 and p_2 in motor ports measured by a strain gauge pressure transducer with range -1÷3 bar and class 0.3;
b) the pressure p_{LPC} and p_{HPC} in motor working chambers measured by a strain gauge pressure transducer with range -1÷3 bar and class 0.3;
c) the delivery Q measured by mass flow meter with range 33 lpm and class 0.1;
d) the rotational speed n of the motor shaft measured by the inductive sensor.

During the test the temperature in the inflow port of motor was also controlled in order to know what the viscosity of the liquid is. This temperature was not register.

The laboratory experiment was carried out using the following liquids:

a) tap water ($\nu = 0{,}9$cSt and $\rho = 999.8$ kg/m^3);
b) mineral oil ($\nu = 40$cSt and $\rho = 862.0$ kg/m^3).

5 Experimental Data

The recorded results of pressure p_1, p_2, p_{HPC}, p_{LPC} and flow rate Q in the motor as a function of the shaft speed n are shown in Fig. 5 and Fig. 6.

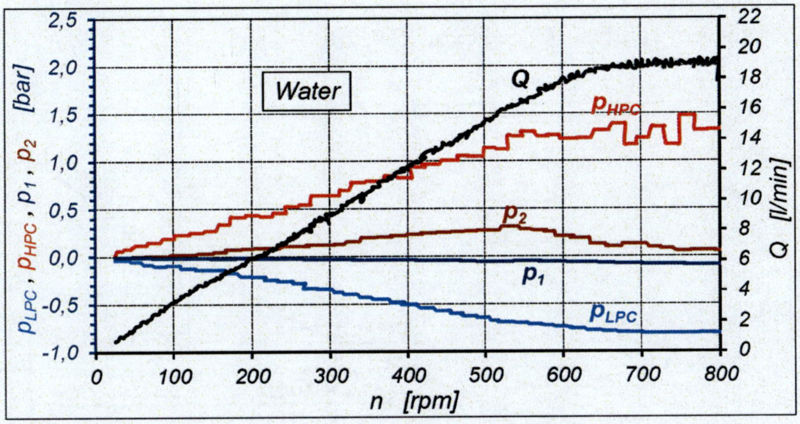

Fig. 5. Characteristics of $p_1 = f(n)$, $p_2 = f(n)$, $p_{HPC} = f(n)$, $p_{LPC} = f(n)$ and $Q = f(n)$ in the motor supplied with water [19, 20].

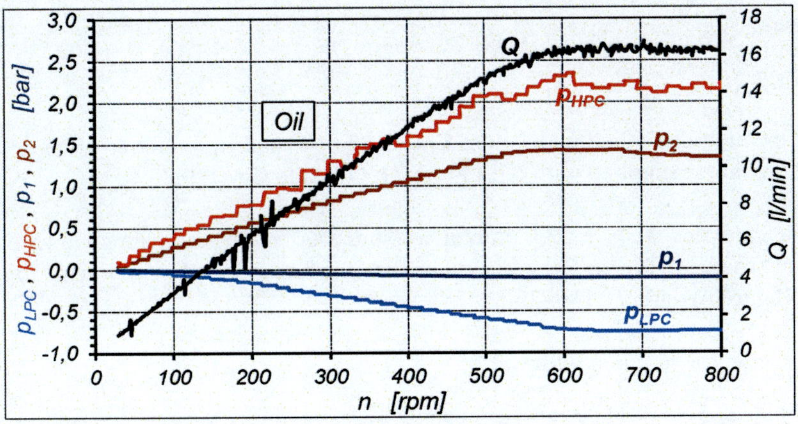

Fig. 6. Characteristics of pressure $p_1 = f(n)$, $p_2 = f(n)$, $p_{HPC} = f(n)$, $p_{LPC} = f(n)$ and flow rate $Q = f(n)$ in the motor supplied with oil [19, 20].

Cavitation occurs (the flow rate Q = const.) at rotational speeds above 630 rpm.

The experimental data (Fig. 5 and Fig. 6) shows that in a low pressure working chamber LPC (Fig. 2), during cavitation, the minimum value of the pressure p_{LPC} is – 0,75 bar for oil and –0,85 bar for water [19].

6 Pressure Drop in Internal Channels

Characteristics of $\Delta p_{ic1} = f(Q)$ whose values were calculated according to formula (7), are shown in Fig. 7. This characteristics Δp_{ic1} and Δp_{ic2} are shown at speeds up to 630 rpm for water and up to 550 rpm for oil. The cavitation does not occur at these speeds. Characteristics of $\Delta p_{ic1} = f(Q)$, shown in Fig. 7 can be described by Eq. (6). Then the values of constants C_t and C_l are [19, 20]:

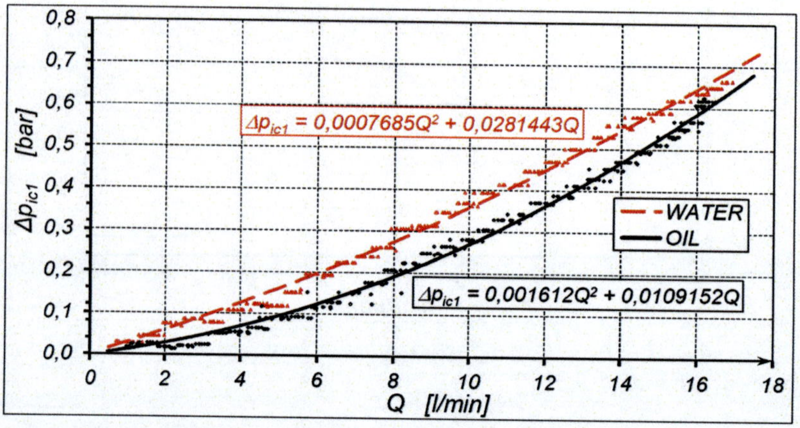

Fig. 7. Characteristics of the pressure drop $\Delta p_{ic1} = f(Q)$ in inflow channels of the motor (no cavitation).

a) for the motor supplied with water: $C_t = 7{,}88 \cdot 10^8$ and $C_l = 409{,}8 \cdot 10^8$;
b) for the motor supplied with oil: $C_t = 6{,}73 \cdot 10^8$ and $C_l = 19{,}2 \cdot 10^8$.

Characteristics $\Delta p_{ich} = f(Q)$ of total pressure drop in the motor supplied with water and oil are shown in Fig. 8. These characteristics can be described by following formulas (according to (6) and (9)) [19, 20]:

a) for the motor supplied with water:

$$\Delta p_{ich} = \left(15{,}76 \cdot \rho \cdot Q^2 + 819{,}6 \cdot \nu \cdot \rho \cdot Q\right) \cdot 10^2 \, MPa \tag{10}$$

b) for the motor supplied with oil:

$$\Delta p_{ich} = \left(13{,}46 \cdot \rho \cdot Q^2 + 38{,}4 \cdot \nu \cdot \rho \cdot Q\right) \cdot 10^2 \, MPa \tag{11}$$

The ratio of the pressure drops $\Delta p_{ich,W}/\Delta p_{ich,O}$ in the internal channels of the motor supplied with water and supplied with oil is shown in Fig. 9.

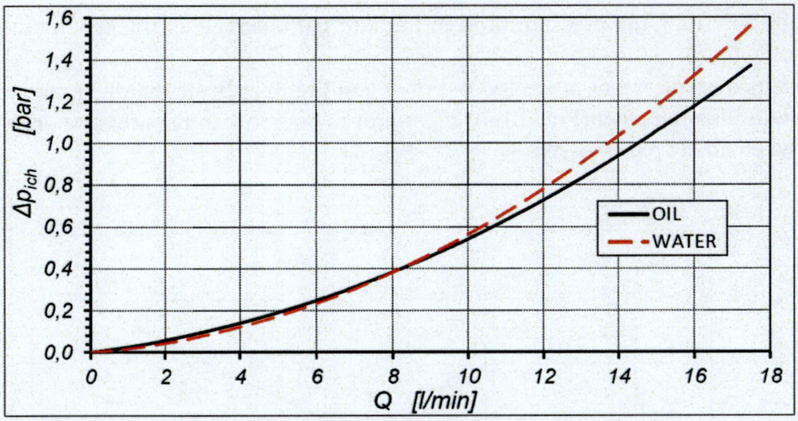

Fig. 8. Characteristics $\Delta p_{ich} = f(Q)$ of the total pressure drop in the motor supplied with oil and water.

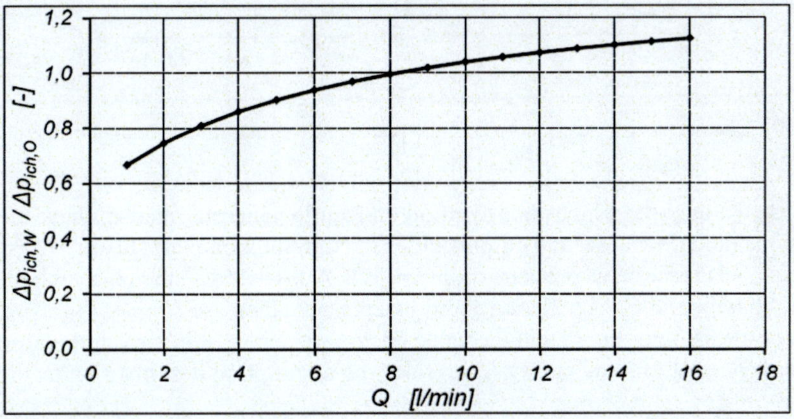

Fig. 9. Characteristic of $\Delta p_{ich,W}/\Delta p_{ich,O} = f(Q)$: $\Delta p_{ich,W}$ and $\Delta p_{ich,O}$ – pressure drop in the internal channels of the motor supplied with water and supplied with oil – result of experiment [19, 20].

7 Discussion

It can be observed that a smaller pressure drop in the internal channels of the motor supplied with a lower flow rate of water (up to 8 l/min) is observed. In the area of a high flow rate, a smaller pressure drop occurs in the motor supplied with oil. This phenomenon should be explained by the fact that, at a low flow rate, the share of the laminar flow component is larger than at a high flow rate. The viscosity of water is about 40 times lower than the viscosity of oil. Thus, at a low flow rate, there is a smaller pressure drop in the motor supplied with water than in the motor supplied with oil.

At a high flow rate, the share of the laminar flow component is very small. In other words, the density of the liquid has the biggest influence on the pressure drop. Therefore, at a high flow rate, the pressure drop in the internal channels is smaller in the motor supplied with oil.

Results of experiment presented in Fig. 5 and Fig. 6 indicate that the pressure drop Δp_{ic2} in outflow channel (Fig. 10) of the motor is bigger than the pressure drop Δp_{ic1} in inflow channel ($\Delta p_{ic1} < \Delta p_{ic2}$).

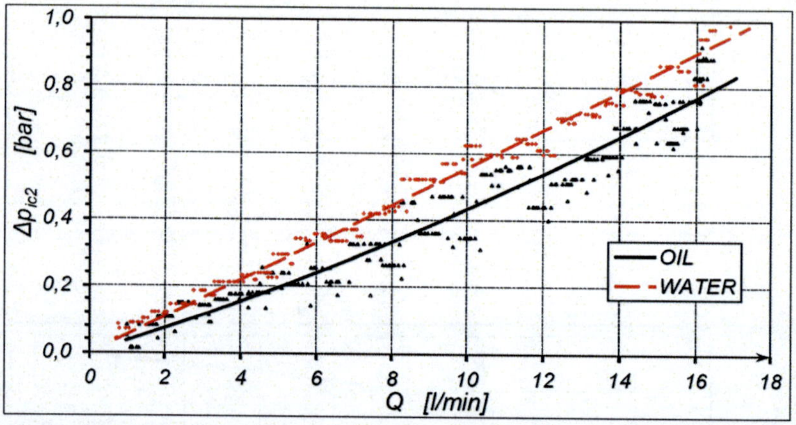

Fig. 10. Characteristics $\Delta p_{ic2} = f(Q)$ of the pressure drop in outflow channel of the motor.

In tested motor, the dimensions of inflow and outflow internal channels have the same dimensions (and the inflow and outflow holes in commutation plates also). Therefore in booth channels is the same pressure drop Δp_{ic1}. If $\Delta p_{ic1} < \Delta p_{ic2}$ then in a high pressure working chamber HPC (Fig. 2) is an additional pressure increase Δp_{HPC} (Fig. 12) caused by the compression of liquid and the value of Δp_{HPC} depends on the rotational speed n of a motor but not on flow rate Q. Therefore, the assumption described by formula (8)

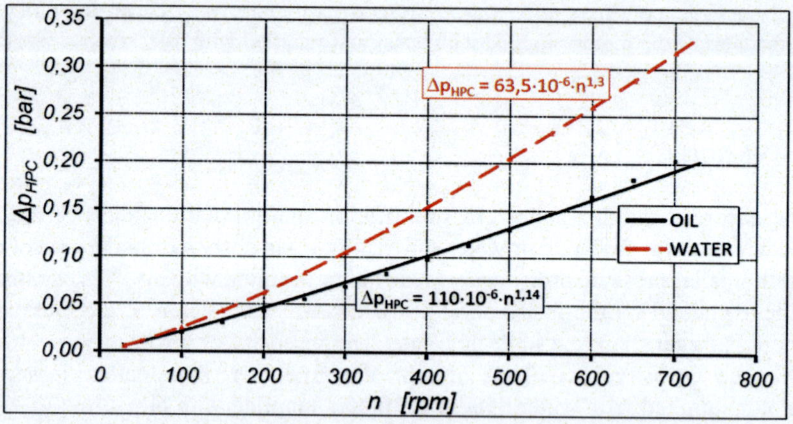

Fig. 11. Characteristics of the pressure increase $\Delta p_{HPC} = f(n)$ in the closed working chamber of a motor supplied with oil and water (calculated according to formula (12)).

is not true. The Δp_{HPC} can be calculated from experimental data as:

$$\Delta p_{HPC} = \Delta p_{ic2} - \Delta p_{ic1} \tag{12}$$

Characteristics of $\Delta p_{HPC} = f(n)$ in the motor are shown in Fig. 11.

According to the above, the progression of pressure in the hydraulic motor should look like Fig. 12. The increase of pressure Δp_{HPC} in the high pressure working chamber only influences the mechanical losses in the working mechanism of the motor. In this way, the Δp_{HPC} is not the component of the pressure losses in the internal channels of the motor.

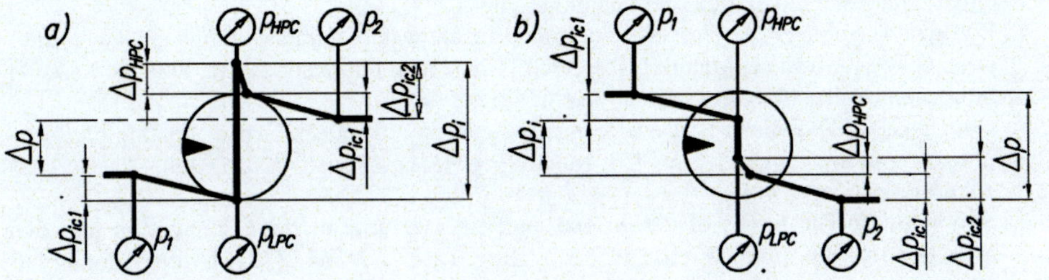

Fig. 12. The illustrative course of pressure in a hydraulic motor including the increase of pressure Δp_{HPC}: a) pump work, b) motor work.

8 Conclusions

The results of the experimental test have shown that:

a) the pressure drop in the internal channels of the hydraulic motor depends on the type of liquid and, in particular, on the parameters of the liquid (kinematic viscosity and density);
b) the flow of both the water and the oil in the internal channels of the motor is not a fully developed turbulent flow;
c) at a low flow rate (up to 8 l/min), there is a lower pressure drop in the motor supplied with water than in the motor supplied with oil. But in the area of a high flow rate, a smaller pressure drop occurs in the motor supplied with oil;
d) the type of liquid has influence on the value of the pressure p_{LPC} in the suction chamber (low pressure chamber LPC (Fig. 2)) of the satellite motor. In the motor supplied with oil, the pressure p_{LPC} is −0.75 bar, but, in the motor supplied with water, the p_{LPC} is −0.85 bar [19, 20];
e) the pressure drop in both the inflow internal channels and in the outflow channels in the hydraulic motor can be described with sufficient accuracy with formula (6);
f) when the volume of the working chamber nears the minimum, additional pressure increase Δp_{HPC} (compression of liquid) occurs in this chamber. This phenomenon depends on the overlap in the commutation unit and the rotational speed of the motor shaft. Therefore, the Δp_{HPC} only has influence on mechanical losses in the working

mechanism of the motor and is not the component of pressure losses in the internal channels of the motor.

References

1. Backe, W.: Water- or oil-hydraulics in the future. In: The Sixth Scandinavian International Conference on Fluid Power. Proceedings of the Conference, Tampere, Finland (1999)
2. Rokala, M.: Analysis of slipper Structures in Water Hydraulic Axial Piston Pumps. Tampere University of Technology (2012)
3. Schoemacker, F., Schmitz, K.: Forschung in der wasserhydraulik - herausforderungen und einsatzgrenzen. Olhydraulik und Pneumatik, pp. 1–2 (2020)
4. Pobedza, J., Sobczyk, A.: Properties of high pressure water hydraulic components with modern coatings. Advanced Materials Research. Trans Tech Publications Ltd, 849/2014 (2014). https://doi.org/10.4028/www.scientific.net/AMR.849.100
5. Dymarski, C., Dymarski, P.: Developing methodology for model tests of floating platforms in low-depth towing tank. Arch. Civ. Mech. Eng. **16**(1), 159–167 (2015). https://doi.org/10.1016/j.acme.2015.07.003
6. Sliwinski, P.: The basics of design and experimental tests of the commutation unit of a hydraulic satellite motor. Arch. Civ. Mech. Eng. **16**(4), 634–644 (2016). https://doi.org/10.1016/j.acme.2016.04.003
7. Kollek, W., Osiński, P., Stosiak, M., Wilczyński, A., Cichoń, P.: Problems relating to high-pressure gear micropumps. Arch. Civ. Mech. Eng. **14**(1), 88–95 (2013). https://doi.org/10.1016/j.acme.2013.03.005
8. Kollek, W., Radziwanowska, U.: Energetic efficiency of gear micropumps. Arch. Civ. Mech. Eng. **15** (2015). https://doi.org/10.1016/j.acme.2014.05.005
9. Osinski, P., Warzynska, U., Kollek, W.: The influence of gear micropump body asymmetry on stress distribution. Pol. Marit. Res. **24** (2017). https://doi.org/10.1515/pomr-2017-0007
10. Jasinski, R.: Problems of the starting and operating of hydraulic components and systems in low ambient temperature (Part III). Pol. Marit. Res. **4**(62) (2009). https://doi.org/10.2478/v10012-008-0052-2
11. Jasinski, R.: Problems of the starting and operating of hydraulic components and systems in low ambient temperature (Part IV). Pol. Marit. Res. **3**(95) (2017). https://doi.org/10.1515/pomr-2017-0089
12. Paszota, Z.: Energy losses in hydrostatic drive. LABERT Academic Publishing (2016)
13. Skorek, G.: Badania laboratoryjne strat ciśnieniowych w pompie wyporowej. Hydraulika i Pneumatyka 2 (2005)
14. Balawender, A.: Physical and mathematical model of losses in hydraulic motors. Developments in mechanical engineering. Gdansk University of Technology Publishers (2005)
15. Maczyszyn, A.: Energy analysis of rotary positive displacement machines used in hydrostatic transmissions. Ph.D. thesis. Gdansk University of Technology (2014)
16. Stosiak, M., Zawislak, M., Nishta, B.: Studies of resistances of natural liquid flow in helical and curved pipes. Pol. Marit. Res. **3**(99) (2018). https://doi.org/10.2478/pomr-2018-0103
17. Balawender, A.: Energy analysis and methodology for testing low-speed hydraulic motors. Mechanika 54, Scientific books of the Gdansk University of Technology, Gdansk (1988)
18. Sliwinski, P.: Pressure losses and power balance in the unloaded satellite pump. Hydraulika a Pneumatika, pp. 1–2 (2013)
19. Śliwiński, A., Patrosz, P.: The methodology for determining pressure losses in a hydraulic motor supplied with water and mineral oil. Chin. J. Mech. Eng. Article in Proceedings (2020)
20. Sliwinski, P.: Satellite displacement machines. Basic of design and analysis of power loss. Gdansk University of Technology Publishers (2016)

CFD Simulations and Tests of a Prototype Flow Control Valve

Marta Zaleska-Patrosz[1(✉)], Piotr Patrosz[2], and Paweł Śliwiński[2]

[1] BIBUS MENOS, Spadochroniarzy 18, 80-298 Gdansk, Poland
mza@bibusmenos.pl
[2] Faculty of Mechanical Engineering, Gdansk University of Technology, Narutowicza 11/12, 80-233 Gdansk, Poland

Abstract. In this paper a prototype of a flow control valve is described and numerically simulated. The flow control valve is used in hydraulic systems to maintain constant fluid flow despite changing loads of a receiver. The standard construction of this type of valves is modified mainly by eliminating the spring. The prototype consists the hydrostatically unloaded throttle valve and pressure ratio valve substituting pressure difference valve. The article concentrates on numerical simulation conducted for different positions of pressure ratio valve's spool and various throttle valve settings. Additionally the rotation of the spool is included in simulation and its influence on valve's characteristics is evaluated. The article also describes the methodology of determining flow characteristics of control valve. Results from numerical simulation are compared to results of experimental research.

Keywords: Flow control valve · Numerical simulation · CFD · Hydraulics · Fluid dynamics

1 Introduction

The flow control valve is used in hydraulic systems to maintain constant fluid flow rate through the receiver, for example hydraulic motor or cylinder, despite pressure changes. The flow control valve consists of an adjustable throttle valve and a differential valve [1].

There are two types of flow control valves: two- or three-way flow control valve (Fig. 1).

In both types of flow control valves the differential valve maintains constant pressure drop on the throttle valve. However, in two-way flow control valve the differential valve is placed in-line with the throttle valve, whereas in three-way flow control valve the differential valve is parallel to the throttle valve [2]. Due to the features mentioned above the two-way flow control valve is always used with pressure relief valve and the three-way control valve can be used as standalone valve.

This article concentrates on two-way flow control valves. The differential valve can be placed either in front or after the throttle valve. The configuration of these two valves does not change the principal of operation [3] therefore only one configuration is presented in Fig. 2 and taken into further considerations.

© The Editor(s) (if applicable) and The Author(s), under exclusive license
to Springer Nature Switzerland AG 2021
J. Stryczek and U. Warzyńska (Eds.): NSHP 2020, LNME, pp. 123–134, 2021.
https://doi.org/10.1007/978-3-030-59509-8_11

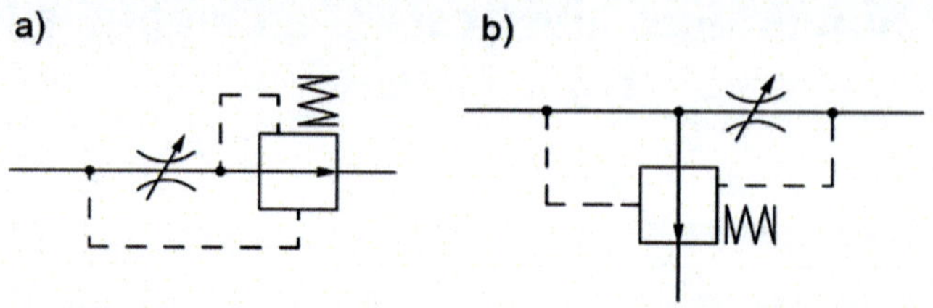

Fig. 1. Scheme of: a) the two-way flow control valve, b) three-way flow control valve.

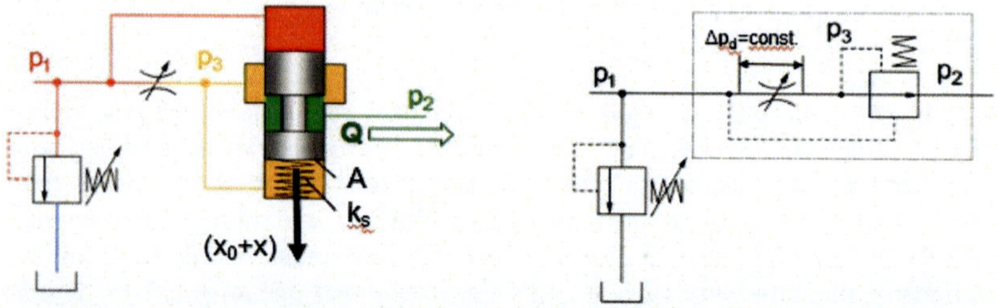

Fig. 2. Scheme of the two-way flow control valve.

Pressure p_1 in front and pressure p_3 behind throttle valve are used as control signals and act on surfaces of the spool. The spool is supported by the spring therefore the pressure drop equation on the throttle valve is described by formula:

$$\Delta p_d = p_1 - p_3 = \frac{k_s(x_0 + x)}{A} \tag{1}$$

where:

p_1 – pressure at the inlet to the flow control valve, adjusted by pressure relief valve,
p_3 – pressure after the throttle valve,
A – area of frontal surfaces of the spool,
k_s – spring constant,
x_0 – initial deflection of the spring,
x – spool's displacement.

As it is shown in Fig. 3 the real characteristic of flow rate through the flow control valve is not constant as it is expected. It is caused by the fact that spool's displacement x increases the force in the spring, which changes the pressure drop Δp_d and affects the flow rate.

The flow control valve without this fault was designed at Gdansk University of Technology [5].

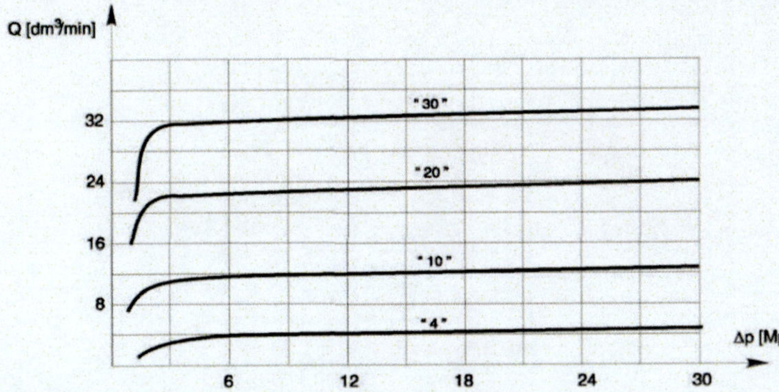

Fig. 3. Characteristics of flow rate Q in function of Δp for valves type UDRN6, produced by Ponar Wadowice, for different valve sizes [4].

1.1 Description of the Prototype Flow Control Valve

Design of the prototype flow control valve is based on a substitution of a standard pressure differential valve with a pressure ratio valve (Fig. 4). The main difference between pressure ratio valve and differential valve is the fact that pressure ratio valve does not have spring and the constant pressure difference Δp_d is obtained thanks to difference between area of surface A_1 and A_2. Higher pressure p_1 acts on the smaller surface A_1 and lower pressure p_3 acts on the bigger surface A_2.

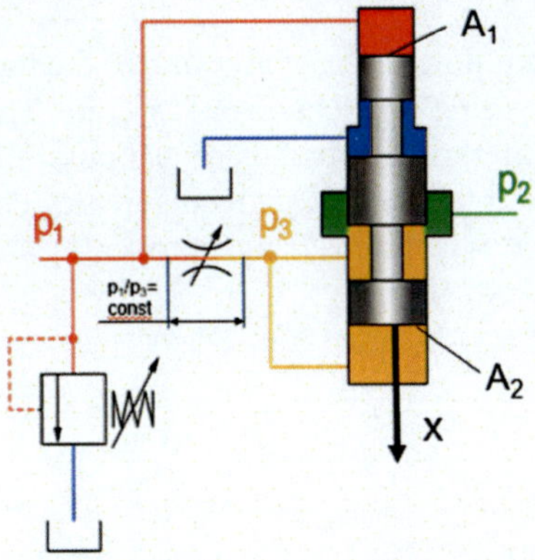

Fig. 4. Scheme of the prototype flow control valve [5].

Due to the fact that the ratio A_1/A_2 is always constant, Δp_d is given by the equation:

$$\Delta p_d = p_1\left(1 - \frac{A_1}{A_2}\right) \tag{2}$$

As long as pressure p_1, set by relief valve, remains constant, Δp_d also remains constant, independently of the spool displacement.

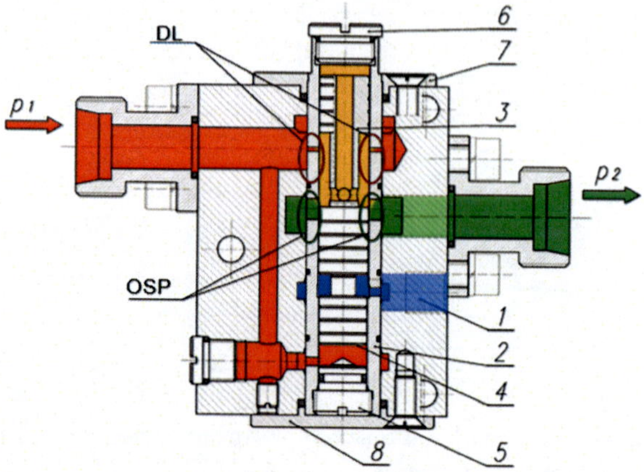

Fig. 5. Flow control valve with pressure ratio valve [6].

The prototype flow control valve shown in Fig. 5 consists of:

- valve housing (1),
- sleeve (2),
- main spool (3),
- auxiliary spool (4),
- throttle valve orifice (DL),
- pressure ratio valve orifice (OSP).

In comparison to standard flow control valve one of the most important advantage is to prevent the impact of the spring's deflection on flow rate. Thanks to using the pressure ratio valve instead of the differential valve, there is a possibility to gain more constant characteristic, which is similar to theoretical characteristic. The throttle valve is a rotary tube with bean-shaped orifice, which can change its opening area when the tube is rotated by an angle between 0° and 68° (Fig. 6).

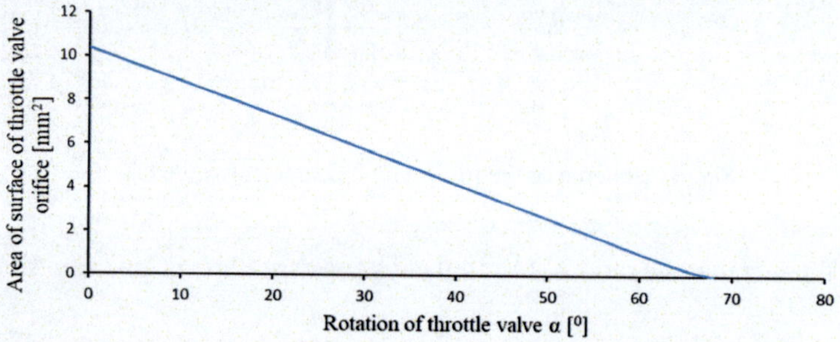

Fig. 6. Relation between area of surface of throttle valve orifice and rotation of throttle valve α.

The prototype is characterized by simple construction and small size. The prototype flow control valve could be used in these branches of industry, where constant input pressure is provided for ex. in mining.

2 Laboratory Tests of the Prototype Flow Control Valve

Laboratory tests were conducted at Gdansk University of Technology under supervision of prof. Paweł Śliwiński. Scheme of a test bench is shown in Fig. 7.

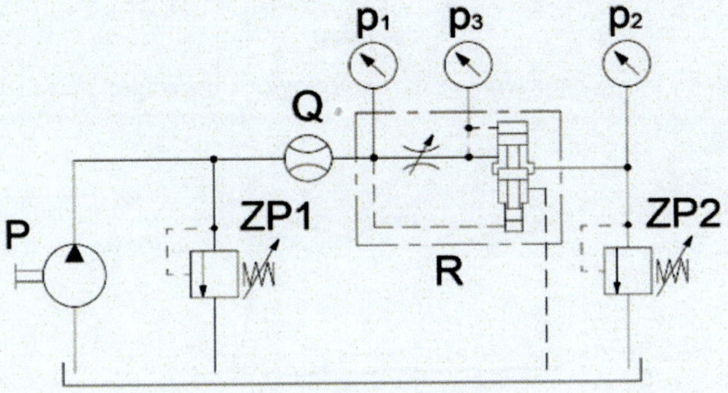

Fig. 7. Scheme of the test bench.

During laboratory tests, the flow rate was measured depending on:

- increase (in characteristics described as "up") or decrease (in characteristics described as "down") of load pressure p_2,
- adjustment of inlet pressure p_1,
- different throttle valve settings α.

Results of laboratory tests (Fig. 8 and 9) show, that there are two main problems in operation of the prototype flow control valve. First of all, the flow characteristics is not stable in the whole range of load pressure (Fig. 8). Characteristic collapse is repeatable but cause of its occurrence is so far unknown. Secondly, the hysteresis between increase and decrease of load pressure p_2 is noticeable and significant (Fig. 9). Additionally, after disassembly of the prototype it was discovered that surfaces of the spool had peripheral scratches, which means that the spool rotates [7]. This is a positive phenomenon but until CFD simulations were conducted, the impact of rotation on the flow rate had been unknown (Fig. 14).

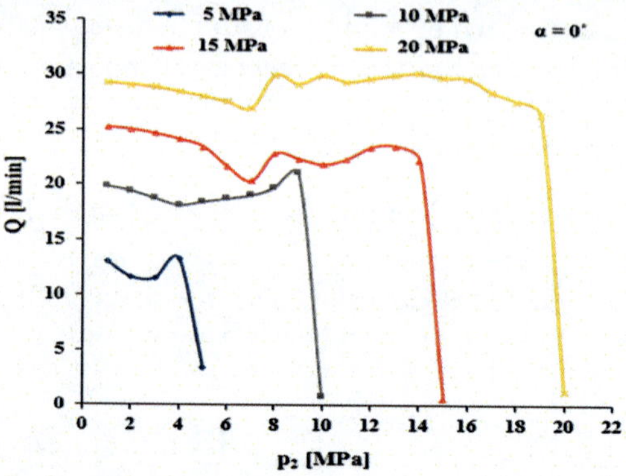

Fig. 8. Flow characteristics of the prototype flow control valve [8].

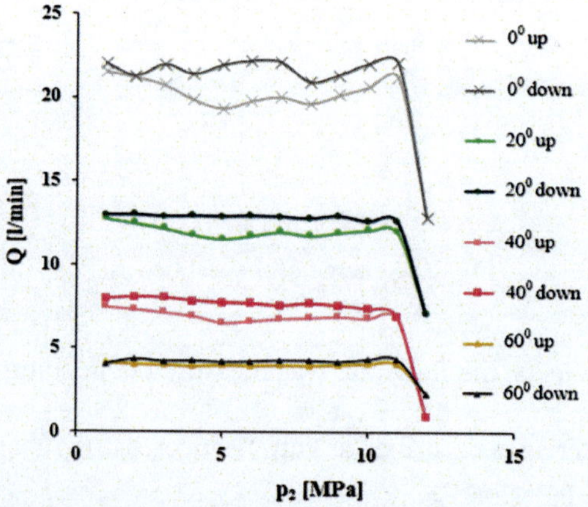

Fig. 9. Flow characteristics during increase(up) and decrease (down) of load pressure p_2 [8].

3 CFD Simulations

CFD (Computational Fluid Dynamics) analysis are used to investigate flow phenomena inside valves. Since the valve presented in this article is a prototype, there are no publications considering it. However, there are several articles, which consider modeling flow control valves with differential valve [9, 10]. Additionally articles presenting CFD models of spool valves [11, 12] proved to be helpful in understanding the phenomena, which occur during operation of researched flow control valve.

3.1 Flow Control Valve Mesh

In order to investigate the prototype flow control valve, Ansys CFX software was used. During preparation of computational model it was necessary to create an appropriate mesh of fluid filling channels of the flow control valve.

One of the problems which always appears in CFD simulation is how to get a compromise between size of elements, hence accuracy of simulation, and time of simulation. In this case it was essential to get as dense mesh as possible in some parts of model, e.g. near walls and orifices, but in other parts it was not so crucial. Model consists of tetrahedron elements, which size depends on position of the spool and rotation of the sleeve. It means that the smaller orifices DL and OSP (Fig. 5) were, the more precise mesh was. The overall quantity of elements varied from 765188 to 934298 depending on analyzed configuration.

Figure 10 presents a mesh of the interior of the prototype flow control valve.

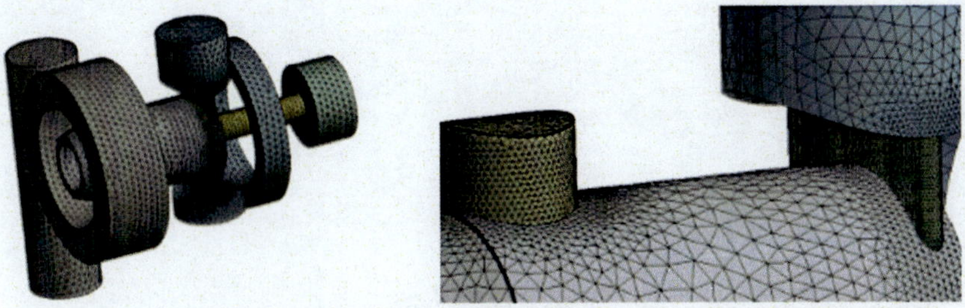

Fig. 10. Mesh of fluid inside the flow control valve.

3.2 Simulation Assumptions and Boundary Conditions

In order to simplify the calculations, initial assumptions were made as follows:

- irrelevant geometric details, for ex. drill cones and screw-threads, were excluded,
- temperature inside the prototype flow control valve was constant,
- fluid was incompressible,
- single-phase flow was set.

Following boundary conditions and solver configurations were set:

- inlet/outlet: opening type,
- turbulence model: Shear Stress Transport (SST),
- type of analysis: Steady State,
- rotation of selected walls: 2 000 rpm,
- fluid: hydraulic oil: density 860 [kg/m^3], viscosity $\nu = 46$ [mm^2/s].

Opening boundary condition for inlet and outlet was selected. It allowed the fluid to flow in both directions in case of vortex near to boundary.

Chosen turbulence model SST which is described in [13–15] combines two models: k-ω and k-ε. Near walls k-ω is used. Away from walls k-ε is used. The transition between these models is defined using blending function. The SST model was selected because it proved to be more stable and converge faster than k-ω and k-ε models.

Rotation of chosen walls was intentionally set so high (2000 rpm) in order to investigate if that rotation has significant influence on flow rate.

Figure 11 presents location of inlet (red surface), outlet (yellow surface), interfaces between domains (green surfaces) and location of rotating walls (blue surfaces) in computational model.

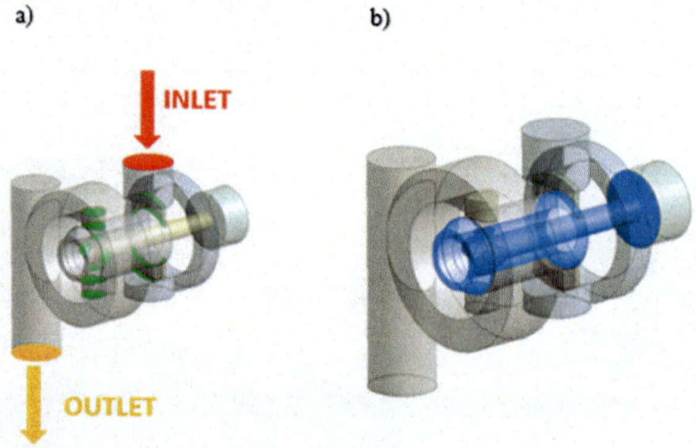

Fig. 11. Boundary conditions: a) location of inlet, outlet and interfaces, b) location of rotating walls.

3.3 Rotation of the Spool

Due to the fact that laboratory tests showed that the spool rotates [7], the CFD analysis were conducted to determine, if the rotation of the spool can have influence on the flow rate.

Figure 12 shows shear stresses on the walls of the spool. An asymmetrical distribution of shear stresses causes rotation and determines the direction of rotation.

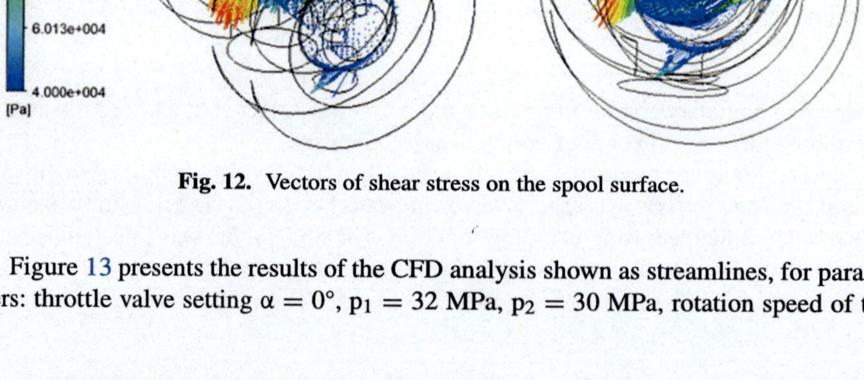

Fig. 12. Vectors of shear stress on the spool surface.

Figure 13 presents the results of the CFD analysis shown as streamlines, for parameters: throttle valve setting $\alpha = 0°$, $p_1 = 32$ MPa, $p_2 = 30$ MPa, rotation speed of the spool n = 2000 rpm and n = 0 rpm. As it is shown rotation does not have significant influence on the flow rate. However, the streamlines and velocity in both cases are very similar.

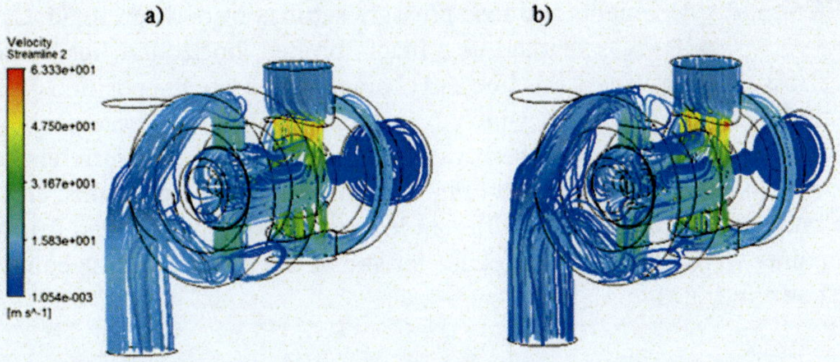

Fig. 13. Results of the CFD analysis – streamlines: a) for n = 2000 rpm, b) for n = 0 rpm.

Figure 14 shows a comparison between results obtained while the spool is rotating and when it is stationary.

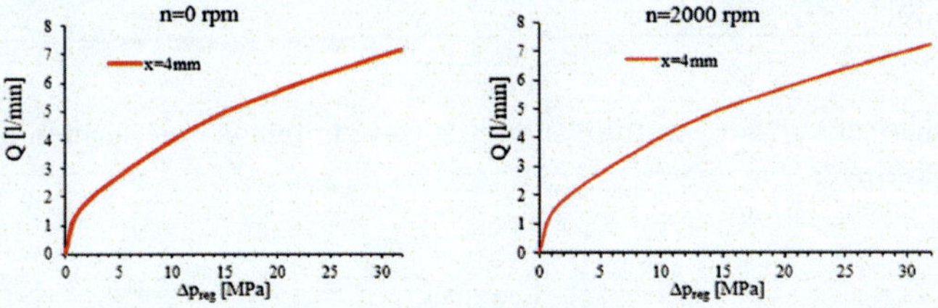

Fig. 14. Characteristics of flow rate in function of pressure drop on flow control valve.

As it is mentioned above the rotation speed was intentionally set to 2000 rpm, which is very high or even impossible. Since such high rotation speed had negligibly small impact on the flowrate, smaller rotation speeds would not have significant influence, either. Therefore to simplify further simulations the spool was considered as stationary.

4 The Methodology of Determining Flow Characteristics of the Prototype from CFD Simulations

For the purpose of determining flow characteristics, it is necessary to know parameters, which define the operation of the prototype. As the position of the spool x during tests is unknown and it is changing depending on load pressure p_2 and adjustment of the throttle valve α, many simulations were conducted for these parameters.

The methodology of determining flow characteristics is based on several steps. Firstly, constant inlet pressure $p_1 = 20$ MPa was applied. Diameter of surface A_1 (Fig. 4) is equal 11.6 mm and diameter of surface A_2 is equal 12 mm [5]. By using Eq. (2) it was calculated that pressure drop on throttle valve Δp_d should equal 1.31 MPa.

CFD simulations were conducted for different load pressures p_2 and different displacements of the spool x. Secondly, characteristics of pressure drop on throttle valve Δp_d, at different displacements and load pressure settings were determined. Calculated value $\Delta p_d = 1.31$ MPa was marked as a black, broken, horizontal line in that chart (Fig. 15). Cross points given by this line and characteristics, are the position of the spool for different load pressures p_2, at which Δp_d equals 1.31 MPa. These positions were marked as dashed, vertical lines on characteristics of flow rate for different positions of the spool x and load pressures p_2 in Fig. 16. Basing on these positions, cross points with flow rate Q characteristics for different load pressures p_2 were read from Fig. 16. Obtained points were used to create characteristic of the prototype flow control valve, which is shown in Fig. 17.

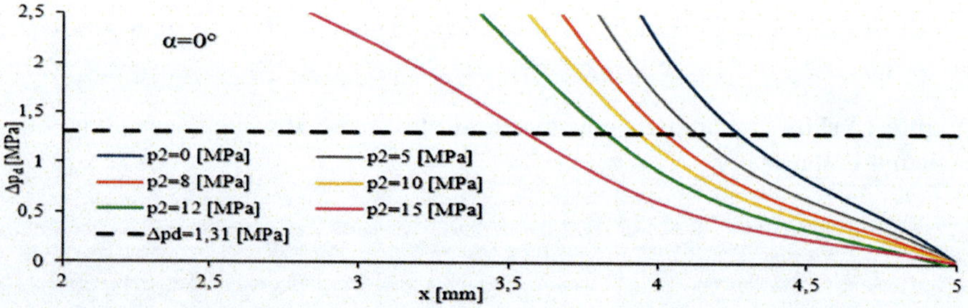

Fig. 15. Characteristics of drop pressure at throttle valve in function of position of the spool for different load pressure.

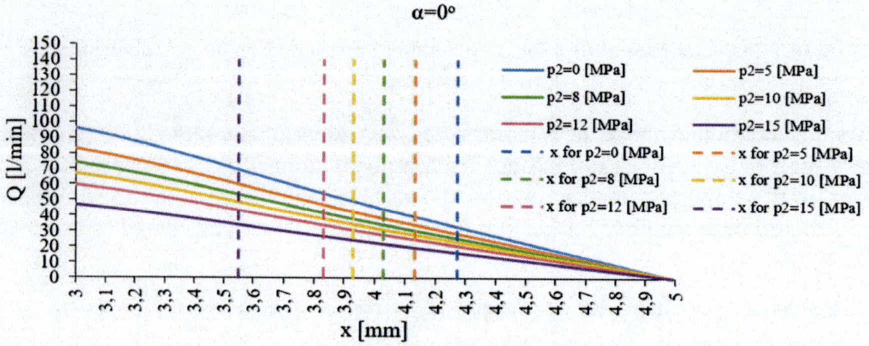

Fig. 16. Characteristics of flow rate in function of position of the spool for different load pressures.

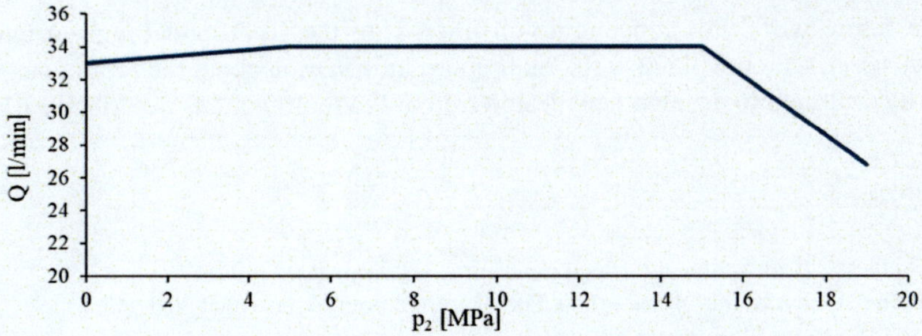

Fig. 17. Characteristic of the prototype flow control valve.

5 Conclusion

In the article experimental and computational method was used to determine the behavior of the prototype flow control valve. CFD simulations were conducted for 610 configurations and results were used to determine the flow characteristic of the prototype flow control valve. For example, Fig. 18 presents comparison between results of laboratory test and CFD simulations for throttle valve setting $\alpha = 0°$.

Both methods have given valuable results. However, separately they did not give the full information about valve operation, but together they proved to be complementary. Thanks to the experimental data the computational model was improved and then validated. Results of laboratory tests and CFD simulations have satisfactory convergence, what is shown in Fig. 18. It is assumed that the convergence can be further improved by including hydrodynamic forces in CFD model. The main problem, that is so far unresolved is the fact that the results of computational simulations do not show characteristic flow rate drop (Fig. 8) which occurs during laboratory tests. Since CFD simulations did not show relation between flow phenomena and flow rate drop, it is very probable that phenomenon is related with friction or hydrodynamic forces, which were not included in CFD analysis.

The methodology of determining characteristics of the flow control valve from simulation proved to be accurate. It has also given very valuable information about spool position during the operation of the valve, which so far has not been reliably measured.

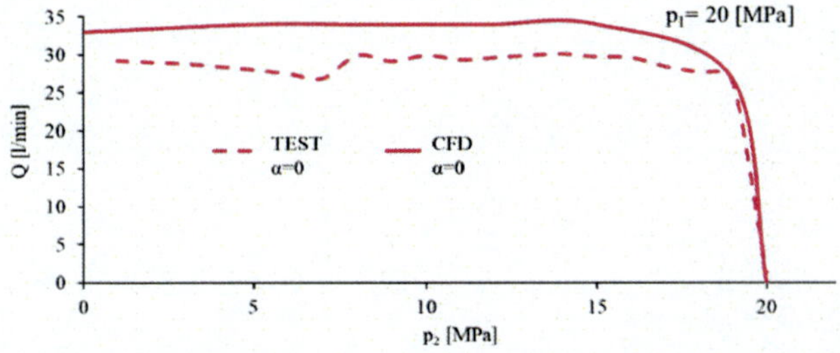

Fig. 18. Comparison between results from laboratory test and CFD simulation [8].

The future work will concentrate on improving the CFD model and on building new test stand, which will allow to obtain more information about the prototype, which hopefully will help to develop new design free of the malfunctions described earlier.

References

1. Osiecki, A.: Hydrostatyczny napęd maszyn. WNT Warszawa (1998)
2. Vademecum hydrauliki: Hydraulika. Podstawy, elementy konstrukcyjne i podzespoły. Tom I. Bosch Rexroth (2007)
3. Stryczek, S.: Napęd Hydrostatyczny. Tom I. Elementy. WNT Warszawa (2005)
4. Datasheet Flow control valves cartridge 2-way typ UDRN6, Ponar Wadowice (2000)
5. Śliwiński, P.: Regulator przepływu// W: Badanie, konstrukcja, wytwarzanie i eksploatacja układów hydraulicznych/ ed. Prof. dr hab. inż. Adam Klich, Dr inż. Antoni Kozieł ul. Pszczyńska, vol. 37, pp. 44–101. Instytut Techniki Górniczej KOMAG, Gliwice, pp. 82–93 (2015)
6. Śliwiński, P.: Regulator przepływu, Polish patent PL 224522 B1
7. Lipski, A.: Badania laboratoryjne regulatora przepływu. Praca dyplomowa magisterska. Promotor: dr inż. Paweł Śliwiński. Politechnika Gdańska (2013)
8. Ellwart, T.: Badania eksperymentalne regulatora przepływu. Praca dyplomowa inżynierska. Promotor: dr inż. Paweł Śliwiński. Politechnika Gdańska (2014)
9. Domagała, M., Momeni, H., Fabis-Domagała, J., Krawczyk, M., Filo, G., Bikass, S.: Numerical simulation of fluid flow inside the flow control valve. Qual. Prod. Improv. **1**(1), 394–399 (2019)
10. Suzuki, K., Urata, E.: Development of a water hydraulic pressure-compensated flow control valve. Int. J. Fluid Power **9**(3), 25–33 (2008)
11. Lisowski, E., Czyżycki, W., Rajda, J.: Three dimensional CFD analysis and experimental test of flow force acting on the spool of solenoid operated directional control valve. Energy Convers. Manag. **70**, 220–229 (2013)
12. Lisowski, E., Rajda, J.: CFD Analysis of Flow Forces Acting on the Spool of Directional Control Valve Type WE10J, Technical Transactions Mechanics, 2-M (2015)
13. Menter, F.R.: Two-equation eddy-viscosity turbulence models for engineering applications. Am. Inst. Aeronaut. Astronaut. J. **32**(8), 1598–1605 (1994)
14. Menter, F.R., Kunts, M., Langtry, R.: Ten years of industrial experience with the SST turbulence model. Turbul. Heat Mass Transf. **4** (2003)
15. Technical Brief Innovative Turbulence Modeling: SST Model in ANSYS CFX **11**(4) (2004)

Experimental Research of an Axial Piston Pump with Displaced Swash Plate Axis of Rotation

Paweł Załuski(✉)

Faculty of Mechanical Engineering, Gdansk University of Technology, Gdańsk, Poland
pawzalus@pg.edu.pl

Abstract. The article describes the influence of displacement of the swash plate rotation axis on the efficiency of axial piston pumps. The dead space volume was defined as the volume of the working chamber in the extreme position of the piston at the end of the pumping phase and its variability was determined as a function of the position of the axis of rotation of the swash plate and the swing angle of swash plate. The influence of swash plate rotation axis displacement on leaks between piston and cylinder was determined. It has been proven that displacement of the swash plate rotation axis reduces the dead space volume and leaks, thus improving the volumetric efficiency visible at small swash plate swing angles. The results of a comparative study of the volumetric and total efficiency of a modified design of a pump that allows the rotation axis to be repositioned are presented. The efficiency curves for the swash plate rotation axis crossing the shaft rotation axis and for the displaced swash plate axis of rotation are compared. The article also presents the author's construction of an axial piston pump with displaced swash plate rotation axis with a follow-up mechanism of capacity change controlled by a stepper motor. The diagram and view of the test stand are shown, as well as the results of tests of volumetric and hydromechanical pump efficiency.

Keywords: Hydraulics · Piston pump · Efficiency

1 Introduction

Piston pumps with swivel swash plate are often used in modern hydraulic systems. The design of such pump is schematically shown in Fig. 1. Shaft (1) is connected to drum (2) by means of a spline. In the drum there are pistons (4) in cylindrical chambers, finished with slippers (5). Pistons, during the rotation of the drum, due to the contact of slippers with the swash plate (6), perform the reciprocating movement in the cylindrical chambers. The face of the rotating drum contacts a fixed timing plate (3) with two kidney holes. One of these openings is connected to the suction port and the other to the pump delivery port. Thus, the cylinder chambers from which the pistons slide out are connected to the suction channel by means of a timing plate and the liquid is sucked in. As the shaft continues to rotate and the pistons pass through the dead-end, they begin to approach the timing plate - then the chambers are connected to the delivery channel and liquids

© The Editor(s) (if applicable) and The Author(s), under exclusive license
to Springer Nature Switzerland AG 2021
J. Stryczek and U. Warzyńska (Eds.): NSHP 2020, LNME, pp. 135–145, 2021.
https://doi.org/10.1007/978-3-030-59509-8_12

are pressed out of the chambers [1, 3]. Figure 1 also indicates the volume of dead space, i.e. the volume occupied by the fluid in the cut-off cylinder chamber when the piston is in its extreme position at the end of the pressing phase [4, 5].

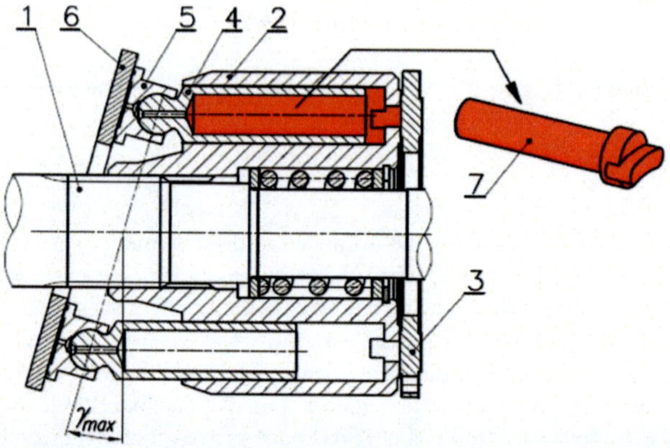

Fig. 1. Working mechanism of axial piston pump [8]: 1 - shaft; 2 - cylinder drum; 3 - timing plate; 4 - piston; 5 - slipper; 6 - swash plate; 7 - dead space volume; γ_{max} - max swing angle.

2 Materials and Methods

2.1 Displacement of Swash Plate Rotation Axis

The standard axial piston pumps have one thing in common: the swash plate axis of rotation crosses the shaft axis of rotation. The volume of the dead space increases as the swing angle of swash plate decreases (see Fig. 2a). If the axis of rotation of the swash plate is displaced as shown in Fig. 2b, the volume of this space will be constant regardless of the swing angle of swash plate. The displacement of the pump V_s is a function of swash plate swing angle γ, piston spacing D and piston diameter d_t [1–3]:

$$V_s = \frac{\pi d_t^2}{4} \cdot D \cdot tan\gamma \tag{1}$$

If the axis of rotation of swash plate intersects the axis of rotation of the shaft, as in Fig. 2a, then for the angle γ_{max} the volume of dead space is minimal and increases with the reduction of the swing angle of swash plate γ according to the following relation [8]:

$$V_D = V_{Dmin} + \frac{\pi d_t^2}{8} \cdot D(tan\gamma_{max} - tan\gamma) \tag{2}$$

V_{Dmin} is the minimum dead space volume for maximum swing angle of swash plate γ_{max}. This space includes a part of the cylinder, the window of the cylinder drum and the empty space inside the piston.

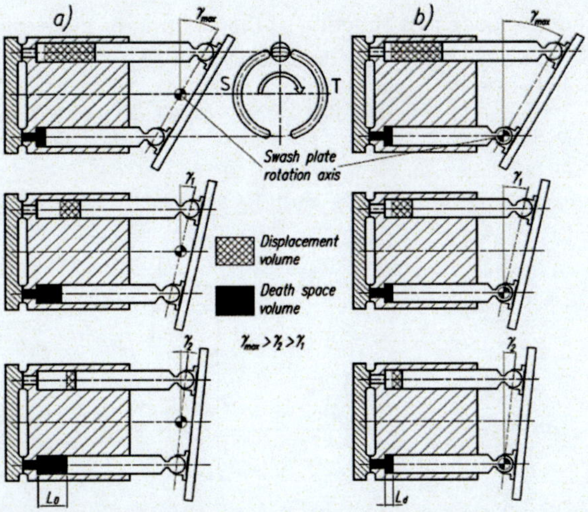

Fig. 2. Comparison of changes in the volume of dead space as a function of the swing angle of swash plate for a pump with the axis of rotation of the swash plate intersecting with the axis of rotation of the shaft a), and a pump with shifted axis of rotation of the swash plate b) [7].

2.2 Influence of Displacement of Swash Plate Rotation Axis on the Dead Space Volume

Figure 3 shows the situation when the axis of rotation of the swash plate is shifted in relation to its original position by a_x and a_y. The boundary condition is that the swash plate is positioned at its maximum swing angle to obtain a minimum volume of dead space V_{Dmin}. Thereby, regardless of the position of the axis of rotation of the swash plate, its position at the maximum swing angle will be the same. Consider the situation of the rotation of the swash plate relative to the displaced center of rotation from the angle γ_{max} to the angle γ in a system of coordinates with the center in the place where the center of rotation of the swash plate was originally located.

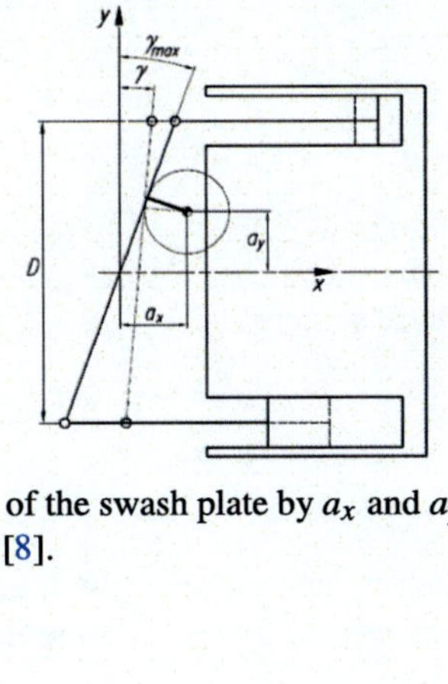

Fig. 3. Shift of the rotation axis of the swash plate by a_x and a_y in relation to the axis of rotation intersecting the axis of the shaft [8].

The volume of dead space as a function of the position of the center of rotation and the swing angle of the swash plate may be described as [8]:

$$V_D = V_{Dmin} + \frac{\pi d_t^2}{4}\left[\frac{D}{2}tan\gamma_{max} - \left(\frac{D}{2} - C\right)tan\gamma\right] \quad (3)$$

The coefficient C depending on a_x and a_y will be [8]:

$$if\ a_x \geq a_y\ tan\gamma_{max} \qquad C = \frac{\left|\frac{a_x}{tan\gamma_{max}} - a_y\right|\sqrt{\frac{1}{tan^2\gamma} + 1}}{\sqrt{\frac{1}{tan^2\gamma_{max}} + 1}} - \frac{a_x}{tan\gamma} + a_y \quad (4)$$

$$if\ a_x \leq a_y\ tan\gamma_{max} \qquad C = -\frac{\left|\frac{a_x}{tan\gamma_{max}} - a_y\right|\sqrt{\frac{1}{tan^2\gamma} + 1}}{\sqrt{\frac{1}{tan^2\gamma_{max}} + 1}} - \frac{a_x}{tan\gamma} + a_y \quad (5)$$

Example results of simulation of a change in the dead space volume as a function of the position of the center of rotation of the swash plate is presented in Fig. 4. The greatest influence on changes in the dead space volume has the movement of the axis of rotation in the direction a_y.

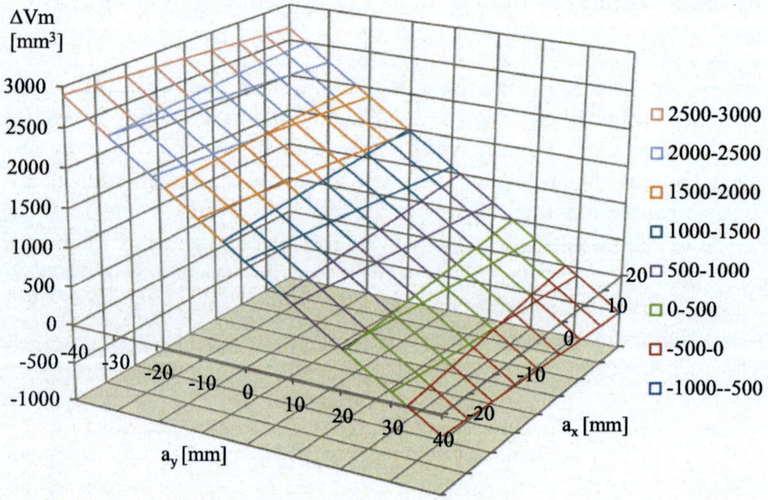

Fig. 4. Change of the dead space volume as a function of the position of the swash plate rotation axis for the pump with parameters: $D = 58$ mm; $d_t = 14{,}15$ mm; $\gamma = 1°$; $\gamma_{max} = 15{,}65°$; $V_{Dmin} = 4{,}27\ cm^3$; ΔVm - change of the dead space volume [8].

2.3 Influence of the Dead Space Volume on the Volumetric Flow Losses

At the end of pumping phase, the residual oil under high pressure is located in the dead space. Thus, when the suction manifold opens, the medium cannot be sucked at the beginning of cycle, because the oil contained in the dead space expands, causing back-flow to the suction duct. Therefore, the volume of actually sucked oil is reduced, thus decreasing the volumetric efficiency. The higher delivery pressure and the larger dead space, there is the higher influence on the phenomenon mentioned above.

The change in the volume of the expanded oil at a pressure drop Δp will be [5]:

$$\Delta V_c = V_D \cdot \frac{\Delta p}{K_I - \Delta p} \tag{6}$$

Since the phenomenon of liquid expansion in the pump is adiabatic, the secant isentropic bulk modulus K_I should be used to describe it. The resulting flow rate losses for a pump driven at speed n with the number of pistons z can be determined as:

$$\Delta Q_s = \Delta V_c \cdot n \cdot z \tag{7}$$

These losses directly affect the reduction of the flow generated by the pump and thus the volumetric efficiency.

2.4 Influence of Displacement of Swash Plate Rotation Axis on the Length of Piston - Cylinder Gap

Figure 2 shows that the displacement of the swash plate rotation axis is carried out under the condition that the position of the swash plate for the maximum angle is the same as for a pump with a non-translocated rotation axis, causing the pistons to slide more into the cylinder drum, hence for the disc rotation angle γ_2 the distance of the piston face from the bottom of the cylinder chamber changes from L_0 to L_d. The difference is:

$$s_d = L_o - L_d = -C \cdot tan\gamma \tag{8}$$

The length of the piston-cylinder gap depends on the angle of rotation of the shaft [6], so greater insertion of pistons into the cylinder drum will result in an increase in the average length of the piston-cylinder gap and will reduce the leakage occurring in this node.

2.5 Influence of Displacement of Swash Plate Rotation Axis on the Volumetric Efficiency

Comparative tests were carried out in the Laboratory of Hydraulics of the Gdansk University of Technology on a pump with the possibility of moving the swash plate axis of rotation. The basic parameters of the pump were as follows:

- piston spacing $D = 58$ mm
- piston diameter $d_t = 14{,}15$ mm
- maximum swing angle of swash plate $\gamma_{max} = 15{,}65°$
- minimal volume of dead space $V_{Dmin} = 4{,}27\ \text{cm}^3$

The results of the tests shown in Fig. 5 showed a noticeable increase in volumetric efficiency for small swing angles of swash plate. These tests confirmed the influence of the position of the swash plate axis of rotation on the volumetric efficiency and were an impulse to design a variable displacement pump with a displaced axis of rotation.

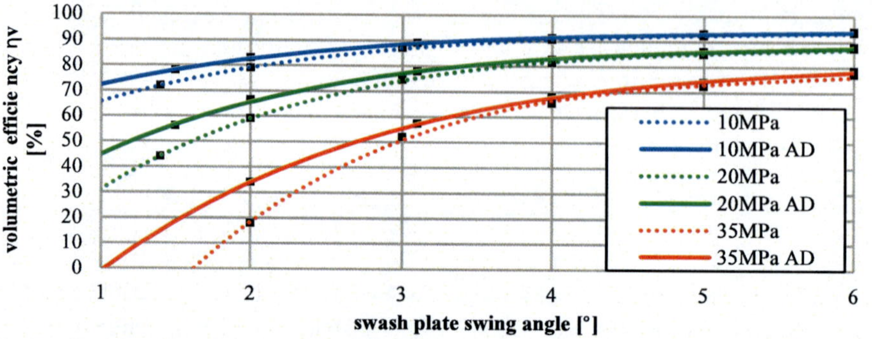

Fig. 5. Volumetric efficiency as a function of the swash plate swing angle for the pump with axis of rotation intersecting the axis of rotation of the shaft and for a pump with the axis displaced by $a_x = a_y \tan\gamma_{max}$ and $a_y = D/2$ (AD); $\nu = 40$cSt, $n = 1500$ rpm [8].

3 Results

3.1 Prototype Pump Construction

The pump shown in Fig. 6 is a prototype construction where the axis of rotation of the swash plate is displaced so that the volume of the dead space remains constant when the swash plate swing angle is changed. The swash plate (2) is fixed in the body by two bolts (3), which form its axis of rotation. The capacity is changed by using a piston (7) ended with a roller. The torque that loads the swash plate has the same turn every time, so there is no need for additional spring support of the plate. The change of the pump capacity is done by moving the control spool (10) by means of the stepper motor (11). This spool controls the flow of liquid through the moving cylinder (12) to the chambers of the actuator (7). Feedback is realized mechanically. Pusher (8) through lever (9) acts on cylinder (12) by moving it in the direction of the control slider movement. The movement of the cylinder continues until it reaches the appropriate position in relation to the control spool. The kinematic ratio of the mechanism, i.e. the ratio of piston stroke (7) to control spool stroke (10), depends on the length and position of the center of rotation of lever (9). A spring (13) acts on cylinder (12), which eliminates clearance in the mechanism and presses cylinder (12) against lever (9). This pump can have two directions of pressing and can be used for hydraulic closed systems. In such a construction, the pump has a symmetrical design, has two displacement pistons and two follow-up mechanisms. When changing the direction of flow, it is necessary to move the axis of rotation by means of hydraulically controlled bolts (3). Main parameters of tested pump were as follows:

- geometric displacement $q = 40$ cm^3/rev
- pistons spacing $D = 67$ mm
- piston diameter $d_t = 15{,}17$ mm
- number of pistons $z = 9$
- dead space volume $V_D = 3{,}244$ cm^3
- max swash plate swing angle $\gamma_{max} = 20°$
- nominal pressure $p_{nom} = 35$ MPa

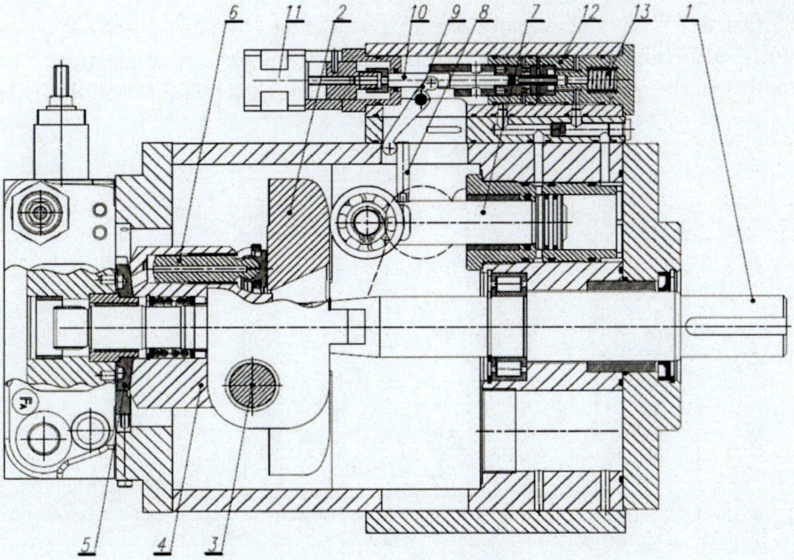

Fig. 6. Axial section of prototype piston pump: 1 - shaft; 2 - swash plate; 3 - bolt; 4 - cylinder drum; 5 - timing plate; 6 - piston; 7 - capacity change piston; 8 - pusher; 9 - lever; 10 - spool; 11 - stepper motor; 12 - cylinder; 13 - spring.

Figure 7 shows the operation of the mechanism of changing capacity when increasing the swash plate swing angle. The follow-up mechanism is delivered straight from the pump pressure collector.

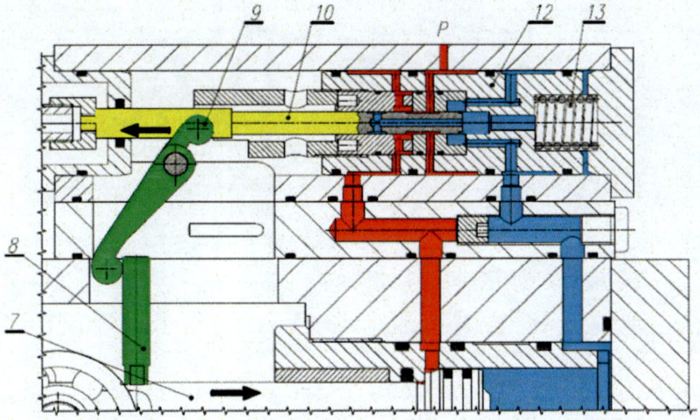

Fig. 7. The functioning of the follow-up mechanism: color red - powered; color blue - drain.

3.2 Test Stand

A scheme of the test stand is shown in Fig. 8. A view of the test stand is shown in Fig. 9. The tested pump was supplied with a supply pump. The suction pressure is regulated by a relief valve (5) and the delivery pressure by a valve (6). During the tests, pressures in front of and behind the pump, oil temperature in the suction connection, flow rate, torque and speed were recorded. Used working fluid was mineral oil of viscosity class VG 46.

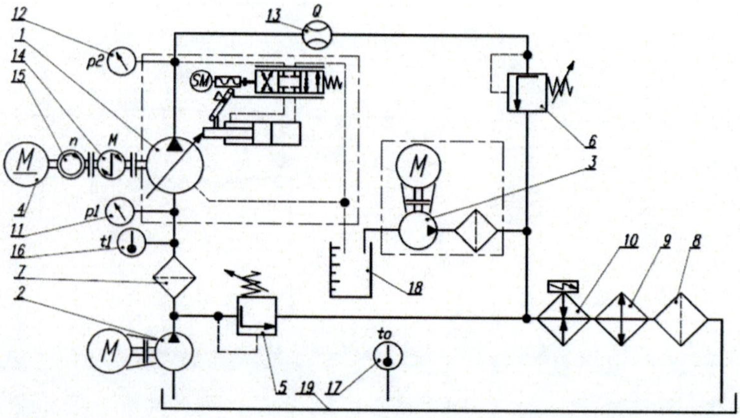

Fig. 8. Scheme of the test stand: 1 - tested pump; 2 - supply pump; 3 - filtering unit; 4 - 70 kW DC electric motor; 5 - suction pressure regulator relief valve; 6 - relief valve 0–40 MPa; 7, 8 - oil filters; 9 - cooler; 10 - 12 × 1,6 kW electric heaters; 11, 12 - pressure gauges; 13 - piston flow meter PT-200; 14 - torque meter HBM T1 500 Nm; 15 - optical speed sensor; 16, 17 - temperature transmitters; 18 - leakage measurement tank; 19 - oil tank 1700 l [9].

Fig. 9. View of the test stand: 1 - electric motor; 2 - speed measurement; 3 - torque measurement; 4 - clutch; 5 - tested pump; 6 - pressure line; 7 - temperature measurement; 8 - suction line; 9 - leakage line [9].

A list of used test apparatus is presented in Table 1.

Table 1. List of test apparatus.

Instrument	Range	Accuracy class	Max error
Manometer MPS-02	0–1,6 MPa	0,2	0,006 MPa
Manometer KFM	0–40 MPa	0,6	0,25 MPa
Thermocouple	0–70 °C	1	1,7 °C
Piston flow meter PT-200	0,3–200 l/min	0,2	0,4 l/min
Torque meter HBM T1	0–500 Nm	0,2	1,01 Nm
Incremental encoder	0–3000 rpm	–	rpm

3.3 Efficiency Characteristic

Volumetric and hydromechanical efficiency as a function of differential pressure and swash plate swing angle for viscosity 20 and 40cSt and rotational speed 2000 rpm are shown in Fig. 10 and 11.

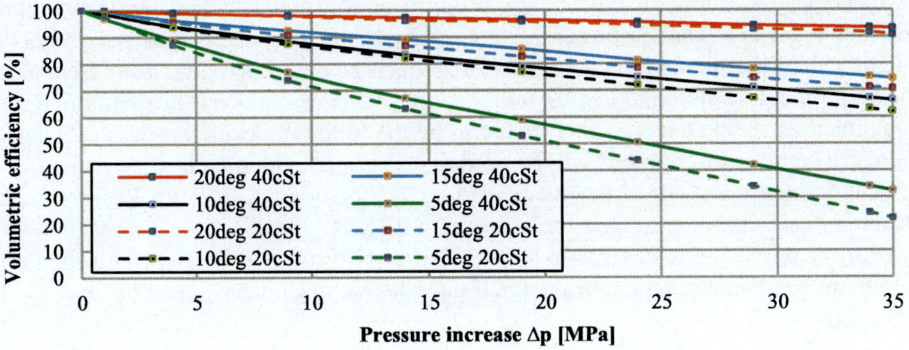

Fig. 10. Volumetric efficiency as a function of differential pressure for different swash plate swing angles and viscosities 20 and 40cSt for 2000 rpm [9].

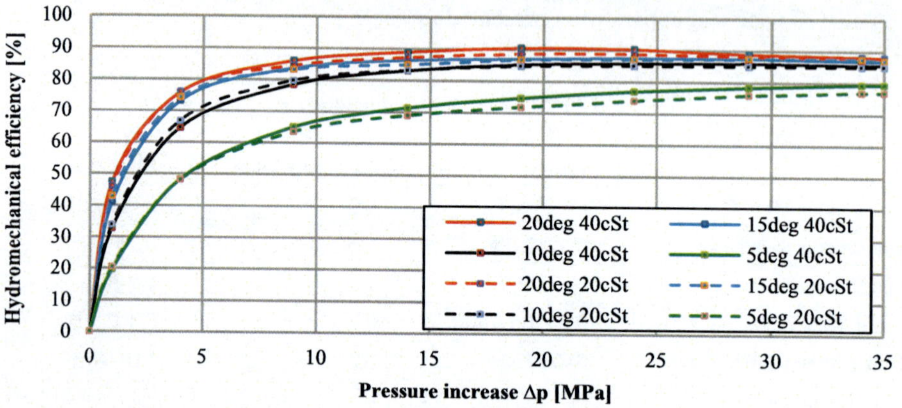

Fig. 11. Hydromechanical efficiency as a function of differential pressure for different swash plate swing angles and viscosities 20 and 40cSt for 2000 rpm [9].

4 Results

The prototype of an axial piston pump with a displaced axis of rotation of the swash plate, based on a commercial pump working mechanism (pistons with slippers, cylinder barrel, timing plate), has achieved a satisfactory volumetric efficiency, comparable to that of commercial pumps. Due to the fixed dead space volume, independent of the swash plate swing angle, the volumetric efficiency was improved for small swash plate swing angles and high pressures. The reason for the reduced hydromechanical efficiency is probably due to a shaft that is too long and can vibrate during pump operation. Tests have also shown problems with the follow-up mechanism, such as too much leakage through the directional valve and too high friction forces on the follow-up cylinder seals, causing it to lock. The results of the tests will be used to create a second, upgraded prototype.

The article is a result of project funded by The National Centre for Research and Development within the framework of program LIDER.

Project no.: LIDER/22/0130/L-8/16/NCBR/2017 Project title: Hydro-mechanical automatic gearbox for agricultural vehicles and heavy machinery. Value of funding: 1 197 500,00 PLN.

References

1. Ivantysynova, M., Ivantysyn, J.: Hydrostatische Pumpen und Motoren. Konstruktion und Berechnung. Vogel, Wurzburg (1993)
2. Manring, N.D., Johnson, R.E.: Modeling and designing a variable-displacement open-loop pump. J. Dyn. Syst. Meas. Control **118**, 267–271 (1996)
3. Osiecki, A.: Hydrostatyczny Napęd Maszyn. WNT, Warszawa (2004)
4. Osiecki, L.: Mechanizmy rozrządu hydraulicznych maszyn wielotłoczkowych osiowych. Wydawnictwo Politechniki Gdańskiej, Gdańsk (2006)
5. Osiecki, L.: Wpływ przestrzeni martwej na straty energetyczne w pompach wielotłoczkowych. Hydraulika i Pneumatyka 3/2007 (2007)

6. Scharf, S., Murrenhoff, H.: Measurement of friction forces between piston and busching of an axial piston displacement unit. Int. J. Fluid Power **6**(1), 7–17 (2005)
7. Załuski, P.: Influence of the position of the swash plate rotation axis on the volumetric efficiency of the axial piston pumps. Machines, Technologies, Materials issue 11/2014, Sofia (2014)
8. Załuski, P.: Wpływ położenia osi obrotu wychylnej tarczy na sprawność objętościową pomp wielotłoczkowych osiowych. Ph.D. dissertation, Gdansk University of Technology (2017)
9. Załuski, P.: Prototype pump testing, Research report in project LIDER/22/0130/L-8/16/NCBR/2017 Gdańsk (2019)

Hydraulic Systems

Check for updates

Intelligent Real-Time Control System for Forging Process Control

Ryszard Dindorf(✉), Jakub Takosoglu, and Piotr Wos

Kielce University of Technology, Kielce, Poland
{dindorf,qba,wos}@tu.kielce.pl

Abstract. The study presents the proposed intelligent real-time control system (IRCS) used to the forging process control on the 80 MN hydraulic forging press (HFP). It is a new control system applied to the industrial open-die forging in the process of elongation heavy large steel forgings. An innovative, energy-saving power supply system for the 80 MN HFP was presented. IRCS based on the parameter prediction of the forging process was developed. Compared to the previous work [1], the authors developed an IRCS for the forgings process control, which includes a model predictive control (MPC) of the multiple-input and multiple-output (MIMO) system, a predictive forging force model (PFFM), a recursive polynomial model estimator (RPME) and a discrete-time non-linear state-space model of the hydraulic system (MHS). The main goal of these studies was to develop and implement an innovative industrial process for hot open-die forging heavy, large, and hard-deformable steel forgings on the 80 MN HFP. Expectations that the application of IRCS will improve the quality and durability of forged products as well as increase the energy efficiency of the forging process have been met.

Keywords: Hydraulic forging press · Forgings process · Energy-saving power supply · Real-time control system

1 Introduction

The study aimed to develop and implement a ground-breaking open die forging process for heavy large steel forgings with the use of an innovative, energy-saving power supply and intelligent real-time control system (IRCS) for the forging process control on the 80 MN hydraulic forging presses (HFP). Heavy large forgings find buyers mainly in the shipbuilding, machinery, energy, and metallurgical industries [2–4]. This study will contribute to significant quality and endurance parameters improvement of forged products, energy-saving, reducing natural gas consumption, and reducing CO_2 emissions. The results of this study will be implemented at the Forged Products Department (FPD) of Celsa Huta Ostrowiec (CHO), Celsa Grup™ steel plant in Poland, in the form of a pilot manufacturing technology. This will allow the company to improve its competitive position in the international market, as well as acquire new recipients of hardly

© The Editor(s) (if applicable) and The Author(s), under exclusive license to Springer Nature Switzerland AG 2021
J. Stryczek and U. Warzyńska (Eds.): NSHP 2020, LNME, pp. 149–158, 2021.
https://doi.org/10.1007/978-3-030-59509-8_13

deformable heavy forgings. The CHO manufactures ingots, forgings, and final products on its own. The production process includes steel melting, forging, heat treatment, and machining. The CHO uses three innovative solutions for steel smelting and casting [5]. Figure 1 shows a manufacturing cycle of forged products at the CHO.

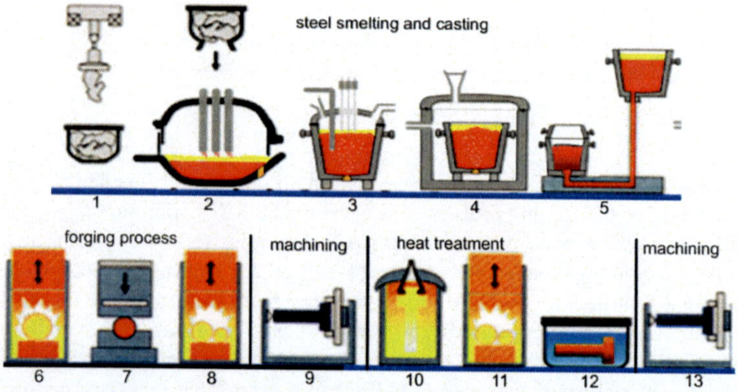

Fig. 1. Manufacturing cycle of forged products at the CHO: 1 – scrap processing, 2 – scrap melting, 3 – heating in a ladle furnace, 4 – vacuum degassing, 5 – molding, 6 – initial heat treatments, 7 – forging press, 8 – inter-operative heating, 9 – mechanical pre-treatment, 10, 11 – heating, 12 – quenching, 13 – mechanical post-treatment [5]

Fig. 2. View of the 80 MN HFP when forging hot ingot (photo FPD)

In FPD operates an industrial HFPs with the forces of 20, 32, and 80 MN for hot open die forging. The view of 80 MN HFP and feedstock (ingot) during hot open-die forging is shown in Fig. 2.

The FPD is equipped with a melt shop, hydraulic forging presses, and modern machines producing finished products with mechanical and thermal treatment. The FPD is a leading European manufacturer of open die forgings for strategic industries: power generation, oil & gas, engine, tool steel, and metal processing. Forged components are made for the world's largest suppliers of wind turbines, high power plant turbines, and global corporations in the energy and mining sectors. The FPD operates an industrial HFPs, which is widely used for hot free forging and hot open die forging in the process of shaping heavy large steel components. Figure 3 shows the machining of a monolithic shaft of a marine engine, as well as the final product of the machine drive shaft.

Fig. 3. Machining of a monolithic shaft of a marine engine and final product of the machine driveshaft (photo FPD)

2 Energy-Saving Power Supply System

The main goal of the research project is to use an innovative, energy-saving power supply and control system in HFP in order to optimize the forging process of heavy, large, and hard-deformable forgings. The technical level of such a solution is not available from competitors of FPD but is wanted in the forging industry [6]. Therefore, it is necessary to solve the following problems that are typical of industrial HFPs:

1. Adjustment of supply power during the operating cycle of the press, depending on the resistance forces of deformable material.
2. Limitation of hydraulic leaks in plunger cylinders.
3. Reduction of pressure pulsations in the high-pressure pipes.
4. Reduction of pressure drops in the hydraulic pipelines.
5. Adaptation of pump performance to the speed of working plungers.
6. Maintaining high pressure in the working plungers during the forging process.

In the previous work [1], the authors analysed industrial HFPs power supply systems by Wepuko Pahnke, Oil Gear, Hauhinco, Inoxihp, Schäfer & Urbach, SMS Meer. As a result, an innovative, parallel energy-saving power supply system of the 80 MN HFPwas used (see Fig. 4). Such a power supply system consists of two parallel hydraulic circuits: one with non-adjustable piston water pumps (pressure 32 MPa and flow 2000 l/min) and the other with an adjustable radial water pump (pressure 42 MPa and flow 500 l/min). The speed and accuracy position of the plunger cylinder is more accurately controlled compared to the standard solutions.

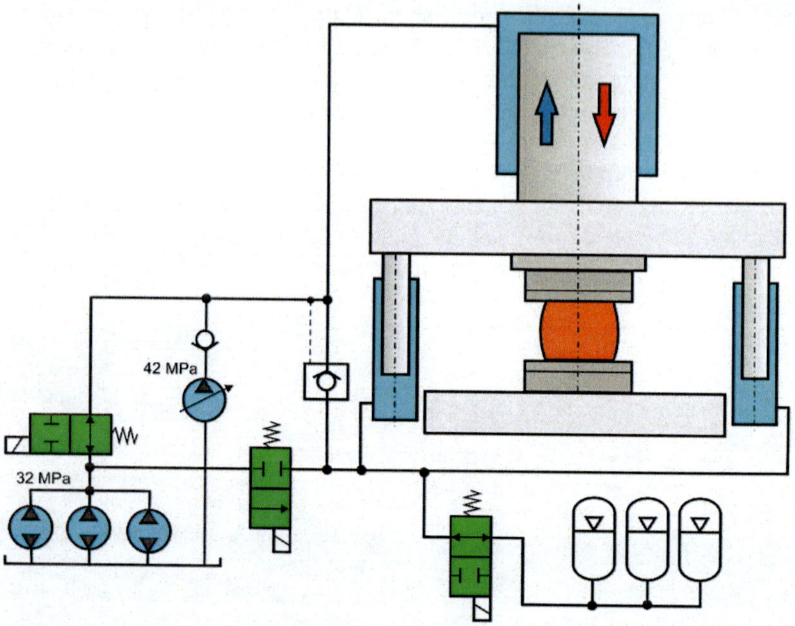

Fig. 4. Diagram scheme of the energy-saving power supply system of the 80 MN HFP

3 Forging Process Simulation

The study concerns an industrial open-die forging process for elongation heavy large steel forgings. The geometry of the forging during the elongation forging process is shown in Fig. 5.

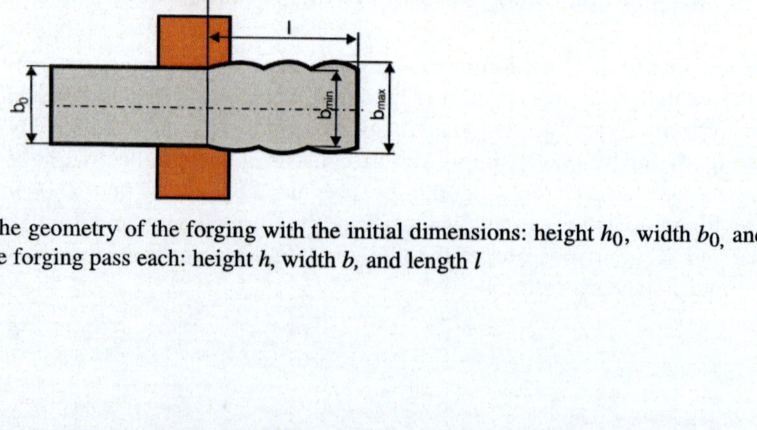

Fig. 5. The geometry of the forging with the initial dimensions: height h_0, width b_0, and length l_0 and one forging pass each: height h, width b, and length l

A schematic diagram of the 80 MN HFP control system is shown in Fig. 6.

Figure 6 shows the parameters used in the dynamic simulation of the 80 MN HFP. It is common practice to simulate industrial deformation processes of forgings before executing industrial trials. In the 80 MN HFP dynamic simulation, a discrete-time nonlinear state-space model was used:

$$\begin{cases} x(k+1) = f(x(k)) + g(x(k))\, u(k) \\ y(k) = C\, x(k) \end{cases} \tag{1}$$

with constraints: $u_{min} \leq u(k) \leq u_{max}$, $u_{min} \leq y(k) \leq y_{max}$.
where $x(k) \in \Re^n$ is the state vector, $u(k) \in \Re^m$ is the input vector, $y(k) \in \Re^p$ is the output vector, $f(x(k))$ and $g(x(k))$ are the nonlinear functions, $C \in \Re^{p+n}$, u_{min} and y_{min} are the vectors of lower bounds, u_{max} and y_{max} are the vectors of upper bounds.

During simulation and control of the 80 MN HFP, it takes into account three phases of the industrial forging process:

1. The drop of the upper die to contact the forging.

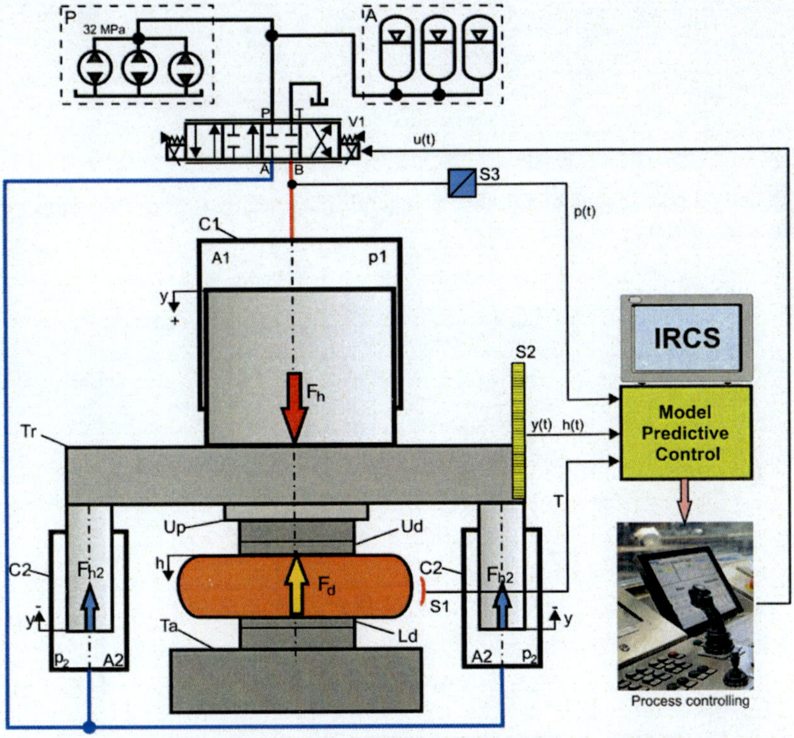

Fig. 6. Schematic diagram of the 80 MN HFP control system: C1 – main (working) plunger cylinders, C2 – return plunger cylinders, V1 – proportional directional 4/4/control valve, P – water pump station, A – high-pressure water accumulators, Tr – movable traverse, Ud – upper die, Dd – down die, Up – upper plate for fixing the upper die, Ta – table for fixing the down die, S1 – infrared temperature sensor T, S2 – position transducer $y(t)$, S3 – pressure transducer $p(t)$

2. Deformation process of the forging.
3. Return of the upper die above the forging,

The following parameters were taken into account in the simulation modeling of the forging process: plunger displacement $y(t)$, forging deformation $h(t)$, plunger speed $v(t)$, working pressure $p_1(t)$, and return pressure $p_2(t)$ in the press cylinders. Forging process simulation results are shown in Figs. 7 and 8.

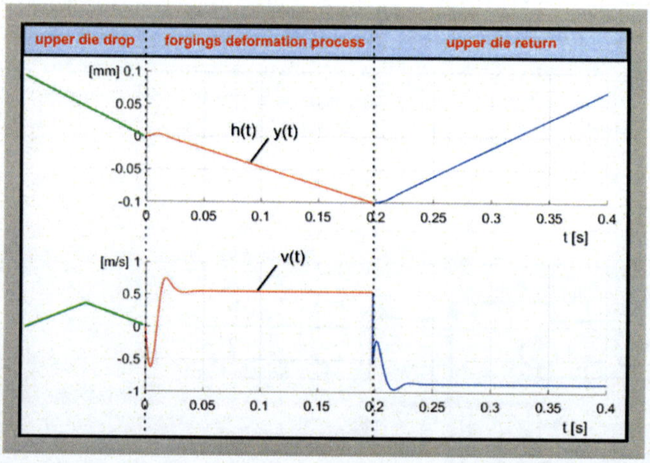

Fig. 7. Forging process simulation results: plunger displacement $y(t)$, forging deformation $h(t)$, and plunger speed $v(t)$

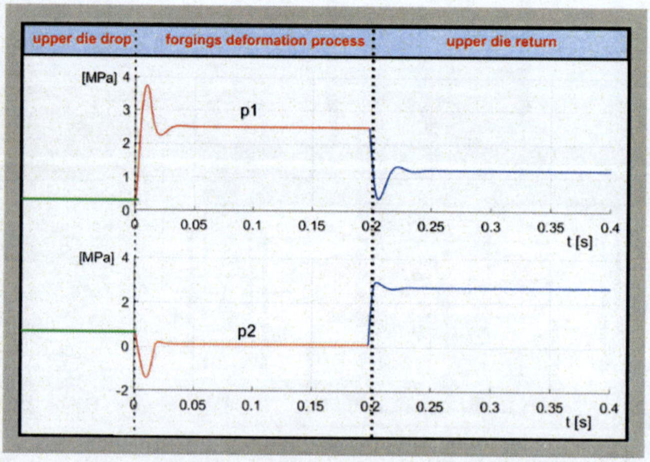

Fig. 8. Forging process simulation results: working pressure $p_1(t)$ and return pressure $p_2(t)$ in the press cylinders

4 Forging Process Control

In the study, IRCS with a prediction of manufacturing results (PMO) as an innovative control method of the deformation forgings process was used. It is a method of predicting the further course of the forging deformation process. The use of predictive methods in the manufacturing industry is met with skepticism. However, Industry 4.0 and the constant drive to optimize the cost of forged products prompts many companies to consider using prediction methods to improve the quality of forged products. This especially applies to the forging process of large and heavy forgings, which is characterized by a high deviation in the quality of the forgings. A large number of rejected and reworked forgings have a serious impact on the cost of production. The control of the forging process should be adapted to the environmental conditions, measurement of parameters, and time-varying events occurring during the deformation of the forgings. In subsequent studies, the use of an adaptive control system for forging presses will be considered [7].

The main task of the IRCS is to control the HFP cycle, taking into account the phases of the open-die forging process. The static parameters are known, but the time-varying dynamic parameters having a direct impact on the forging process and the quality and strength of the final product are unknown. Despite simulating the parameters of the forging process with the use of various advanced programs, such as FORGE, it is difficult to avoid deformation errors of forgings. The HFP control takes into account the parameters of the forging process, such as forging/deformation force, deformation strain rates, and deformation temperature. The quality of the forging process is also greatly influenced by uncontrolled hydraulic transients, such as volumetric elasticity (compressibility modulus) of the working fluid, as well as leakage in the plunger cylinder [8]. The basis of IRCS operation is the time-varying parameters of the forging process, such as the compressive force $F(t)$, the pressure $p(t)$, and the plunger displacement $y(t)$. These parameters are recorded in real-time, directly during the forging process. While the temperature T of the forgings is always measured at the beginning of each pass of the forging process. A schematic diagram of the IRCS for the forgings process control on the 80 MN HFP is shown in Fig. 9.

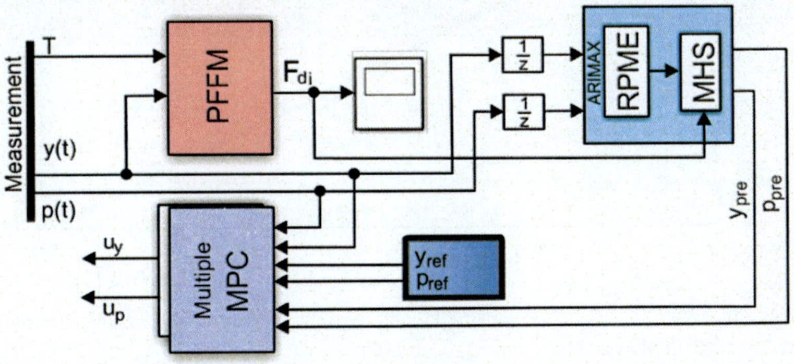

Fig. 9. Schematic diagram of the IRCS for the forgings process control on the 80 MN HFP

The IRCS used to the forgings process control on the 80 MN HFP contains the following modules (see Fig. 9): model predictive control (MPC) of the multiple-input and multiple-output (MIMO) system, the predictive forging force model (PFFM), the recursive polynomial model estimator (RPME) and the nonlinear state-space equations containing a dynamic model of the hydraulic system (MHS). The MPC is an optimization control method that uses the current state and a predictive model to obtain the optimal input vector by solving the optimization problem [9, 10]. The RPME estimates the discrete-time of polynomial models of ARIMAX (autoregressive integrated moving average with explanatory variable) structures. The multiple MPC was used to predict control of the forging process, which includes two independent algorithms: deformation/forging forces $F(t)$ algorithm and plunger displacement $y(t)$/forgings height $h(t)$ algorithm. The predictive control (PC) is based on the measurement of input parameters recorded in real-time. The plunger displacement $y(t)$ is measured using a position transducer, the pressure $p(t)$ in the working cylinder is measured using a pressure transducer, and the forgings temperature T is measured by an infrared temperature sensor. The goal of the MPC is to minimize the difference between the reference control signals $y_{ref}(t)$ and $p_{ref}(t)$ and predicted control signals $y_{pre}(t)$ and $p_{pre}(t)$ for optimal input signals u_y and u_p in the control valve [1]:

$$\begin{cases} \min u_y \;\; J_h(y_{ref}, y_{pre}) \\ \min u_p \;\; J_p(p_{ref}, p_{pre}) \end{cases} \tag{2}$$

where J_h and J_p are the objective functions.

The results of the predictive response for the plunger displacement $y(t)$ and the working pressure $p(t)$ are presented in Figs. 10 and 11.

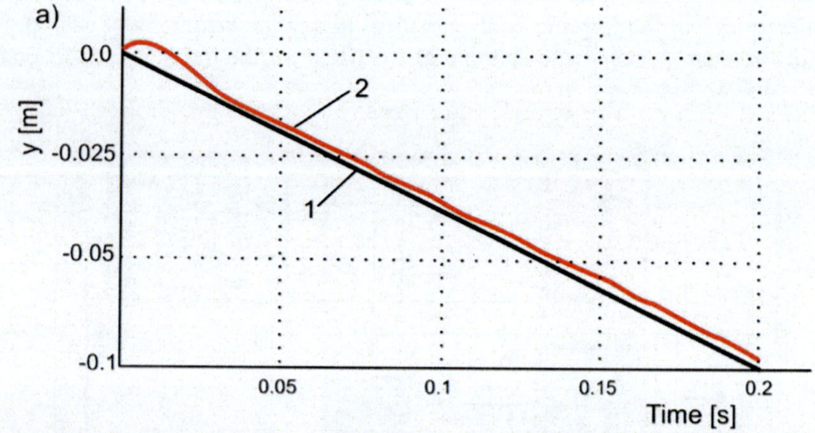

Fig. 10. Predictive response for plunger displacement $y(t)$: 1 – reference control signal $y_{ref}(t)$, 2 – predicted control signal $y_{pre}(t)$

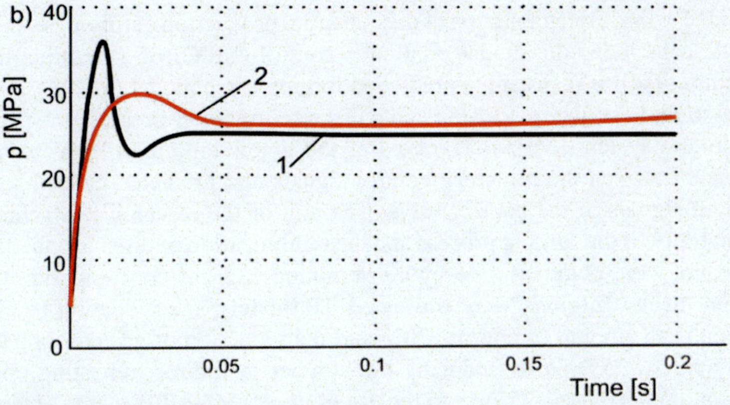

Fig. 11. Predictive response for working pressure $p(t)$: 1 – reference control signals $p_{ref}(t)$, 2 – predicted control signal $p_{pre}(t)$

The used IRCS to the forging process control on the 80 MN HFP gives greater possibilities of accurate shaping of heavy, large forgings during the open-die forging process. Its main advantage is the repeatability of the forging process and the minimization of size deviations δh of the forgings (see Fig. 12).

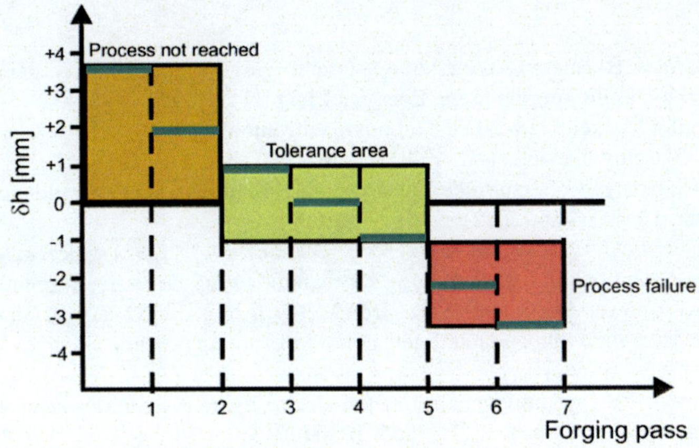

Fig. 12. The size deviation δh of the forgings

5 Conclusion

The paper contains completely new, previously unpublished results on the use of IRCS to control the forging process on 80 MN HFP. The presented test results are of practical importance for the industrial forging process in the elongation of heavy, large, and hard-deformable steel forgings. The research was carried out in cooperation with FPD CHO and was aimed at the implementation of the industrial production of forgings. The applied IRCS solution includes the model predictive control (MPC) of a multiple-input and multiple-output (MIMO) system, a predictive forging force model (PFFM), a recursive polynomial model estimator (RPME) and a discrete-time non-linear state-space model of the hydraulic system (MHS). The goal of the prediction control was to minimize the difference between the reference control signals and predicted control signals for optimal input signals in the control valve. The aim of the research was achieved, the high repeatability of the forging process and the minimization of dimensional deviations of the forgings were obtained. Energy consumption and gas consumption during the forging large heavy forgings were estimated. Ultimately, the reduction in electricity consumption is to amount to approx. 20% and the reduction of gas consumption is to amount to approx. 25%. After reducing the number of indirect reheating operations, gas consumption decreased. Thanks to the use of the 80 MN HFP energy-saving power supply system, the energy consumption per ton of forgings is to be reduced by 45%, and the water savings are to be approx. 10% by reducing leakages in hydraulic plunger cylinders. By reducing pressure pulsation and limiting the hydraulic strike effect, it is expected to reduce the noise level to 80 dB. Limiting the influence of hydraulic transients not only extends the life of the hydraulic components but also affects the quality of forged products. Increases the energy efficiency of the forging process is environmentally advantageous.

References

1. Dindorf, P., Wos, P.: Energy-saving hot open die forging process of heavy steel forgings on an industrial hydraulic forging press. Energies **13**(7), 1–17 (2020)
2. Global Forging Market 2018–2022. Technovio, Toronto (2019)
3. Gregor, M.: Forging. Didactic text. VŠB TU Ostrava, Ostrava (2014)
4. Kukuryk, M.: Analysis of deformation and damage evolution in hot elongation forging. Arch. Metal. Mater. **12**, 417–424 (2012)
5. Zdonek, B., Szypuła, I., Binek, S., Kowalski, J., Dudkiewicz, P., Karbowniczek, M., Barański, J.: Innovative solutions in the technology for manufacturing large-size forgings from ultra-clean steels for power equipment. Works Iron Metals Inst. **4**, 53–59 (2013). (In Polish)
6. Dindorf, R.: Energy saving and intelligent control of forging process. Forgings Today **10**(4), 6–10 (2020)
7. Wos, P., Dindorf, R.: Adaptive control of the electro-hydraulic servo-system with external disturbances. Asian J. Control **15**(4), 1065–1080 (2013)
8. Dindorf, R., Wos, P.: Control of integrated electro-hydraulic servo-drives in a translational parallel manipulator. J. Mech. Sci. Technol. **33**(11), 5437–5448 (2019). https://doi.org/10.1007/s12206-019-1038-y
9. Ganesh, U.G., Krishna, M., Hariprasad, A.P.: Review on models for a generalized predictive controller. In: CCSEA 2011, First International Conference on Computer Science, Engineering and Applications, Chennai, India, pp. 418–424 (2011)
10. Navratil, P., Ivanka, J.: Recursive estimation algorithms in matlab & simulink development environment. WSEAS Trans. Comput. **13**, 691–702 (2014)

Digital Pumping System with Electromechanical Repartition

Drumea Petrin[1], Dumitrescu Catalin[1(✉)], Barbu Valentin[1], Opruta Dan[2], and Banyai Daniel[2]

[1] INOE 2000 IHP, Cutitul de argint Street, No.14, sector 4, Bucharest, Romania
dumitrescu.ihp@fluidas.ro
[2] Technical University of Cluj-Napoca, Memorandumului Street, No. 28, Cluj-Napoca, Cluj, Romania

Abstract. Digital fluid power means a hydraulic plant having one or more discrete valued components actively controlling the system output. In this paper authors present a solution of parallel connected of six fixed pumps. The proposed digital pumping system consists of an electrical single speed motor, fixed displacement pumps, an electromechanically repartition element and an electronic controller. This complex system which can achieve 63 flows was made in our Research Institute and in present is in the Digital hydraulic laboratory for tests.

Keywords: Digital hydraulics · Digital hydraulic pump · Pumping system · Electromechanical distributor

1 The Concept of Digital Hydraulics

In recent years, hydraulic systems have an increasingly important competition from the electric and electromechanical ones, especially in terms of energy efficiency. The basic criterion, on which the systems comparison is made, consists on a closer value for the spent energy to the necessary energy, useful for actuating any machine, but especially in the case of the mobile ones in agriculture and construction. Hydraulics can no longer be competitive with overall efficiency of 35%–50%, so with a huge energy waste that exceeds half of the consumed one.

Among the solutions of recent years, where specialists in hydraulic actuators try to reduce energy losses, the use of digital hydraulics is also included. It should be noted that digital hydraulics does not mean traditional, classical or modern hydraulics with analog-type elements that we drive with digital electronics. Digital hydraulics means hydraulic systems that have at least one hydraulic element that ensures at the output, accurate discrete values, actively controlled with appropriate electronics, so a good computerization of the system. At European level it seems that the idea of digital hydraulics with parallel connection and digital hydraulics based on switching technology is accepted. An interesting synthesis is made by Prof. Scheidl (Linz) [1] starting from an article by Professor Linjama (Tampere) [2] and presented in this material in Fig. 1. Although it

© The Editor(s) (if applicable) and The Author(s), under exclusive license
to Springer Nature Switzerland AG 2021
J. Stryczek and U. Warzyńska (Eds.): NSHP 2020, LNME, pp. 159–168, 2021.
https://doi.org/10.1007/978-3-030-59509-8_14

seems that, the digital solution with parallel connection, is complicated and involves many elements, things are simpler and easier to achieve in reality.

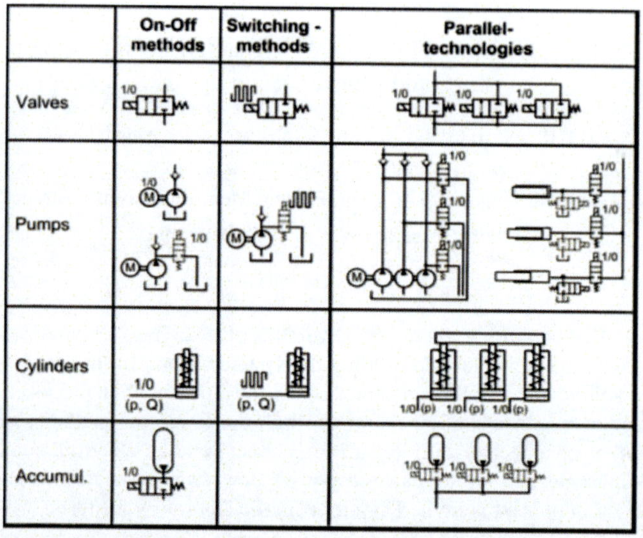

Fig. 1. The examples of digital hydraulics with parallel connection and based on switching technology [1, 2].

2 Digital Hydraulic Pumps

From all the elements of digital hydraulics that have passed the prototype phase, the only ones entered in series production are the piston pumps developed by the company ARTEMIS. Luke Wadsley from Sauer proposes a functional scheme (Fig. 2) of the pump in which the basic solution of a simplified pumping element is presented [3]. Finally, the pump is made from the assembly of several basic elements. Each element is

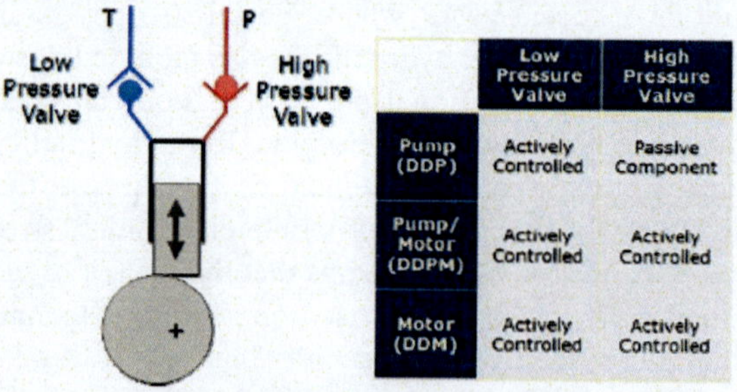

Fig. 2. Digital pumping element [3].

independently controlled, with specific electronics, being thus realized a digital pump, as a whole, that can ensure the flow required by the load, reducing the energy losses.

There are many elements involved in the reduction of energy losses in the case of digital pumps compared to the classic pumps with axial pistons, which, for example, in addition to the fluid flow changes the reduced noise produced, advantage is added, which helps the faster implementation of the new technology in complex machines. The digital pumping element is based on a cylinder with piston and two check valves integrated with a fast acting repartitor and a suitable electronics.

The only company that produces digital hydraulic pumps in series is ARTEMIS, which has managed to provide for flow changes a time response below 30 ms, which allows a good functioning for hydraulic installations on mobile machines. An industrial version is shown in Fig. 3.

Fig. 3. Digital pump made by ARTEMIS.

3 Digital Pumping Systems with Electromechanical Repartitor

The pumping system with multiple pumps proposed by the authors is a fairly old general idea, the novelty referring to the fact that in this situation it is a complex pumping system, which ensures the variation of flowrates through a special selection of fixed displacement pumps. Through the construction of the repartitor and the choice of fixed pumps, a digital delivery of the flow is ensured, being obtained flowrates of different values at different orders. Since the size of the fixed flows delivered by the fixed pumps of the system are realized in a binary scheme it can be accepted that the whole system becomes a digital pumping system.

A similar solution in principle, only as a hydraulic scheme, and with a summary analysis, was presented at HERVEX 2012 by the German specialist H. Theissen from the Technical University of Aachen [4, 5].

The solution presented in this article consists of a single drive motor, six pumps with different displacement, respecting the increase in size under binary numbers conditions, a single simple and robust mechanical part, a simple system for selecting the pumps by driving them with electromagnetic disc couplings, controlled by the digital electronic block, which selects according to the indications from the Table 1, which pumps are switched on and which pumps remain inactive.

The advantages of the digital pumping system are:

- Simplify the mechanical block simultaneously with increasing the level of compaction.
- With a number of six fixed pumps, three of the same displacement size and driven at three different speeds, a range of 63 flowrates is easily accessible through an extremely simple digital electronic control. The 63 values of flows finally ensure a precision of approximately 2%.
- The fixed pumps, that are not involved in achieving the chosen flow, are not driven, the electromagnetic coupling being inactive and as a result reduces the power consumption to the level of the necessary value and also reduces the wear.
- The pressure level taken into account is 210 bar, provided by a motor with a power around 68 kW with a speed of 1450 rpm.

Figure 4 shows the hydraulic diagram of the system, which includes as basic elements the six fixed pumps, chosen in two dimensional sizes, together with the three speeds obtained through the electromechanical distributor will deliver six different, fixed flows, located in a binary sequence. A very important role is played by the six electromagnetic couplings with discs, that will drive or not, the six fixed pumps, realizing the usable flow to the required size. The electrical control panel contains the electric drive system and the electronic block for selecting and controlling the flow by actuating the couplings,

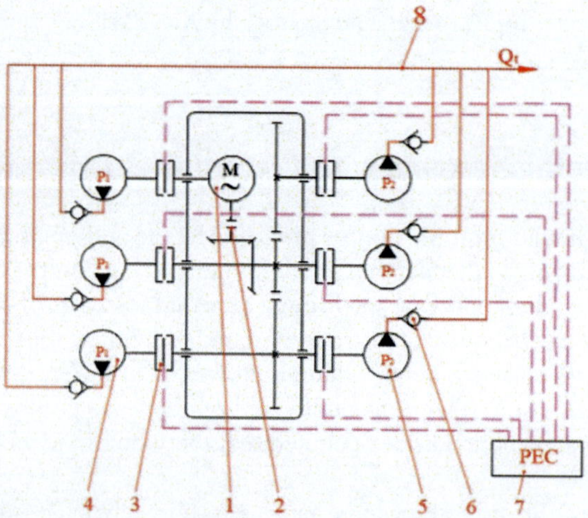

Fig. 4. Digital pumping system with electromechanical repartitor: 1 - electrical or thermal motor, 2- elastic coupling with bolts, 3 - electromagnetic coupling with discs, 4- fixed flow pump (PRD size 1), 5- fixed flow pump (PRD size 2), 6 - one-way valves, 7 - electrical control panel, which also includes the electronic control unit, 8 - pressure line.

according to the specifications from Table 1. The role of the check valves is to prevent the flow from an active pump to return to an inactive pump and turn it into a hydraulic motor.

The electromechanical repartitor (EMR) is made in a conical-cylindrical construction. The electric motor (ME) transmits the movement to the axis 1 of the repartitor through a conical gear that amplifies the axis speed from 1450 min^{-1} to the speed $n_1 = 1600\ \text{min}^{-1}$. The movement is transmitted from axis 1 to axis 2 and 3 through cylindrical gears, thus obtaining the speeds $n_2 = 800\ \text{min}^{-1}$ at axis 2 and $n_3 = 400\ \text{min}^{-1}$ at axis 3. Each axis has at the ends electromagnetic couplings with discs through which transmits the movement to the pump on the respective axis.

In the example presented in Figs. 4 and 5 we have:

EMR – electromechanical repartitor;
P1- hydraulic pump with 8 cm^3/rot displacement;
P2- hydraulic pump with 64 cm^3/rot displacement;

The speeds of the three axes:

$n1 = 1600\ \text{min}^{-1}$;
$n2 = 800\ \text{min}^{-1}$;
$n3 = 400\ \text{min}^{-1}$.

Table 1. The fluid flow generated by the system based on the active pumps.

	Active pump combination	Ax. 1	Ax. 2	Ax .3	Ax .1	Ax . 2	Ax. 3	Flow [l/min]	Necc. Power [kW]	Motor axis torque [daNm]	N axis 1 [kW]	N axis 2 [kW]	N axis 3 [kW]
		Q6	Q5	Q4	Q3	Q2	Q1						
1	P1						3.2	3.2	1.07	0.68	0	0	1.07
2	P2					6.4		6.4	2.13	1.36	0	2.13	0
3	P1+P2					6.4	3.2	9.6	3.2	2.04	0	2.13	1.07
4	P3				12.8			12.8	4.27	2.72	4.27	0	0
5	P3+P1				12.8		3.2	16	5.33	3.4	4.27	0	1.07
6	P3+P2				12.8	6.4		19.2	6.4	4.07	4.27	2.13	0
7	P3+P2+P1				12.8	6.4	3.2	22.4	7.47	4.75	4.27	2.13	1.07
8	P4			25.6				25.6	8.53	5.43	0	0	8.53
9	P4+P1			25.6			3.2	28.8	9.6	6.11	0	0	9.6
10	P4+P2			25.6		6.4		32	10.67	6.79	0	2.13	8.53
11	P4+P2+P1			25.6		6.4	3.2	35.2	11.73	7.47	0	2.13	9.6
12	P4+P3			25.6	12.8			38.4	12.8	8.15	4.27	0	8.53

(*continued*)

Table 1. (*continued*)

	Active pump combination	Ax. 1	Ax. 2	Ax. 3	Ax. 1	Ax. 2	Ax. 3	Flow [l/min]	Necc. Power [kW]	Motor axis torque [daNm]	N axis 1 [kW]	N axis 2 [kW]	N axis 3 [kW]
		Q6	Q5	Q4	Q3	Q2	Q1						
13	P4+P3+P1			25.6	12.8		3.2	41.6	13.87	8.83	4.27	0	9.6
14	P4+P3+P2			25.6	12.8	6.4		44.8	14.93	9.51	4.27	2.13	8.53
15	P4+P3+P2+P1			25.6	12.8	6.4	3.2	48	16	10.19	4.27	2.13	9.6
16	P5		51.2					51.2	17.07	10.87	0	17.07	0
17	P5+P1		51.2				3.2	54.4	18.13	11.54	0	17.07	1.07
18	P5+P2		51.2			6.4		57.6	19.2	12.22	0	19.2	0
19	P5+P2+P1		51.2			6.4	3.2	60.8	20.27	12.9	0	19.2	1.07
20	P5+P3		51.2		12.8			64	21.33	13.58	4.27	17.07	0
21	P5+P3+P1		51.2		12.8		3.2	67.2	22.4	14.26	4.27	17.07	1.07
22	P5+P3+P2		51.2		12.8	6.4		70.4	23.47	14.94	4.27	19.2	0
23	P5+P3+P2+P1		51.2		12.8	6.4	3.2	73.6	24.53	15.62	4.27	19.2	1.07
24	P5+P4		51.2	25.6				76.8	25.6	16.3	0	17.07	8.53
25	P5+P4+P1		51.2	25.6			3.2	80	26.67	16.98	0	17.07	9.6
26	P5+P4+P2		51.2	25.6		6.4		83.2	27.73	17.66	0	19.2	8.53
27	P5+P4+P2+P1		51.2	25.6		6.4	3.2	86.4	28.8	18.34	0	19.2	9.6
28	P5+P4+P3		51.2	25.6	12.8			89.6	29.87	19.02	4.27	17.07	8.53
29	P5+P4+P3+P1		51.2	25.6	12.8		3.2	92.8	30.93	19.69	4.27	17.07	9.6
30	P5+P4+P3+P2		51.2	25.6	12.8	6.4		96	32	20.37	4.27	19.2	8.53
31	P5+P4+P3+P2+P1		51.2	25.6	12.8	6.4	3.2	99.2	33.07	21.05	4.27	19.2	9.6
32	P6	102						102.4	34.13	21.73	34.13	0	0
33	P1+P6	102					3.2	105.6	35.2	22.41	34.13	0	1.07
34	P2+P6	102				6.4		108.8	36.27	23.09	34.13	2.13	0
35	P1+P2+P6	102				6.4	3.2	112	37.33	23.77	34.13	2.13	1.07
36	P3+P6	102			12.8			115.2	38.4	24.45	38.4	0	0
37	P3+P1+P6	102			12.8		3.2	118.4	39.47	25.13	38.4	0	1.07

(*continued*)

Table 1. (*continued*)

	Active pump com-bi-nation	Ax. 1	Ax. 2	Ax. 3	Ax. 1	Ax. 2	Ax. 3	Flow [l/min]	Necc. Power [kW]	Motor axis torque [daNm]	N axis 1 [kW]	N axis 2 [kW]	N axis 3 [kW]
		Q6	Q5	Q4	Q3	Q2	Q1						
38	P3+P2+P6	102			12.8	6.4		121.6	40.53	25.81	38.4	2.13	0
39	P3+P2+P1+P6	102			12.8	6.4	3.2	124.8	41.6	26.49	38.4	2.13	1.07
40	P4+P6	102		25.6				128	42.67	27.16	34.13	0	8.53
41	P4+P1+P6	102		25.6			3.2	131.2	43.73	27.84	34.13	0	9.6
42	P4+P2+P6	102		25.6		6.4		134.4	44.8	28.52	34.13	2.13	8.53
43	P4+P2+P1+P6	102		25.6		6.4	3.2	137.6	45.87	29.2	34.13	2.13	9.6
44	P4+P3+P6	102		25.6	12.8			140.8	46.93	29.88	38.4	0	8.53
45	P4+P3+P1+P6	102		25.6	12.8		3.2	144	48	30.56	38.4	0	9.6
46	P4+P3+P2+P6	102		25.6	12.8	6.4		147.2	49.07	31.24	38.4	2.13	8.53
47	P4+P3+P2+P1+P6	102		25.6	12.8	6.4	3.2	150.4	50.13	31.92	38.4	2.13	9.6
48	P5+P6	102	51.2					153.6	51.2	32.6	34.13	17.07	0
49	P5+P1+P6	102	51.2				3.2	156.8	52.27	33.28	34.13	17.07	1.07
50	P5+P2+P6	102	51.2			6.4		160	53.33	33.96	34.13	19.2	0
51	P5+P2+P1+P6	102	51.2			6.4	3.2	163.2	54.4	34.63	34.13	19.2	1.07
52	P5+P3+P6	102	51.2		12.8			166.4	55.47	35.31	38.4	17.07	0
53	P5+P3+P1+P6	102	51.2		12.8		3.2	169.6	56.53	35.99	38.4	17.07	1.07
54	P5+P3+P2+P6	102	51.2		12.8	6.4		172.8	57.6	36.67	38.4	19.2	0
55	P5+P3+P2+P1+P6	102	51.2		12.8	6.4	3.2	176	58.67	37.35	38.4	19.2	1.07
56	P5+P4+P6	102	51.2	25.6				179.2	59.73	38.03	34.13	17.07	8.53
57	P5+P4+P1+P6	102	51.2	25.6			3.2	182.4	60.8	38.71	34.13	17.07	9.6
58	P5+P4+P2+P6	102	51.2	25.6		6.4		185.6	61.87	39.39	34.13	19.2	8.53
59	P5+P4+P2+P1+	102	51.2	25.6		6.4	3.2	188.8	62.93	40.07	34.13	19.2	9.6

(*continued*)

Table 1. (*continued*)

	Active pump combination	Ax. 1	Ax. 2	Ax .3	Ax .1	Ax .2	Ax. 3	Flow [l/min]	Necc. Power [kW]	Motor axis torque [daNm]	N axis 1 [kW]	N axis 2 [kW]	N axis 3 [kW]
		Q6	Q5	Q4	Q3	Q2	Q1						
	P6												
60	P5+P4+P3+P6	102	51.2	25.6	12.8			192	64	40.75	38.4	17.07	8.53
61	P5+P4+P3+P1+P6	102	51.2	25.6	12.8		3.2	195.2	65.07	41.43	38.4	17.07	9.6
62	P5+P4+P3+P2+P6	102	51.2	25.6	12.8	6.4		198.4	66.13	42.1	38.4	19.2	8.53
63	P5+P4+P3+P2+P1+P6	102	51.2	25.6	12.8	6.4	3.2	201.6	67.2	42.78	38.4	19.2	9.6

A first condition for EMR sizing is that the distance between the axes 1 and 2 is greater than the dimensions of the pump P2, in horizontal plane and the space required for the intervention at the pump and for fixing the hydraulic network connections.

$$a_{12} \geq \frac{G_{P2} + G_{P2}}{2} + S \tag{1}$$

where:

G_{P2} - dimensions of the pump P2, in horizontal plane;

S - the space needed to fix the hydraulic network connections.

The speeds of the shafts are realized by the transmission ratio of this gears. From the speed of $n1 = 1600\ \text{min}^{-1}$ of axis 1 to $n2 = 800\ \text{min}^{-1}$, the transmission ratio is u = 2.

The number of teeth of the gear wheels is:

On axis 1 the recommendation for the number of gear teeth is at least $z_1 = 25$;

The number of gear teeth on axis 2 is given by the transmission ratio $u_{12} = 2$ resulting in 50 teeth we choose $z_2 = 51$ teeth (z_1 and z_2 it should not have common divisors).

The distance between axes 1 and 2 is given by the relation:

$$a_{12} = \frac{m_n(z_1 + z_2)}{2} \tag{2}$$

replacing in relation (2) the value obtained from relation (1) results in a value for the normal gear module:

$$m_n = \frac{2a_{12}}{(z_1 + z_2)} \tag{3}$$

For dimensioning the EMR from the stress point of view, to which the respective gear is subjected, we determine:

The power required to drive the pumps on the respective axis at the maximum flow and pressure values, as follows:

$$N_{ax2} = Q_1 \frac{P_1}{612} + Q_2 \frac{P_2}{612} \tag{4}$$

The torque:

$$M_{taxis2} = \frac{N_{ax2}}{n_2} [daNm] \tag{5}$$

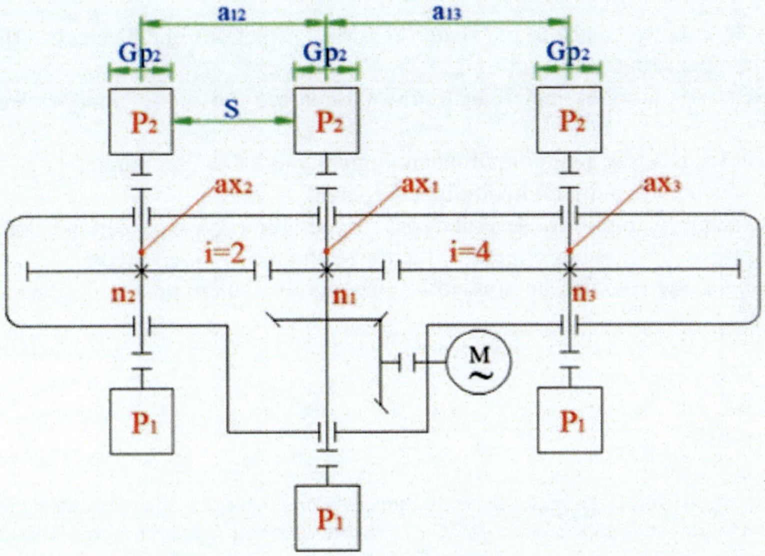

Fig. 5. Mechanical drive scheme

For the minimum distance between the axes can be written, depending on the torque [6]:

$$a_{min} = (u+1) \sqrt[3]{\frac{M_{tax2} K_H (Z_H Z_E Z_\varepsilon Z_\beta)^2}{2u^2 \psi_a \sigma_{HP}^2}} \tag{6}$$

where:

M_{tax2} - the torque on gear from axis 2;
K_H - load correction factor;
Z_H – contact area factor;
Z_E – material factor;

Z_ε- teeth coverage factor;
Z_β - tilt of the teeth factor;
σ_{HP}- admissible stress limit at contact;
Ψ_a – gears width coefficient;
u – transmission ratio;

Comparing the values obtained with the relations (1) and (6) there can be two variants:
$a_{12} < a_{min}$ then the value for the distance between the axes is chosen: a_{min};
$a_{12} \geq a_{min}$ then the value for the distance between axes is adopted: a_{12}.
The mechanical diagram of the pumps driving system is shown in Fig. 5.

4 Conclusion

- Digital hydraulics is still in the search for work directions and as a result there are only sporadic series manufacturing.
- The basic solutions of digital hydraulic pumps are those with independent axial pistons.
- The article proposes a simplified technological way of digital pumping system that can be accepted as a digital hydraulic equipment.
- On the same technology, several types of pumping systems can be composed, depending on the number of pumps, their size and the driving speeds.
- The technology can also be applied to turbopumps, where flow variations are also required.

References

1. Scheidl, R., Kogler, H., Winkler B.: Hydraulic switching control - objectives, concepts, challenges and potential applications. In: The 20th International Conference of Hydraulics and Pneumatics - HERVEX, 7–9 November 2012, Calimanesti-Caciulata, Romania (2012)
2. Linjama, M.: Digital fluid power-state of art. In: The 12th Scandinavian International Conference of Fluid Power, 18–20 May 2011, Tampere, Finland (2011)
3. Wadsley, L.: Optimal system solutions enabled by digital pumps. In: International Exposition for Power Transmission and Technical Conference, Proceedings of the 52nd National Conference on Fluid Power, March 23–25, 2011, Las Vegas, Nevada, USA (2011)
4. Theissen, H.: Fluid power for sustainability. In: The 19th International Conference of Hydraulics and Pneumatics - HERVEX, 9–11 November 2011, Calimanesti-Caciulata, Romania (2011)
5. Pavel, I., Rădoi, R.I., Chiriţă, A., Hristea, M., Tudor, B.A.: Technical solutions for digital hydraulic cylinders and test methods. Hidraulica (3), 41–49 (2017)
6. Palade, V., Constantin, V., Hapenciuc, M.: Reductoare cu roţi dinţate, Editura ALMA Galaţi (2003)

Control of the Test Rig with Hydraulic Integrated Actuator for Spring Stiffness Measurement

Petr Noskievič(✉), Ahmed Al Zaid, and Yadhu Swaroop Chandra Mohan

Faculty of Mechanical Engineering, VŠB-Technical University of Ostrava, Ostrava-Poruba, Czech Republic
petr.noskievic@vsb.cz

Abstract. The paper deals with the control of the test rig for measurement of the stiffness of the spring. The test rig is equipped with the integrated hydraulic actuator which allows to control the piston velocity and position and to produce the force needed for the spring compression and stiffness measurement. The integrated hydraulic actuator was developed at the department, it and has an integrated manifold with four proportional flow control valves connected as the full hydraulic resistance bridge. Each control valve is controlled separately using the input signal. That makes it possible to realize different connections and control strategies of the hydraulic cylinder using the different control algorithms. The hydraulic circuit was modelled and simulated, and the achieved simulation results are presented in the paper. The test rig is controlled using the microprocessor control system and is equipped by different transducers – position transducer, pressure transducers, force transducer to be able to collect data important for the stiffness characteristic evaluation. The experimentally obtained results and evaluated stiffness characteristic of the tested spring are presented in the following chapters.

Keywords: Integrated hydraulic actuator · Simulation model · Spring stiffness · Control · Measurement

1 Introduction

The most important properties of the hydraulic drives determining their use in the production machines, manipulators, robots and testing equipment are high power to weight ratio, good controllability of pressure and flow, high dynamics, fast time response, simple transformation of the hydraulic power in the mechanical power of the longitudinal motion. The control of the power transmission in the hydraulic driven machines is realized using the controllable pumps and flow control valves. In recent years, the efficiency of the power transmission in fluid power systems has become the goal of the development of the new components and systems. The power to weight ratio of the axial piston pumps increased approx. three times, and of the pilot-operated valves approx. fifteen times [12]. Decreasing the weight of the hydraulic components is one way how to be

© The Editor(s) (if applicable) and The Author(s), under exclusive license to Springer Nature Switzerland AG 2021
J. Stryczek and U. Warzyńska (Eds.): NSHP 2020, LNME, pp. 169–178, 2021.
https://doi.org/10.1007/978-3-030-59509-8_15

competitive with electrical drives. Another way is the application of the control systems and the design of the hydraulic systems with controllable components. The use of the modern pumps with variable geometric volume allows to realize the displacement control, or the fix-displacement pump can be controlled by the speed controlled AC-motor and the similar control strategies leading to the higher efficiency can be applied [1, 3]. High dynamic hydraulic drives and control of different quantities – velocity, pressure, force – are very often done using the resistance control and flow control valves. The high dynamics, accuracy and easy integration into the control system is the advantage of the valve technology [2, 8]. The resistance control is typically realized using the spool, which is placed in the control valve and creates four synchronously operated hydraulic resistances. The disadvantage of the valve-controlled drives is the losses due to the pressure drop across the valve – the control resistances and also the asymmetric behaviour of the drive by the use of the differential cylinder [6].

In order to eliminate the disadvantages of the valve-controlled cylinder, the full hydraulic resistance bridge was realized by means of the four proportional flow control poppet valves [5]. The manifold with these four poppet valves was designed as compact as possible to the hydraulic cylinder, and in this way, the integrated hydraulic actuator was developed [7]. The use of the full hydraulic bridge realized using the independently controlled four flow control valves, and the application of the integrated hydraulic actuator in the test rig for spring stiffness measurement is described in more details in the next chapters.

2 Test Rig Description

2.1 Structure of the Test Rig

The developed test rig with hydraulic integrated actuator for spring stiffness characteristic measurement is shown in Fig. 1. The test rig has a frame welded from the I-sections, which is holding the hydraulic actuator in a vertical position [10]. The piston of the actuator is going down by the extending motion and after achieving the upper end of the spring starts to compress the tested spring.

The compression of the spring is done very slowly to obtain correct data for the followed evaluation of the spring characteristic and stiffness calculation. The cylinder is equipped with the integrated linear position sensor and with the pressure sensors for the measurement of the pressures in the cylinder chambers. The force transducer is installed on the piston rod.

The structure of the hydraulic circuit and the connection of the actuator to the control system is shown in Fig. 2. The movement of the piston rod is controlled using the full hydraulic resistance bridge realized using four proportional solenoid – operated valves. These valves are installed in the manifold which is placed on the hydraulic cylinder very close to the chambers of the hydraulic actuator to make the hydraulic capacities as small as possible. The control valves and installed sensors are connected to the control system ECU-0710, Fig. 2. The ECU-0710 is a general-purpose programmable controller and can be used as a stand-alone controller, or for integrating into a CAN network with other devices [9]. The ECU-0710 controller is connected to the PC, which allows to program

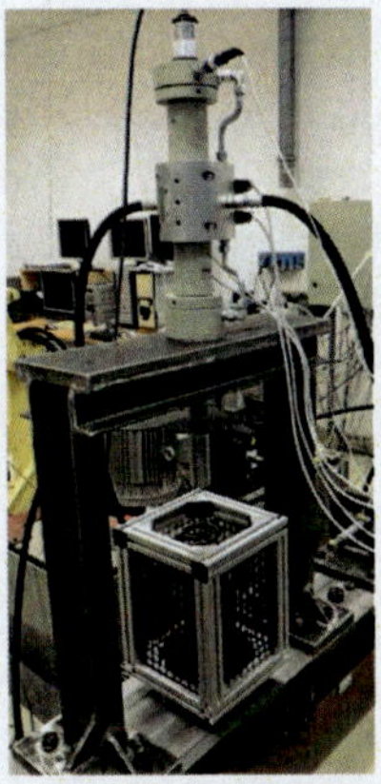

Fig. 1. Test rig with the integrated hydraulic actuator for spring stiffness measurement.

the ECU, collect and evaluate the obtained data and finally to operate with the system in automatic testing mode.

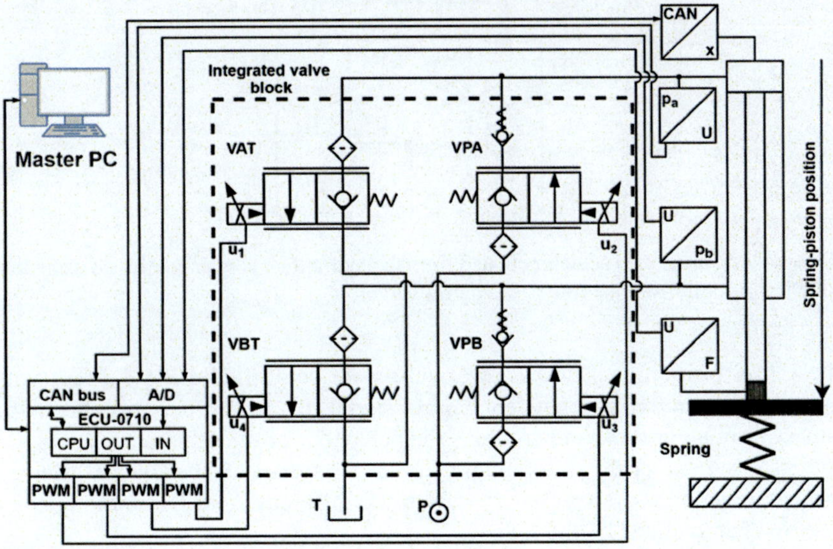

Fig. 2. Connection of the hydraulic actuator and sensors to the control system ECU-0710.

2.2 Testing Cycle

The ECU reads the piston position from the position sensor [9]. The spring compression required is entered by the user as an input. Using this compression required value, the controller calculates the desired piston position. Depending on the actual piston position, the ECU gives the control signal to the integrated valves accordingly. If the piston needs to move from the upper position to the lower position, Fig. 2, then the ECU gives the command signal to input valve VPA and output valve VBT. If the piston needs to move upwards, then the command is given to input valve VPB and output valve VAT. This is how the piston moves to the desired position. Along with position control, we control the velocity of the piston at any given time of spring compression. This is done to reduce the risk of the spring flying away and to get better dynamic results from the sensor.

3 Hydraulic Drive

3.1 Hydraulic Actuator with Integrated Control Valves

The actuator creates a differential hydraulic cylinder with one side piston rod controlled using the four proportional flow control valves connected in the full hydraulic bridge-type A+A, Fig. 3 [1–5].

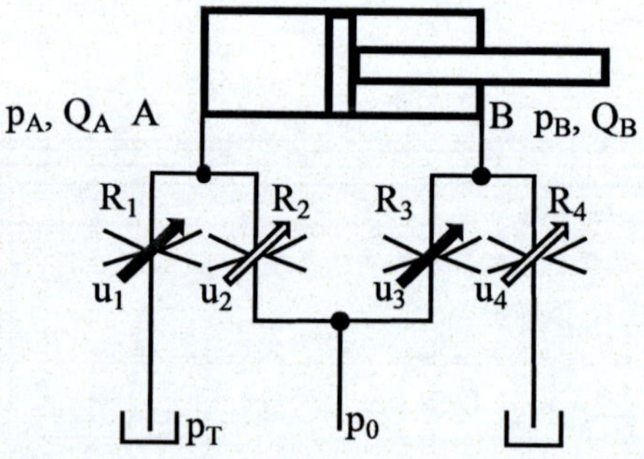

Fig. 3. Hydraulic integrated actuator created from the hydraulic cylinder controlled using the full hydraulic bridge built from the four flow control valves.

These four proportional flow control valves are controlled using the four voltage inputs, i.e. using the four independent control signals u_1, u_2, u_3, u_4. This allows to use different control algorithms and optimize the behaviour of the actuator.

The disadvantage of the differential cylinder is asymmetric behaviour. The reason for that is the geometrical asymmetry due to the different areas S_A in the piston and ringside S_B characterized by the ratio

$$\alpha = \frac{S_A}{S_B} \quad (1)$$

The asymmetric construction of the cylinder leads to the different hydraulic capacities of the cylinder chambers and asymmetric velocity-load force characteristic, which causes the different piston velocity by extending and retracting of the piston of the cylinder controlled in the open loop. It means that the velocity gain of the open-loop system depends on the direction of the motion – piston velocity [1, 3, 6]. Next property which should be taken into account by the design of the drive based on the differential cylinder is the course of the pressures in the cylinder chambers A and B and the limits for the pulling and pushing external force acting on the piston rod during the motion. The maximal allowed values of the load forces depend on the motion direction. The analysis of the static and dynamic properties of the differential actuator and their compensation by the control using the full hydraulic Wheatstone bridge was done and can be found in [1, 3, 6–8]. The fact that each orifice in the hydraulic resistance bridge can be controlled separately allows to achieve the compensation of the pressure asymmetry in the cylinder chambers instead of using the valves with the special spool with different flow areas S_{VA}, S_{VB} characterized by the ratio

$$\beta = \frac{S_{VA}}{S_{VB}} \tag{2}$$

The control of the separate proportional valves using the suitable generated input signals allows to set the valve opening, i.e. cross-section areas of the corresponding valves to the given ratio β or to the similar value or keep some valve fully opened and control the flow and movement only using one resistance. The full hydraulic resistance bridge offers a significant potential for different control algorithms, and improvement of the properties of the resistance controlled differential cylinder.

4 Simulation of the Testing Cycle

4.1 Simulation Model of the Test Rig

The simulation model of the whole test rig was built in the simulation program MATLAB-Simulink. The simulation model is based on the mathematical models of the hydraulic cylinder, proportional flow control poppet valves, connections, models of the transducers and finally on the model of the tested spring. After adding the controller, it allows simulating the whole testing cycle [6–8]. The hydraulic cylinder is placed in a vertical position, and the force of gravity should be taken into account. When the spring force acts on the piston the motion equation for the piston position x has the form

$$m\ddot{x} + b\dot{x} = S_A p_A - S_B p_B - F_s - F_T \operatorname{sgn}(\dot{x}) + F_g \tag{3}$$

Where $F_s = k\Delta x$ is the force of spring depending on the spring stiffness k and spring compression Δx, F_g is the gravitational force, F_T is the Coulomb friction, m is the mass, B viscous friction coefficient, S_A and S_B are the surface area of the piston side and rode side.

Pressures p_A, p_B in the cylinder chambers are given by the differential equations for the hydraulic capacities C_A and C_B

$$\dot{p}_A = \frac{1}{C_A}(Q_A - S_A v - Q_{LA} - Q_L) \tag{4}$$

$$\dot{p}_B = \frac{1}{C_B}(S_B v - Q_B - Q_{LB} + Q_L) \tag{5}$$

$$C_A = \frac{V_{A0} + Sx}{K}, \quad C_B = \frac{V_{B0} + S(H - x)}{K} \tag{6}$$

The poppet valve is modelled as a variable orifice created by a cylindrical sharp-edged stem and conical seat [11]. The flow through the poppet valve is proportional to the valve opening and nonlinear depending on the pressure drop p_i across the valve

$$Q_i = C_D * A(h_i) * \sqrt{\frac{2}{\rho}} * \left(\frac{\Delta p_i}{(\Delta p_i^2 - p_{cr}^2)^{1/4}} \right) \tag{7}$$

Where i = PA, PB, AT, BT, $A(h_i)$ is the passage area depending on h_i. The minimum pressure for turbulent flow p_{cr} is given by

$$\text{pcr} = (\text{pavg} + \text{patm})(1 - \text{Blam}) \tag{8}$$

where p_{avg} is average pressure from pressures on the inlet and outlet of the valve, p_{atm} is atmospheric pressure, and B_{lam} is constant [11]. The poppet dynamics was modelled using the second-order differential equation for the valve opening h_i in dependence on the input voltage u_i.

The parametrization of each component of the simulation model was done using the available data from the datasheet of the producer.

4.2 Simulation of the Testing Cycle

Using the built and verified simulation model, the whole testing cycle was simulated. The courses of the important state variables of the hydraulic actuator – piston position x,

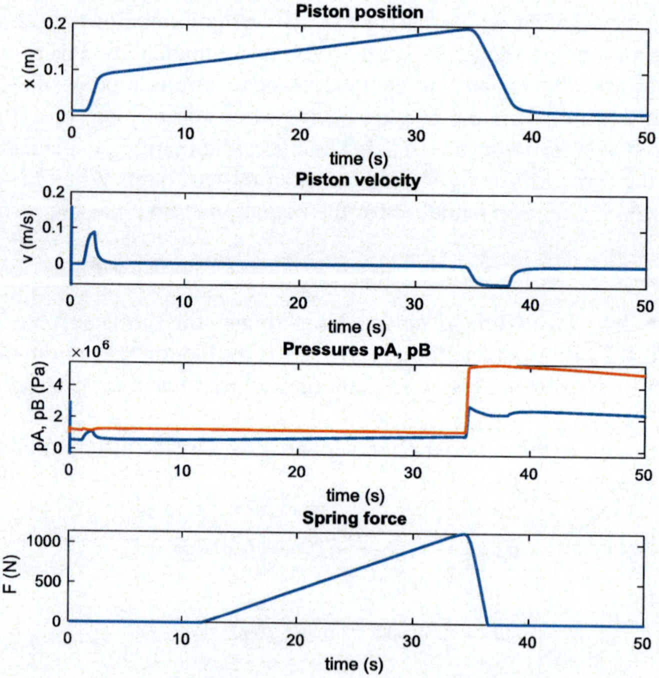

Fig. 4. Simulated state variables during the testing cycle, $\beta = 1$.

piston velocity v, pressures p_A, p_B in the cylinder chambers and the spring force during the testing cycle by control using the same four control signals, i.e. the coefficients β are equal to one for each valve, are shown in Fig. 4.

The courses of the pressures p_A and p_B are typical for the differential cylinder and extension motion of the piston. Especially at the beginning, when the force of gravity is pulling the piston rod and later when the additional force of the compressed spring acting in the opposite direction is small, the pressures are very low and close to zero. Figure 5 shows the simulation results for β equal to $\alpha = 0.48$.

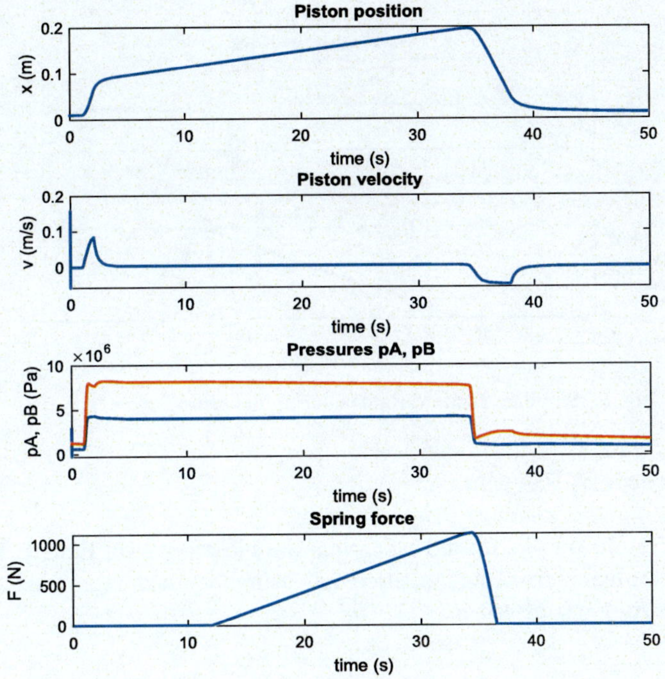

Fig. 5. Simulated state variables during the testing cycle, $\beta = 0.48$.

The impact of the different control signals to the valves VPB and VBT is observable from the pressures p_A, p_B, whose values are significantly higher and are getting closer to the value of the system pressure.

Figure 6 shows the simulation results for β equal to 0.7. The opening of the valves VPB and VBT is about 70% of the other valves, and the chamber pressures are higher than in the first case for β equal to one, but the pressures did not achieve the values as in the case for β equal to $\alpha = 0.48$.

Separate control of each proportional flow control valve makes it possible to use different valve opening for each valve, i.e. to set different values of the ratio β in specific phases of the working cycle of the machine and in this way to optimize the behaviour and also the energy efficiency of the installed hydraulic system.

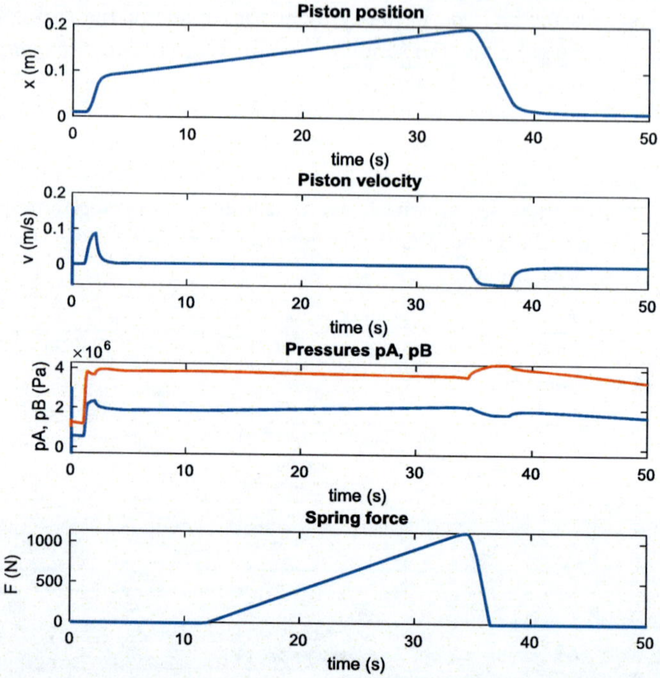

Fig. 6. Simulated state variables during the testing cycle, $\beta = 0.7$.

5 Measurement Results

The obtained results for three tested car springs are summarized in Table 1. The stiffness constant was evaluated from the measured data using the linear regression as a slope of the obtained spring characteristic.

Table 1. List of calculated spring constant stiffness

Stiffness constant of spring No. 1 (N/m)	29410
Stiffness constant of spring No. 2 (N/m)	14350
Stiffness constant of spring No. 3 (N/m)	18890

The achieved results by the testing of the car spring No 2 are shown in Fig. 7 and Fig. 8. It is observable that the part of the test during the spring compression between the spring-piston position is equal to approximately 0.42 m and 0.5 m on the measured course Fig. 7 (values are in a different coordinate system than the piston position x, see Fig. 2). This part corresponds to the simulated process and piston displacement x in Figs. 4, 5 and 6. The graph in Fig. 8 is showing the values of the piston displacement (spring compression) and spring force and the evaluated spring characteristic.

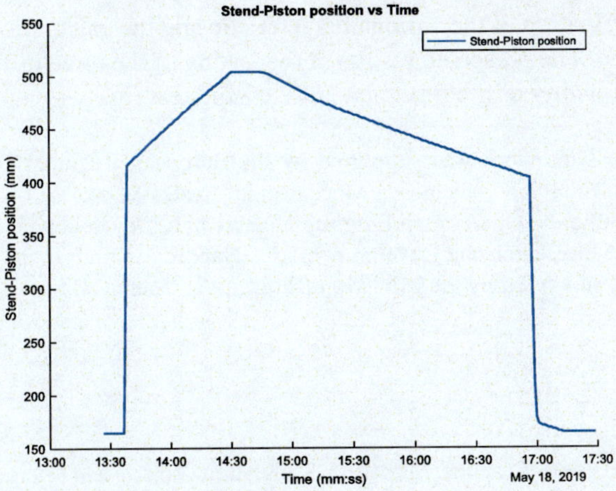

Fig. 7. Measured spring-piston position in the test rig due to the testing cycle.

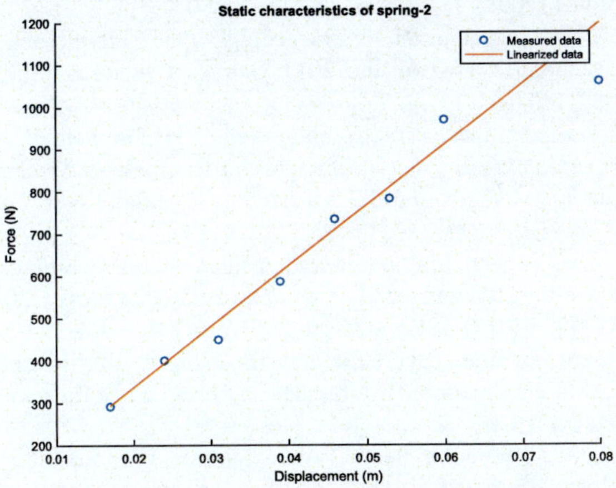

Fig. 8. Measured values of the spring force in the dependence on the spring compression and calculated stiffness characteristic of the spring.

6 Conclusions

The test rig and the automatic testing cycle for the evaluation of the spring characteristic and the spring stiffness constant was introduced in the paper. The tested spring is compressed using the hydraulic actuator with integrated full hydraulic resistance bridge realized using the proportional flow control valves. The fact that the control of the movement of the piston is done using the four independent control signals instead of one input signal in case of the use of the traditional 4/3 proportional flow control valve allows to optimize the allowed pulling force and the course of the pressures in the cylinder chambers. This is the advantage of the use of the full hydraulic bridge of A-A type. The control of the drive was realized using the ECU-071 controller, which was connected to the PC where the final evaluation of the spring characteristic and the calculation of the stiffness constant were done. The presented results of the testing of the three different car springs confirm the functionality of the designed and realized test rig.

Acknowledgement. This work was supported by the European Regional Development Fund in the Research Centre of Advanced Mechatronic Systems project, CZ.02.1.01/0.0/0.0/16_019/0000867 within the Operational Programme Research, Development and Education and the project SP2020/57 Research and Development of Advanced Methods in the Area of Machines and Process Control supported by the Ministry of Education, Youth and Sports.

References

1. Watton, J.: Fluid Power Systems. Modeling, Simulation, Analog and Microcomputer Control. Prentice Hall International (UK) Ltd. (1989). ISBN 0-13-323197-6
2. Backé, W.: Servohydraulik. RWTH Aachen (1984)
3. Murrenhoff, H.: Servohydraulik - Geregelte hydraulische Antriebe. IFAS RWTH Aachen. Shaker Verlag GmbH (2008). ISBN 978-3-8322-7067-4
4. Gundlach, M.: Efficient regeneration. In: Proceedings of the 21st International Conference on Hydraulics and Pneumatics. 1st–3rd June 2011, Ostrava, Czech Republic, pp. 21–28 (2011). ISBN 978-80-248-2430-7
5. Koňařík, P.: Syntéza řízení mechatronického systému s využitím simulačních metod a prostorových modelů virtuální reality. (Control Algorithm Synthesis of Mechatronic System with the use of Simulation Methods and 3D Virtual Reality Model's. In Czech). Dissertation. VŠB-Technical University of Ostrava (2009)
6. Noskievič, P.: Control of asymmetric hydraulic actuator. In: Proceedings of the 22nd International Conference on Hydraulic and Pneumatics. 23–25 October 2013, Prague, Czech Republic, pp. 113–120 (2013). ISBN 978-80-248-3136-7
7. Noskievič, P.: Control of linear hydraulic actuator using the full hydraulic bridge, 6p. In: Proceedings of the 19th International Conference on Research and Education in Mechatronics, 7–8 June 2018, Delft (2018)
8. Jelali, M., Kroll, A.: Hydraulic Servo-systems: Modelling, Identification and Control. Springer, Heidelberg (2004). ISBN 1852336927
9. Zaid, A.A: Control design of the hydraulic actuator with integrated control valves. Diploma thesis, Faculty of Mechanical Engineering, VŠB-Technical University of Ostrava (2019)
10. Chandra Mohan, Y.S.: Design of the test rig of the integrated actuator. Diploma thesis, Faculty of Mechanical Engineering, VŠB-Technical University of Ostrava (2019)
11. Mathworks. Poppet valve. https://www.mathworks.com/help/physmod/hydro/ref/poppetvalve.html. R2020a
12. Garbacik, A., Stecki, J.: Developments in Fluid Power Control of Machinery and Manipulators. Fluid Power Net Publication. Cracow (2000). ISBN 83-86219-71-8

Hydrostatic Actuator Drive Control with Pump Leakage Compensation

Łukasz Stawiński(✉), Andrzej Kosucki, and Adrian Morawiec

Institute of Machine Tools and Production Engineering, Lodz University of Technology, Stefanowskiego 1/15, 90-924 Łódź, Poland
lukasz.stawinski@p.lodz.pl

Abstract. The paper presents research of the hydraulic cylinder drive system equipped with gear pump driven by asynchronous motor fed by the frequency inverter using a developed method of pump leakage compensation (LPC) to control the flow. The pump parameters at various pressures and various angular speeds of the asynchronous motor, including in the speed range below the minimum recommended by the pump manufacturer, have been tested. The obtained pump characteristics showed the presence of two components of internal pump leakage - independent and dependent of pressure. The linear internal leakage model was developed for the pump. This model has been implemented to the verified model of the drive system of the variable ratio mechanism driven by a hydraulic cylinder - scissor lift. The pump flow correction algorithm has been added to the previously developed hydraulic cylinder speed control system. This correction was aimed at improving the mapping of the assumed speed of the hydraulic cylinder. The compensation block has been added to the hydrostatic drive of the scissor lift equipped with an open control system - developed in the LabView environment. Experimental tests confirmed the improvement in mapping of the hydraulic cylinder assumed speed regardless of its variable load resulting from the scissor system geometry and the weight of the load on the scissor lift platform. The developed leakage compensation system on the pump can be used in hydrostatic systems with fixed or variable ratio, regardless of the load on the hydraulic cylinder.

Keywords: Pump leakage compensation · Volumetric efficiency · Flow rate · Velocity control

1 Introduction

Hydraulic actuators are used in mobile machines and stationary drives, due to their good capabilities in positioning, fast and smooth response, high power density and compact size. Using hydraulic actuators for constant speed operation is highly required in many applications, such as lifting, pressing, cutting, automatic production and assembly, etc. However, maintaining a constant hydraulic cylinder speed can be problematic due to many factors. These include, among others, variable load on the hydraulic cylinder, pressure drop, changes in hydraulic oil viscosity due to temperature and leakage in

© The Editor(s) (if applicable) and The Author(s), under exclusive license
to Springer Nature Switzerland AG 2021
J. Stryczek and U. Warzyńska (Eds.): NSHP 2020, LNME, pp. 179–191, 2021.
https://doi.org/10.1007/978-3-030-59509-8_16

hydraulic components. To minimize the impact of these factors on the speed of the hydraulic actuator, two control methods are used - flow control or pump control. The flow control method most often uses electronically controlled valves. These are proportional directional valves or servo valves. Flow regulators are also used to eliminate the impact of pressure changes on the speed of the hydraulic cylinder. The pump control method uses variable displacement pumps or fixed displacement pumps with a variable speed motor (VSPC – variable speed pump control).

An example of flow control is application of a variable displacement pump driven by a motor with inverter is used in hydraulic excavator to meet the requirements of high dynamic, high energy efficiency and zero emission of hydraulic excavator under different working conditions, following [1, 2].

Electrohydraulic Fuzzy-PID position control system by compensating for leakage through proportional flow control valve was presented in [3]. The main goal for this is reducing the energy consumption and improving the position control of the cylinder. However, it is not the simplest and low-cost solution.

A cheaper option is the VSPC and appropriate change of leakage through the proportional throttle valve, described in [4]. The method combines pump control and throttle valve control using valve-pump weight ratio. Authors mentioned that, in order to make the system have higher response while keeping high efficiency, the valve-pump weight ratio should be reasonably set.

The combination of these two solutions is the use of a variable displacement pump driven by motor with inverter and a proportional valve. The authors [5, 6] show the behavior of the pump and the system for changes in the flow rate in the system. The leakage compensation has been shown to be achieved by either controlling the proportional electro-hydraulic valve or controlling the pump speed in system with constant load.

The applicability of a low-cost drive using a speed-controlled motor in combination with a fixed displacement pump applied in a load-sensing control strategy is described in [7]. The main achievements were low energy losses, low noise emission, improved control dynamics, reduced steady-state error.

In mobile machinery, when we are dealing with combustion engine authors proposed a new lift control strategy based on the hydraulic supply pressure and flow control [8]. The hydraulic pump speed is precisely controlled using a two-input gearbox mechanism. This eliminates the need for precision control of the solenoid valves opening interval within every cycle.

Analyzing the literature presented above, it was found that the cheapest and most effective method of maintaining a constant speed of a hydraulic cylinder is to use the VSPC. The fixed displacement gear pumps are very often used in hydrostatic drive systems. They are installed in systems with throttled control of the movement of actuators. A few research projects have been conducted to define volumetric efficiency of gear pump models [9–11]. However, the studies were limited only to the rotational speeds of pumps recommended by manufacturers, mainly in the range from 500–700 to 3000 rpm. Even in newly designed test stands the adjustment of pump speed is not considered which can be confirmed by [12], where the authors designing the test rig for the volumetric efficiency of the gear pump, did not foresee the possibility of alternation of pump speed.

During modelling of hydrostatic drives, according to [13, 14], the flow is calculated using the theoretical flow decreased by leakage which is defined as proportional to the pressure drop only.

Nowadays almost every inverter has a vector control mode (FOC) which provides constant motor speed regardless of its load. However, this function is not often used in industrial solutions and electric motors are running in V/f mode, which is sensitive to low speed and may not maintain a desired torque. The result is a decrease in motor speed due to increased load. This will not provide the desired flow from the pump and will affect the speed of the hydraulic cylinder.

The research on motor-inverter systems has been conducted for many years in the team of Working Machines, Drives and Controls at the Technical University of Lodz. Some of them concern hydrostatic systems, such as hydrostatic transmissions, hydraulic-rope lifts, scissor lifts and variable load mechanisms. Research includes the impact of inverters on reducing vibration and overloads, pressure drops and maintaining constant speed of receivers, regardless of their load [15, 16].

The use of inverters in hydraulics is increasingly common. The objective of [17] is to study possibilities of reducing the flow ripple by adjusting the speed of the pump by utilizing the fast torque control of the electric motor. The results show that a proper control method of the electric drive reduces the flow ripple in a hydraulic system and open-loop control of the hydraulic system is possible.

This paper presents experimental tests of the constant speed of a hydraulic actuator with the use of a system consisting of the VSPC using a pump leakage compensation system (LPC). Leakage from the pump have been eliminated by shaping the speed of the electric motor using an inverter and external controller. The scissor lift mechanism was used for testing, in which hydraulic cylinders are subjected to changing force during lifting. The main objective is to compensate the leakage of the pump that leads to sudden change in the hydraulic actuator speed and to maintain constant operation parameters, regardless the pressure. The concept of the system is shown in Fig. 1.

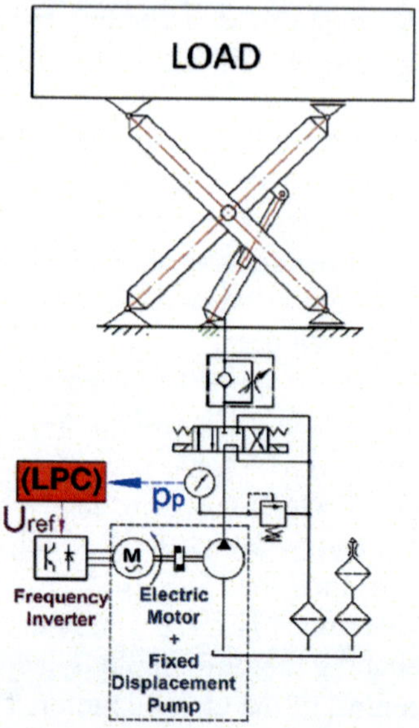

Fig. 1. Concept of the system with LPC leakage pump compensation system.

2 Leakage Compensation System Description

To develop LPC for the system, a series of pump (XV-0R/0.98, were displacement q_P is equal to 0.92 cm^3/rev) tests, have been conducted. The hydraulic scheme of the pump test stand has been presented in Fig. 2. The measurements were registered for rotational speed of the pump n_P between 100 and 1500 rpm, what corresponds to the theoretical flow from 0.092 to 1.38 l/min, following the formula:

$$Q_t\left[\frac{l}{min}\right] = \frac{q_P \cdot n_P}{1000} \tag{1}$$

The load of the pump was set from 0 to 16 MPa. The experiment was performed using HL-46 hydraulic oil, operating in temperature 35–40 °C. The temperature effect was omitted because of its low influence (experimentally confirmed) on pump efficiency in range 35 to 60 °C. The theoretical pump flow Q_t was compared with actual measurements Q and a flow loss $\Delta Q = Q_t - Q$ diagram for the pump was obtained (Fig. 3). Dashed lines indicate speeds below the pump manufacturer's recommended speed (below 700 rpm). These tests have shown that the pump has significant internal leakage and is dependent on two factors. One of them is the load due to the pressure at the outlet of the pump, and the other is a constant value, depending on the design of the pump.

It was assumed that the knowledge of the nature and the level of leakage in the pump allows to correct the control input signal in order to improve mapping of the actual flow. The equation of flow losses as a function of pump pressure was determined for synchronous speed. Factors A and B for the 1500 rpm are shown in Fig. 4, following by the formula in SI system:

$$\Delta Q\left[\frac{m^3}{s}\right] = f(p_P) = A\left[\frac{m^5}{N \cdot s}\right] \cdot p_P[Pa] + B\left[\frac{m^3}{s}\right] \tag{2}$$

The correction consisted in adding to the inverter control signal the certain value of voltage, calculated according to the formula (3), where ΔU is a value of the control signal correction, U_{max} [V] is a maximum value of the control signal corresponding to the maximum pump speed, Q_{max} [l/min] is theoretical maximum pump flow corresponding to its maximum speed and ΔQ [l/min] is a constant component of leakage.

$$\Delta U[V] = \frac{U_{max}}{Q_{max}} \cdot \Delta Q \tag{3}$$

The diagram of the LPC is shown in Fig. 5. The block 1 represents a computer equipped with control-measurement card. The block was used to determine the inverter input signal U(t) [V] and to generate the corrected input signal U(t) + ΔU. The block 2 shows the vector-controlled inverter with feedback from the motor driving the pump (Uω). The block 3 represents the unit consisted with the electric motor and the gear pump – the flow Q generator. The software used to control the system was developed in the LabView environment. The same software allowed the acquisition of data necessary to present the following charts.

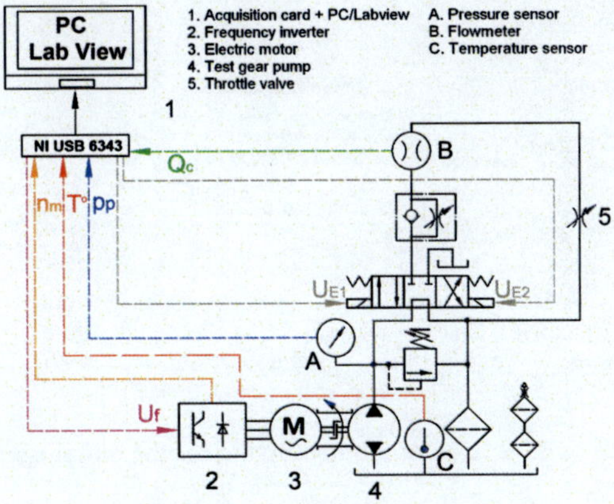

Fig. 2. The pump test stand.

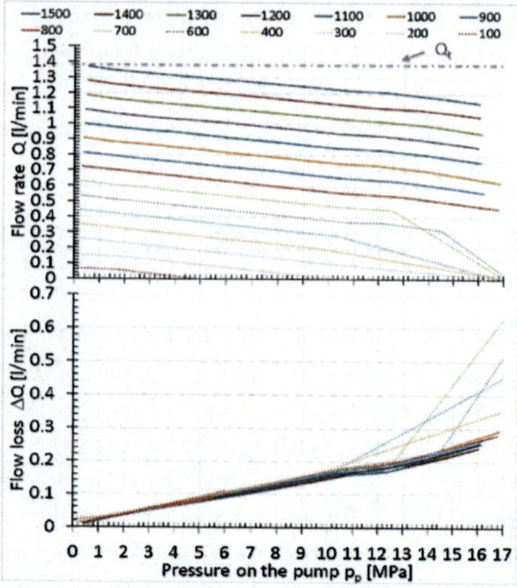

Fig. 3. Theoretical flow Qt, actual Q and leakage ΔQ of gear pump PGP505 for various rotational speeds and pressures.

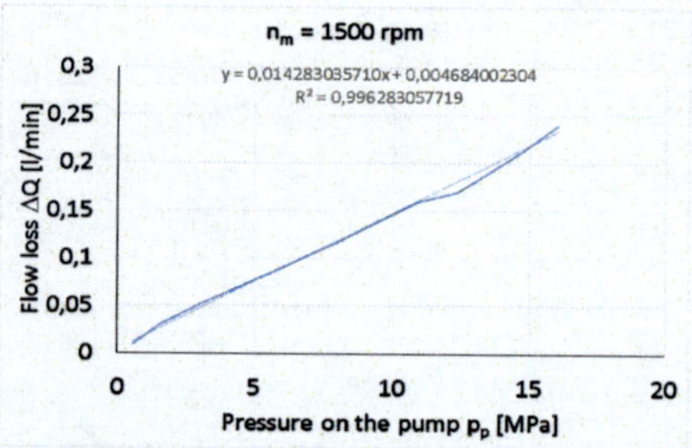

Fig. 4. Flow losses as a function of pump pressure for the motor speed of 1500 rpm.

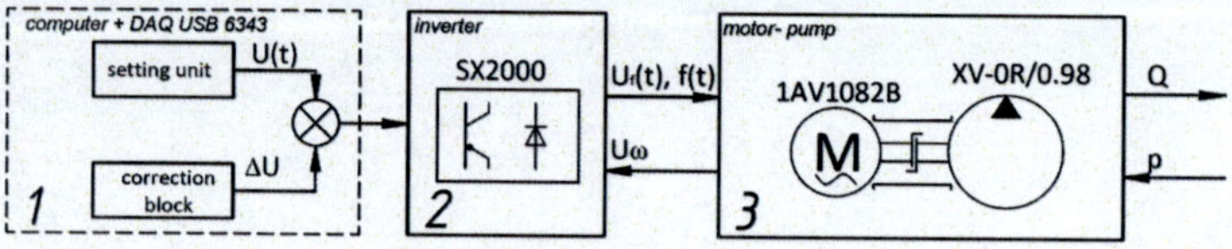

Fig. 5. Diagram of the LPC.

3 Experimental Test Rig

Experimental tests were carried out on a specially modernized scissor lift, which in addition to the conventional one, was equipped with elements of the proposed LPC (Fig. 6). A set of sensors gives the possibility of registering the operating parameters of the systems in different conditions. The signal from the pressure sensor is directed to the LPC. The technical parameters of the main elements of the test stand in Table 1.

Table 1. Test stand setup.

No	Component	Parameters
1	PC + LabVIEW Software + NI USB 6343	16 AI, 2 AO, 24 DIO USB
2	Frequency Inverter SX2400-0R7G-2	0.75 kW/0–10 V
3	Motor Simotics GP 1AV1082B	0.55 kW/1385 rpm
4	Pump XV-0R/0.98	0.92 cm^3/rev
6	Hydraulic cylinders 40-25-300	40-25-300 mm
A	Pressure sensors P3297B084001	25 MPa/4–20 mA
B	Flowmeter GFM-5	0.05–2 l/min/5.250 pulses/l
C	Length sensor EMAX-000-01.5-2-CAO	1.5 m/125 kBit/s

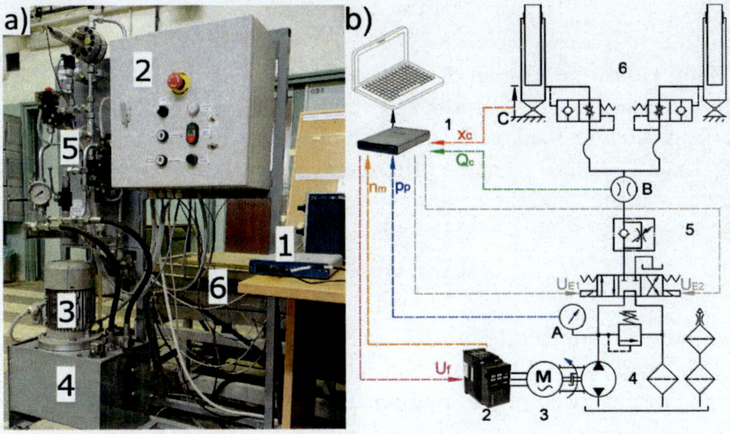

Fig. 6. Experimental test rig: a) test stand, b) schematic diagram of the hydraulic and sensor circuits.

4 Experimental Results

The objective of the experiments is to validate the LPC for different test cases covering the operating range of the drive. To reduce oil pulsation, the start ramp on the inverter has been set to 5 s. The voltage reference functions are carried out in four cases:

Uf = 10 V which is equal to 1500 rpm (157 rad/s) of electric motor rotational speed without leakage compensation – index NO.
Uf = 10 V + ΔU_A which is responsible for the pressure dependent correction factor in the leakage compensation system – index A.
Uf = 10 V + ΔU_B which is responsible for the pressure independent correction factor in the leakage compensation system – index B.
Uf = 10 V + ΔU_{AB} which corresponds with the total leakage compensation system – index AB.

All test compared the behavior of flow rate in the system Q, the hydraulic actuator speed v_c and calculations of changes in volumetric efficiency η_v during correction.

The research was divided into two stages. The first assumed comparing the behavior of the pump flow rate, and the second was the behavior of the hydraulic cylinder.

4.1 Pump Flow Rate Tests

The tested pump is XV-0R/0.98 with a displacement $q_p = 0.92\ cm^3/rev$. The pump manufacturer states that the flow rate of this pump Q_p at speed $n_p = 1500$ rpm and pressure in the pump $p_p = 0,05$ MPa is equal to 1.311 l/min. Theoretically, if the volumetric efficiency of the pump would be 100%, the flow rate Q_t should be equal to 1.38 l/min, follow the formula (1).The volumetric efficiency of the pump η_v in this case is equal to 95% according the formula below:

$$\eta_v[\%] = \frac{Q_p}{Q_t} \cdot 100\% \tag{4}$$

If the pump is powered by an electric motor supplied directly from the mains, it should be remembered that it will work according to its mechanical characteristics. This means that the greater the load on the motor shaft, the lower its speed. The torque on the motor/pump shaft T_p comes from the differential pressure at the pump inlets $\Delta p_p = p_p - p_0$ (where p_0 – is the max. inlet suction pressure 0.1 bar), according to the formula:

$$T_p[Nm] = \frac{q_p \cdot \Delta p_p}{2 \cdot \pi \cdot \eta_{hm}} \tag{5}$$

With the motor manufacturer's data such as synchronous speed $n_s = 1500$ rpm (50 Hz), rated speed $n_m = 1385$ rpm and rated torque $T_m = 3,79$ Nm, it is possible to calculate the actual speed of the motor and pump shaft n_p according to the formula below:

$$n_p[rpm] = n_s - \frac{T_p}{T_m} \cdot (n_s - n_m) \tag{6}$$

If the pressure on the pump changes, the pump speed will change accordingly, which will change the pump flow rate Q_v. Referring to the theoretical volumetric efficiency of the pump, a significant decrease in the efficiency of the pump driven by a motor powered directly from the mains has been noticed, as depicted in Fig. 7.

The FOC alone maintains a constant flow rate, almost regardless of load, but this level only indicates the value provided by the pump manufacturer, as depicted in Fig. 8.

Comparison of the flow rate from the pump, in different cases, is shown in Fig. 9. The pump has been tested with speed 1500 rpm on the experimental test rig, described above, where the pressure changes when the actuator moves. This is due to the variable load on the actuator. It is caused by the variable ratio of the scissor mechanism speed. The scissor lift has no load on the platform, however the weight of the structure caused changes in pressure in the system during lifting. This can be seen in the variable Q_v waveform, where the motor was powered directly from the mains. The flow value in this case is much lower than that given by the manufacturer. The LPC increases the value of the flow rate in relation to the system without compensation. The factor B gives the least impact. This is due to the small size of the pump. The factor A, depending on the changing pressure, strongly increases the flow value.

The LPC significantly increases the flow rate from the pump (Fig. 10). Factor A increases the flow value, reducing the effect of pressure impact on pump efficiency. The factor B raises the flow by a constant value. By using the LPC, it was possible to virtually increase the volumetric efficiency of the pump. Comparison of pump volumetric efficiency under the load with various variants of pump compensation is shown in Fig. 11. The volumetric efficiency of the pump decreases as the load increases. When using the LPC, the higher the load, the closer to the theoretical pump flow rate. This is mainly due to the influence of the A factor, which depends on the pressure.

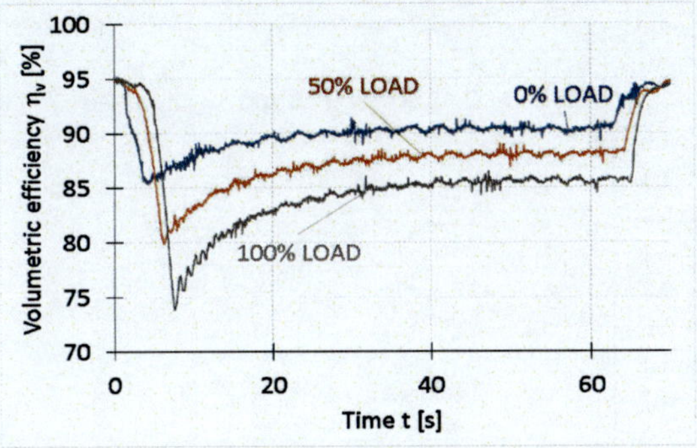

Fig. 7. Pump volumetric efficiency with different loads. Motor driven directly from the mains.

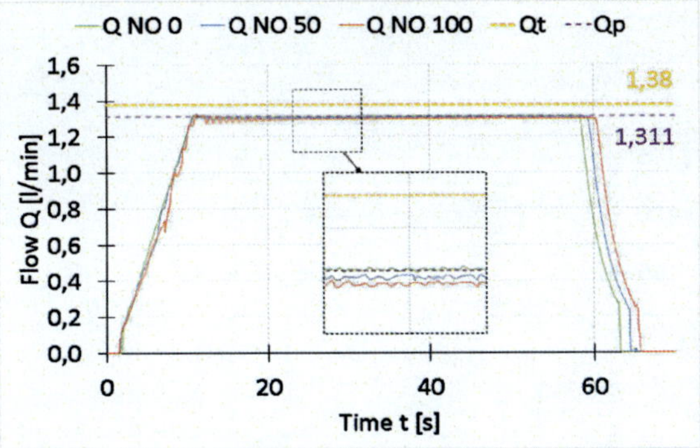

Fig. 8. Impact of load on pump flow rate. Motor fed by FOC, without LPC.

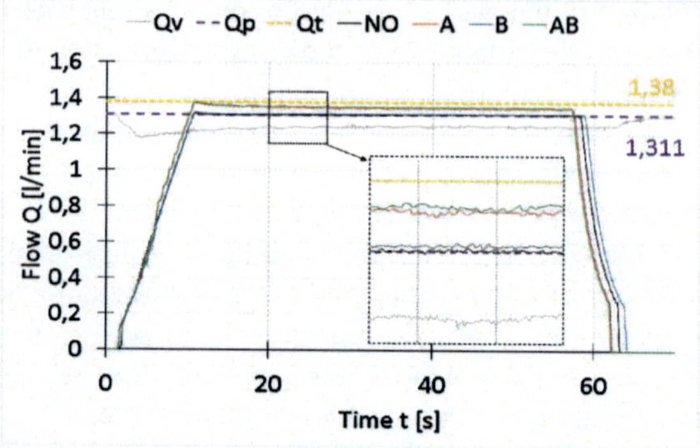

Fig. 9. Comparison of the pump flow in different cases.

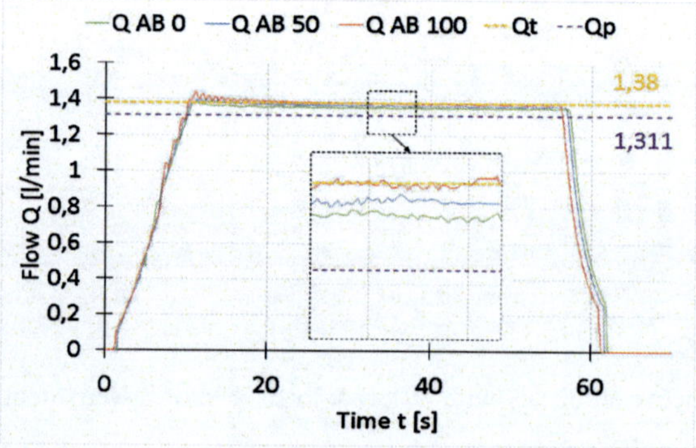

Fig. 10. Impact of load on pump flow rate with full LPC.

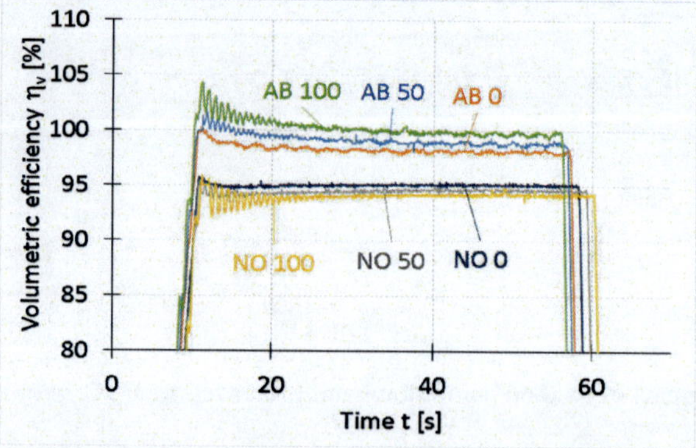

Fig. 11. Comparison of pump volumetric efficiency for different load with and without LPC.

4.2 Hydraulic Actuator Velocity Tests

The tests were carried out without and for three types of compensation (A, B and AB) with the hydraulic actuators loaded with the weight of the structure and the load on the platform in the discrete interval $load_{ext}$ = [0, 50, 100]% while lifting with the same stroke $x_s = 0.24$ m. The ramps on the inverter has been set on 5 s.

The velocities of the actuator depending on the flow from the pump with no load on the platform has been depicted in Fig. 12. The theoretical value of velocity was marked as vt, where pump would not have losses. The line marked as vp is the value that the actuator should reach according to the given pump flow for the engine speed of n_m = 1500 rpm, according to the pump manufacturer's data. Both values have been calculated for 2 cylinders with nominal diameter of the piston each $D_c = 40$ mm and displacement of the pump q_p = 0,92 cm^3/rev following the formula:

$$v_c = \frac{Q_c}{A_c} = \frac{q_p \cdot n_m}{1000 \cdot \frac{\pi \cdot (D_c)^2}{4}} [m/s] \tag{7}$$

Other speed values were measured by a displacement sensor located on the actuator. There are no major differences in velocity levels between the system with and without compensation. The only difference appears in the actuator velocity when the pump is driven by a motor connected directly from the mains (without FOC).

The actuator velocity in the system with LPC, with different loads, has been presented in Fig. 13. The actuator velocity curve without LPC at full load has been added to highlight the difference (NO 100). The use of LPC increases the actuator velocity. Despite the variable load, the velocity of the actuator in the system with LPC is higher than calculated based on the pump manufacturer data (vp). Furthermore, the actuator velocities exceed the theoretical velocity vt. The reason for this may be the inaccurate actuator design. A small change in the diameter of the actuator significantly changes its speed.

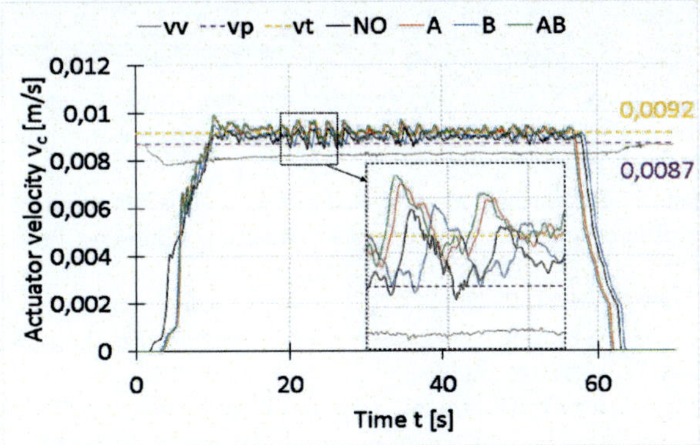

Fig. 12. Comparison of actuator velocities in different cases with no load on the platform.

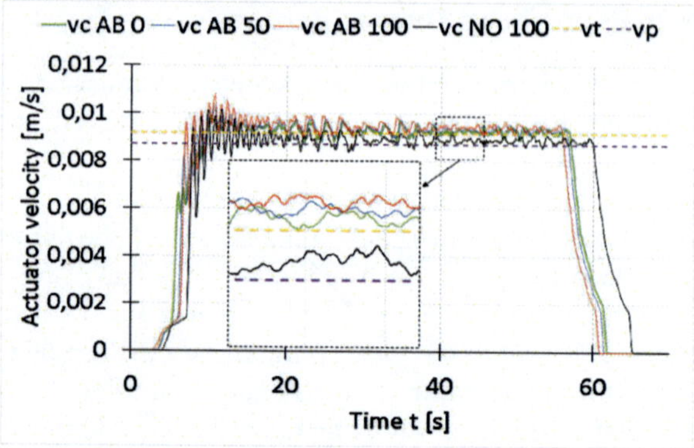

Fig. 13. Comparison of actuator velocities with LPC and different load.

5 Conclusion

In this paper, the compensation method of pump leakage and its impact on the operation of the hydrostatic system with an actuator loaded with variable force is presented. The characteristics of gear pump covering full range of speed and allowable pressure have been obtained. It allowed to determine two factors (pressure dependent and independent) responsible for the pump's leakage.

These factors were used to develop the LPC system, correcting oil flow using a simple control function of speed of the pump driven by the electric motor fed by an inverter.

Experimental studies of the behavior of a gear pump driven by a motor connected directly from the mains and fed by an inverter with a vector system (FOC) were carried out. The use of the FOC allows to maintain a constant pump speed regardless of pressure. In addition, the use of LPC increases the flow rate to the theoretical value of the pump, maintaining a constant actuator velocity.

References

1. Ge, L., Quan, L., Zhang, X., Zhao, B., Yang, J.: Efficiency improvement and evaluation of electric hydraulic excavator with speed and displacement variable pump. Energy Convers. Manag. **150**, 62–71 (2017)
2. Wrat, G., Ranjan, P., Bhola, M., Mishra, S.K., Das, J.: Position control and performance analysis of hydraulic system using two pump-controlling strategies. Proc. Inst. Mech. Eng. Part I: J. Syst. Control Eng. **233**(9), 1093–1105 (2018)
3. Wrat, G., Bhola, M., Ranjan, P., Mishra, S.K., Das, J.: Energy saving and Fuzzy-PID position control of electro-hydraulic system by leakage compensation through proportional flow control valve. ISA Trans. (2020, in press)
4. Ding, H., Zhao, J., Cheng, G., Wright, S., Yao, Y.: The influence of valve-pump weight ratios on the dynamic response of leaking valve-pump parallel control hydraulic systems. Appl. Sci. **8**(7), 1201 (2018)
5. E.-Azm Aly, A.A.E., Pascal, M., Taher, R.: Experimental study of leakage compensation for actuator speed control in electro-hydraulic systems. Int. J. Eng. Res. Technol. **4**(12), 129–138 (2015)

6. Pascal, M.A., Abou El-Azm, A.S., Taher, R.: Experimental study of leakage compensation on dynamic characteristics of hydraulic power systems. In: Proceedings of the 16th International Conference on Applied Mechanics and Mechanical Engineering, Cairo, Egypt, pp. 179–195 (2014)
7. Lovrec, D., Kastrevc, M.: Electro-hydraulic load sensing with a speed-controlled hydraulic supply system on forming-machines. Int. J. Adv. Manuf. Technol. **41**, 1066–1075 (2009)
8. Pournazeri, M., Khajepour, A.: An efficient lift control technique in electro-hydraulic camless valvetrain using variable speed hydraulic pump. SAE Int. J. Engines **4**(1), 1247–1259 (2011)
9. Zhou, Y., Hao, M.: The study of leakage of circular arc–involute–circular arc gear pump. Adv. Mech. Eng. **9**(9), 1–9 (2017)
10. Inauguma, Y.: A practical approach for analysis of leakage flow characteristics in hydraulic pumps. Proc. Inst. Mech. Eng. Part C: J. Mech. Eng. Sci. **227**(5), 980–991 (2012)
11. Rundo, M.: Models for flow rate simulation in gear pumps: a review. Energies **10**, 1261 (2017). https://doi.org/10.3390/en10091261
12. Elshorbagy, K., Kandil, H., Latif, M.: Development of a Multi-functional Hydraulic Test Stand. J. Sci. Eng. Res. **5**(1), 123–132 (2018)
13. Xu, M., Ni, J., Chen, G.: Dynamic simulation of variable-speed valve-controlled-motor drive system with a power-assisted device. Strojniški vestnik – J. Mech. Eng. **60**(9), 581–591 (2014). https://doi.org/10.5545/sv-jme.2013.1532
14. Willkomm, J., Wahler, M., Weber, J.: Potentials of speed and displacement variable pumps in hydraulic applications. In: Proceedings of 10th International Fluid Power Conference, pp. 379–392 (2016)
15. Kosucki, A., Stawiński, Ł.: Badania układów hydrostatycznych zasilanych falownikami (Investigations of hydrostatic systems powered by inverters - in Polish). In: The International Scientific and Technical Conference: Hydraulic and Pneumatic Drives and Control, vol. 1, pp. 5–16 (2016)
16. Stawinski, L., Kosucki, A., Morawiec, A., Sikora, M.: A new approach for control the velocity of the hydrostatic system for scissor lift with fixed displacement pump. Arch. Civil Mech. Eng. **19**(4), 1104–1115 (2019)
17. Minav, T.A., Laurila, L.I.E., Pyrhöen, J.J.: Axial piston pump flow ripple compensation by adjusting the pump speed with an electric drive. In: The 12th Scandinavian International Conference on Fluid Power, Tampere, Finland (2011)

Design Rules for Fuzzy Logic Controllers for Pneumatic Systems

Jakub Takosoglu(✉), Ryszard Dindorf, and Piotr Wos

Faculty of Mechatronics and Mechanical Engineering, Kielce University of Technology, Al. Tysiaclecia Panstwa Polskiego 7, 25-314 Kielce, Poland
qba@tu.kielce.pl

Abstract. Position control is particularly important in systems with pneumatic drives. Many centres around the world are looking for an optimal and effective method especially for rod and rodless pneumatic cylinders. Position control of pneumatic drives is a difficult process due to the compressibility of the working medium, friction occurring in the drive system, stick-slip phenomenon, clearances in the cylinder, sticking of the piston to the cylinder in final positions. The problem is to set and maintain a constant speed of motion, especially in the low speed range, and to precisely stop the piston of the cylinder in intermediate positions between the end positions of the cylinder.

The article presents the theoretical foundations of fuzzy logic controllers and the principles of their design for pneumatic drive systems. A fuzzy control system for an electro-pneumatic servo-drive made of a rodless cylinder controlled by a proportional flow valve has been analysed. An experimental stand for testing designed fuzzy logic controllers and assessing the quality of control has been presented. Experimental tests have been conducted for the implementation of changeover control, tracking control and following up the curvilinear trajectory.

Keywords: Fuzzy logic controller · Electro-Pneumatic Servo-Drive

1 Introduction

Problems related to the optimal control of pneumatic devices contributed to the search for new AI-based control techniques methods – fuzzy logic [1]. Use of FLC fuzzy logic controllers enables the transition from a quantitative description to a qualitative description of the process [2]. In traditional control systems, control algorithms are developed intuitively by operators based on their own experience. Applying fuzzy logic methods, knowledge gained during process operation and handling, or the knowledge of operators, i.e. intuitive algorithms for controlling control objects, can be saved with verbal logic converted into mathematical operation and used in the control process. Fuzzy logic systems can also be used in processes where there are non-linearities, uncertainties about their parameters or other adverse features of a control object. Fuzzy logic systems are intelligent control methods [3–5] in which the knowledge encoded in the rule base

© The Editor(s) (if applicable) and The Author(s), under exclusive license to Springer Nature Switzerland AG 2021
J. Stryczek and U. Warzyńska (Eds.): NSHP 2020, LNME, pp. 192–204, 2021.
https://doi.org/10.1007/978-3-030-59509-8_17

results from experience and intuition as well as theoretical and practical understanding of process dynamics [5, 6]. The use of fuzzy logic in controllers results from the fact that the knowledge of the process dynamics is not required for the correct tuning of the controller. Fuzzy control is also popular because the actual control objects are non-linear and therefore require special control techniques, which are usually very difficult, laborious and sometimes impossible to design. For such objects, designing fuzzy controllers can be much easier, and they can replace standard or status controllers [7].

2 Fuzzy Logic Controllers

The fuzzy control algorithm is a knowledge-based algorithm, described by fuzzy logic methods [8]. Fuzzy control systems are some kind of expert systems based on knowledge. The knowledge base contains a control algorithm stored as a rule base. The control strategy and knowledge contained in the rule base are derived from human experience and intuition and from a theoretical and practical understanding of the dynamics of a controlled object. The main difference between conceptually fuzzy and conventional control is the lack of analytical description [8].

The Control Law (1) of the fuzzy controller is described by a knowledge-based system containing ***IF-THEN*** rules with unspecified predicates and a control mechanism with fuzzy logic [8]:

$$u(k) = F[e(k), e(k-1), \ldots, e(k-v), u(k-1), u(k-2), \ldots, u(k-v)] \quad (1)$$

where F is a non-linear function representing the FLC fuzzy controller described by the rule base. The FLC describes the relationship between the $u(k)$ control signal on the one hand and the $e(k)$ error and its change $\Delta e(k)$ on the other. Therefore, the overall relationship of the FLC control (2) for v = *1* can be presented as follows:

$$u(k) = F[e(k), \Delta e(k)] \quad (2)$$

In the case of a fuzzy logic controller, its type – P, PI, PD, PID – determines the selection of process status variables, control variables and the content of the predecessor and successor of each rule. For process status variables (input variables) that represent the rule predecessor content, i.e. the ***IF*** rule part, the following is obtained:

- control error e,
- control error change Δe,
- control error sum δe,

For process control variables (output variables) that represent the rule predecessor content, i.e. the ***THEN*** rule part, the following is obtained:

- for-loop u,
- for-loop change Δu.

By analogy to the standard controller, it can be presented as follows:

$$\begin{aligned}
e(k) &= y_0(k) - y(k) \\
\Delta e(k) &= e(k) - e(k-1) \\
\Delta(\Delta e(k)) &= e(k) - 2e(k-1) + e(k-2) \\
\delta e(k) &= \sum_{i=1}^{k} e(i) \\
\Delta u(k) &= u(k) - u(k-1)
\end{aligned} \tag{3}$$

where:
$y_0(k)$ – set-point (input function),
$y(k)$ – command (response).

PID fuzzy logic controllers PID implemented in two versions: direct and incremental (see Fig. 1).

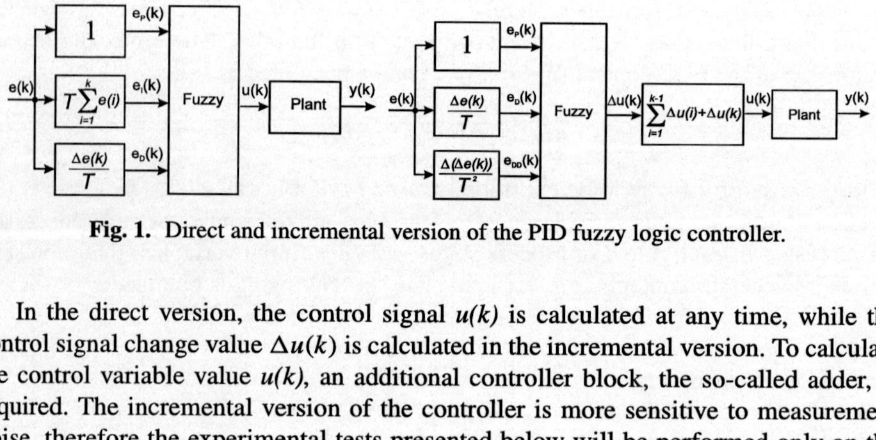

Fig. 1. Direct and incremental version of the PID fuzzy logic controller.

In the direct version, the control signal *u(k)* is calculated at any time, while the control signal change value $\Delta u(k)$ is calculated in the incremental version. To calculate the control variable value *u(k)*, an additional controller block, the so-called adder, is required. The incremental version of the controller is more sensitive to measurement noise, therefore the experimental tests presented below will be performed only on the basis of the direct version.

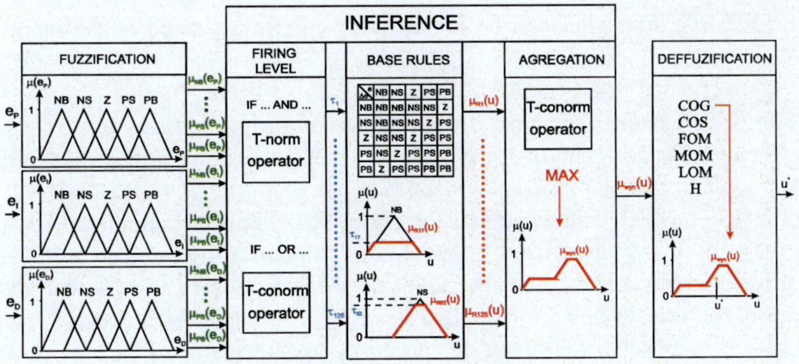

Fig. 2. Fuzzy structure.

The structure of the fuzzy part for the direct version is shown in Fig. 2. Five main stages of fuzzy inference were taken into account (fuzzification, determination of rule firing levels, launching of rules, aggregation of individual rule outputs, defuzzification).

Due to the complexity of the fuzzy inference process, the structure of the fuzzy part presented in Fig. 2 contains only some of the results of the full process. Fuzzification was carried out in three e_P, e_I, e_D tracks using seven fuzzy sets of NB, NM, NS, Z, PS, PM, PB. The firing level, depending on the design of the premise in the rule base, can be determined using T-norm or T-conormy operators. The complete rule base contains 147 rules including all combinations of NB, NM, NS, Z, PS, PM, PB inputs. The launching of rules was performed using the Mamdani implication and the aggregation of the fuzzy sets $\mu_{Ri}(u)$ of launched rules using the *MAX* T-conormy operator. Defuzzification was performed using the COG (Center of Gravity) method.

2.1 Design Rules for Fuzzy Logic Controllers

Due to the complexity of fuzzy logic instrument, designing FLC fuzzy controllers can be a complex and time consuming process. The problem is both the determination of the FLC parameters, i.e. the linguistic labels and fuzzy sets, and the structure of the FLC – rule base with a fuzzy inference mechanism. To perform the fuzzy operation, the membership function must be precisely defined in terms of quantity, i.e. parameters and function coefficients, as well as quality, i.e. function types.

The following rules are recommended for determining parameters and tuning fuzzy logic controllers [9]:

- The most commonly used membership functions are triangular functions,
- number of intersection points of two membership functions – the intersection coefficient usually takes the value of 1 (one intersection point),
- the membership function value at the intersection point of the membership function – the intersection level usually assumes the value 0.5,
- symmetrical membership functions are most often applied,
- A large range of cut-off values in the membership function diagram means low gain, and narrow membership functions mean high sensitivity of the controller,

- The fuzzy logic controller can be tuned by changing the shape of the membership function.

Building a rule base is the most difficult stage of designing fuzzy logic controllers. It is also a cause for criticism of fuzzy logic control because, in principle, there are no systematic tools to create a FLC rule base [8].

The fuzzy control algorithm reflects the control mechanism implemented by people, without the use of any formalised knowledge of the controlled object in the form of mathematical models and without the analytical description of the control algorithm [8]. One way to derive a linguistic control strategy is human experience and reasoning, another way is to observe the phase surfaces of conventional PI, PD, PID controllers, from which fuzzy logic control rules can be derived.

Another, more formal approach to FLC rule base construction is to use a template rule base. The template rule base can be considered a basic tool that combines engineering common sense and fuzzy logic control experience [8]. This standard rule base is the one proposed by Mac Vicar-Whelan (Table 1).

Table 1. Mac Vicar-Whelan rule base.

Δe\e	NB	NM	NS	Z	PS	PM	PB
NB	NB	NB	NB	NB	NM	NS	Z
NM	NB	NB	NM	NM	NS	Z	PS
NS	NB	NM	NS	NS	Z	PS	PM
Z	NB	NM	NS	Z	PS	PM	PB
PS	NM	NS	Z	PS	PS	PM	PB
PM	NS	Z	PS	PM	PM	PB	PB
PB	Z	PS	PM	PB	PB	PB	PB

Knowledge of the technical and operational parameters of pneumatic control elements is extremely important when designing control systems. Manufacturers usually give only basic parameters of their products and also present selected theoretical characteristics. Such ideal characteristics can introduce errors into the control system. Position control of pneumatic drives uses proportional valves for pressure and flow. Pneumatic proportional valves (servo-valves) are elements with a high gain of the input signal. They are used to amplify and transform continuously a weak input signal into a strong output signal in the form of flow rate (flow valves) or pressure (pressure valves). The highest gain can be achieved by using electrical signals as input signals, thus proportional valves with electrical control are most often applied. The valve consists of two basic units:

- An electromechanical transducer that converts an electrical voltage or current signal into either linear (angular) motion or force (moment),
- one-, two-, or three-stage amplifier, and
- spool valve.

Proportional valves are used primarily in control systems for fast-changing production processes. The design of electro-pneumatic servo-valves does not differ significantly from electrohydraulic servo-valves. Most often, these are slide servo-valves with four or two distributing edges. Pneumatic servo-valve slide-sleeve assemblies do not contain elastic seals to reduce frictional resistance and the resulting hysteresis. The lack of sealing is a source of significant compressed air leakage, which is prevented by the use of a positive cover of the flow channels in the slider-sleeve assembly, which, however, introduces a servo-valve insensitivity zone in the middle position of the slide. Knowledge of the actual characteristics of the control valve can significantly facilitate the design process of the fuzzy logic controller for the electro-pneumatic servo-drive, as fuzzy logic provides the opportunity to eliminate non-linear phenomena occurring during the control process.

3 Object of Study

The experimental tests focused mainly on the design and implementation of the fuzzy control algorithm for the electro-pneumatic servo-drive for the task of precise positioning (changeover control, tracking control and any motion trajectory). In industrial environments, servo-drives are performing increasingly complex positioning tasks under changing operating conditions [10]. Due to the increasing demands on pneumatic servo drives, the control algorithm must be resistant to variable mass load [11], change of the set piston motion speed of the cylinder, change of the set piston position of the cylinder, change of the frequency of the set sinusoidal trajectory, change of the supply pressure, to the type and accuracy of the applied position transducers, to the parameters of the pneumatic cylinder used for construction. Figure 3 shows a diagram of an electro-pneumatic servo drive, in which a rodless cylinder is used as an actuator.

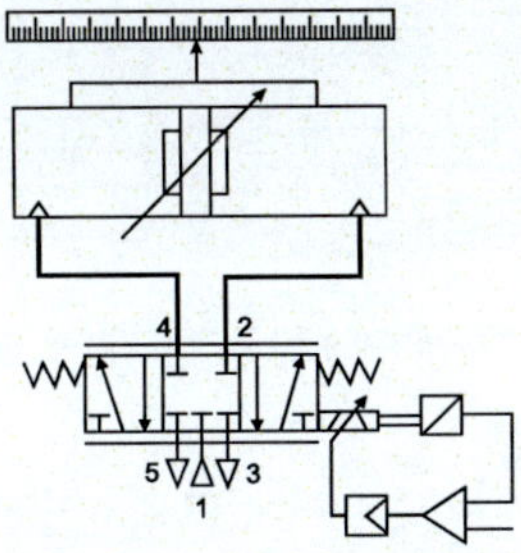

Fig. 3. Electro-pneumatic servo-drive diagram with rodless cylinder.

The main components of the servo drive are:

- DGPL-25-224 rodless cylinder from Festo (Fig. 4a),
- MPYE-5-1/8-HF-010-B proportional valve from Festo (Fig. 4b),
- BTL5-A11-M0600-P-S32 magnetostriction transducer of Balluff (Fig. 4c).

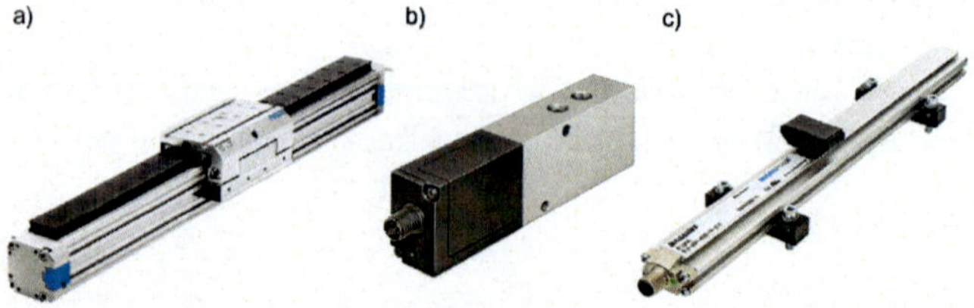

Fig. 4. General view: a) rodless cylinder, b) proportional valve c) magnetostriction transducer.

MPYE-5-1/8-HF-010-B proportional valve is a five-way, three-position, voltage controlled valve ($0 \div 10$ V). It is a valve with a nominal flow rate of 700 l/min and a switching time of 80 Hz [12]. Figure 5a shows the actual flow characteristics [13] of the valve and Fig. 5b shows the actual pressure characteristics of the valve.

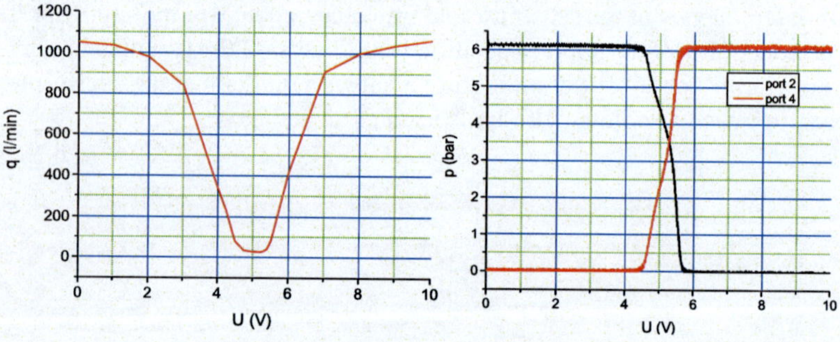

Fig. 5. Pressure and flow characteristics.

A rodless cylinder 224 mm long and a piston diameter of 25 mm was used as the drive. A non-contact magnetostrictive position transducer with a gauge length of 600 mm and resolution of less than 2 μm with an analogue output ($0 \div 10$ V) works basing on the magnetostrictive effect principle. Figure 6 shows the measurement noise of the magnetostrictive position transmitter.

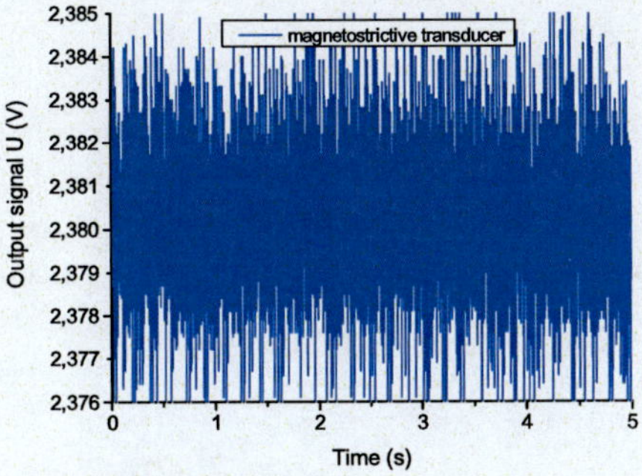

Fig. 6. Measurement noise of the BTL5-A11-M0600-P-S32 position transducer.

4 Fuzzy PD Controller Design

When designing the PD fuzzy logic controller, experimental tests of the proportional valve and the position transducer were taken into account. The controller was designed using the *Fuzzy Logic* toolbox of Matlab-Simulink package. The variable domain error e is defined from -3.7 V to 3.7 V because these values correspond to the operating range of this variable for the DGPL-25-224 rodless cylinder with BTL5-A11-M0600-P-S32 position measurement. The Γ and L-type sets are used for extreme values, which avoids the lack of completeness of the knowledge base for large amplitude values of the error e input signal. In addition, a trapezoid set was used for the near zero error to determine the value of the assumed static deviation 2δ and to avoid oscillations around the near zero deviation and the set was compacted. This reduced the dynamics of the system but allowed achieving "smoothness" of the output signal (i.e. small jumps when passing from one rule to another) near the zero error. For the second input signal, i.e. the change of error Δe, the same was done during the fuzzy process, but its domain was set from – 25 V/s to 25 V/s, because these values of the amplitude of the input signal Δe correspond to the maximum speed of error change (maximum speed of the cylinder). Taking into account the above mentioned assumptions, the input signal of the e error was subjected to the fuzzy process with the distribution of fuzzy sets as in Fig. 7a, while the input signal Δe as in Fig. 7b.

For the magnetostrictive transducer the measurement noise is ± 0.007 V. The determination of the measurement noise was necessary to determine the assumed static deviation and thus to calculate the quality indicators for the control of the fuzzy electro-pneumatic servo drive. The determination of the measurement noise was also used to correctly design the fuzzy logic controller. Due to the position transducer used and the requirements for pneumatic drives under industrial operating conditions, a static deviation $2\delta = \pm 0.022$ V was assumed.

As experimental tests (flow characteristics) have shown, the proportional valve from Festo does not operate symmetrically and the flow curve is similar to the sigmoid shape. Therefore, a sigmoid fuzzy sets shifted to the right (due to the asymmetrical operation of the valve) were used. For extreme fuzzy sets Γ and L-type sets are used, similarly to input signals. The rule base comprises of 49 fuzzy rules presented in Table 1.

5 Experimental Tests

The experimental tests on the electro-pneumatic servo-drive were mainly focused on checking the operation of the PD fuzzy logic controller under changing operating conditions. The fuzzy logic controller had a feedback loop from the position of the cylinder piston using a position transducer. Figure 7 shows a general view of the experimental test stand.

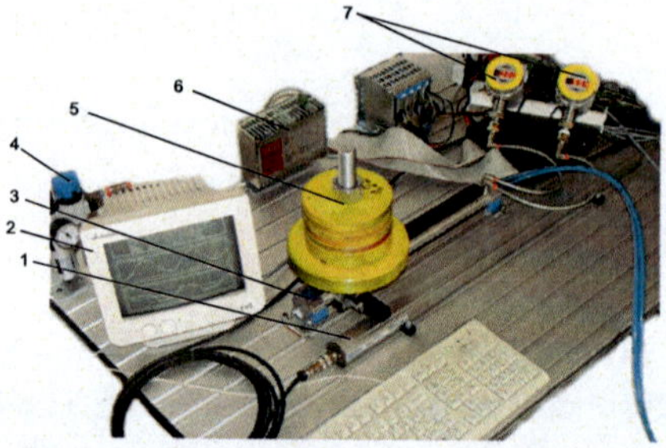

Fig. 7. A general view of the test stand: 1 – position transducer, 2 – screen, 3 – rodless cylinder, 4 – FRL unit, 5 – mass load, 6 – power supply, 7 – pressure transducer.

The problem of pneumatic servo drive positioning results from the complex and complicated process of converting the energy of compressed gas into mechanical energy of the cylinder piston motion. Solving such a problem in an environment of insufficient and incomplete information is a difficult task, but especially important when it comes to the possibility of using the proposed solution in industrial working conditions. The design of servo drives equipped with extra sensors, e.g. pressure transducers or force sensors, is uneconomic under industrial conditions. In the time of rapid development of automation and the growing demands from industrial sectors, the controller faces a number of tasks, mainly concerning the task of changeover control, tracking control and reproduction of any motion trajectories. Due to the use of pneumatic servo drives in the construction of manipulators and industrial robots, one of the most important requirements is the resistance of the control algorithm to alternating mass load and supply pressure fluctuations. Positioning is also required over a wide range of displacements and pre-set piston speeds of the cylinder.

A number of experimental tests have been conducted on the designed and built stand. The graphs below show selected dynamic and quality characteristics (IAE - Integral of Absolute Error, ISC - Integral of Square Control Signal) of electro-pneumatic servo drive piston positioning (Figs. 8, 9, 10, 11 and 12).

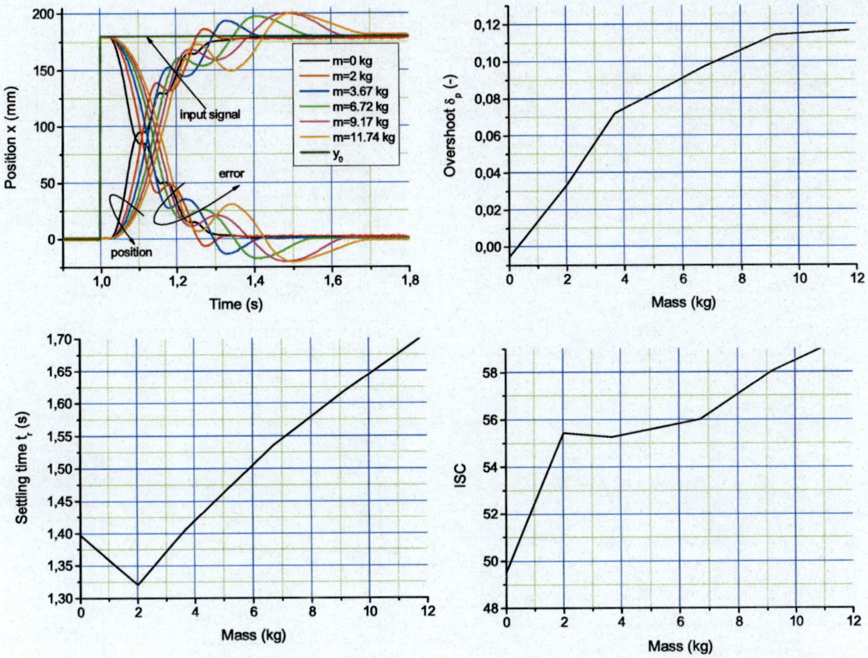

Fig. 8. Changeover control with variable load mass.

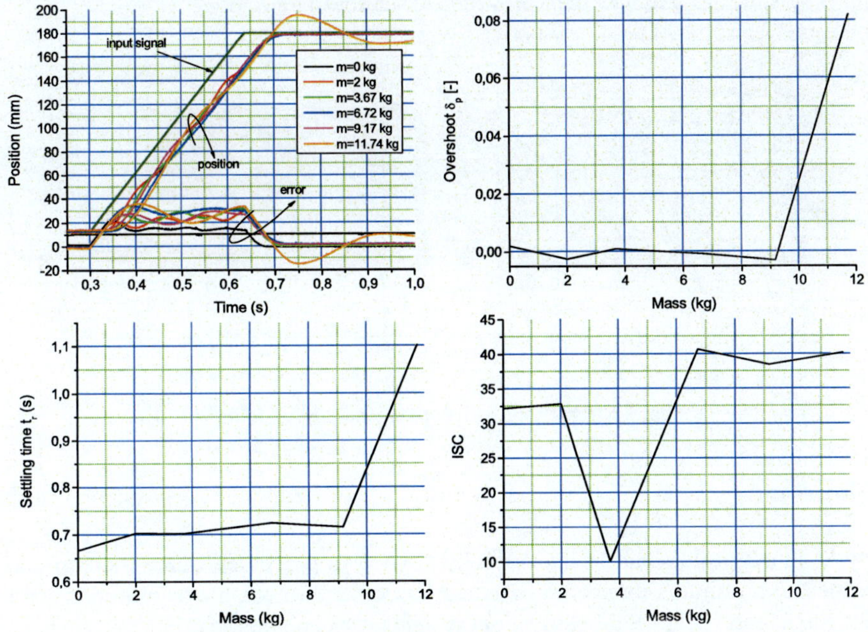

Fig. 9. Tracking control with variable load mass

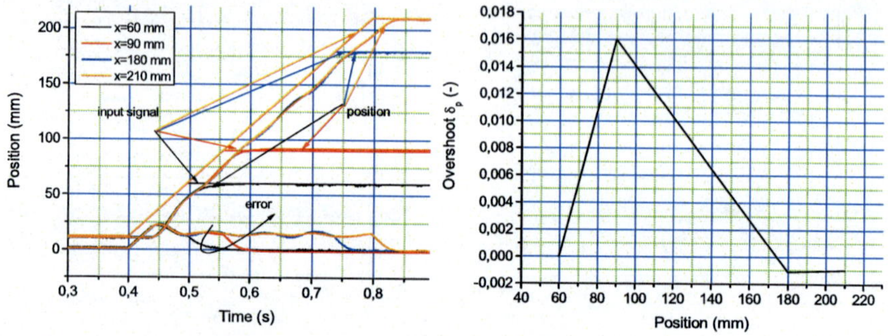

Fig. 10. Tracking control with position variable.

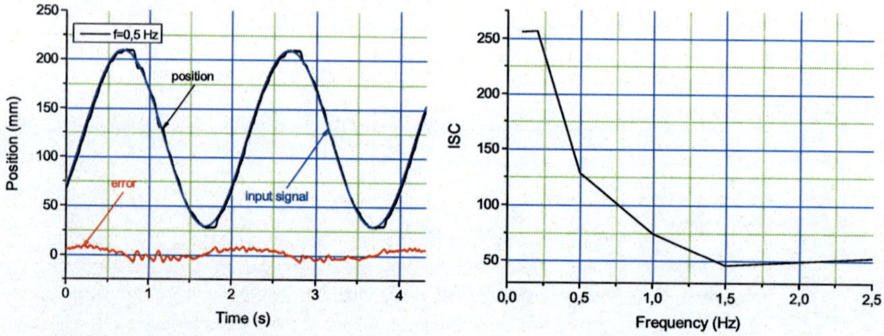

Fig. 11. Implementation of sinusoidal trajectory.

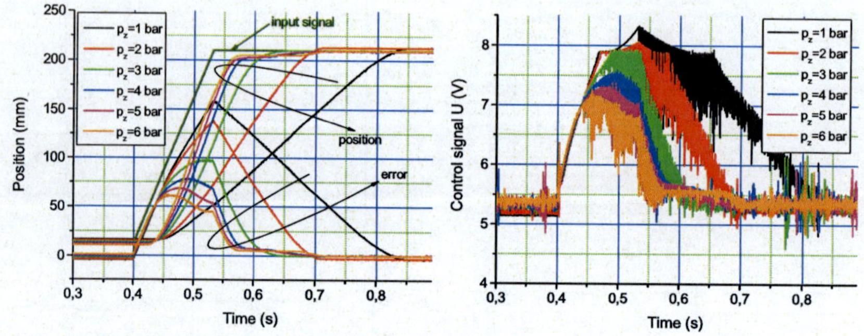

Fig. 12. Tracking with supply pressure variable.

6 Summary

Fuzzy logic controllers can be successfully applied in processes where there are non-linearities, uncertainties about their parameters or other unfavorable features of a control object. Such features are characterized by systems with pneumatic drives and especially analyzed electro-pneumatic servo drive with a rodless cylinder controlled by a proportional flow valve. The tests have been conducted in a wide range of variable operating conditions of the servo drive. The designed PD fuzzy logic controller performed very well the assumed tasks of the changeover control, tracking control and sinusoidal trajectory. The static deviation of $2\delta = \pm 1.32$ mm has been assumed for the tests, which was a result of the measurement noise occurring in the control system and positioning accuracy required in industrial conditions. The aim of the tests was to check the resistance of the fuzzy electro-pneumatic servo control algorithm to variable mass load of the cylinder, change of the pre-set speed of the cylinder piston motion, change of the pre-set position of the cylinder piston, change of the supply pressure, change of the pre-set frequency of the sinusoidal trajectory.

The designed fuzzy logic controller is resistant to variable mass load of the cylinder. The fuzzy logic controller works correctly up to a mass of 6.72 kg in the case of the changeover control and 0.5 m/s in the case of the tracking control up to a mass of 9.17 kg.

For low speeds (0.075 m/s and 0.25 m/s), an oscillating stick-slip (transition phase from solid to fluid friction) can be observed in the initial start-up phase. Above 0.25 m/s, no stick-slip occurs. A step change in the friction force value during this phase results in a significant speed error, which is effectively compensated by the fuzzy logic controller during the further control process. The drive maintains the set point values for motion parameters. An additional advantage is that it is not necessary to tune the controller during changes in operating conditions.

The analysis of the quality characteristics showed that the piston positioning of the cylinder is accurate enough for industrial conditions.

References

1. Zadeh, L.A.: Fuzzy sets. Inf. Control **8**(3), 338–353 (1965). https://doi.org/10.1016/S0019-9958(65)90241-X
2. Schulte, H., Hahn, H.: Fuzzy state feedback gain scheduling control of servo-pneumatic actuators. Control Eng. Pract. **12**, 639–650 (2004). https://doi.org/10.1016/S0967-0661(03)00148-5
3. de Silva, C.W.: Intelligent Control: Fuzzy Logic Applications. CRC Press, Cambridge (1995)
4. Takosoglu, J.E., Dindorf, R.F., Laski, P.A.: Rapid prototyping of fuzzy controller pneumatic servo-system. Int. J. Adv. Manuf. Technol. **40**, 349–361 (2007). https://doi.org/10.1007/s00170-007-1349-5
5. Harris, J.: Fuzzy Logic Applications in Engineering Science. Springer, New York (2005)
6. Ross, T.J.: Fuzzy Logic with Engineering Applications. Wiley, Hoboken (2010)
7. Takosoglu, J.E., Laski, P.A., Blasiak, S.: A fuzzy logic controller for the positioning control of an electro-pneumatic servo-drive. Proc. Inst. Mech. Eng. Part I J. Syst. Control Eng. (2012). https://doi.org/10.1177/0959651812456498
8. Yager, R.R., Filev, D.P.: Essentials of Fuzzy Modeling and Control, 1st edn. Wiley, Hoboken (1994)
9. Driankov, D., Hellendoorn, H., Reinfrank, M.: An Introduction to Fuzzy Control. Springer, Heidelberg (1996)
10. Saravanakumar, D., Mohan, B., Muthuramalingam, T.: A review on recent research trends in servo pneumatic positioning systems. Precis. Eng. **49**, 481–492 (2017). https://doi.org/10.1016/j.precisioneng.2017.01.014
11. Lisowski, E., Filo, G.: Automated heavy load lifting and moving system using pneumatic cushions. Autom. Constr. **50**, 91–101 (2015). https://doi.org/10.1016/j.autcon.2014.12.004

12. Takosoglu, J.E.: Experimental research of flow servo-valve. In: Dancova, P. (ed.) Proceedings of the International Conference Experimental Fluid Mechanics 2016. Technical University Liberec, pp. 819–823 (2016)
13. Takosoglu, J.E., Laski, P.A., Blasiak, S., et al.: Determination of flow-rate characteristics and parameters of piezo valves. In: Dancova, P. (ed.) Proceedings of the International Conference Experimental Fluid Mechanics 2016, pp. 814–818, Technical University Liberec (2016)

Simulation of Transient Flow in Micro-hydraulic Pipe System

Kamil Urbanowicz[1(✉)], Michał Stosiak[2], Krzysztof Towarnicki[2], Huan-Feng Duan[3], and Anton Bergant[4]

[1] Faculty of Mechanical Engineering and Mechatronics, West Pomeranian University of Technology, Szczecin, Poland
kamil.urbanowicz@zut.edu.pl
[2] Faculty of Mechanical Engineering, Department of Technical Systems Operation and Maintenance, Wrocław University of Science and Technology, Wrocław, Poland
{michal.stosiak,krzysztof.towarnicki}@pwr.edu.pl
[3] Department of Civil and Environment Engineering, Hong Kong Polytechnic University, Hung Hom, Kowloon, Hong Kong
hf.duan@polyu.edu.hk
[4] Litostroj Power d.o.o. (Full-Time) and Faculty of Mechanical Engineering, University of Ljubljana (Part-Time), Ljubljana, Slovenia
anton.bergant@litostrojpower.eu

Abstract. This paper presents the modelling and simulation of transient flow in micro-hydraulic pipe systems. Liquid stream energy dissipation occurs mainly as a result of friction losses. Theoretical considerations of water hammer resulting from rapid valve closing, supported by experimental verification, were undertaken. The experimental system incorporated a straight two-meters long section of a steel pipe with an internal diameter of $4 \cdot 10^{-3}$ m. An attempt was made to determine the degree of conformity of the transient flow model (previously verified in conventional pipes) to the experimental results obtained for small-internal-diameter pipes. Shear stress on the pipe wall was modelled using first a simplified quasi-steady approach and then an effective modified unsteady friction model. The pressure waveforms at the valve (at the downstream end of the pipe) were obtained for initial flow velocity, $v_{01} = 2.39$ m/s and $v_{02} = 1.14$ m/s, respectively. Experimental studies were carried out in the region of laminar flows with Reynolds numbers below 100.

Keywords: Water hammer · Wall shear stress · Micro-hydraulic pipe · Laboratory apparatus · Modelling · Method of characteristics · Unsteady friction

1 Introduction

Unsteady flows occur commonly in hydraulic, water supply and transmission systems as well as in living organisms (circulatory systems) [1–4]. Because of the wave character of flow disturbance propagation, they pose a considerable danger to complex systems with

© The Editor(s) (if applicable) and The Author(s), under exclusive license to Springer Nature Switzerland AG 2021
J. Stryczek and U. Warzyńska (Eds.): NSHP 2020, LNME, pp. 205–215, 2021.
https://doi.org/10.1007/978-3-030-59509-8_18

many branches (water supply and hydraulic systems) since superposition of waves can occur in these systems [5]. In hydraulic systems (hydraulic machines, motor vehicles, etc.), where directional valves are commonly used, each switch-over of these devices can lead to unsteadiness. The high pressures accompanying such states can cause leaks and in extreme cases (bursts) they can lead to the total destruction of the overloaded system components [6]. Therefore, elements protecting such systems are commonly employed [7]. But they considerably increase the planned costs of the systems. In order to avoid these costs, it is necessary to better identify the behaviour of such systems under unsteady loads. This knowledge will help to optimize the choice of system components. Mathematical models describing flows with acceptable accuracy can be used for this purpose. There is lack of existing commercial simulation software that will model all phenomena associated with transient flow with sufficient accuracy and/or efficiency. In most of them, the dynamic frictional effect is modelled in a simplified 1-D quasi-steady way. In practice, 3-D simulations are not performed because of their very long computation time and complexity. The models which enjoy increasing popularity are characterized by essential simplicity, whereby they can be simply applied to any considered system. The group of such models includes 1D models having a highly simplified structure [8]. They are described by a set of partial differential equations and are mostly solved with the method of characteristics [9]. One can use them to calculate (in close to real time) changes of two basic flow parameters, i.e. flow rate (the mean flow velocity in the cross section) and pressure, in specific cross-sections of the analysed system. Research has been conducted for many years to increase the efficiency of such models [10–12]. Recently, it was possible to avoid a serious error that was responsible for incorrect modelling of wall shear stress [13–15]. In this paper a corrected 1-D efficient numerical model [16] will be further verified and compared with new experimental results received for a short length and very small diameter oil pipe system. Fluid hammer experiments in a micro scale been not presented before. Correct knowledge and description of the phenomenon of transient flow is important in such micro-hydraulic systems. Micro-hydraulic systems are widely used in medical applications, manufacturing industry, mobile machines, aircraft and vessels. To this end, the relevance of transient friction effect to the pipe scales will be examined [17] and verified through experimental work. Since in the experiments directional valve switching time was not instantaneous it was necessary to use a modified boundary condition.

2 Test Stand

An abrupt closure of the flow in a hydraulic conduit results in an instantaneous pressure rise in the hydraulic line. In order to determine the effect of a micro-hydraulic system switch-over the test stand shown in Fig. 1 was built.

A steel hydraulic pipe with an outer diameter of 0.006 [m] and an inner diameter of 0.004 [m] is the principal test stand member constituting the basis for the mathematical model. Pressure transducers (13) and (14) were installed at the pipe's ends. The pressure transducers can measure pressure up to max. $25 \cdot 10^6$ [Pa]. A Casapa PLP 10.1D0 external gear pump (3), with a unit delivery of $1 \cdot 10^{-6}$ [m^3/rotation], was used to generate liquid flow and pressure. A cut-off valve (17) was installed on the pipe's end to rapidly

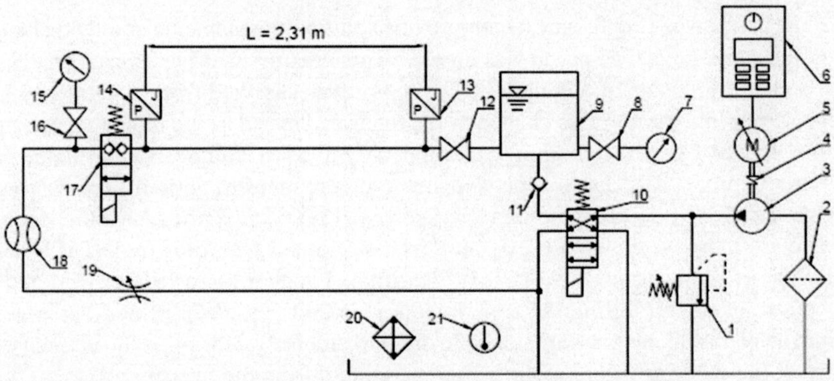

Fig. 1. Schematic of tests stand and instrumentation: 1 – safety valve, 2 – suction filter, – displacement gear pump, 4 – flexible coupling, 5 – electric drive motor, 6 – power supply box with built-in frequency converter, 7 – pressure gauge, 8 – cut-off valve, 9 – accumulator, 10–4/2 directional control valve controlled by electromagnetic coil, 11 – check valve, 12 – cut-off valve, 13, 14 – pressure transducer, 15 – pressure gauge, 16 – cut-off valve, 17 – cutting off valve controlled by electromagnetic coil, 18 – flowmeter, 19 – throttle valve, 20 – cooler, 21 – thermometer.

close the flow through the pipe. The valve's closing element is controlled by the coil of a conventional magnet (acc. to the manufacturer specifications, the leakage from the valve amounts to $5.5{\cdot}10^{-7}$ [m^3/s]. In the vicinity of the pressure gauge a flowmeter (with measuring accuracy $\pm 5\%$ of the mean value) was installed to measure the rate of flow in the pipe. A pre-set throttle valve (19) was connected to the flowmeter's outlet to obtain the mean pressure in the investigated hydraulic line. A flexible hydraulic conduit with an inner diameter of 0.004 [m] was connected to the throttle valve whereby the liquid was transported back to the tank. A suction filter (2) was installed on the pump's inlet port to ensure the proper filtration of the working liquid. Further on the main delivery line a conventionally electrically controlled directional valve (10) was connected in series. Its function was to change the direction of the liquid flow generated by the pump at the instant when the flow was closed by the cut-off valve situated at the pipe's end. A check valve (11) was installed to prevent the flow of the liquid along the section from the directional valve (10) to the investigated flow at the instant of directional valve switch-over. A hydraulic accumulator (9) was installed after the check valve on the main supply line. The accumulator had the form of a cylindrical tank with a capacity of 0.013 [m^3]. For system servicing purposes a cut-off valve (12) was placed between the accumulator and the investigated pipe. On the other side, at the connection of the investigated pipe, a pressure gauge (7) and a cut-off valve (8) were installed to measure mean pressure in the accumulator (9). The accumulator's supply connection was made at its bottom axis. The hydraulic system was equipped with a thermometer (21) and a cooler (20) – a basic system temperature stabilization component. Owing to the system's thermal inertia and the temperature stabilization component the set temperature could be attained with an accuracy of ± 2 [K]. The assumed water hammer could be produced by simultaneously changing over the cutting off valve (17) and the directional valve (10). The electromagnetic coils whose function was to control the above elements were connected in parallel to the electric power supply unit. Owing to this solution the elements could be changed over at the same time. The pump was driven by connecting its drive shaft to an electric motor's shaft by means of a coupling. The motor was supplied from a frequency converter (6) enabling the stepless change of the rotational speed of the motor's drive shaft. The investigated hydraulic conduit was immovably fixed to a foundation insulated from ambient vibrations. The distance between the pressure measuring points (pressure sensors) was $L = 2.31$ [m], the pressure wave speed $c = 1180$ [m/s], the kinematic viscosity of the liquid in the system $\nu = 1 \cdot 10^{-4}$ [m^2/s] (determined according to ASTM D7042), the density of the oil $\rho = 869$ [kg/m^3] (determined according to ISO 12185) and the temperature of the oil during the experiments $T = 292$–294 [K]. Mineral oil was the medium used during the measurements. In order to precisely determine the viscosity and density of the working liquid at the set temperature during the measurements a sample of the oil was sent to a specialist laboratory. The laboratory tests conclusively showed that oil ISO32 was the oil used in the experiments.

3 Mathematical Model (of Transient Flow in Micro-hydraulic Pipe Systems)

The unsteady flow in pressure conduits is described in this paper by the system of basic equations [8] of continuity (1) and motion (2):

$$\frac{\partial p}{\partial t} + c^2 \rho \frac{\partial v}{\partial x} = 0 \tag{1}$$

$$\frac{\partial v}{\partial t} + \frac{1}{\rho}\frac{\partial p}{\partial x} + \frac{2\tau}{\rho R} = 0 \tag{2}$$

where:
$v = v(x, t)$ – mean velocity of the liquid in the pipe cross section,
$p = p(x, t)$ – pressure in the pipe cross section,
R – inner diameter of the pipe,
τ – wall shear stress,
ρ – density of the liquid,
g – acceleration due to gravity,
c – pressure wave speed,
t – time,
x – axial coordinate along the pipe.

Above system of equations is solved with use of method of characteristics [8, 9] in time domain. The distributed model was chosen in this study because it could provide full picture of the dynamic and propagating phenomena that often accompany with this type of flow (e.g. retarded strain in plastic pipes, cavitation, etc.). It is noted that another type of lumped parameters model in the frequency domain – transmission line method [18, 19] has been well developed and also widely applied for the simulation and analysis of transient flows in a pipeline. However, the lumped model may not fully capture the details of complex transient flow process due to the potential simplification and approximation used in its mathematical expressions such as linearized friction effect and constant wave speed [20]. It is expected that, with the detailed transient flow results achieved by the distributed model in this study, the lumped model will be further investigated in the future work for comparison.

The wall shear stress τ is a function of the mean flow velocity and its derivative $\tau = \tau\left(v, \frac{\partial v}{\partial t}\right)$ [21]:

$$\tau = \frac{f\rho}{8} v|v| + \frac{2\mu}{R} \int_0^t w(t-u) \cdot \frac{\partial v(u)}{\partial t} du \tag{3}$$

The first term of the above equation represents a quasi-steady quantity in which f is the Darcy-Weisbach friction factor. The second term, which is a convolution integral, represents a time dependent quantity in which $w(t-u)$ is the so-called weighting function. Please note that in practice, the order of magnitude of pressure wave speed ranges from 100 to 1400 [m/s] and the flow velocity is of order 1 to 10 [m/s]. Therefore, the Mach number, in hydraulic/water hammer applications is often in the range from 10^{-3} to 10^{-2} [22]. This allowed to reject the convective term of velocity in the main hydrodynamic Stokes equation of parallel axisymmetrical flow that was Zielke starting point when he derived the solution in form of Eq. 3. In this paper the convolution integral was solved in an effective manner using the Urbanowicz's solution [16]:

$$\tau_{(t+\Delta t)} = \frac{\rho f_{(t+\Delta t)}}{8} v_{(t+\Delta t)} \left| v_{(t+\Delta t)} \right| + \frac{2\mu}{R} \sum_{i=1}^{3} \frac{\left[A_i y_{i,(t)} + \eta B_i \left[v_{(t+\Delta t)} - v_{(t)}\right] + C_i [1-\eta] \left[v_{(t)} - v_{(t-\Delta t)}\right]\right]}{y_{i,(t+\Delta t)}} \tag{4}$$

Coefficients A_i, B_i, C_i and weighting function correction factor η are calculated from the formulas discussed in detail in paper [16]. The fact that the above relation currently consists of only three expressions is the result of the filtration of the effective weighting functions used in the calculations. Details of this filtration are discussed in paper [23].

3.1 Boundary Condition

Since the directional valve switching time was not instantaneous, the pressure inside the valve during the switch-over will be calculated here similarly as for slow valve closing. As opposed to the formulas found in the literature [8, 24, 25], the formulas presented below were derived from the orifice equation [8] for the basic flow parameters, i.e. averaged flow and pressure velocities.

During a switching time one may calculate the instantaneous velocity value at $t < T_c$ using the method of characteristics and the following formula:

$$v_{(1,t)} = \frac{0.5c\rho\theta^2 v_{(1,t=0)}^2}{p_{(1,t=0)}} - \sqrt{\left(\frac{0.5c\rho\theta^2 v_{(1,t=0)}^2}{p_{(1,t=0)}}\right)^2 + \frac{c\rho\theta^2 v_{(1,t=0)^2}}{p_{(1,t=0)}} \left(\frac{p_{(2,t-\Delta t)}}{c\rho} - v_{(2,t-\Delta t)} + \frac{4\Delta t \tau_{(2,t-\Delta t)}}{\rho D}\right)} \tag{5}$$

After calculating the above value inside the directional valve one can then determine the pressure prevailing in the same place from the following relation:

$$p_{(1,t)} = c\rho\left(v_{(1,t)} - v_{(2,t-\Delta t)}\right) + p_{(2,t-\Delta t)} + \frac{2c\Delta t \tau_{(2,t-\Delta t)}}{R} \tag{6}$$

In the above formulas the characteristic of the directional valve being closed is modelled by the time function $\theta(t)$. In this paper the following form of this function was assumed:

$$\theta = 1 - \left(\frac{t}{T_c}\right)^2 \tag{7}$$

The shape of the function is shown in Fig. 2.

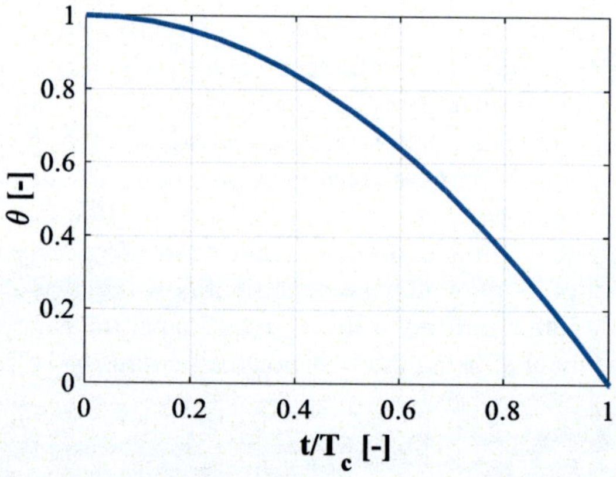

Fig. 2. Hydraulic valve closure relation characteristic.

4 Simulation Results

For the purposes of this study a series of experiments were carried out on the test stand described in details in Fig. 1. The Oil ISO32 was the liquid flowing in the analysed hydraulic system. The tests were conducted on November 12, 2019. During the tests the temperature inside the hydraulic system increased slightly from 292 to 294 [K]. The change in temperature affected the values of the two key mathematical model parameters: viscosity and density, whose detailed values are presented in Table 1.

The method of characteristics (MOC) [8, 9] was used to solve the system of partial differential Eqs. (1) and (2) presented in Sect. 3. This numerical method interprets well the physical nature of the unsteady flow and is characterized by quick convergence, the ease of taking different boundary conditions into account, and a high accuracy of the calculation results. Using it one can solve the analysed system of equations in a simple way. The solution consists of finding an equivalent system of four ordinary differential equations, which is then solved by means of finite differences. Approximation with first-order differential schemes yields satisfactory results. The detailed solution can be found in [10]. The calculated pressure wave speed in this short pipe ($L = 2.31$ [m]) was estimated from the obtained experimental results and its value was $c = 1180$ [m/s]. In all the investigations presented below the conduit was discretized into $N = 16$ [-] elements (the number of reaches in the numerical calculation). This axial discretization guaranteed the fulfilment of the computational compliance criteria [26] and forced the following time step $\Delta t = 1.22 \cdot 10^{-4}$ [s]. All the other parameters needed for simulations are contained in Table 1. Selected results of the tests are presented in Figs. 3 and 4 below, where the experimental results (the broken line) are compared with the simulation results (the solid line). For each of the investigated flows the simulation results were obtained in four ways:

a) by rapidly (instantaneously) switching over the directional valve ($T_c < \Delta t$) – hydraulic resistance calculated in the quasi-steady way (denoted as IC+QSF);
b) by spreading the switching-over time of the directional valve (the switching time of directional valve was assumed in numerical calculations to be a quarter of a period of hydraulic hammer $T_c = \frac{L}{c} = \frac{2.31}{1180} = 1.96 \cdot 10^{-3}$ [s]) – hydraulic resistance calculated in the quasi-steady way (denoted as RC+QSF);
c) by rapidly (instantaneously) switching-over the directional valve – hydraulic resistance calculated using the unsteady model (denoted as IC+UF);
d) by spreading the switching-over time of the directional valve – unsteady resistance (denoted as RC+U).

For clarity, the abbreviations used in the above four cases and the following figures stand for: IC – instantaneous closing time, QSF – quasi-steady friction, UF – unsteady friction, RC – realistic closing time.

Table 1. Parameters of analysed flows.

Case	Q [m^3/s]	v_0 [m/s]	Re [-]	p_R [MPa]	ρ[kg/m^3]	ν [m^2/s]	T [K]
C01	$3.00 \cdot 10^{-5}$	2.39	86.86	5.25	870	$1.1 \cdot 10^{-4}$	292
C02	$1.44 \cdot 10^{-5}$	1.14	45.73	5.00	869	$1 \cdot 10^{-4}$	293

The comparisons for case C01 (characterized by the highest initial flow velocity) clearly indicate the effect of unsteady friction and the boundary condition (real closing) on the modelled graphs. Both the boundary condition and unsteady friction may indirectly affect the duration of the successive pressure amplitudes [27]. When RC and UF are used, this duration will be extended. Owing to this, when considering the effect of unsteady friction, the simulation conformity can be regarded as satisfactory (Fig. 3d). As shown in Fig. 3a, the use of only quasi-steady resistance with a rapid instantaneous switching-over of the directional valve can result in a simulation graph significantly different from the real one. The discrepancy is most visible for the first amplitude, where the experimental graph gradually increases until it reaches its maximum value (at $t =$ 0.005 [s]), and then gradually decreases over time until it reaches its minimum at $t =$ 0.009 [s]. Whereas in the case of the simulated graph, the pressure rise and drop are instantaneous and out of phase (e.g. the sharp pressure drop at $t = 0.005$ [s]). These figures show that the device registering pressure jumps (the pressure sensor) captured the first three amplitudes. The next amplitudes are not visible in the graph, which is

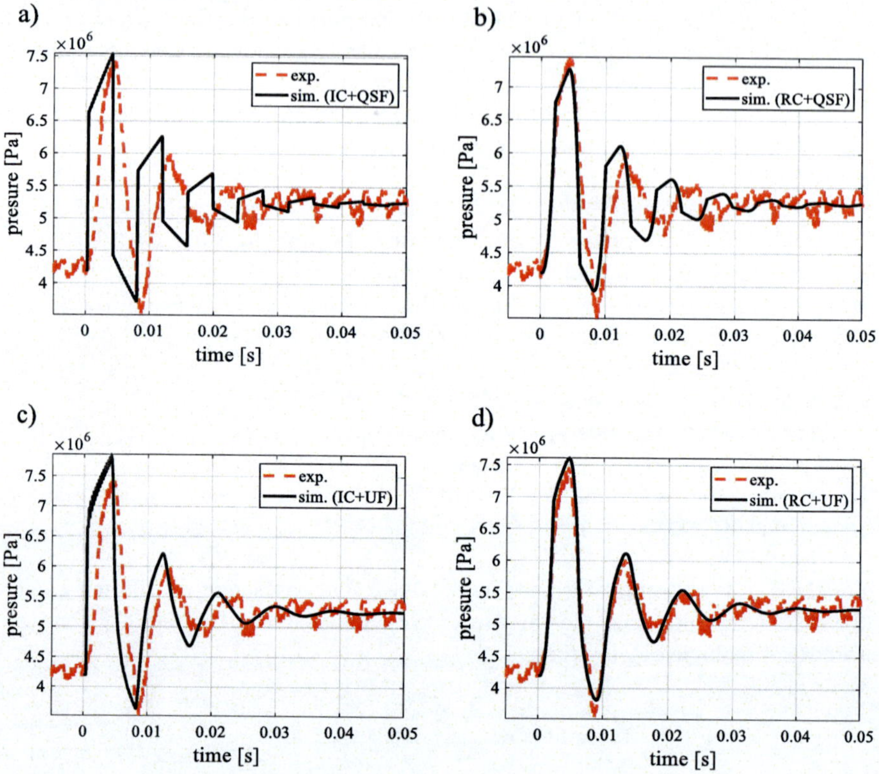

Fig. 3. Pressure variation in case C01 ($v_{01} = 2.39\,[\mathrm{m/s}]$).

due to the high noisiness of the experimental results performed in this micro-hydraulic pipeline system.

The second case, i.e. C02, was characterized by relatively low flow velocity $v_{02} = 1.14\,[\mathrm{m/s}]$ that has a lower Reynolds number than former case C01. Similarly, the measurement noise is also significant at this rate. Fortunately, the first three pressure amplitudes have been correctly registered, while the next amplitudes are very noisy. Meanwhile, the use of solely the quasi-steady model approach (Fig. 4a) can result in significant discrepancies (e.g., the third amplitude is modelled in counter phase) and that the use of a correct boundary condition (described by Eqs. (5) to (7)) coupled with unsteady friction (Fig. 4d) would be able to provide better fit of the simulated result to the experimental one. Furthermore, the comparison of results in Figs. 3 and 4 demonstrates that the system and measurement noise can provide more significant influence in case C02 than case C01, which indicates the resistance of initial Reynolds number flow state to the influence of system and measurement noise (i.e., noise tolerance).

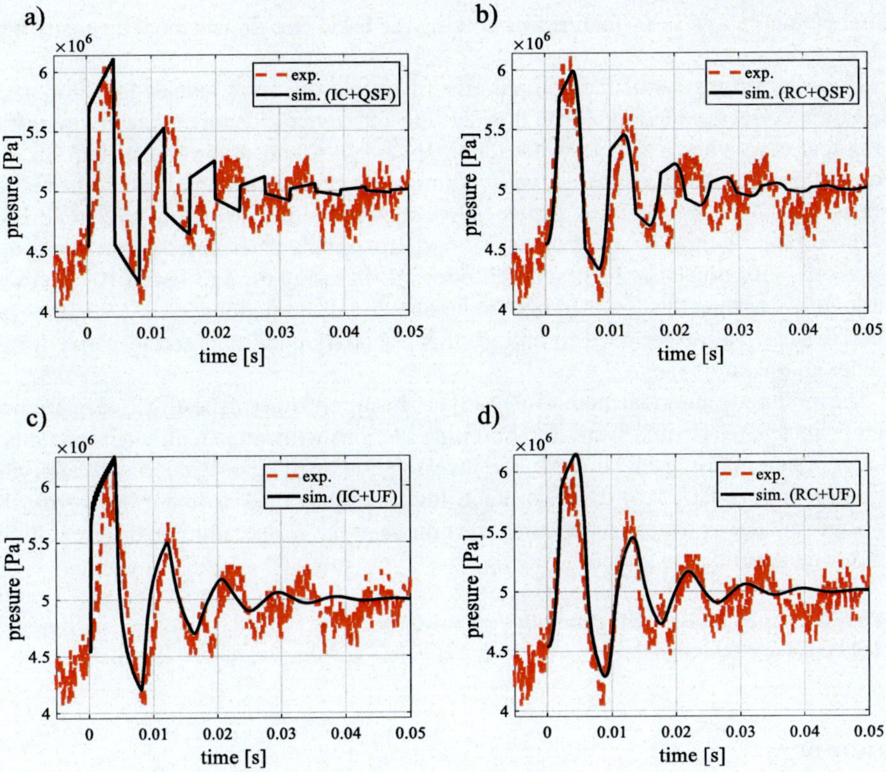

Fig. 4. Pressure variation in case C02 ($v_{02} = 1.14\,[\mathrm{m/s}]$).

Summing up, the comparisons of these two cases show that:

- in microscale conduits, in which the closing time of the flow cutting off element (a valve, a slide, etc.) is greater or equal to ¼ of the water hammer period $T_c \geq L/c$, the effect of this closure can be regarded as significant;
- when considering the real valve closing time in the modelling, this has a significant effect on the results of simulations (the shape of the modelled graphs is consistent with that of the experimental graphs);
- in the case of such microscale conduit sections, it is recommended to take unsteady friction into account since it also has an important influence on the final shape of the modelled amplitudes (e.g., comparing Fig. 3b and Fig. 4b (RC+QSF) with the results obtained using the model presented in this paper – Fig. 3d and Fig. 4d (RC+UF));
- the conventional model, commonly used to simulate flows in pipes with inner diameter $D > 10{\cdot}10^{-3}$ [m], is also suitable for modelling water hammers in pipes with relatively small diameters $D \leq 4 \cdot 10^{-3}$ [m] (usually regarded as microtubes [28, 29]).

5 Conclusions

Numerous comparative investigations were carried out as part of this research on fluid hammer in a micro-hydraulic pipeline system. The experimental results were compared with the numerical results obtained by solving the basic one-dimensional unsteady pipe flow equations.

The selected results of the comparative investigations have revealed an important role of the correct modelling of the flow cutting-off device closing characteristic (close to the real one) when a simple water hammer ($T_c \leq \frac{2L}{c}$) occurs in the system. In real microscale systems, directional valve switching time is never instantaneous. The results discussed in this paper and the analysis have shown that the switching-over time is of key importance in this kind of systems. The experimentally determined (from the results of pressure pulsations – at the first amplitude) valve closing time $T_c = 2.6{\cdot}10^{-3}$ [s] was a little different than the assumed one in the numerical simulations $T_c = 1.96{\cdot}10^{-3}$ [s], which indicates a further need to modify this boundary condition and to verify it in a broader comparative range.

The modified numerical method [16, 23] has been confirmed to be effective in predicting transient states in such systems (short pipe sections with very small inner diameter). As a next step in our research work, the investigations will be continued with applying the other fluid (oil ISO10) that has much lower viscosity, so as to further verify the developed model as well as to examine the effects of the temperature on different fluids in micro-hydraulic pipe systems.

Acknowledgements. Bergant gratefully acknowledges the support of Slovenian Research Agency conducted through the research project L2-1825 and the programme P2-0162.

References

1. Karadzić, U., Janković, M., Strunjaš, F., Bergant, A.: Water hammer and column separation induced by simultaneous and delayed closure of two valves. Strojniški vestnik – J. Mech. Eng. **64**(9), 525–535 (2018)
2. Iffa, R.B., Triki, A.: Assessment of inline technique-based water hammer control strategy in water supply systems. J. Water Supply: Res. Technol.-Aqua **68**(7), 562–572 (2019)
3. Jiang, D., Lu, Q., Liu, Y., Zhao, D.: Study on pressure transients in low pressure water-hydraulic pipelines. IEEE Access **7**, 80561–80569 (2019)
4. Mei, C.C., Jing, H.: Effects of thin plaque on blood hammer - an asymptotic theory. Eur. J. Mech. - B/Fluids **69**, 62–75 (2018)
5. Jung, B.S., Karney, B.: A practical overview of unsteady pipe flow modelling: from physics to numerical solutions. Urban Water J. **08**, 502–508 (2016)
6. Stosiak, M.: The impact of hydraulic systems on the human being and the environment. J. Theoret. Appl. Mech. **53**(2), 409–420 (2015)
7. Moghaddas, S.M.J., Samani, H.M.V., Haghighi, A.: Transient protection optimization of pipelines using air-chamber and air-inlet valves. KSCE J. Civil Eng. **21**(5), 1991–1997 (2017)
8. Wylie, E.B., Streeter, V.L.: Fluid Transients in Systems. Prentice-Hall Inc., Englewood Cliffs (1993)
9. Streeter, V.L., Lai, C.: Water-hammer analysis including fluid friction. J. Hydraul. Division **88**(3), 79–112 (1962)
10. Trikha, A.K.: An efficient method for simulating frequency–dependent friction in transient liquid flow. ASME J. Fluids Eng. **97**(1), 97–105 (1975)
11. Schohl, G.A.: Improved approximate method for simulating frequency – dependent friction in transient laminar flow. ASME J. Fluids Eng. **115**(3), 420–424 (1993)

12. Vítkovský, J.P., Stephens, M.L., Bergant, A., Simpson, A.R., Lambert, M.F.: Efficient and accurate calculation of Vardy–Brown unsteady friction in pipe transients. In: Proceedings of the 9th International Conference on Pressure Surges, Chester, UK, pp. 405–419 (2004)
13. Vardy, A.E., Brown, J.M.B.: Evaluation of unsteady wall shear stress by Zielke's method. J. Hydraul. Eng. **136**(7), 453–456 (2010)
14. Meniconi, S., Duan, H.F., Lee, P.J., Ghidaoui, M.S., Ferrante, M.: Further developments in rapidly decelerating turbulent pipe flow modeling. J. Hydraul. Eng. – ASCE **140**(7), 04014028 (2014)
15. Duan, H.F., Meniconi, S., Lee, P.J., Brunone, B., Ghidaoui, M.S.: Local and integral energy based evaluation for the unsteady friction relevance in transient pipe flows. J. Hydraul. Eng. – ASCE **143**(7), 04017015 (2017)
16. Urbanowicz, K.: Fast and accurate modelling of frictional transient pipe flow. J. Appl. Math. Mech. **98**(5), 802–823 (2018)
17. Duan, H.F., Ghidaoui, M.S., Lee, P.J., Tung, Y.K.: Relevance of unsteady friction to pipe size and length in pipe fluid transients. J. Hydraul. Eng. – ASCE **138**(2), 154–166 (2012)
18. Johnston, N.: The transmission line method for modelling laminar flow of liquid in pipelines. Proc. Inst. Mech. Eng. Part I: J. Syst. Control Eng. **226**(5), 586–597 (2012)
19. Lee, P.J., Duan, H.F., Ghidaoui, M.S., Karney, B.W.: Frequency domain analysis of pipe fluid transient behavoirs. J. Hydraul. Res. – IAHR **51**(6), 609–622 (2013)
20. Duan, H.F., Che, T.C., Lee, P.J., Ghidaoui, M.S.: Influence of nonlinear turbulent friction on the system frequency response in transient pipe flow modelling and analysis. J. Hydraul. Res. – IAHR **56**(4), 451–463 (2018)
21. Zielke, W.: Frequency-dependent friction in transient pipe flow. ASME J. Basic Eng. **90**(1), 109–115 (1968)
22. Ghidaoui, M.S., Zhao, M., McInnis, D.A., Axworthy, D.H.: A review of water hammer theory and practice. Appl. Mech. Rev. **58**(1), 49–76 (2005)
23. Urbanowicz, K.: Analytical expressions for effective weighting functions used during simulations of water hammer. J. Theoret. Appl. Mech. **55**(3), 1029–1040 (2017)
24. Pérez-Sánchez, M., López-Jiménez, A.P., Ramos, H.M.: PATs Operating in water networks under unsteady flow conditions. Control Valve Manoeuvre and Overspeed Effect. Water **10**(4), 529 (2018)
25. Jensen, R.K., Larsen, J.K., Lassen, K.L., Mandø, M., Andreasen, A.: Implementation and validation of a free open source 1D water hammer code. Fluids **3**(3), 64 (2018)
26. Urbanowicz, K.: Computational compliance criteria in water hammer modelling. In: E3S Web of Conferences, vol. 19, pp. 302–315 (2017)
27. Bergant, A., Karadžić, U.: Developments in valve-induced water hammer experimentation in a small-scale pipeline apparatus. In: Proceedings of the 12th International Conference on Pressure Surges, Dublin, Ireland, pp. 639–652 (2015)
28. Vekariya, P.B., Subbaiah, R., Mashru, H.H.: Hydraulics of microtube emitters: a dimensional analysis approach. Irrig. Sci. **29**, 341–350 (2011)
29. Ji, H., Li, H., Huang, Z., Wang, B., Li, H.: Measurement of gas-liquid two-phase flow in micro-pipes by a capacitance sensor. Sensors **14**, 22431–22446 (2014)

Check for updates

The Electro-Hydraulic Lifting and Leveling System for the Bricklaying Robot

Piotr Wos(✉), Ryszard Dindorf, and Jakub Takosoglu

Faculty of Mechatronics and Machine Design,
Kielce University of Technology, Kielce 25-314, Poland
wos@tu.kielce.pl

Abstract. The leveling process for the hydraulic lifting mechanism of the bricklaying robot is complicated. The article presents the designing process and method of controlling a four-cylinder electro-hydraulic (EH) servo drives system. As a part of the research, a mathematical model was established and the process of leveling a mobile support platform of the bricklaying robot was optimized. A kinematic analysis of the lifting mechanism was carried out. The mathematical model of the hydraulic system was presented. The work scheme was designed on the base of the strategy of leveling and planning the displacement trajectory of individual EH servo drive. The designed control system takes into account not only position errors for individual drive axis, but also synchronization errors with neighboring axes. The synchronization function for the control system, which takes into account the positioning errors of individual drive axis, was specified.

Keywords: Bricklaying robot · Electro-hydraulic servo drive · Leveling system

1 Introduction

Masonry works are one of the very early human craftsmanship. However, this discipline has not achieved a high degree of automation. A number of attempts were made to develop mobile construction works, the most advanced of which were the projects [1, 2]. These studies date back to the early 90s of the 20th century, when the motivation for designers was to improve the efficiency and savings of building construction, mainly through the use of the ability of machines to carry larger loads [4]. These works, although very advanced, did not find full practical application because they were not flexible enough to adapt and respond to various situations occurring on the construction site. Thanks to the introduction of digital technologies in architecture, research centers such as Harvard GSD, Carnegie Mellon and the University of Stuttgart have created research laboratories for the construction of construction works [8]. These centers have developed various technologies and prototype elements aimed at increasing the role of robots in construction works, in particular bricklaying. In addition, some commercial masonry works were developed, such as Hadrian and Semi Automated bricklaying robot SAM, which perform masonry processes. The common feature of these projects is that they

© The Editor(s) (if applicable) and The Author(s), under exclusive license
to Springer Nature Switzerland AG 2021
J. Stryczek and U. Warzyńska (Eds.): NSHP 2020, LNME, pp. 216–227, 2021.
https://doi.org/10.1007/978-3-030-59509-8_19

have all tried to replace building materials with new materials that may be more suitable for the use of robots [5].

Contemporary works intended for construction works can be divided into several categories:

- remote controlled robots, which are controlled by a human using a remote control, where decisions are made by the operator based on the collected information.
- programmable construction machines in which the operator can choose from a function menu or teach a robot to perform a new function.
- intelligent systems in which unmanned, fully automated construction machines perform tasks without direct human intervention. This category may also include semi-automatic machines that are not completely unmanned and some level of human control is present.

Because bricklaying is a repetitive process, we can easily automate it. To carry out the masonry process, the robot must detect and then take a brick from the warehouse. Then the brick should be placed in the right position on a properly prepared base - mortar. During capturing and laying bricks, the robot should avoid collisions with obstacles. For such a dynamic area as a construction site, obstacles detection requires the active motion planning techniques based on real-time sensory data. In addition, the bricklaying machine must be "aware" of the progress of wall evolution. To introduce this technology during masonry work, the following basic technical problems must be solved: robot movement along the erected wall, accurate reference of the robot working area axis relative to the entire wall (building), creating and programming complex interactions within the working area for the process collecting and laying of individual bricks, creating and programming interactions for the process of filling joints with mortar [3].

As a part of the research and development works, a design of the Bricklaying Robotic System (BRS) prototype was made (Fig. 1), meeting the expectations of construction companies in the field of robotization of time-consuming and arduous manual construction works.

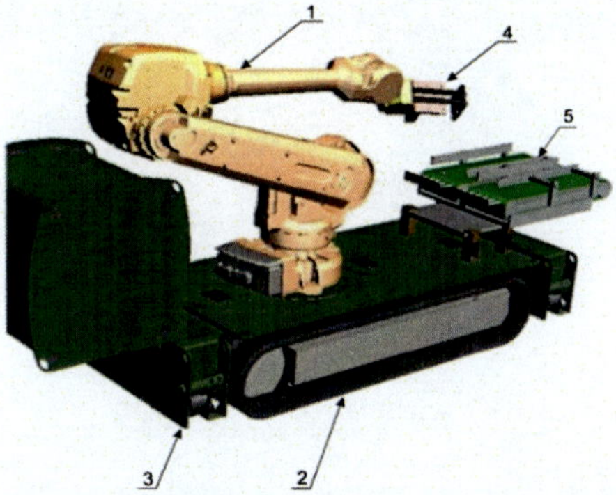

Fig. 1. Model view Bricklaying Robotic System (BRS) 1 - basic robot (BR), 2 - mobile hydraulic module (MHM) 3 - support module (SM), 4 - gripper 5 - bricks feeder.

The Bricklaying Robotic System (BRS) consists of a basic robot (BR) (1) equipped with a gripper (4). Basic robot (BR) is set on a mobile hydraulic module (MHM) (2). The BRS is equipped with a lifting and positioning control system as well as a control system for working movements of the basic robot. Concept of the control system for leveling as well as the system of indicators for measuring and controlling BRS work parameters were developed. BRS will have a feeder with a reservoir (5) for various building materials used during bricklaying, and a mortar feeder. Figure 2 and 3 respectively show: the process of bricklaying simulation and the working range of the designed BRS prototype.

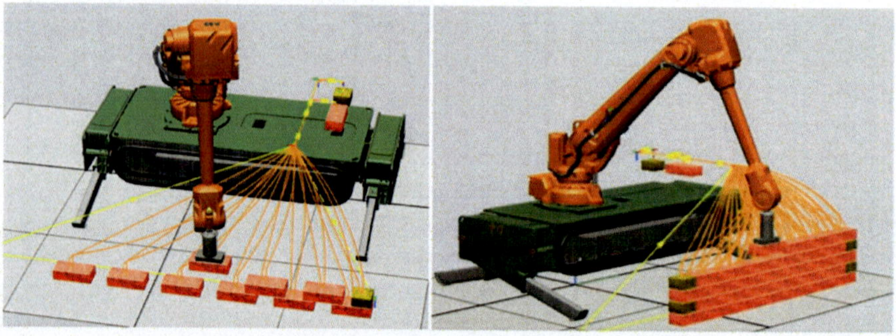

Fig. 2. Simulation tests of the bricklaying process.

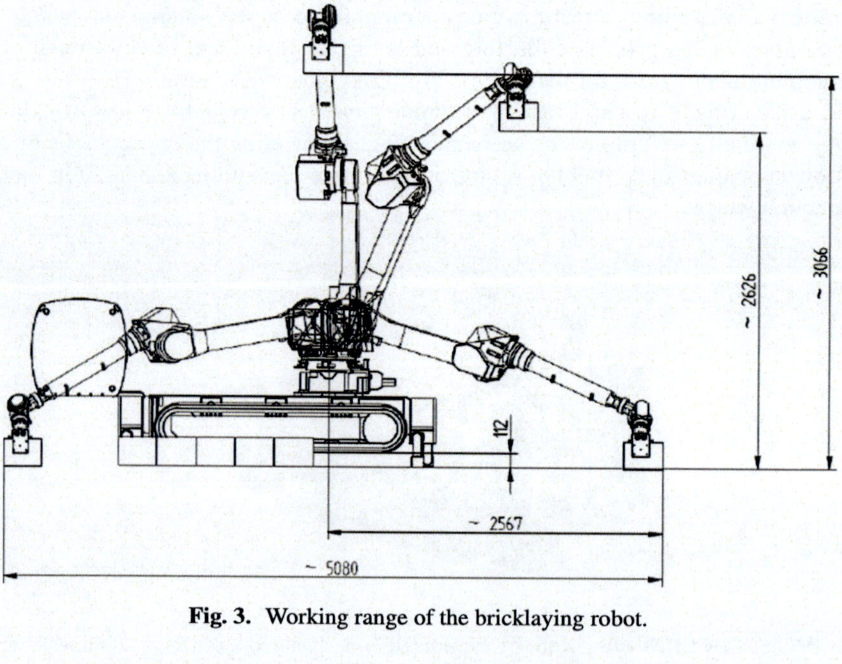

Fig. 3. Working range of the bricklaying robot.

2 Dynamic Model of the MHM

Mobile hydraulic module (MHM) is a support platform for further development of devices used in cooperating construction works, such as: construction manipulator, material warehouse or lifting platform. It is made of a mobile platform, which ensures the movement of MHM and two supporting modules (SM). Each of the supporting modules has extendable legs with a cross structure, which are lowered by means of two electro-hydraulic servo drives. The main task of the platform is to move the BRS to the designated working area and then precisely raise the BRS to the desired height along with its leveling. Both of these tasks are significantly influenced by the dynamic behavior of the mobile module itself and the mechanisms mounted on it. Figure 4 shows the model of the mobile hydraulic module (MHM).

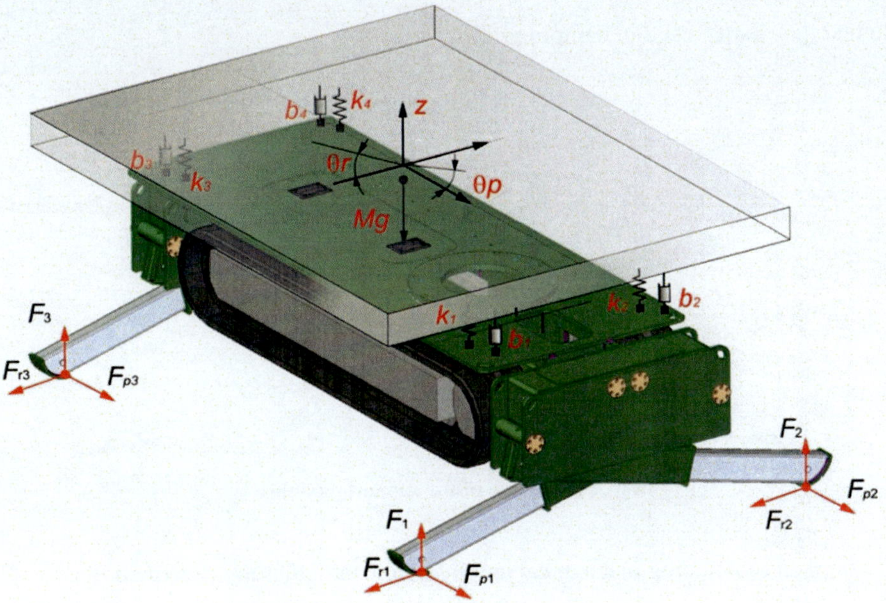

Fig. 4. Mobile hydraulic module (MHM).

Calculation diagram of the single supporting module (SM) consisting of the double-acting cylinder with one-sided piston rod and proportional directional control valve is presented in the Fig. 5. The mass of the load M was placed centrally on a plane $2a$ wide and $2b$ long for simplicity.

For the module (MHM) (Fig. 4) the following equations of motion along the direction of gravity and rotation around the axis x (r – roll) and y (p – pitch) can be obtained:

$$M\ddot{z}_M = \sum_{i=1}^{4} F_i + \sum_{i=1}^{4} F_{ri}\sin(\theta_r) + \sum_{i=1}^{4} F_{pi}\sin(\theta_p) - Mg$$

$$J_r\ddot{\theta}_r = \sum_{i=1}^{4}((F_i + F_{ri}\sin(\theta_r))a_{ri}$$

$$J_p\ddot{\theta}_p = \sum_{i=1}^{4}((F_i + F_{pi}\sin(\theta_p))b_{pi} \tag{1}$$

where: z_M – module center of gravity position, M – total mass of the load, g – gravitational constant, F_i, F_{ri}, F_{pi} – reaction forces for electrohydraulic servo drives for $i = 1, 2, \cdots n$, a_{ri}, b_{pi} – arms length for moments of forces, J_r, J_p – moments of module inertia, θ_r, θ_p – angles of rotation. If we assume, that θ_r, θ_p is small then $\sin(\theta_r) \approx 0$ and $\sin(\theta_p) \approx 0$.

In contact with the hydraulic supports with the ground, the following occurs:

$$G_i = k_i(z_0 - z_i) + b_i(\dot{z}_0 - \dot{z}_i), \quad i = 1, 2, \cdots, n \tag{2}$$

where: k_i, b_i – stiffness and damping ratio.

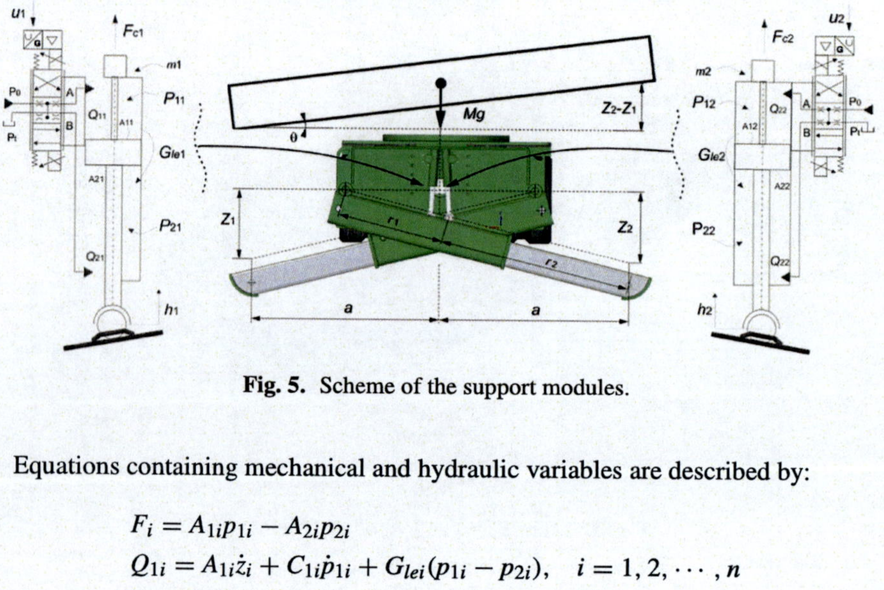

Fig. 5. Scheme of the support modules.

Equations containing mechanical and hydraulic variables are described by:

$$\begin{aligned} F_i &= A_{1i}p_{1i} - A_{2i}p_{2i} \\ Q_{1i} &= A_{1i}\dot{z}_i + C_{1i}\dot{p}_{1i} + G_{lei}(p_{1i} - p_{2i}), \quad i = 1, 2, \cdots, n \\ Q_{2i} &= A_{2i}\dot{z}_i - C_{2i}\dot{p}_{2i} + G_{lei}(p_{1i} - p_{2i}) \end{aligned} \tag{3}$$

where: Q_{1i} – volumetric flow rate, C_{1i}, C_{2i} – fluid capacitances in the cylinder chambers, G_{lei} – coefficients of leakages in the cylinder, A_{1i}, A_{2i} – effective area of cylinder's piston, p_{1i}, p_{2i} – pressure in individual cylinder's chambers.

The hydraulic forces developed by the actuators are given by:

$$[p_{1i} \cdot A_{1i} - p_{2i} \cdot A_{2i}] = \left[A_{1i} \cdot p_0 - A_{2i} \cdot p_T - \frac{A_{1i}^3 + A_{2i}^3}{(K_1 \cdot u + K_{0.1})^2} \cdot h_i^2\right], \quad i = 1, 2, \cdots n \tag{4}$$

where: u_i – control inputs, $K_1, K_{0.1}$ – constants factors, p_0 – supply pressure, p_T – pressure in the tank which correspond to the main and leakage valve flow paths.

The equation of motion of each hydraulic cylinder piston is:

$$m_i\ddot{z} = (P_{1i} - P_{2i})A_{2i} - b_i\dot{z} - F_{ci} \tag{5}$$

and the equations of motion of the support module are:

$$\begin{cases} F_1 + F_2 - Mg = M\,\ddot{z}_M \\ (F_1 + F_2)a = J\ddot{\theta} \end{cases} \tag{6}$$

where: $x_M = (z_1 + z_2)/2$, $F_i = F_{ci}\frac{r_1}{(r_1+r_2)}$ and $\theta = (z_1 - z_2)/2a$.

Due to the symmetrical shape of MHM, it can be assumed that the center of gravity of the platform is in the middle. The system can be described by the formulas:

$$\begin{cases} \boldsymbol{M}_L\ddot{\boldsymbol{z}}_q = \boldsymbol{L}_m\boldsymbol{G} - \boldsymbol{M}_g \\ \boldsymbol{G} = \boldsymbol{K}_i(z_0 - z_i) + \boldsymbol{B}_i(\dot{z}_0 - \dot{z}_i), \quad i = 1, 2, \cdots, n \\ m\ddot{z}_L = \boldsymbol{F} - \boldsymbol{G} \end{cases} \tag{7}$$

where:

$\mathbf{G} = [G_1, G_2, \cdots, G_n]^T$, is vector loading forces,
$\mathbf{z}_q = [z_1, z_2, \cdots, z_n]^T$, is platform displacement vector,
$\mathbf{z}_h = [z_{h1}, z_{h2}, \cdots, z_{hn}]^T$, is hydraulic cylinder displacement vector,
$\mathbf{L}_m = \begin{bmatrix} 1 & 1 & \cdots & 1 \\ a_{r1} & a_{r2} & \cdots & a_{rn} \\ b_{p1} & b_{p2} & \cdots & b_{pn} \end{bmatrix}$, is an arm matrix with a fixed moment relative to the axis r (roll) and p (pitch),
$\mathbf{M}_g = \begin{bmatrix} M \cdot g \\ 0 \\ 0 \end{bmatrix}$, is a mass load matrix,
$\mathbf{F} = MA \cdot \begin{bmatrix} A_{11} \cdot p_{11} - A_{21} \cdot p_{21} - m_1 g \\ A_{12} \cdot p_{12} - A_{22} \cdot p_{22} - m_2 g \\ A_{13} \cdot p_{13} - A_{23} \cdot p_{23} - m_3 g \\ \cdots \\ A_{1n} \cdot p_{1n} - A_{2n} \cdot p_{2n} - m_n g \end{bmatrix}$, is the matrix of forces generated by hydraulic cylinders, $MA = \frac{r_1}{(r_1+r_2)}$ is mechanical advantage.
$\mathbf{M}_L = diag([M\ J_r\ J_p])\mathbf{L}_{rp}^{-1}$, is the load inertia matrix,
were: $\mathbf{L}_{rp} = \begin{bmatrix} 1 & a_{r1} & b_{p1} \\ 1 & a_{r2} & b_{p2} \\ \vdots & \vdots & \vdots \\ 1 & a_{rn} & b_{pn} \end{bmatrix}$,
$\mathbf{K}_i = diag([k_1\ k_2\ \cdots\ k_n])$, is the contact stiffness matrix,
$\mathbf{B}_i = diag([b_1\ b_2\ \cdots\ b_n])$, is the contact viscous damping coefficient matrix.

3 Lifting Process MHM

The process of extending the support module (SM) takes place in two stages: the extension of the module legs to the maximum elongation, followed by the MHM lifting process. The lifting process can be divided into three stages. The first stage occurs when the cross-drive cylinders quickly extend to the ground without load. After placing the legs on the ground, the second stage occurs, when the support cylinders synchronously extend to a certain height. This is followed by a leveling step when the MMH actuators move according to the horizontal slope relationship to complete the leveling. After leveling, the MHM platform is locked mechanically.

3.1 Synchronous Movement

The problem of synchronization of many linear hydraulic cylinders arises in heavy, hydraulically controlled applications such as lifting devices, construction and agricultural machinery [7, 9]. This problem is most often solved by introducing to each parallel actuator branch an additional hydraulic resistance adjusted manually or automatically. In the case of manually adjustable hydraulic resistance, this boils down to the use of a throttle valve, and in the case of automatically adjustable hydraulic resistance, the use of a flow controller. In both cases, the idea of solving the problem of motion synchronization boils down to creating additional pressure drops in the power supply branches of individual drives so that the pressure in the pump discharge line, determined from the equilibrium pressure equations arranged for the power branches of individual drives, is the same [6, 7].

During synchronization, the differential position errors between the hydraulic actuators converge to zero. The position error of i-th hydraulic actuator is defined as follows:

$$e_i(z) = z_i^d(t) - z_i(t), \tag{8}$$

where: $z_i^d(t)$ is the desired generated length and $z_i(t)$ is the actual measured length of i-th hydraulic actuator.

The position error vector of four actuators is:

$$\mathbf{e}(t) = [e_1(t), e_2(t), e_3(t), e_4(t)]^T \tag{9}$$

If the ratio of the actual position of each hydraulic actuator at each sampling time is equal to that of all other actuators, then the MMH moves in a synchronous manner. The synchronization goal is understood to be as follows:

$$e_1(t) = e_2(t) = e_3(t) = e_4(t) \tag{10}$$

Thus, the synchronization error is defined as:

$$\begin{cases} \varepsilon_1(t) = e_1(t) - e_2(t) \\ \varepsilon_2(t) = e_2(t) - e_3(t) \\ \varepsilon_3(t) = e_3(t) - e_4(t) \\ \varepsilon_4(t) = e_4(t) - e_1(t) \end{cases} \tag{11}$$

where: $\boldsymbol{\varepsilon}(t) = [\varepsilon_1(t), \varepsilon_2(t), \varepsilon_3(t), \varepsilon_4(t)]^T$ and $\mathbf{T} = \begin{bmatrix} 1 & -1 & 0 & 0 \\ 0 & 1 & -1 & 0 \\ 0 & 0 & 1 & -1 \\ -1 & 0 & 0 & 1 \end{bmatrix}$ is 4×4 synchronization transformation matrix.

The position control system is designed by combining the position error and synchronization error. Herein, in conceiving a cross-coupled controller (CCC), the cross-coupling error c_i includes the position error e_i and the synchronization error ε_i:

$$c_i(t) = e_i(t) + \beta_i \int_0^t \varepsilon_i(\omega)d\omega, \tag{12}$$

where β_i is positive coupling parameter which determines the weight of the synchronization error, wherein as β_i increases, the synchronization error ε_i decreases, ω is a parameter at a variable time, from $t = 0$ to t.

The position synchronization error looks for the difference between the position errors of many axes and is based on the synchronization function (12).

The block example diagram of the synchronization function designed in Matlab/Simulink is shown in the Fig. 6 [10].

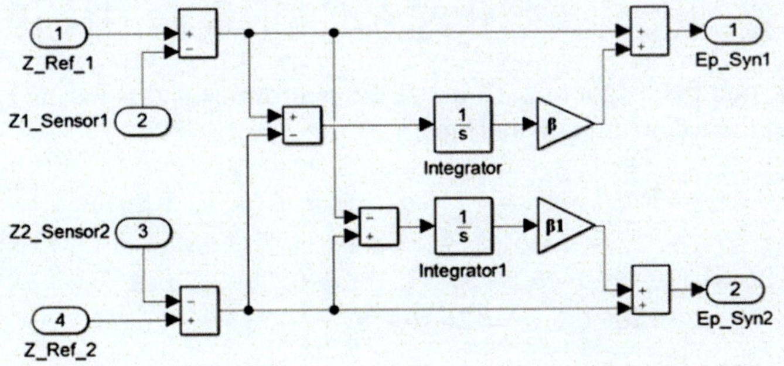

Fig. 6. Scheme a subsystem of the synchronization controller in Matlab/Simulink.

3.2 Leveling

To obtain the relationship between the slope and displacement angles of hydraulic cylinders, it was assumed that the angle θ_p (pitch) is measured from the X axis and the angle θ_r (roll) from the Y axis in the direction of the Z axis. The purpose of leveling is to set the surface of the robot platform so that these angles are zero. The determined coordinate system is shown in Fig. 7.

$$R_x(\theta_r) = \begin{bmatrix} 1 & 0 & 0 \\ 0 & \cos\theta_r & \sin\theta_r \\ 0 & -\sin\theta_r & \cos\theta_r \end{bmatrix}, R_y(\theta_p) = \begin{bmatrix} \cos\theta_p & 0 & \sin\theta_p \\ 0 & 1 & 0 \\ -\sin\theta_p & 0 & \cos\theta_p \end{bmatrix} \tag{13}$$

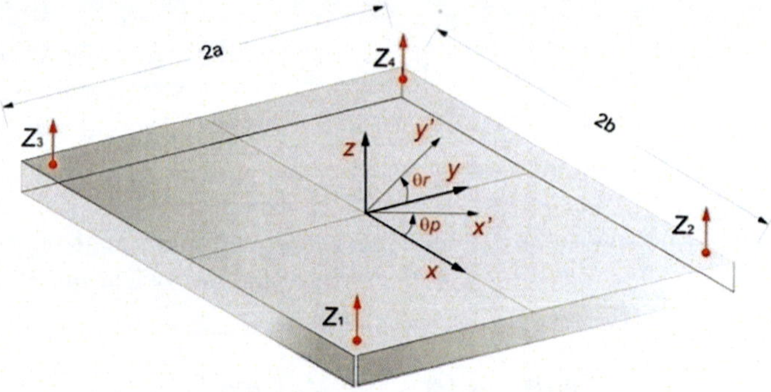

Fig. 7. Coordinate system for MHM.

In fact, during the leveling process the deviation in the platform level is small. With this assumption, we can determine the following relationships: $\cos\theta_r = \cos\theta_p = 1$ and $\sin\theta_r = \theta_r$, $\sin\theta_p = \theta_p$.

The equation for the coordinate transformation of a leveling plane is obtained as follows:

$$\begin{bmatrix} x' \\ y' \\ z' \end{bmatrix} = R_x(\theta_r)R_y(\theta_p)\begin{bmatrix} x \\ y \\ z \end{bmatrix};\begin{bmatrix} x' \\ y' \\ z' \end{bmatrix} = \begin{bmatrix} 1 & 0 & \theta_p \\ 0 & 1 & \theta_r \\ -\theta_p & \theta_r & 1 \end{bmatrix}\begin{bmatrix} x \\ y \\ z \end{bmatrix} \tag{14}$$

Assuming that the height of each leg of the platform supports means z_1, z_2, z_3 and z_4, we can get the following relationships:

$$\begin{cases} z_1 = 0,\ z_2 = 2a \cdot \theta_r,\ z_3 = -2b \cdot \theta_p,\ z_4 = -2b \cdot \theta_p + 2a \cdot \theta_r, \\ z_1 = -2a \cdot \theta_r,\ z_2 = 0 \cdot \theta_r,\ z_3 = -2b \cdot \theta_p - 2a \cdot \theta_r,\ z_4 = -2b \cdot \theta_p, \\ z_1 = 2b \cdot \theta_p,\ z_2 = 2b \cdot \theta_p + 2a \cdot \theta_r,\ z_3 = 0,\ z_4 = 2a \cdot \theta_r, \\ z_1 = 2b \cdot \theta_p - 2a \cdot \theta_r,\ z_2 = 2b \cdot \theta_p,\ z_3 = -2a \cdot \theta_r,\ z_4 = 0. \end{cases} \tag{15}$$

Each zero value of offsets z_1, z_2, z_3 and z_4 means the highest position in relation to the other legs of the supports. Figure 8 shows the control system diagram for MHM with synchronization and leveling.

For the simulation purposes constant values of parameters were accepted and implemented (Table 1). Figures 9, 10 and 11 show graphs results of the synchronization and leveling processes.

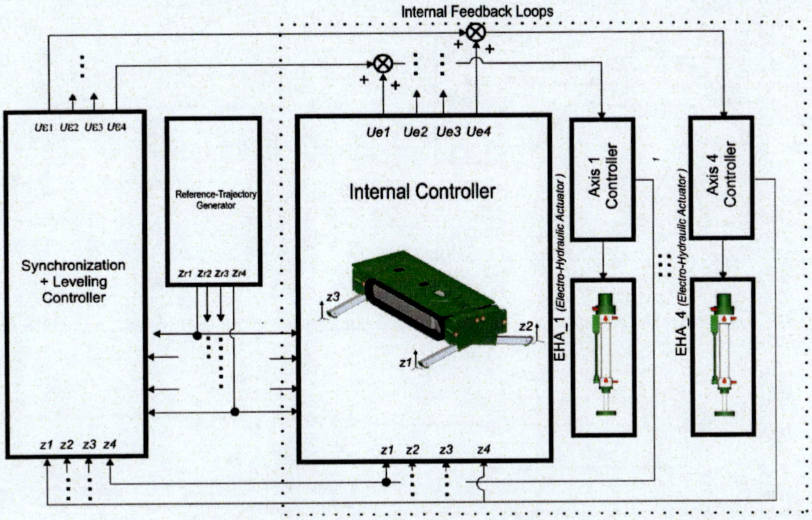

Fig. 8. The controller block diagram of the MHM with synchronization.

Table 1. Simulation parameters.

p_s	16 MPa	M	1000 kg
A_{1i}	1.256×10^{-3} m^2	A_{2i}	1.134×10^{-3} m^2
$2a$	1568 mm	$2b$	2358 mm

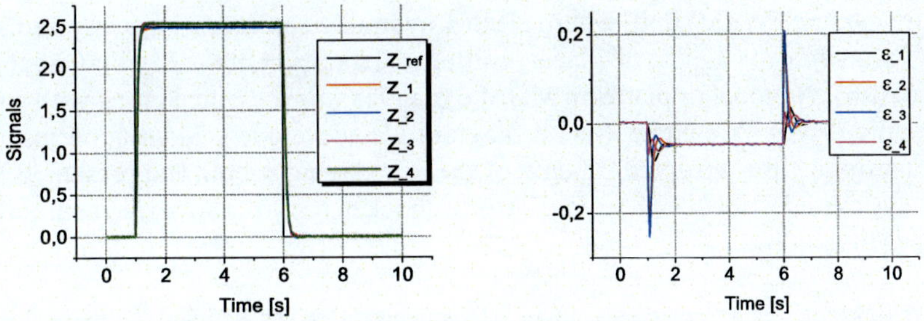

Fig. 9. Synchronization error ε_i for references rectangular input signals u_{iref}.

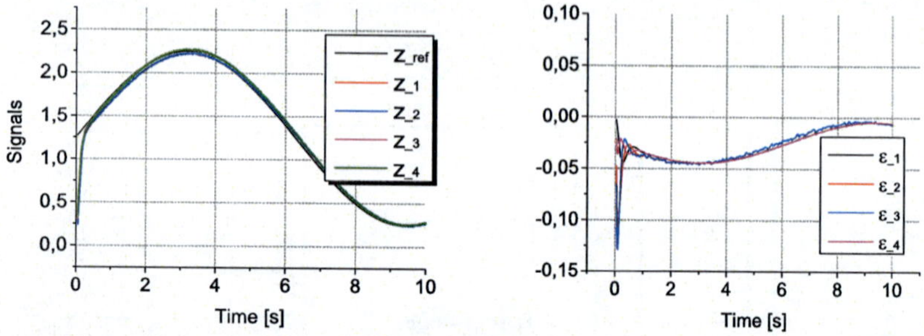

Fig. 10. Synchronization error ε_i for references sinusoidal input signals $u_{iref} = 0.08$ Hz.

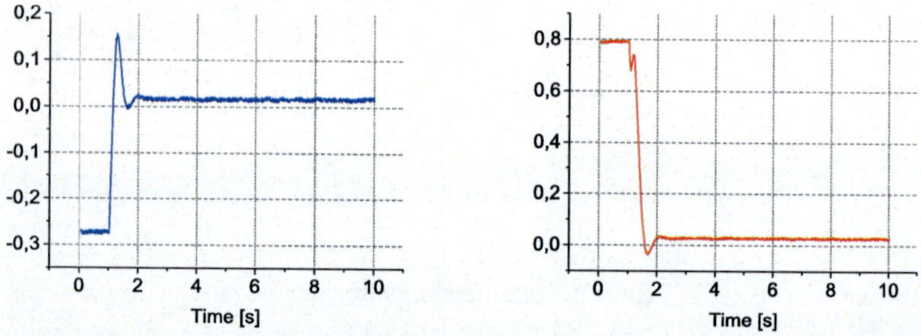

Fig. 11. Leveling angles signals: θ_r – roll and θ_p – pitch.

4 Summary

A technical solution was presented to the lifting and leveling system of the mobile hydraulic module (MHM) unit, where electro-hydraulic control and drive systems were used. This assembly is used to displace, position and transport of the Bricklaying Robotic System (BRS). A model of platform MHM dynamics was presented along with a model of hydraulic drives. A control system diagram was developed. Simulation tests were carried out to test the accepted scheme of the bricklaying robot. The presented MHM control system uses feedback from platform bracket position errors, synchronization errors and tilt angles.

Acknowledgment. This research was financially supported by The National Centre for Research and Development in Poland (Grant No. POIR.04.01.02-00-0045/18-00).

References

1. Aguair, M.L., Behdinan, K.: Design prototyping and programming of a bricklaying robot. J. Student Sci. Technol. **8**(3), 1–10 (2015)
2. Andres, J., Bock, T., Gebhart, F.: First results of the development of the masonry robot system ROCCO. In: Proceedings of the 11th International Symposium on Automation and Robotics in Construction, pp. 87–93 (1994)

3. Ardiny, H., Witwicki, S., Mondada, F.: Are autonomous mobile robots able to take over construction? A review. Int. J. Robot. **4**(3), 10–21 (2015)
4. Chamberlain, D., Ala, S., Watson, J., Speare P.: Masonry construction by an experimental robot. In: Proceedings of 9th International Symposium on Antenna and Robotics in Construction, Tokyo (1992)
5. Helm, V., Ercan, S., Gramazio, F., Kohler, M.: Mobile robotic fabrication on construction sites DimRob. In: IEEE International Conference on Intelligent Robotics System, pp. 4335–4341 (2012)
6. Kassem, S., El-Din, T.S., Helduser, S.: Motion synchronization enhancement of hydraulic servo cylinders for mould oscillation. Int. J. Fluid Power **13**(1), 51–60 (2012)
7. Ning, S., Wang, S.: Synchronization control to the hydraulic width system in hot rolling. In: 2010 International Conference on Measuring Technology and Mechatronics Automation, pp. 710–713. IEEE (2010)
8. Spath, I.D., Schmidt, I.J., Andres D.I.J.: Robot systems for residential building construction: integration and operation. In: Proceedings of the 14th ISARC, pp. 185–197 (1997)
9. Sun, D.: Position synchronization of multiple motion axes with adaptive coupling control. Automatica **39**(6), 997–1005 (2003)
10. Wos, P., Dindorf, R.: Synchronized trajectory tracking control of 3-DoF hydraulic translational parallel manipulator (2015)

Operating Problems of Lubrication of Friction Nodes in Mining Machines Working in an Aggressive Environment

Grzegorz Wszelaczyński[1], Dymitry Capanidis[2], Maciej Paszkowski[2], and Tadeusz Leśniewski[2](✉)

[1] Systemy Smarowania Maszyn, (A.A.L.S. S.A), Legnica, Poland
[2] Department of Fundamentals of Machine Design and Mechatronic Systems, Wroclaw University of Science and Technology, 27 Wybrzeże Wyspiańskiego Street, 50-370 Wrocław, Poland
tadeusz.lesniewski@pwr.edu.pl

Abstract. The paper presents problems related to the lubrication of moving machine nodes that are operating in an aggressive environment occurring in copper mines. The analysis was based on the exploitation of self-propelled wheeled machines, such as transport vehicles, loaders, drilling and anchoring cars. In self-propelled machines there are from 15 to 40 moving parts exposed to high impact loads and stresses. These are mainly bolted joints performing swinging motion as well as rotary motion at low rotational speeds up to 10 rpm. Observations and operational tests of several dozen manifolds and pumps operating in the environment occurring in KGHM Polska Miedź copper mines were carried out during two years. It was stated that elements made of aluminum or its alloys should not be used, and should be replaced with bronze (the proposal results in the increase of weight and costs of the system, but observational studies have shown that the durability and reliability of pumps increased more than three times, and the durability of the manifolds with the bronze body has increased more than five times). It was recommended to avoid the use of electronic systems on actuators in central lubrication systems (control systems should be encapsulated) and to use pumps with a hydraulic drive, which is commonly used in the working elements of machines used in a copper mine. The proposed lubrication system made of bronze and with hydraulic drive is several times more durable and cheaper to operate compared to the central lubrication systems used so far.

Keywords: Mining machinery · Exploitation · Grease · Central multi-circuit lubrication · Progressive lubrication

1 Introduction

Skillful and effective lubrication of machines is very important from the profitability of modern industry point of view. Primarily, it is associated with the extension of the service life of lubricated equipment, reduction of production interruptions and reduction

© The Editor(s) (if applicable) and The Author(s), under exclusive license to Springer Nature Switzerland AG 2021
J. Stryczek and U. Warzyńska (Eds.): NSHP 2020, LNME, pp. 228–238, 2021.
https://doi.org/10.1007/978-3-030-59509-8_20

of production costs. Appropriate selection of lubricant is the easiest way to extend the life of machines [1–5]. In many cases, a slight modification or change of the lubricant [6, 7] extends the life of the element, slightly (or not at all) increasing the cost of use. It is much cheaper procedure than changing materials from which the cooperating elements are made.

Another and even the main condition for effective lubrication is its automation. Lubrication systems primarily enabled the precise supply of the right amount of lubricant to all tribological nodes, and also minimized the possibility of omission by the lubricating person hardly accessible and invisible at first glance friction pairs. Lubrication systems, by rational dosing of plastic greases and lubricating oils in terms of the quantity and frequency of lubrication of friction pairs, have contributed to more effective protection of the natural environment. This is of great importance considering the amount of lubricants used in highly industrialized countries. For example, in 2015 global consumption of lubricants was recorded at 35.6 million m^3, while only in the United States about 9 million m^3 was used at the time [8]. In special cases, centralized lubrication systems compared to traditional lubrication have reduced lubricant consumption by up to 90% [9].

An optimally designed lubrication system should reliably distribute lubricant to specific collection points. Designing such a system is particularly difficult if it supplies heavily loaded machinery with lubricant, containing a large (up to several hundred) number of various types of friction nodes, arranged non-linearly, at a large distance from each other. Extremely important problem, often ignored by engineers, are difficult environmental conditions in which lubrication systems are operated (e.g. high humidity, pollution with coal dust, water, brine). All these factors makes it a big challenge to design a lubrication system that works well even for an experienced designer specializing in this particular field.

The aim of this study is to present the main problems encountered during the exploitation of a fully automated lubrication systems of friction nodes of mining machines working in underground mines. Problems related with lubrication of machine moving nodes operating in an aggressive environment occurring in copper mines, with particular emphasis on the mining plants Polkowice-Sieroszowice and Rudna, belonging to the KGHM Polska Miedź S.A. group were presented. The analysis was based on the operation of self-propelled wheeled machines, such as transport vehicles, loaders, drilling and anchoring cars. In these machines there are from 15 to 40 moving elements that are subjected to high impact loads and stresses (high pressure with concentrated contact, which is the worst type of contact of cooperating elements). These are mainly bolted joints, which perform swinging motion, as well as rotary motion at low rotational speeds up to 10 rpm. Centralized lubrication systems of self-propelled wheeled machines supplying lubricants from a dozen to several dozen tribological nodes have been analyzed. The work contains the test results of structural components of lubrication systems, including manifolds and valves, which are elements of the hydraulic system deciding on the value of basic parameters that significantly affect the work of the energy receiver and the entire hydraulic system [10, 11]. The paper also provides valuable tips for future designers and constructors of central lubrication systems for mining machines.

2 Machines and Environmental Parameters in Selected Copper Mines

Mining machines working in copper mines have a specific design and size. For example, the total height of the machine does not exceed the average human height and is a maximum of 1800 mm. This parameter and other construction and operational requirements set for mining machines result from many factors. They consist of geological factors determining the dimensions of excavations, safety and employee protection recommendations, as well as a number of construction and material guidelines determining many parameters of machines. Additionally, recommendations and limitations resulting from the experience of employees and broadly understood human factors should be remembered. Figures 1 and 2 show examples of self-propelled mining machines operating in copper mines.

Fig. 1. Self-propelled loading vehicle [12].

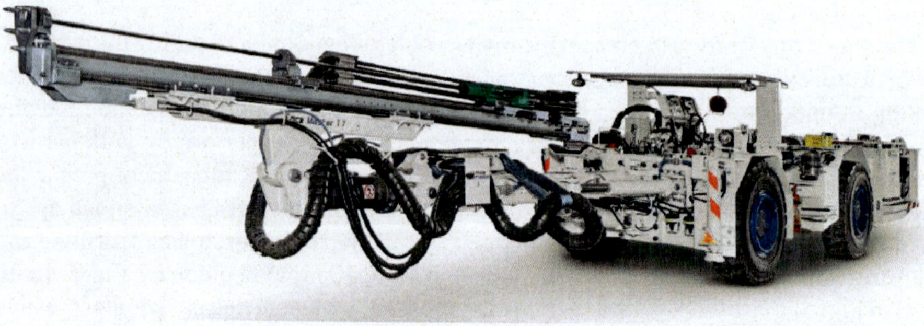

Fig. 2. Self-propelled drilling and anchoring vehicle Face Master 1.7 [13].

Observation and analysis of lubrication systems were carried out in the mining plants of ZG Polkowice-Sieroszowice and ZG Rudna belonging to KGHM Polska Miedź S.A. The operational tests of these elements depended on the ambient conditions, which are not identical in both mines. Environmental and operating conditions of these elements are presented in Table 1 [14].

Copper ore deposits in these mining plants are located at a depth of 500 m to over 1100 m. The depth of copper deposits and a high geothermal degree in this region of Poland on average 32.8 m/1 °C [15] (the average geothermal gradient is therefore

Table 1. Working conditions of machines in the analyzed coal mines KGHM Polska Miedź S.A.

	Department of ZG Rudna	Department of Polkowice-Sieroszowice
1. Climatic conditions:		
a. air temperature [°C]	47	41
b. air humidity [%]	95%	97
c. dust [mg/m^3]	20	19.31
d. air composition [mg/m^3]:		
- Carbon monoxide - nitric oxide - Sulphur dioxide - hydrogen sulfide	180 10 5 (only for NDS-10) 10	120 3 0.2 20
2. Water pollution [g/l]		
a. index pH	7	7.81
b. total dissolved substances	270	311.2
c. dissolved mineral substances	250	297.2
d. salt content	250	297.2
e. chlorides	145	193.23
f. sulfate content	2.5	5.59
g. sodium content	85	126.7
h. potassium content	1.2	2.140
i. magnesium content	1.5	2.04
j. calcium content	9	2.108
k. mineralization	250	311.2
3. Type of work:	a. work in four-shift system b. working time during the shift is about 5 h c. working time with a maximum allowed efficiency of about 85%	

3 °C/100 m) imposes high working temperature of machines and people. The ambient temperature assuming the degree of geothermal and heat generated by the operating machines underground is from 25 °C to 40 °C. It can be concluded that the lubricating nodes described in this article operate at elevated temperatures (in fact, the operating temperature of the node is higher than the ambient temperature). In addition, copper ore deposits are irregularly located, and sometimes arranged in bands about 1 m high, with a tendency to 'fall down'. It should be noted that the geological structure is complex. Above the copper deposits at a depth of 300–500 m there are rich salt beds after the previous presence of a lake or salt sea.

Taking into account the geological factors and temperature it should be noted that working conditions of machinery and equipment are very unfavorable underground. The geological structure of copper ore, the presence of NaCl salts, high humidity and the emission of other chemical compounds cause that pumped from the earth's surface air contains a lot of compounds that negatively affect mining machines and people working there. The content of these compounds in the atmosphere is shown in Table 1.

Analyzing the presented data, it should be clearly stated that mining machines work in a chemically aggressive environment and at elevated ambient temperature. Due to these difficult operating conditions, since the 90 s in over 80% of all self-propelled machines central lubrication systems have been used in their moving nodes. Most solutions use progressive multi-circuit systems with automatic steering and control. The lubricant is plastic greases with various thickeners and base oil. Lubricants with a consistency of NLGI 1 or 2 are used [9, 16, 18–20].

3 Central Lubrication Systems

The purpose of using central lubrication systems (CUS) is to extend the machine's operating time by preventing seizure of bolt connections in their operating systems. The simultaneous use of CUS eliminates the human factor for carrying out maintenance and conservation activities in the these machines.

3.1 Characteristics of Research Objects and Operational Tests

Figure 3 shows a diagram of a typical central lubrication system used in the observed mining machines [17].

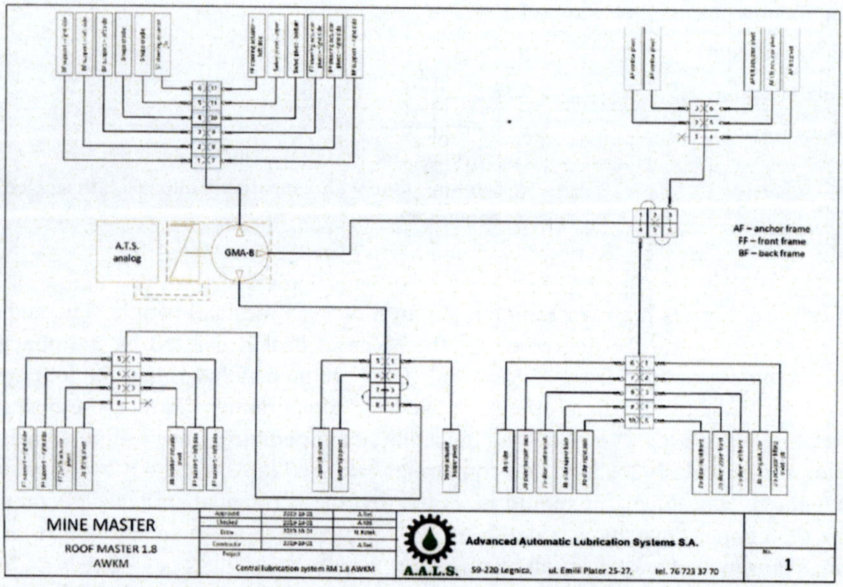

Fig. 3. Diagram of central lubrication system [17].

The presented central lubrication system is a three-circuit progressive system. It uses a pump with an electric drive with a 24 V DC motor, where the pump body is made of aluminum alloys. The progressive distributors installed in it also had aluminum bodies. On grease supply lines polyurethane (PUR) with a textile braid and a polyethylene (PE), polyurethane or polypropylene (PP) sheath is used.

The solution with the use of aluminum and their alloys allowed the installation of lubrication systems with lower weight than other metals and their alloys. This approach is right, because taking into account the criterion of minimum emissions from combustion engines driving machines and their operating systems, the design of these machines strives to keep their total weight as low as possible with maximum reliability and efficiency.

Figure 4 presents photos of pumps, distributors and accessories installed in mining machines equipped with lubrication systems.

a) b)

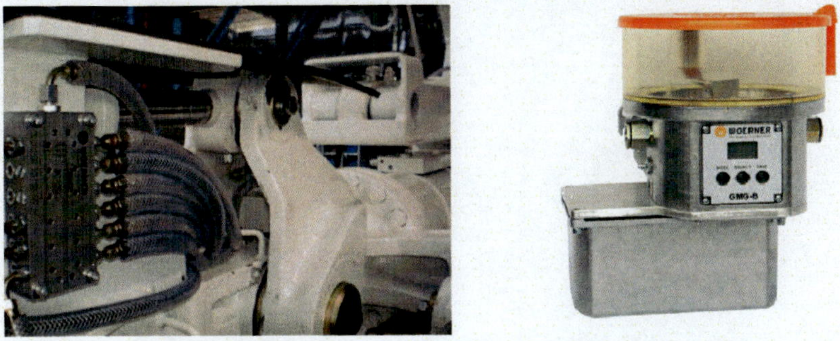

Fig. 4. Elements of the central lubrication system: a - distributor, b - pump aggregate [17].

In the years 2017–2018 at the mining departments of the plants belonging to the company KGHM Polska Miedź S.A. operational observations were carried out and an analysis of the failures of these lubricating system components was carried out. Table 2 contains the type and number of lubrication system components evaluated for durability and reliability during their operation.

Table 2 presents a summary of the assessment results of the operational reliability of lubrication system components. Together, 36 manifolds with an aluminum housing were used for the tests, which were subject to frequent failures during the tests and observations. The main cause of damage was corrosion of housings, resulting in a total failure of the machine's lubrication system. Distributor housing and its condition prevented further repairs, which necessitated the replacement of these elements with new ones. It was similar with fixing elements, which together with the distributor were also damaged and had to be replaced. An example of manifold corrosion is shown in Fig. 5, where numerous pits, body defects and breaking off of fixing elements of the lubrication system are visible. They perfectly show how destructively aggressive mine environment affects the machine and its components - and only after less than 3 months of work.

Table 2. Description of failure of analyzed lubrication system elements during exploitation.

Item	Item name	Quantity	Description of wear or failure
1	Distributor progressive type VPB-B/6/with housing made of AL	21	Housing material loss due to corrosion. Corrosion of steel structural elements of the manifold (elements sealing the flow of grease in the manifold housing) with few defects in the components. Few leakage of grease between the distributor housing and connecting elements with the central lubrication system
2.	Distributor progressive type VPB-B/8/with housing made of AL	9	
3.	Distributor progressive type VPB-B/10/with housing made of AL	6	
4.	GMG-L series lubrication pump with built-in electronic controller with system operation control	5	Numerous failures of the control system. Corrosion of electronic systems. Corrosion of the pump's aluminum components. Electric drive failure

Fig. 5. Grease distributors after the service life.

Five GMG-L lubrication system pumps were also selected for operational tests. These are electrically driven pumps with a built-in pump control and monitoring system. This hybrid is widely used in mobile central lubrication systems. However, as shown by operational observations, such a system is not suitable for use in a copper ore mine. The mine's environment had a destructive effect on the aluminum components of the pump housing and on the controller electronics. The average trouble-free operation of the pump controller was two months. After this period, it was necessary to replace it with a new one. On average, every 3 months, the pump housing components were cleaned, maintained and repaired.

Figure 6 shows an example of the pump after about three months of operation. Pay attention to manifolds and their corrosion throughout (Fig. 8). These elements worked many times directly in the water on the mine floor.

Fig. 6. Grease pump during operation.

3.2 Design Recommendations and Operational Tests of Modified Lubrication Systems

After analyzing the wear, efficiency and failure rate of UCS machines and components, and determining the permissible time of their operation (durability), new construction solutions were proposed. Instead of aluminum manifolds, new manifolds of similar design and performance have been used, but with a bronze housing. The capacity, dimensions and connections of the manifolds have been the same as for the manifolds with aluminum housing.

A hydraulic drive pump with a carbon steel body, protected with a varnish coating (urea paint) and progressive distributors with a bronze body were also introduced.

Figure 7 shows a GAT hydraulic drive pump, and Fig. 8 shows a photograph of a progressive distributor which body is made of bronze.

The results of operational research are shown in Table 3.

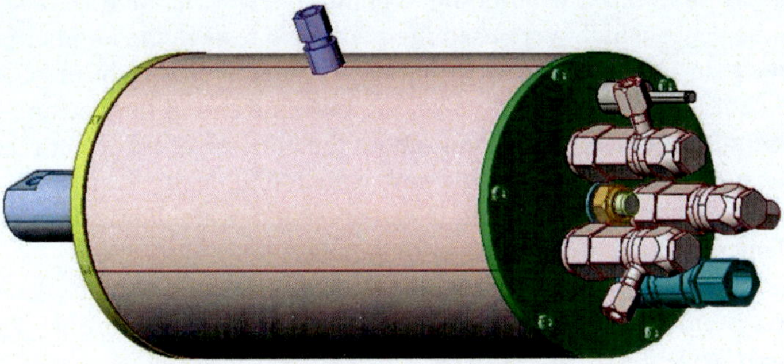

Fig. 7. Pump with GAT hydraulic drive.

After operational tests of 11 new manifolds, it turned out that the bronze housing is very resistant to corrosion caused by the aggressive environment in the copper ore mine. The distributors, after some time, about 2 months, were covered with a corrosion layer of copper sulphide, but the corrosion did not continue. A thin layer of green copper sulphide and copper compounds acted as a natural anti-corrosion agent and the distributors were

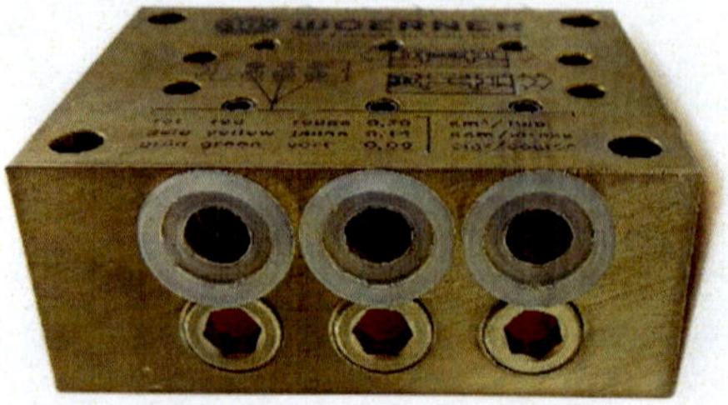

Fig. 8. Progressive distributor with a bronze body.

Table 3. Description of failures of modified lubrication system elements.

Item	Item name	Quantity	Description of wear or failure
1.	Distributor progressive type VPB-H/6/with a bronze housing	4	Few corrosion of the housing. The change of the housing color to green due to the mine environment - separation of the copper sulphide layer limited further corrosion of the housing
2.	Distributor progressive type VPB-B/8/with a bronze housing	5	
3.	Distributor progressive type VPB-B/10/with a bronze housing	2	
4.	Pump with hydraulic drive, GAT series	2	Sparse failures sensor lubricant level in the pump

working properly. Only a few lubricant leaks appeared in the housing connections with the grease lead connector, but it was possible to quickly eliminate leaks or replace these fixing elements.

The situation was similar with modified pumps. New generation GAT series pumps with a hydraulic drive, which were used in operational tests in the number of 2, did not show any defects in electrically driven pumps. The pump control of the GAT series is placed in the central controller of the working machine and in the housing resistant to the atmosphere of the mine. Only a few, about three, cases of incorrect reading of the location of the grease level sensor in the tank were found, which were quickly repaired and it did not have significant meaning in the operation of the lubrication system. This is the only element mounted outside the pump body.

However, The introduced modifications also have disadvantages. The design solution of lubrication systems using a pump with a hydraulic drive and manifolds with a bronze body, increased the weight of the entire UCS system by over 100%. However, operational observations showed that the durability and reliability of the pumps increased more than 3 times. Electrical failure caused by corrosion or interruption of the electric wires supplying and controlling the pump have been eliminated. The durability of distributors with a bronze body compared to those with an aluminum body increased more than five times.

3.3 Summary and Conclusions

The paper presents problems related to the lubrication of moving nodes of self-propelled wheeled machines, operated in an aggressive environment and at elevated temperatures occurring in copper mines. Observations and operational tests of several dozen manifolds and pumps operating in the environment occurring in KGHM Polska Miedź copper mines during two years allowed to draw a number of conclusions:

- elements made of aluminum or its alloys should not be used, and should be replaced with bronze (the design solution of lubrication systems using a pump with a hydraulic drive and manifolds with a bronze body resulted in an increase in the weight of the entire UCS system by over 100%, however, operational observations showed that pump durability and reliability increased more than three times, and manifolds with bronze body increased more than five times),
- the use of electronic systems on actuators in central lubrication systems should be avoided (eliminated electrical failures caused by corrosion or interruption of the electric wires supplying and controlling the pump by enclosing the control systems),
- hydraulic drive, which is widely used in the working elements of machines used in a copper mine, turned out to be the optimal drive in central lubrication pumps.

The above observations are very promising, because even considering the fact that elements made of aluminum are 3 times cheaper than those made of bronze, and the hydraulic drive is about 2 times more expensive than the electric drive, the proposed lubrication system made of bronze and with hydraulic drive is several times more durable and cheaper in operation compared to the central lubrication systems used so far. It should be remembered that the cost of production alone does not affect the final price of the device, because the cost of downtime and repairs is an important economic component.

During the work, further issues emerged that should be considered during subsequent work aimed at improving the lubrication systems of mining machinery friction nodes. The problems related to the flow and thixotropy of lubricants should be solved and the design of the pin and bearing (operating node) should be adjusted in terms of optimizing the lubricant supply to the node. A separate issue is the impact of individual types of grease on the durability of pins and bushes in working nodes, and the impact of the materials from which these elements were made.

References

1. Krawiec, S.: Kompozycje smarów plastycznych i stałych w procesie tarcia stalowych węzłów maszyn. Oficyna Wydawnicza Politechniki Wrocławskiej, Wrocław (2011)
2. Capanidis, D., Wieleba, W., Kowalewski, P.: Wpływ wybranych smarowych preparatów eksploatacyjnych na właściwości tribologiczne materiałów polimerowych podczas tarcia ze stalą. Tribologia **6**(41), 11–23 (2010)
3. Lugt, P.: Grease Lubrication in Rolling Bearings. Wiley, Chichester (2014)
4. Bruce, R.: Handbook of Lubrication and Tribology. CRC Press, Boco Raton (2012)
5. Bruce, R.: Handbook of Lubrication and Tribology, vol. II. CRC Press, Boco Raton (2012)
6. Krawiec, S., Krawiec, M., Leśniewski, T.: The influence of the type of grease on the synergy of its composition with PTFE powder and copper powder. Tribologia **1**(47), 45–55 (2016)

7. Leśniewski, T., Krawiec, S.: The effect of ball hardness on four-ball wear test results. Wear **264**, 662–670 (2008)
8. Paszkowski, M.: Przepływy smarów plastycznych w układach smarowniczych i węzłach tarcia. Wybrane zagadnienia, Oficyna Wydawnicza Politechniki Wrocławskiej, Wrocław (2017)
9. Czarny, R.: Smarowanie układów jezdnych maszyn roboczych i pojazdów. Oficyna Wydawnicza Politechniki Wrocławskiej, Seria: Navigator 11, Wrocław (1998)
10. Stosiak, M.: Ways of reducing the impact of mechanical vibrations on hydraulic valves. ACME **15**, 392–400 (2015)
11. Stosiak, M.: The modelling of hydraulic distributor slide-sleeve interaction. Arch. Civil Mech. Eng. **12**(2), 192–197 (2012)
12. KGHM Zanam Homepage. https://www.kghmzanam.com/produkty/maszyny-gornicze/ladowarki/lkp0405. Accessed 11 Feb 2020
13. MineMaster Homepage. https://www.minemaster.eu/pl/wiertnice/face-master-1-7. Accessed 11 Feb 2020
14. Operating information in mining plants of KGHM Polska Miedź S.A
15. Majorowicz, J.: Zakład Geofizyki Instytutu Geologicznego w Warszawie (1971)
16. Lawrowski, Z.: Technika smarowania. Wydawnictwo Naukowe PWN, Warszawa (1996)
17. Company materials Woerner Zentralschmieranlage i Elektroluber S.C. (previously A.T.S. Electro-Lube Polska)
18. Wysocki, M.: Systemy smarownicze w przemyśle ciężkim. Wydawnictwo "Śląsk", Katowice (1971)
19. Zwierzycki, W.: Oleje, Paliwa i Smary dla przemysłu. Wydawnictwo i Zakład Poligraficzny Instytutu Technologii Eksploatacji, Radom (2001)
20. Totten, G., Gresham, R.: Lubrication and Maintenance of Industrial Machinery: Best Practices and Reliability. CRC Press/Taylor & Francis Group, Boca Raton (2009)

Cavitation, Dynamics, Noise and Vibration

Reduction of Noise Emission of Hydraulic Power Units

Wiesław Fiebig[1(✉)] and Piotr Rosikowski[2]

[1] Faculty of Mechanical Engineering, Wroclaw University of Science and Technology, Lukasiewicza Street 7/9, 51-370 Wroclaw, Poland
wieslaw.fiebig@pwr.edu.pl

[2] Ponar Wadowice S.A, R&D Center, Sw. Jana Pawla II 10a Street, 43-170 Laziska Gorne, Poland
piotr.rosikowski@ponar-wadowice.pl

Abstract. The article presents the results of investigations for reducing noise emissions from a hydraulic power unit. The main sources of noise like electric motor and bellhousing in the power unit were determined. As a measure for the noise reduction the sound-insulating capsulation for the electric motor with mounting bracket (bellhousing) was used. With that measure a significant reduction in sound power from the power unit has been achieved.

Keywords: Hydraulic power units · Noise reduction

1 Introduction

Hydraulic power units are in many industrial applications the noise sources with highest intensity. This noise is mainly caused by vibrations originating from positive displacement pumps, which are transferred to elements mechanically or hydraulically connected to the pumps. The process of noise generation in hydraulic power units is very complex and to obtain noise reduction it is necessary to know the main sources of noise. Hydraulic power units in some applications should have low noise emission. For this reason, this paper presents a method of reducing their noise level by using a sound absorbing and insulating capsulation for an electric motor, which, as stated in [1–6], is one of the main sources of noise.

2 Noise Generation in Fluid Power Units

The reasons for the development of noise in hydraulic systems are presented in Fig. 1. In case of positive displacement pumps, the vibrations are mainly caused by variable forces occurring in the cross section processes. The flow, pressure pulsation [7–9] and mechanical causes (impacts between pump elements) have also an influence on pump noise generation. It should be noted that the noise accompanying the flow is particularly

© The Editor(s) (if applicable) and The Author(s), under exclusive license to Springer Nature Switzerland AG 2021
J. Stryczek and U. Warzyńska (Eds.): NSHP 2020, LNME, pp. 241–251, 2021.
https://doi.org/10.1007/978-3-030-59509-8_21

intense in places with a very small flow areas where the flow rate reaches very high values above 10 m/s. The noise caused by cavitation occurs often in the cross-section areas. In vane pumps the flow noise caused by cavitation may be decisive for certain operational parameters. The electric motor vibrations are caused mainly by the pumps but also due to the forces of electrodynamic and aerodynamic origin.

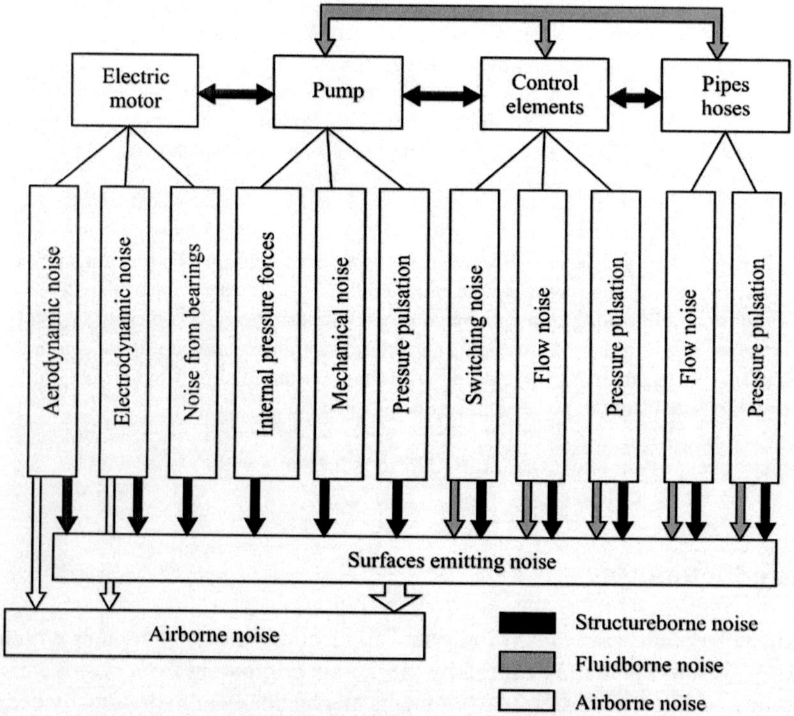

Fig. 1. Causes of noise in hydraulic power units (HPU) [6].

Hydraulic elements, such as valves and hydraulic pipe elements and hoses are excited to vibrations due to pressure pulsation. The vibrations of these elements are excited by a pump that is hydraulically and mechanically connected to them.

The vibrations of electric motors, valves, tank walls, and other elements are coupled with vibrations of the pumps. The amount of sound power emitted to the environment by the fluid power units (HPU) is often up to 10 dB (A) higher than the sound power level emitted by the pumps. This fact indicates that displacement pumps cannot be treated as the main noise sources in such HPU's. Displacement pumps, due to their discontinuous nature of operation, are among the most important sources of structure and fluid borne noise, which will be transferred to the elements connected to the pumps mechanically or hydraulically. The share of the noise of pumps themselves is usually smaller than the share of elements of HPU with larger surface radiating sounds like electric motors or tank walls. The answer to the question whether internal pressure forces or pressure pulsation is more important for the generation of noise depends on the design of the power unit and cannot be given without tests. The relationship between these basic causes of noise in fluid power units is very complex. In general the pressure pulsation is more responsible for the noise development in mobile applications.

It should be noted that the mechanisms of noise generation in HPU with the horizontal motor-pump assembly installed on the tank significantly differ from the HPU with the vertical located motor pump group [2, 6]. In case of horizontal HPU the main noise source is usually an electric motor. For the motor noise - as confirmed by many tests – the pump is responsible, because the fundamental frequency of the pump (resulting from the rotational speed and number of displacement elements) and its harmonics, are dominant in the spectral analysis of the motor.

Figure 2 presents, in an illustrative way, the noise and vibration transmission paths in a power unit. The pump's mechanical vibrations are transmitted to the electric motor by path 1. On the one hand, by means of a rubber ring and a console (1a) and on the other hand by means of a flexible coupling (1b).

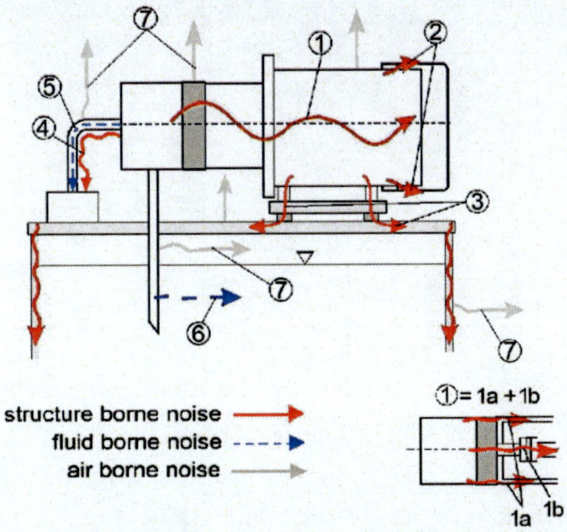

Fig. 2. Noise generation in a hydraulic power unit with horizontally mounted electric motor [6].

The vibrations are transferred to the fan cover (path 2) and to the tank structure (path 3) via the motor. Part of the vibration energy is transferred from the pump to the tank through the lines connecting it to the valves (path 4). The vibrations in the form of pressure pulsations (path 4) are also transmitted through the fluid. As a result of vibrations of the pump body and the suction pipe, the sounds are excited in the oil tank, which are conducted through the oil to the tank (path 6). The tanks are usually filled with 80% oil and for this reason, airborne sounds are propagated inside them (path 7). As a result, noise is emitted to the environment by an electric motor (path 3) and a tank (path 7). It should be noted that some of the vibration energy coming from the pump is transferred to the tank structure through fluid. The influence of the fluid inside the tank on transmission of vibrations in liquid inclusive FSI – Fluid Structure Interactions is presented in paper [6]. The investigations show that the presence of liquid affects both dampening of vibrations of the tank structure and its natural frequencies.

3 Investigated Power Unit

The power unit, Fig. 3, consists of a 15 kW electric motor, a axial-piston pump with a swing-out disc with a geometric displacement of 28 cm^3/rev, nominal pressure of 280 bar and a tank with a capacity of 250 dm^3 [12].

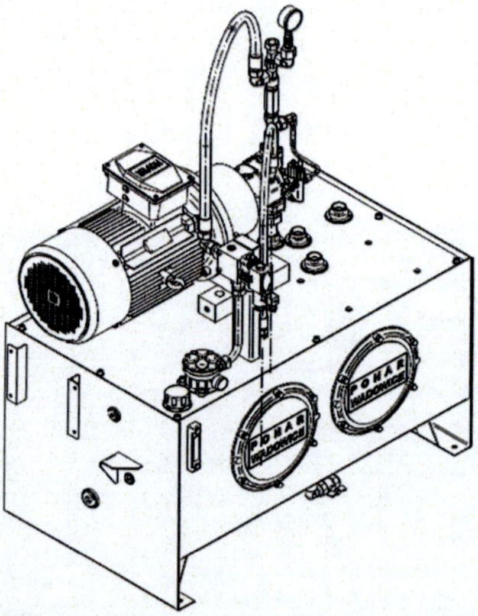

Fig. 3. CAD model of power unit.

Fig. 4. View of the power unit.

The pump was equipped with a pressure, flow and constant power regulator, and the pump load was obtained using an electrically controlled pressure relief valve.

The tank has been equipped with the drain filter, breathing and flood filter, optical oil level indicator, two inspection covers and a drain valve. During the tests, HLP46 hydraulic oil was used. The pump set has been built on the top cover of the tank to have easy access to the pump regulators - as usual in this type of power units. In Fig. 4 the view of the investigated hydraulic power unit has been shown.

4 Measurement Methods

The measurements of the sound power emitted from the power supply were made in a semi-anechoic acoustic chamber based on the PN ISO 3745 standard. During the measurements, the results of the sound pressure measurement were averaged from 9 measurement points arranged as in Fig. 5.

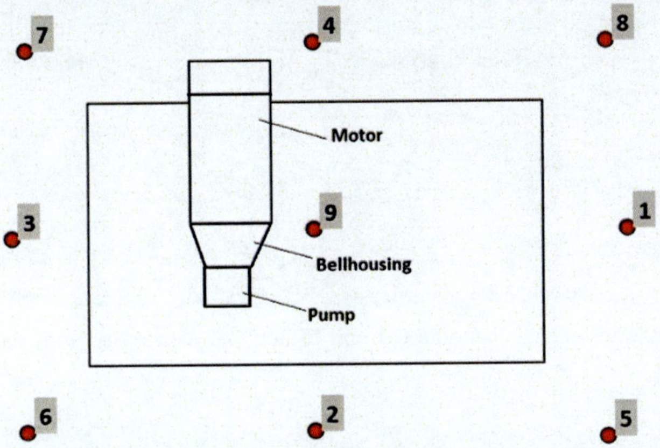

Fig. 5. Arrangement of measuring point around the power unit.

The noise sources of power units can be also identified with an acoustic camera and Beamforming Method. An acoustic camera (Fig. 6) uses an array of 112 microphones coupled with a video camera, digital signal processing hardware and computing algorithms. The microphone array pattern, number of microphones and the size of the array define parameters such as resolution, frequency range and map's dynamic [10]. The video camera is positioned in the center of the microphone array. Acoustic pressure of each microphone was registered and computing algorithms like Beamforming and Acoustic Holography were used [10, 11]. As the result of the measurement an acoustic photo or video with color coded acoustic pressure distribution was created.

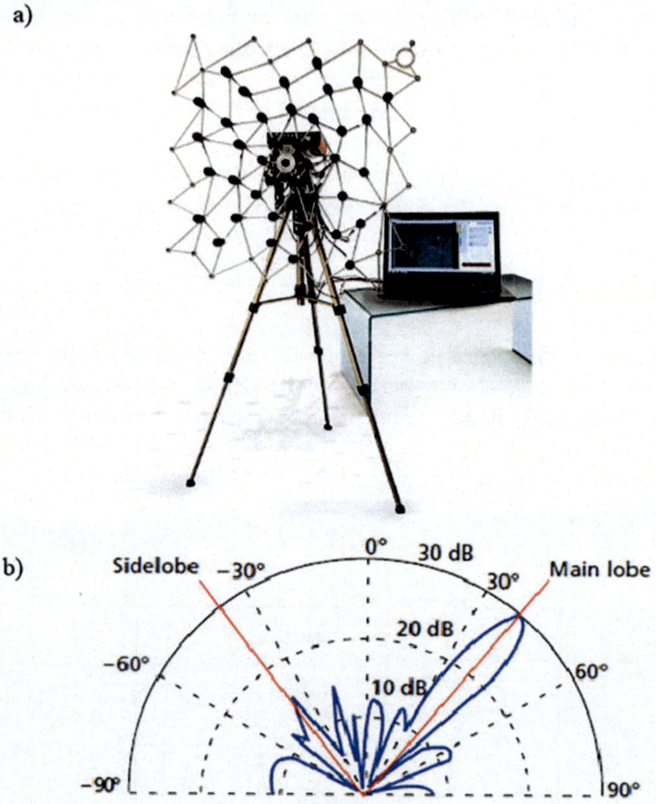

Fig. 6. Acoustic camera CAE Systems (a) and directional sensitivity with the main focused direction (b).

It can be seen that the highest amplitudes of sound pressure level occur at the pump basic frequency and its harmonics. Sound distribution on the surface can be shown for the selected frequency or frequency range. The value of sound pressure level is calculated from the pressure level at the microphone array and the distance of the array from the measured object, which should be measured prior to the measurement and implemented into the data acquisition software [10, 11]. In this way, the noise from different sources can be identified. In Fig. 7 and Fig. 8 the noise sources for different frequency ranges are shown. The measurements with the acoustic camera were carried out in a semi anechoic chamber.

a)

b)

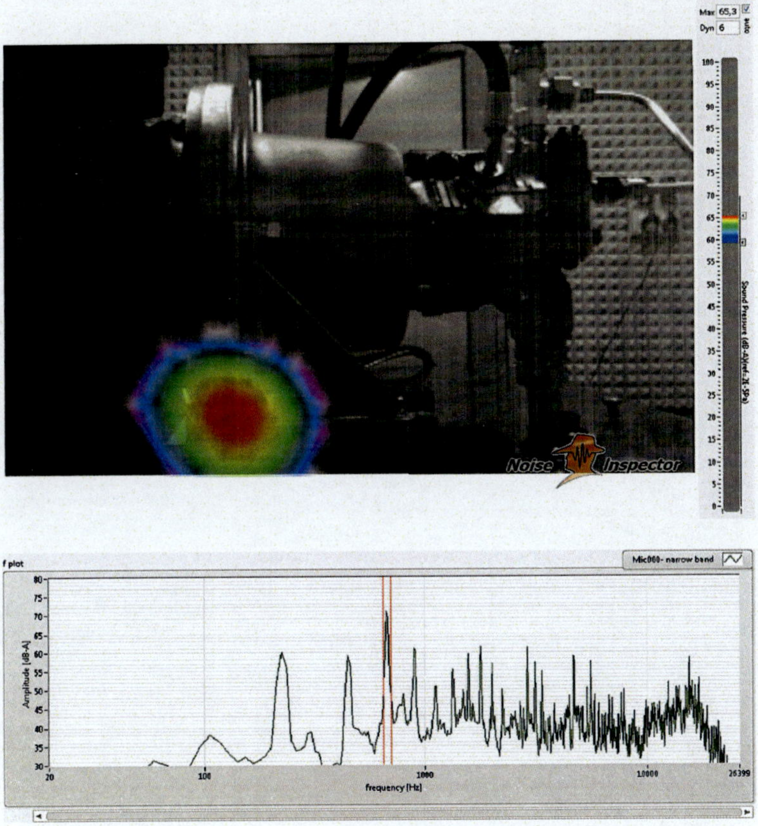

Fig. 7. The noise source on motor feet (a) and FFT analysis of sound pressure signal (b).

In Fig. 7 the noise source on the motor feet and frequency spectrum of the sound pressure signal from a selected microphone has been shown.

In Fig. 8 the second noise source at the fan cover of the electric motor and the frequency spectrum has been shown. The highest levels of the sound pressure occurs for all measurement points at the basic frequency of the pump and its harmonics. That indicates that the pump is responsible for the noise generation in the power unit.

In both cases the highest sound power levels have been observed. That is way the capsulation of the electric motor has been proposed as a possible measure for the noise reduction of the investigated power unit. Also in many other applications [7–16] the electric motor was the one of main noise sources.

a)

b)

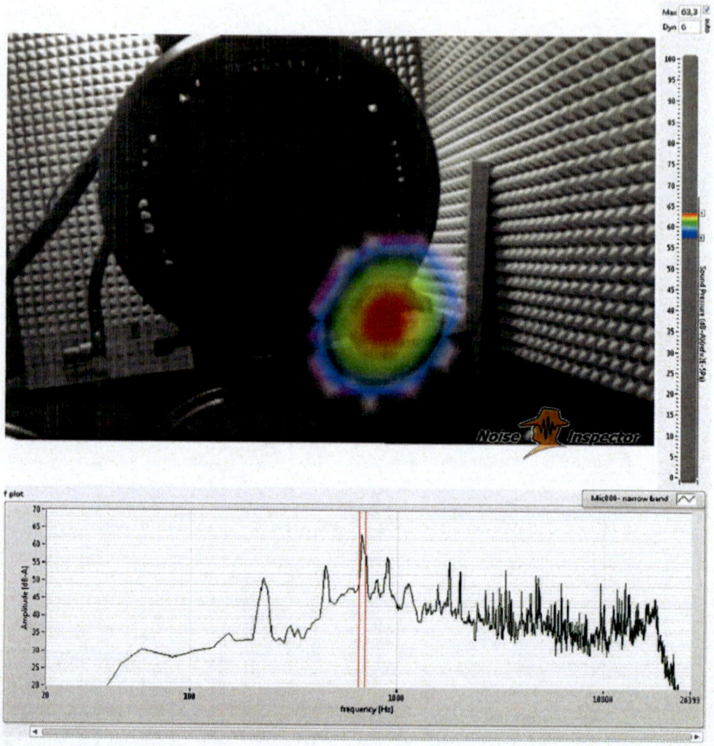

Fig. 8. The noise source at fan cover (a) and FFT analysis of sound pressure signal (b).

5 Fluid Power Unit with Motor Capsulation

In order to reduce the level of noise emitted from the power supply, a sound absorbing and insulating capsulation was made, which has the following features (Fig. 9):

- the capsulation includes main noise sources, namely the electric motor with the
- mounting bracket of the pump,
- the positive displacement pump is located outside the capsulation,
- there is a forced air flow inside the housing for motor cooling,
- the housing is demountable.

Other components of the power unit like pipe elements and hoses, valves, etc. remain the same.

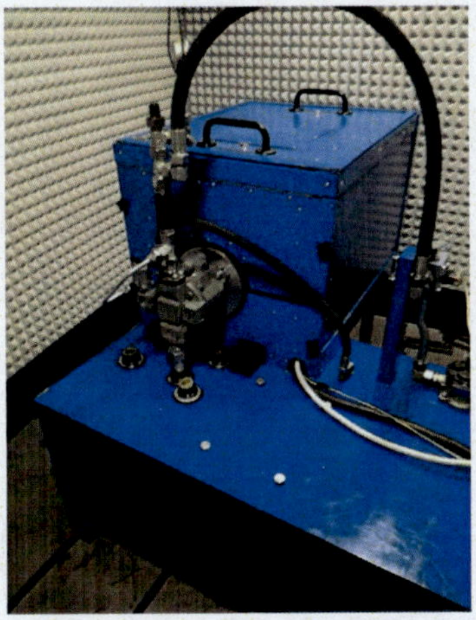

Fig. 9. Investigated power unit with capsulation of the electric motor and the bellhousing.

6 Discussion of Results

The results of the sound power measurements are summarized in Table 1 and Fig. 10. Table 1 shows that the sound pressure values at individual measuring points differ from each other.

Table 1. Comparison of sound pressure and sound power levels with and without motor capsulation

Pressure [MPa]	Measurement point	Without motor capsulation		With motor capsulation	
		Sound pressure level L_{Aeq} [dBA]	Sound power level L_{WA} [dBA]	Sound pressure level L_{Aeq} [dBA]	Sound power level L_{WA} [dBA]
25.0	1	70.9	85.3	67.6	81.1
	2	71.1		67.3	
	3	76.1		69.4	
	4	71.4		63.8	
	5	70.3		64.6	
	6	69.1		67.2	
	7	69.1		67.6	
	8	69.5		60.3	
	9	72.4		71.3	
5.0	1	67.6	83.8	66.6	79.2
	2	68.2		66.2	
	3	72.8		66.7	
	4	69.9		59.6	
	5	65.0		64.2	
	6	66.1		68.8	
	7	70.0		64.8	
	8	72.1		57.2	
	9	72.6		66.3	

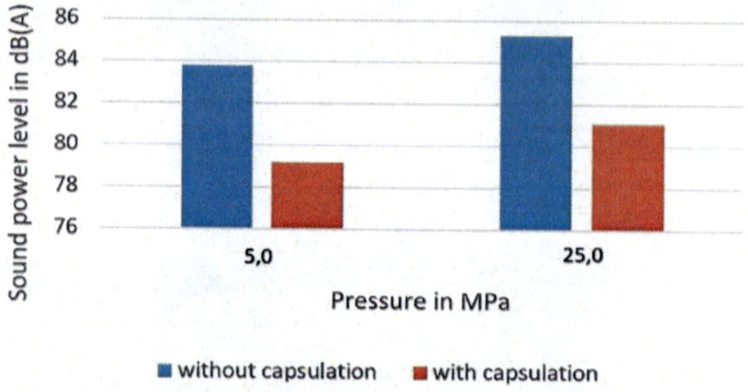

Fig. 10. Sound power levels without and with capsulation of the electric motor.

A direct comparison shows that the reduction of sound power levels is 4.6 dB (A) at 5 MPa and 4.2 dB (A) at 25 MPa. It should be noted that a reduction of the sound power level by 4.2 dB (A) means a reduction of the sound energy emitted to the environment by factor 2.6. Thus, it can be concluded that the achieved noise reduction is significant and the applied measure in the form of a soundproofing enclosure will fulfill its role. In this solution, the axial piston pump is outside the housing and the heat is transferred through it to the environment.

7 Conclusions

This article presents the results of investigations done to reduce the level of noise emitted from a hydraulic power unit. Based on measurements using an acoustic camera, it was established that the electric motor together with the mounting bracket for the pump is one of the main sources of noise in the investigated power unit. It was established that the vibrations and noise from the electric motor is generated by a positive displacement pump. The pump itself generates less noise because it has a much smaller surface than the motor. For this reason and because of the heat balance, the pump is located outside the housing. A significant reduction in the sound power levels has been established and the proposed solution can be used in applications in which reduced noise emission from hydraulic power units is required.

References

1. Dahm, M.: Wirksame Schallreduzierung von Hydraulikaggregaten, O+P, Ölhydraulik und Pneumatik, **41**(2) (1997)
2. Fiebig, W.: Location of noise sources in fluid power machines. Int. J. Occup. Saf. Ergon. (JOSE) **13**(4), 441–450 (2007)
3. Akers, A., Gassman, M., Smith, R.: Hydraulic Power System Analysis, Chapter 4: Noise Control. Taylor & Francis Group, Boca Raton (2006)
4. Skaistis, S.: Noise Control in Hydraulic Machinery. Marcel Dekker Inc., New York (1988)
5. Fiebig, W.: Geraeuschminderung in hydraulischen Systemen, VDI-Fortschritt-Berichte, **11**(337) (2008)
6. Fiebig, W., Wróbel, J.: System approach in noise reduction in fluid power units. In: Proceedings of ASME Bath Conference (2018)

7. Edge, K.A., Johnston, D.N.: New method for evaluating the fluid borne noise of hydraulic components. In: 7th International Fluid Power Symposium, Bath (1986)
8. Longmore, D.K., Schlesinger, A.: Transmission of vibration and pressure fluctuations through hydraulic hoses. Proc. IMechE J. Syst. Control Eng. **205**(2), 105–111 (1991)
9. Chenxiao, N., Xushe, Z.: Study on vibration and noise for the hydraulic system of hydraulic hoist. In: Proceedings of 2012 International Conference on Mechanical Engineering and Material Science (MEMS) (2012)
10. Smith, M.G., Kim, K.B., Thompson, D.J.: Noise source identification using microphone arrays. Proc. Inst. Acoust. **29**(5) (2007)
11. Christensen, J.J., Hald, J.: Beamforming, Bruel & Kjaer Technical Review No. 1 (2004)
12. Jędrzykiewicz, Z., Stojek, J., Rosikowski, P.: Hydrostatic drive and control (in Polish), Vist Sp. z o.o., Kraków (2017)
13. KTR Couplings Limited: Noise reduction in hydraulics. The Engineering Lab on 30th June 2011. http://www.ktrcouplings.co.uk
14. International Organization for Standardization (ISO): Hydraulic fluid power test code for determination of airborne noise levels (Standard No. ISO 4412-1:1991). ISO, Geneva (1991)
15. Fiebig, W.: Noise control of fluid power units. In: 23 rd International Congress on Sound and Vibration, 10-14 July, Athens, Greece (2016)
16. Edge, K.A.: Designing quieter hydraulic systems—Some recent developments and contributions. In: Proceedings of the Fourth JHPS International Symposium on Fluid Power, Tokyo, Japan, 15–17 November 1999, pp. 3–27 (1999)

Assessment of the Effectiveness of Passive and Active Methods in Noise Suppression in Machines and Equipment with the Hydrostatic Drive

Wacław Kollek, Piotr Osiński, and Kacper Leszczyński(✉)

Department of Technical Systems Operation and Maintenance, Faculty of Mechanical Engineering, Wrocław University of Science and Technology, Wrocław, Poland
{waclaw.kollek,piotr.osinski,kacper.leszczynski}@pwr.edu.pl

Abstract. This paper focuses on the noisiness of work machines in the workplace. It assesses the effectiveness of contemporary solutions for lowering the sound level of work machines. Active and passive methods of noise suppression are distinguished. The importance of the two kinds of methods in the process aimed at reducing the noise generated by contemporary equipment and making the latter comply with the current regulations is highlighted. Original solutions designed, made and tested in the Department of Hydraulic Drives and Automation at Wrocław University of Science and Technology are presented.

Keywords: Noise · Passive methods · Active methods · Hydraulic drives

1 Introduction

The rapid advances made in work machine technology in the last century have brought about great changes in many fields of economy and industry. But the rapid development of technology brings not only positive effects. Vibroacoustic processes pose a very serious danger to living organisms. The European programme of environmental protection indicates noise as one of the most arduous phenomena in urban areas and in the work environment.

Due to their daily many-hour exposure to high-intensity sound, manual workers are the group most affected by noise. Many work machines are characterized by high noise emission (Fig. 1).

Hearing organ diseases are high on the list of occupational diseases in Poland [2]. Hearing deteriorates with age and hearing loss is usually irreversible. Data on worker noise exposure indicate that it is necessary to monitor and assess the effect of noise-generating phenomena on the environment and the human body. Then measures should be taken to reduce the sound level of work machines. Currently the control of the acoustic environment covers such aspects as:

© The Editor(s) (if applicable) and The Author(s), under exclusive license to Springer Nature Switzerland AG 2021
J. Stryczek and U. Warzyńska (Eds.): NSHP 2020, LNME, pp. 252–263, 2021.
https://doi.org/10.1007/978-3-030-59509-8_22

Noise source	Sound level A [dB]
Machines constituting energy sources	98 - 130
Fluid-flow equipment	98 - 120
Compressed-air tools and engines	90 - 120
Material handling equipment	98 - 112
Shredding, crushing, sieving, citting and cleaning machines	96 - 111
Wood cutting machine tools	92 - 108

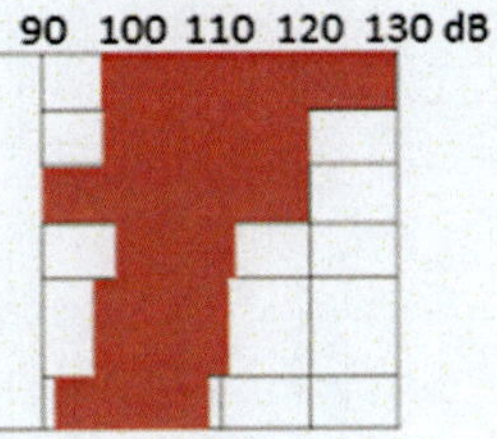

Fig. 1. Sound levels of work machines [1].

- the protection of the environment and people against noise;
- the monitoring of sound sources, aimed at determining (through exploratory investigations) the acoustic state of a facility;
- exploratory investigations enabling signal analyses and the classification of noise emitting objects [3].

Hydraulically driven machinery is an example of work machines characterized by high noise emission. Hydraulic systems offer several advantages, but have a serious drawback – while operating they generate high sound levels, which is often the reason why they are rejected in the case of machines which are to work close to people. Their noisiness is mainly due to:

- hydraulic phenomena connected with the flow of the working medium in the system (cavitation, forcing pressure pulsations, pressure surges, the effect of water in the working medium [12]);
- mechanical phenomena, i.a., vibrations of structural components caused by, e.g., manufacturing and assembly errors, excessive clearances and the unbalance of rotating parts [4].

Figure 2 shows a simple hydraulic system with marked potential sources of sound generation. It also highlights the fact that vibrations can be transmitted by the working liquid or the material of the components contacting the system (lateral transmission). This is a highly adverse phenomenon since, as a result of the vibrations transmitted to the above-mentioned components as well as to the walls or the foundation, all the vibrating bodies emit a certain level of sound.

Hence, the following three basic methods of reducing machine and equipment noisiness are distinguished:

- operational methods – consisting in removing the causes of noise or reducing noise emission through remedies applied to the noise source [4];
- passive methods – limiting the area of propagation of sound waves from the source of their emission, through sound scattering and absorption [4];
- active methods – using a sound generating source to suppress primary source noise.

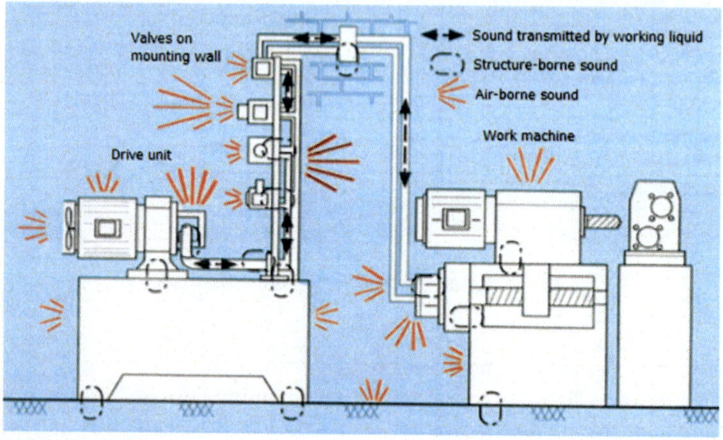

Fig. 2. Schematic of hydraulic system with marked potential sound sources [5].

The noisiness requirements set for contemporary machines cannot be satisfied using solely passive methods. The latter are complementary to the more effective active methods [4].

2 Passive Methods

Passive methods are usually complementary to active methods. They consist in limiting the area of propagation of acoustic waves through absorption and/or scattering [4].

Various passive methods are used in fluid power hydraulics to absorb or dissipate the acoustic waves generated by the hydraulic power unit.

One of the simplest and most often used ways of limiting sound propagation consists in immersing the hydraulic pump or the latter together with the driving motor in oil in the tank, whereby the acoustic waves are dissipated and dampened by the liquid. Thanks to this solution the vibrations transmitted to the walls of the tank have a considerably lower amplitude, whereby the system is characterized by a much lower sound emission.

A major problem in hydraulic drives is the propagation of structure-borne sounds, i.e. vibrations propagating in solid bodies being in contact with the system's components. In order to counteract this, dampeners made of vibration damping materials are used. Special dampeners are placed, e.g., at the contact between the pump and the tank, in places where the system's components contact the surrounding bodies and under the power units, to reduce the transmission of vibrations to the foundation.

The use of such passive methods can significantly affect the sound emission of the whole hydraulic system. Figure 3 shows the effects of lowering the acoustic power level of the hydraulic power unit through the use of several passive methods and by replacing the pump with a less noisy one. When a mounting block alone was used, the sound power level decreased by about 6 dB(A). When, additionally, the pump was replaced with an internal gear pump, characterized by a lower noise emission, which was immersed in the tank and damping elements were employed, the sound power level was reduced by further 16 dB(A).

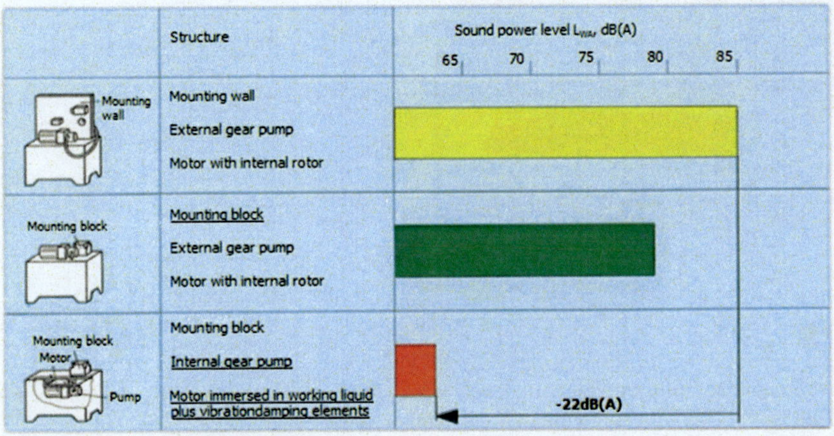

Fig. 3. Effects of using passive methods and replacing pump on acoustic power level of hydraulic power unit [5].

In order to reduce pressure fluctuations and surges broadband pressure pulsation dampeners are used in hydraulic systems. They are especially effective in the low-frequency (infrasound) range and in the range of a few hundred Hz [6]. The basic components of a broadband pressure pulsation dampener are a passive chamber dampener (with an internal conduit) and an active dampener [6]. Figure 4 shows a broadband pressure pulsation dampener installed in a tested system in the Department of Hydraulic Drives and Automation at Wrocław University of Science and Technology. The aim of the tests was to determine the effectiveness of pressure pulsation damping and system noisiness reduction. The sound pressure levels for the system with and without the broadband dampener are compared in Fig. 5.

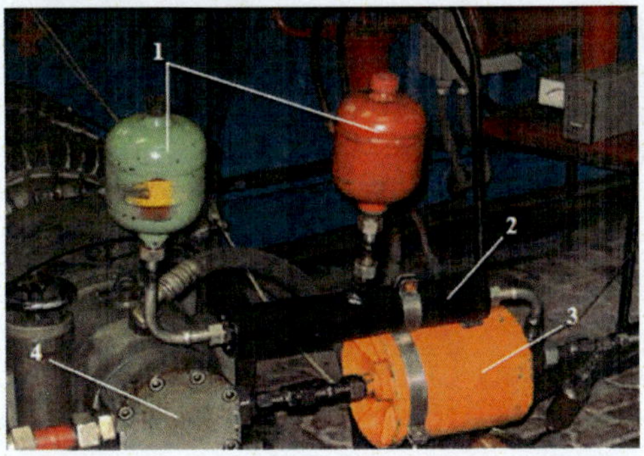

Fig. 4. Broadband pressure pulsation dampener installed in tested system: 1 – hydropneumatic accumulators, 2 – active dampener, 3 – chamber dampener, 4 – hydraulic pump [6].

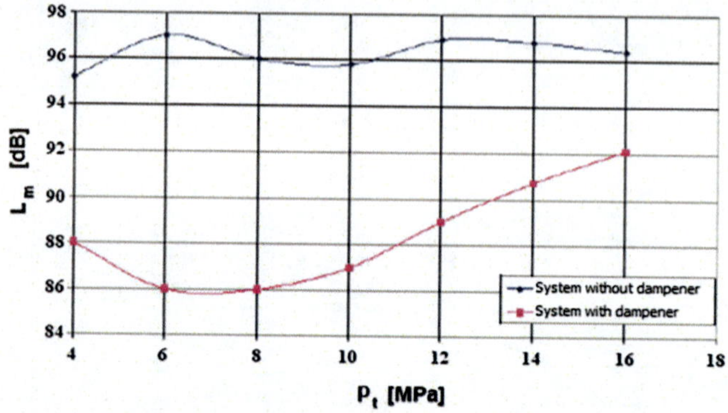

Fig. 5. Comparison of sound pressure levels versus forcing pressure at pump rotational speed of 1500 rpm for system with and without broadband pressure pulsation dampener [6].

The mentioned graphs show that the use of the broadband pressure pulsation dampener results in a few-fold reduction in pressure pulsations and in a reduction in the system's sound power level by 4–11 dB.

Investigations [7], consisting in measuring the sound intensity of an undampened/dampened hydraulic power unit, show the need for the use of pressure pulsation dampeners. As part of the investigations the distribution of sound intensity on the particular surfaces of the hydraulic power unit was determined using a dual-microphone sound intensity probe. In this way the power unit's noisiest components were identified. The frequency and amplitude of the basic pressure pulsation harmonics were determined in order to dimension a pressure pulsation dampener of the branched type and to measure vibrations in the point where sound intensity was highest. A schematic of the tested power unit with marked measuring points is shown in Fig. 6. The dominant frequency in the pressure pulsation spectrum was the second harmonic resulting from displacement pump kinematics, amounting to $f_2 = 340$ Hz. Also the one-third octave spectra of power unit sound intensity before and after installing the branched dampener ($l = \lambda/8 = 920$ mm) and stiffening the top cover of the tank indicate pressure pulsation to be the main cause of the power unit's noisiness. As a result of this upgrading, sound intensity in the particular measuring points decreased by a few to 10–20 dB (Figs. 7, 8).

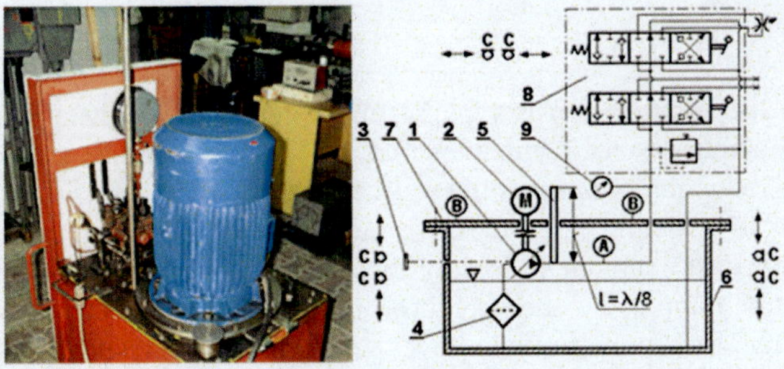

Fig. 6. General view and hydraulic schematic of investigated power unit, with marked measuring points [7]: 1-axial piston pump type PNZ-25, 2-electric motor, 3-screw system for pump delivery change, 4-suction filter, 5-branched pressure pulsation dampener, 6-tank, 7-tank cover, 8-two-way valve block with RBS valve, 9-pressure gauge, A-pressure pulsation transducer, B-vibration transducer, C-sound intensity probe.

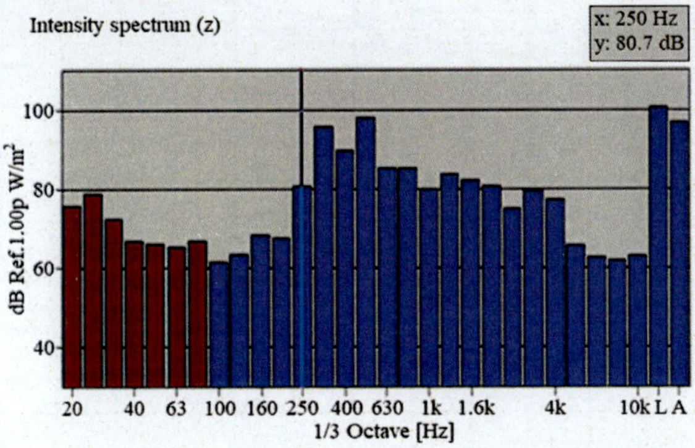

Fig. 7. One-third octave spectrum of sound intensity in point where sound intensity was highest before upgrade [7].

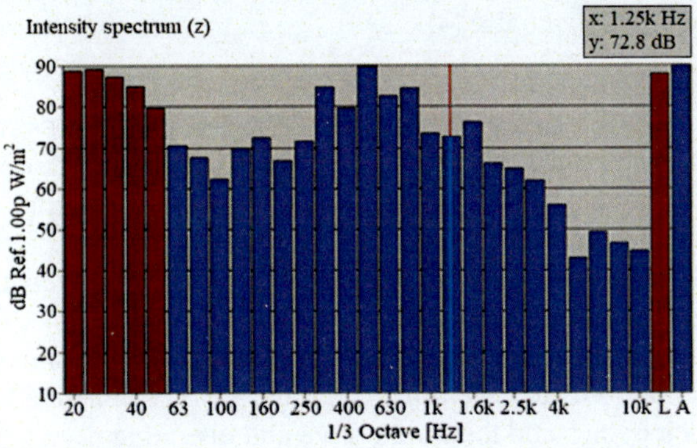

Fig. 8. One-third octave spectrum of sound intensity in point where sound intensity was highest after power unit upgrade [7].

3 Active Methods

The idea of active methods consists in altering the design of a given machine. This is a very costly operation since it involves making many prototypes and their thorough testing, but the reduction of noise at the source gives the best results [4].

In the case of hydraulic drives, active methods come down to alterations in the design of the noisiest component, mainly the hydraulic pump. Sound emission values for a multi-piston pump and a one-way throttling valve of type ZDJ are given in Fig. 9 which shows traces of sound pressure level versus forcing pressure. The difference in sound level values for each of the forcing pressures amounts to about 20 dB [4].

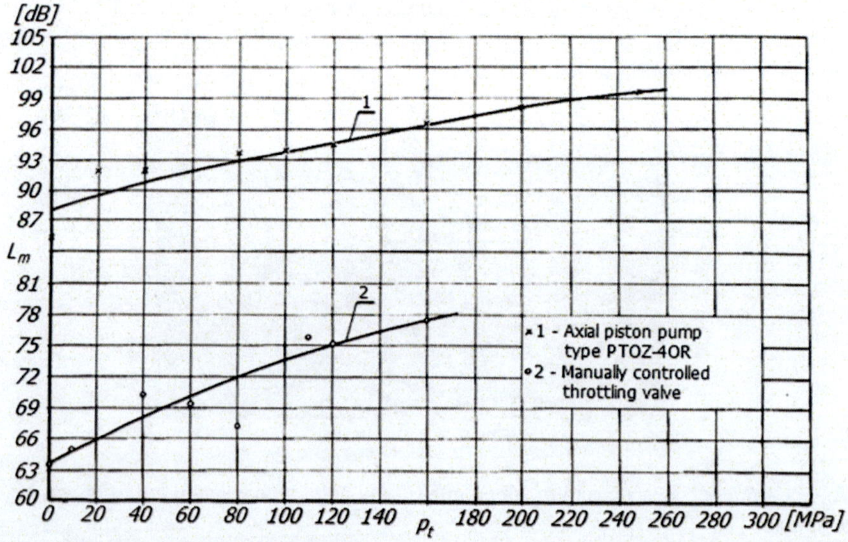

Fig. 9. Comparison of sound pressure level versus forcing pressure generated by respectively multi-piston pump and throttling valve [4].

This is a clear proof that pumps are the main source of the noise emitted by hydraulic systems, while the contribution of the other system components to this emission is much smaller. Among hydraulic pumps, multi-piston pumps and external gear pumps are characterized by the highest sound emission. Output irregularity coefficient values for different types of hydraulic pumps are compared in Fig. 10 [8]. This coefficient describes fluctuations in pump output, which are one of the main causes of noise emission. Therefore one can conclude that the simplest active method is to use a pump characterized by the lowest sound emission. However, this is not usually possible due to the inadequate performance of such pumps and their high price and operating costs.

Owing to its simple design, low price and very good performance the most frequently used pump is the external gear pump [9]. For this reason the largest number of studies and patents aimed at lowering the level of noise of hydraulic systems are devoted to external gear pumps.

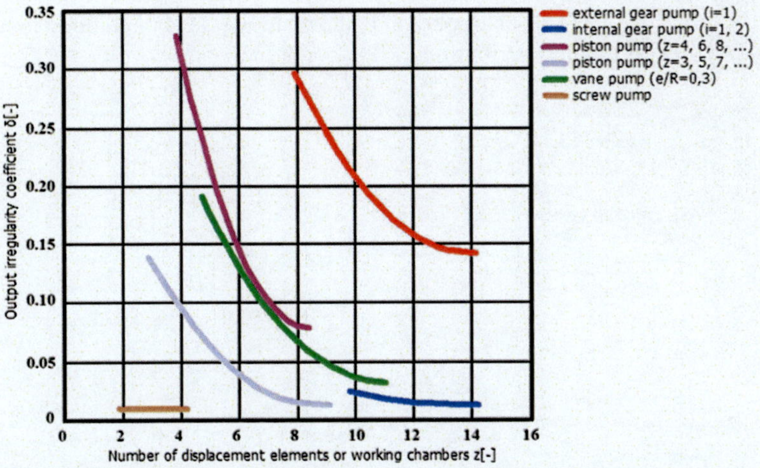

Fig. 10. Comparison of output irregularity coefficients for different types of hydraulic pumps [8].

A trapped volume is widely relieved by grooves made in the housings of slide bearings (Fig. 11). The purpose of the grooves is to connect the trapped volume with the delivery chamber of the pump until the volume decreases [10]. This prevents a detrimental pressure rise in the trapped volume. The rise of the pressure by about 15 MPa results in a sound pressure level rise of over 10 dB(A) [11].

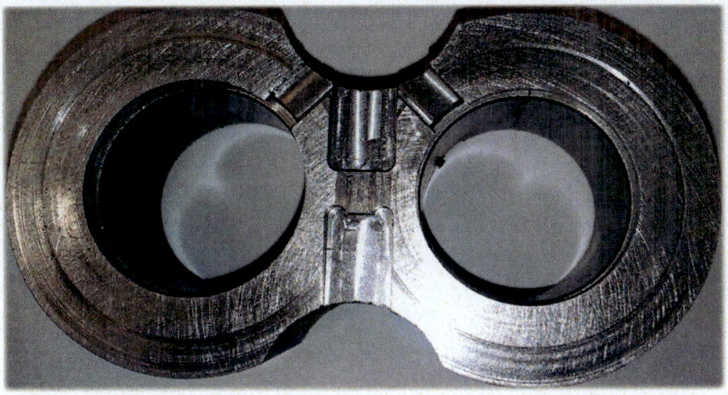

Fig. 11. Relief grooves in slide bearing housing.

The axial piston pump is characterized by very good performance and its output can be adjusted, but of all the hydraulic pumps it generates the highest sound levels. In the design of such pumps the priority is to ensure a proper pressure distribution in the working space. A pressure surge as the cylinder passes from the suction side to the delivery side (this process is called a change-over), and vice versa, should be prevented. This is done by rotating the change-over axis by an appropriate angle β_2 relative to the axis of the dead centre positions of the pistons, in the direction consistent with the direction of rotor rotations [4]. This solution is shown in Fig. 12. For comparison, Fig. 13 shows an incorrect pressure distribution.

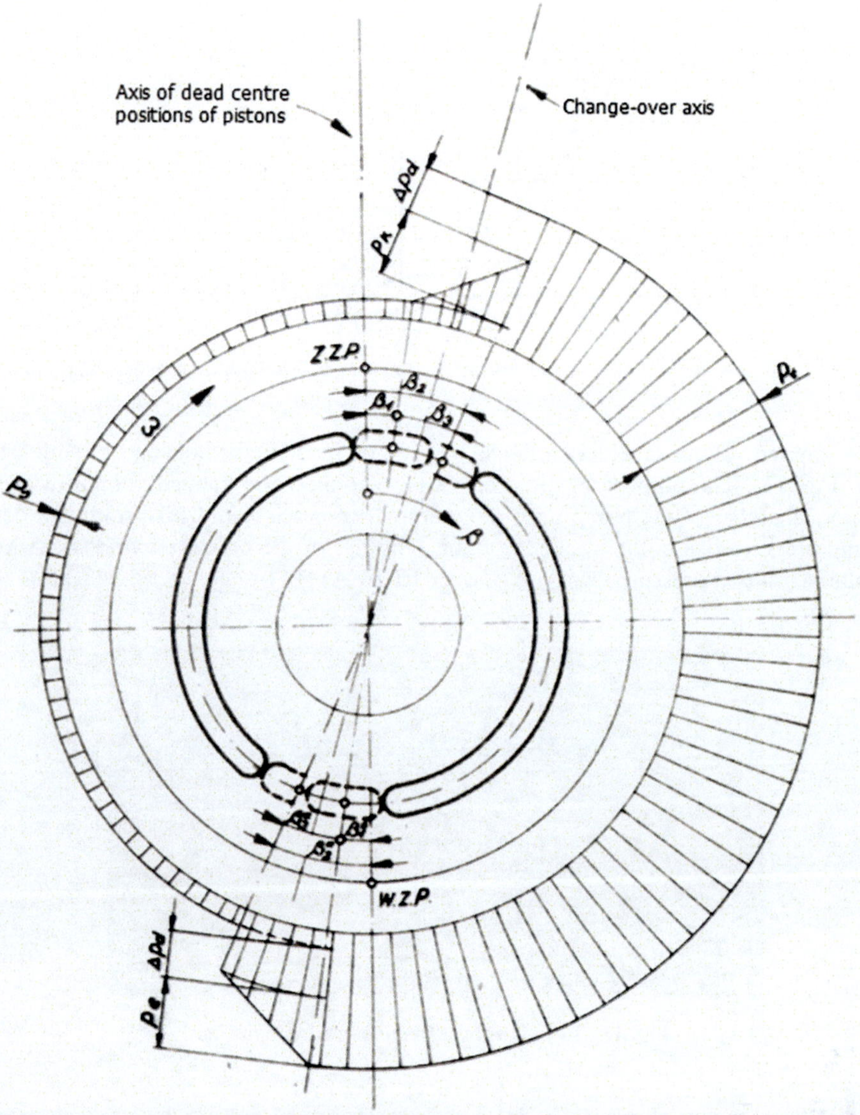

Fig. 12. Correct pressure distribution in axial piston pump working space at change-over axis rotated consistently with direction of rotor rotations [4].

The results of multi-piston pump noise reduction depending on the alterations made are presented in Fig. 14. Curves 1, 2 and 3 represent the noise pressure level of respectively: a mass-produced pump, a pump with a novel distributive disc and a pump with

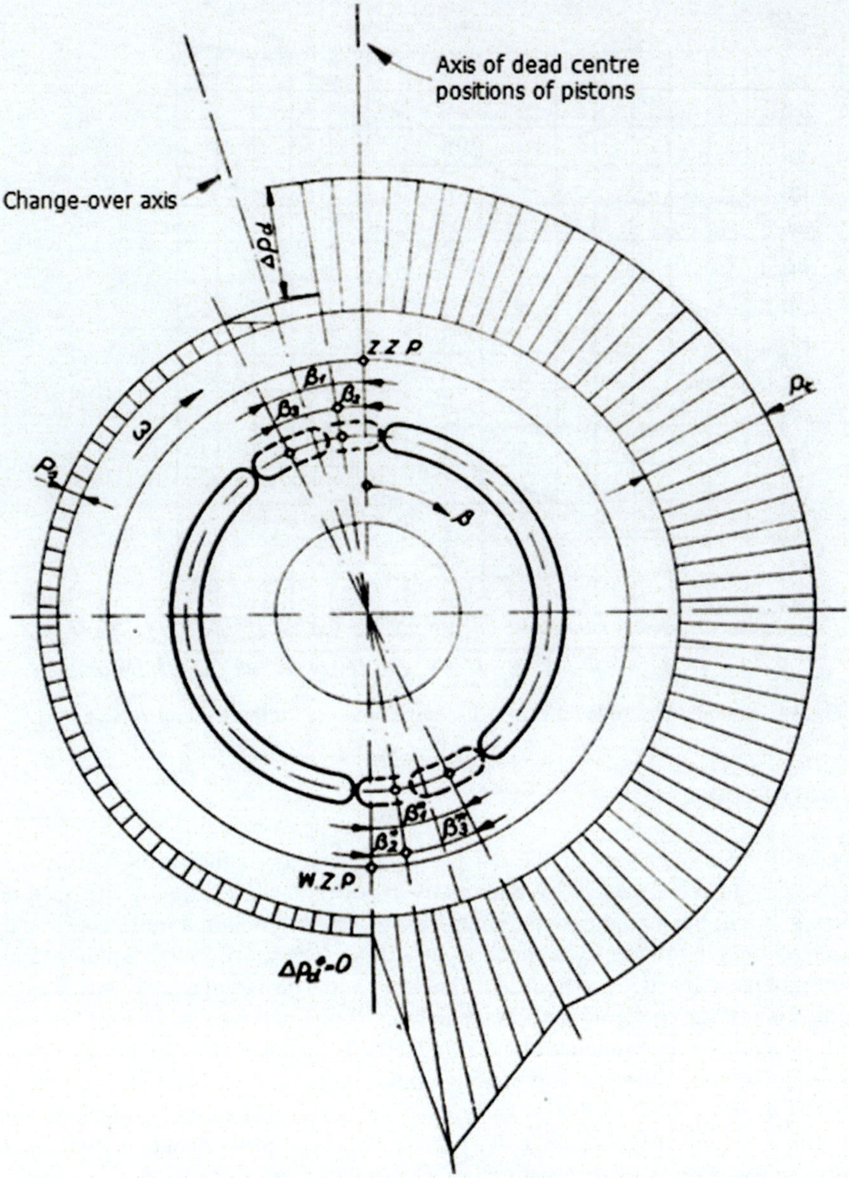

Fig. 13. Incorrect pressure distribution in axial piston pump working space at change-over axis rotated in direction opposite to direction of rotor rotations [4].

further design alterations. Already in the partly modified pump (2) the level of emitted acoustic pressure is lower by 7–13 dB, while in the experimental pump (3) it is lower by as much as 18–20 dB [4].

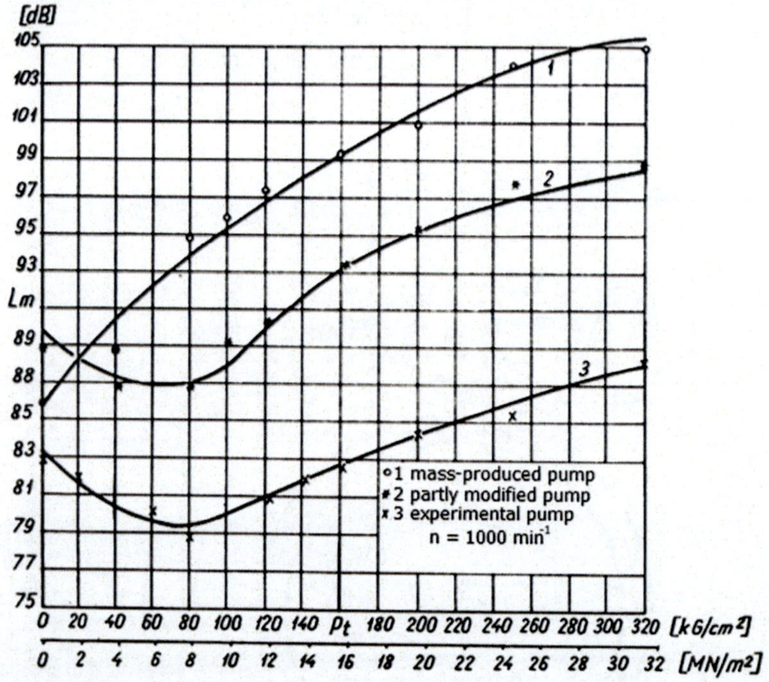

Fig. 14. Sound pressure level versus discharge pressure for three pump versions [4].

4 Conclusion

It appears from the analysis of the noise reduction methods that using only passive methods one can reduce noise by maximally 10 dB(A), while using active methods, i.e. intervening in the sound generation processes, one can obtain a more considerable noise reduction, amounting to as much as 20 dB(A). The most effective solution is to combine passive and active methods. An alteration of the design of a noisy machine, e.g. a modification of the change-over process in a multi-piston pump, will result in a lower sound level. If this is combined with passive methods (consisting in the dissipation and absorption of sound waves), comprehensive noise reduction is achieved. From among the passive methods used to lower the level of the sound generated by machines with a hydraulic drive, the original design of a broadband active pressure pulsation dampener presented in [6], deserves special attention. Using this dampener one can considerably reduce the amplitudes of low-frequency noise (especially below 20 Hz). Moreover, the recently developed investigative methods based on sound intensity measurements, such as the sound intensity probing, acoustical holography and beam forming, are worthy of note. The radiosity methods mentioned are particularly useful for determining the dominant sound generating sources. The correct quantitative identification of noise sources contributes to the high effectiveness of the active methods in reducing the noise emitted to the environment and consequently, to an improvement in the acoustic environment.

Further development directions should be directed towards systemic implementation in industry of active and passive methods of reducing noise of machines and devices. If minimum levels are reached due to physical, technological or economic constraints as a supplement to active and passive methods, wider attempts should be made to implement active methods involving the use of the phenomenon of destructive interference of waves of a noise source and an additional emitter source (anti-noise).

References

1. Central Institute for Labour Protection: Sources of exposure in work and life environment (in Polish). https://www.ciop.pl/CIOPPortalWAR/appmanager/ciop/pl?_nfpb=true&_pageLabel=P62005986134017866l073&html_tresc_root_id=32274&html_tresc_id=3000243&html_klucz=32274&html_klucz_spis=. Accessed 13 Jan 2020
2. Uzarczyk, A.: Harmful and oppressive factors in the work environment (in Polish). Human Resources Consulting and Improvement Center, Gdańsk (2006)
3. Kollek, W., Osiński, P.: Assessment of the usefulness of energy measurement methods for identifying sound sources in hydraulic devices (in Polish). Development of hydraulic machines and device, Editorial Institution of Wrocław Board of Scientific Technical Societies Federation NOT, pp. 5–19, Wrocław (2006)
4. Kollek, W.: Problems of Active Reduction of Axial Piston Pumps Noise (in Polish). WUS&T, Wrocław (1976)
5. Bosch-Rexroth training materials
6. Kudźma, Z.: Damping of Pressure Pulsation and Noise in Hydraulic Systems in Transition and Steady States (in Polish). WUS&T Publishing House, Wrocław (2012)
7. Kollek, W., Kudźma, Z., Osiński, P., Rutański, J.: Localization and attempt of noise reduction in hydraulic feeder (in Polish). Probl. Work Mach. **15**, 101–112 (2000)
8. Kollek, W., Kudźma, Z.: Experience in the silencing machinery with the hydrostatic drive systems (in Polish). Hydraulic drives in machines and vehicles, pp. 59–76. Wrocław University of Science and Technology, Racibórz (2012)
9. Osiński, P.: High-Pressure and Low-Pulsation External Gear Pumps (in Polish). WUS&T Publishing House, Wrocław (2013)
10. Kollek, W., Maćkiewicz, K.: Theory and Calculation of Gear Pumps: Hulls and Preloaded Systems (in Polish). The Ossolineum, Wrocław (1999)
11. Osiński, P.: Gear Pumps with Reduced Noise Emission Level (in Polish). WUS&T Publishing House, Wrocław (2017)
12. Śliwiński, P.: The influence of water and mineral oil on mechanical losses in the displacement pump for offshore and marine applications. Pol. Marit. Res. **25**, 178–188 (2018)

Technology of Ultrasonic Cavitation Cleaning of Elastic Surfaces

Katerina Luhovska, Andrey Movchanuk, Volodymyr Feshich, and Alina Shulha(✉)

«Igor Sikorsky Kyiv Polytechnic Institute», National Technical University of Ukraine, Kiev, Ukraine
{kateryna_lugovska,f_v_p}@ukr.net, movchanuk@rtf.kpi.ua, a.liashok@kpi.ua

Abstract. The possibilities of ultrasonic cleaning of elastic surfaces as exemplified in fabrics are considered. The effect of ultrasonic cavitation processing of fabric for its strength characteristics is analyzed. It has been experimentally proven that cavitation processing of fabric leads to less mechanical destruction compared to traditional drum-type washing. The effect of ultrasonic cavitation processing on the brightness of dyed fabrics faded over time has been experimentally studied. Attention is paid to the inactivation of microorganisms harmful to human health in the ultrasonic cavitation field. Based on the conducted research, the design of an ultrasonic emitter drive and technological equipment for the implementation of technology of ultrasonic cavitation cleaning of elastic surfaces is proposed.

Keywords: Ultrasonic cavitation · Elastic surfaces · Ultrasonic emitter drives

1 Introduction

The technology of ultrasonic cavitation cleaning of hard surfaces is widely known in the technology. In this case, the cleaning effect is achieved due to exposing solid contaminants and greasy film deposits to shock waves, high-energy cumulative jets and intense microflows arising from vibrations and collapse of cavitation bubbles [1–3].

However, the technology of ultrasonic cavitation cleaning of elastic, for example, fabric surfaces is no less well-known and important [4–6]. The need to develop and study such a technology is caused by significant shortcomings of traditional drum-type cleaning of fabrics, associated with mechanical damage of fabrics, the need to use chemical detergents, the drain of which into natural reservoirs is unacceptable in accordance with modern environmental standards, the inability to restore the brightness of fabric color without chemically active substances, as well as poor ability to inactivate harmful and dangerous for human health microorganisms when cleaning medical fabrics. Effective disinfection of fabrics and washing solutions is achieved due to the fact that, in addition to the already mentioned effects accompanying the collapse of cavitation bubbles, there is a local increase of temperature up to 1000 °C and pressure up to 1000 MPa, as well as

© The Editor(s) (if applicable) and The Author(s), under exclusive license to Springer Nature Switzerland AG 2021
J. Stryczek and U. Warzyńska (Eds.): NSHP 2020, LNME, pp. 264–271, 2021.
https://doi.org/10.1007/978-3-030-59509-8_23

activation, in the cavitation area, of oxidative processes that contribute to the inactivation of microorganisms.

The purpose of this article is to analyze the capabilities of the ultrasound technology under consideration, as well as the problems that arise when designing the appropriate technological equipment.

2 Main Part

First, it is necessary to conduct comparative tests of the degree of mechanical damage to fabric threads in a cavitation medium and during drum-type washing by studying the change in tensile strength depending on the number of washes. The experiments were carried out on four types of fabric: cambric, calico, silk and chintz [4, 6]. "Ariel" washing powder was used as a chemical detergent. The temperature during the tests was maintained at 40 °C ... 45 °C. The duration of each wash was 30 min. Identical samples were cut from the selected types of fabrics, 100 mm long and 30 mm wide. In accordance with the experimental design theory, the parameters of the experiment necessary to achieve a maximum degree of disinfection of 99.399% and a maximum rate of contamination removal of 99.473% [7], were chosen. Cavitation processing was carried out in a bath with a low-amplitude ultrasonic emitter with 100 W power with a resonant frequency of 38 kHz, the ultrasound intensity was 2.5 W/cm^2. The samples were completely dried after each wash using two technologies. After drying, part of the samples was tested for strength, and another part was sent for further washing.

Strength tests were carried out in a certified laboratory using the TIRAtest-2151 universal testing machine. The samples were clamped according to the scheme shown in (Fig. 1), using special grips to prevent damage of fibers.

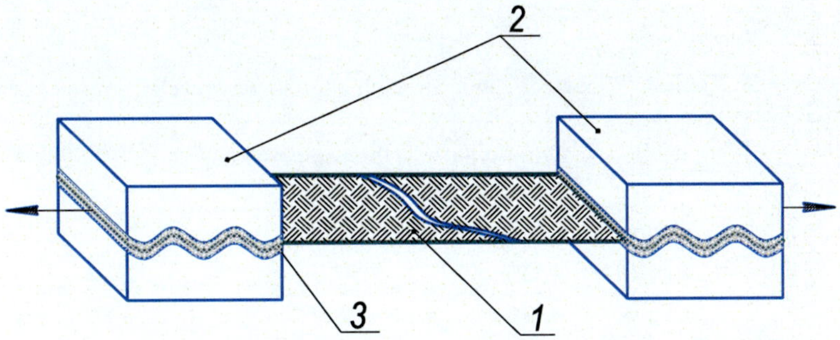

Fig. 1. Scheme of testing of fabric samples for strength (1-test sample; 2-grips; 3-corrugated plates preventing damage of the sample).

The speed of movement of the active grip of a tension testing machine was 50 mm/min at a temperature of 22 °C and relative humidity of 86%.

The results obtained indicate (Fig. 2) that after 30 technological cycles of both drum-type and ultrasonic cavitation washing, the strength of almost all types of fabrics considered has decreased by 11...33% of the initial values (Table 1). At the same time, the comparison of drum-type and cavitation washing shows that both processes under consideration reduce the strength of the fabric. An exception is cambric, which shows high resistance to cavitation. Its strength after 30 cycles of cavitation washing has decreased by only 2.5%. At the same time, after 30 cycles of drum-type washing, the strength of the cambric has decreased by 32%.

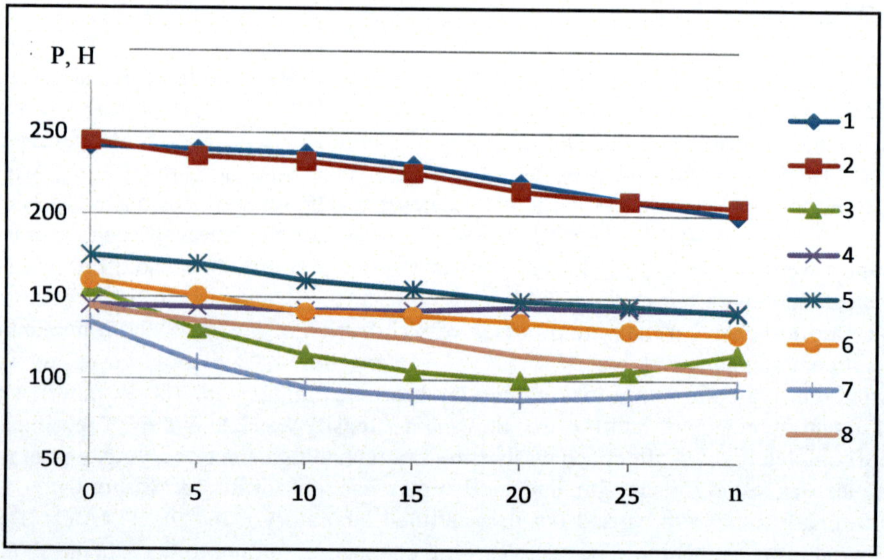

Fig. 2. The relation of the tensile strength of fabric samples and the number of processing cycles (Cleaning in a drum-type machine: 1 - calico; 3 - cambric; 5 - silk; 7 - chintz. Cleaning by ultrasonic cavitation technology: 2 - calico; 4 - cambric; 6 - silk; 8 - chintz).

Table 1. Change in fabric strength after 30 wash cycles.

Fabric	Drum-type washing	Ultrasonic washing
Cotton	−14%	−11%
Batiste	−32%	−2.5%
Chintz	−33%	−27%
Calico	−13%	−13%

Chintz was selected to assess the quality of processing. Blood and sunflower oil were used as pollutants. Three methods of cleaning were investigated: a traditional mechanical drum-type and two cavitation ones using different modes of ultrasonic cavitation, namely the regime with an ultrasonic wave with an intensity of up to 10 W/cm^2 created by a low-amplitude resonant drive-emitter, and the regime with an ultrasonic wave with an intensity of more than 10 W/cm^2 created by high-amplitude resonant drive emitter.

In the case of low-amplitude cavitation in a liquid, small cavitational vapor-gas bubbles are formed in the rarefaction phase, which are almost not observed with an unaided eye. Having made one or several vibrations and accumulated energy, the bubbles in the compression phase collapse. Moreover, if the bubble is removed from neighboring bubbles or from solid surfaces, then its shape is close to spherical and its collapse occurs with the formation of spherical shock waves. Otherwise, the shape of the bubble is distorted and due to the radial imbalance of forces, the bubble collapses with the formation of a high-energy cumulative jet directed towards the nearest surface. Shock waves and cumulative jets destroy hardened impurities, tear off impurities from fibers, overcoming the adhesion and cohesion. Fluctuations of bubbles and diverging shock waves cause microflows in the washing technological solution. Microflows contribute to the mixing of the washing solution at the molecular level. A large number of collapsing bubbles provide high erosion activity to the resulting cavitation medium, activate the liquid due to the formation of free radicals when the bubbles collapse, and lead to intensification of chemical processes in the liquid. At the same time, the action of the chemical components of the washing solution is significantly activated, which reduces the time of the technological process of cleaning and reduces the amount of consumed detergents.

In case of high-amplitude cavitation, in which the intensity of ultrasonic vibrations exceeds 10 W/cm^2, cavitation bubbles formed in the rarefaction phase due to the large amplitude values of sound pressure do not have time to slam in the compression phase, but they continue to fluctuate in time with the ultrasonic wave, gradually increasing in size to those observed with an unaided eye. Fluctuations of such bubbles lead to the occurrence of intense microflows in the volume of the technological washing solution. The washing solution transfers to a shaking dynamic state. In this case, a fixed, contaminated with fat, fabric surface begins to be intensively cleaned due to dissolving and washing off contaminants with intense microflows of the washing solution. Unlike traditional drum-type mechanical cleaning, the fabric, remaining motionless, is not mechanically deformed and not damaged. It should be noted that with the increase in the intensity of ultrasonic vibrations introduced into the liquid, the introduction efficiency decreases due to the formation of a two-phase cavitation layer on the radiating surface that absorbs and scatters ultrasonic energy.

During testing, the increased ultrasound intensity was achieved through the use of a stepwise ultrasonic speed transformer. At that, the amplitude of fluctuations of the radiating surface of the drive was 25 μm, the intensity of the ultrasonic wave introduced into the liquid reached 18…25 W/cm^2.

Studies have shown that the highest quality of cleaning hardened contaminants is achieved by cavitation cleaning with ultrasound of low and medium intensity, and fat ones – by cavitation cleaning with ultrasound of high intensity.

Increasing the productivity of the technological process of ultrasonic cavitation cleaning of fabrics requires increasing the power consumed by ultrasonic emitter drives. In this case, the maintenance of the required fluctuation intensity is achieved by creating a developed radiation surface.

An analysis of the results of an experimental study of the effect of ultrasonic cavitation impact on the color saturation of samples of cotton fabric was carried out optically.

The analysis method consisted in photofixation of samples with subsequent digital processing of the obtained images in RGB palette with the selection of each color and the determination of its content in the color gamut.

The color saturation of the processed samples was determined as a percentage of the color saturation of the original sample. The total color saturation was derived by averaging the color saturation indices of individual samples.

Experimental samples were cut from chintz that had naturally aged. The data obtained indicate that the cavitation effect on the fabric allows to partially restore its color. The effect manifests itself from the first seconds of processing and increases with increasing processing time. In the experiments, a more intense increase in the color saturation of the fabric was observed in the first third of the processing time. With a further increase in processing time, the restoration of color continues, however, the intensity of the effect is relatively low (Fig. 3).

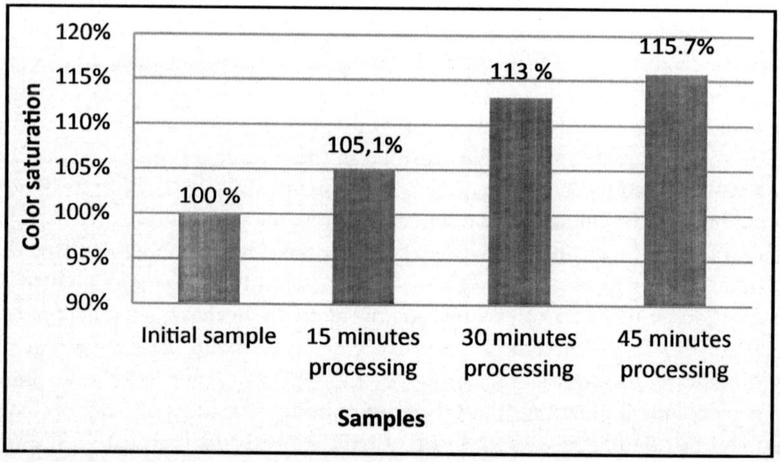

Fig. 3. The results of an experimental study of the effect of ultrasonic cavitation on the color saturation of the fabric.

The restoration of colourity may be due to both cavitation cleaning of the fabric from solid and fatty contaminants, and cavitation erosion of the upper pigment layer of the fabric dye. From physical standpoint, the mechanism of cleaning the fabric from contaminants can be explained by the specifics of the perception by the fibers of the fabric of the energy-force impact caused by cavitation. Due to its elasticity and permeability, the fabric partially dampens the stresses arising on its surface. The high degree of wettability of the fabric due to the sound-capillary effect in the ultrasonic field provides a weakening of the adhesion strength. Cohesion forces in hardened contaminants and bonds holding foreign particulates in interfiber spaces and fiber defects, as well as contaminants sorbed and retained in the surface layer of the fabric, and macromolecular substances and compounds are much more susceptible to intense shock waves and cumulative jets that occur during the collapse of cavitation bubbles. However, coloring principles under the action of cavitation erosion, are destroyed gradually, revealing brighter, not discolored layers, which, in combination with these phenomena, leads to a partial restoration of the color saturation of the fabric.

The performed studies show that for the effective implementation of the technology of ultrasonic cavitation cleaning of the fabric, the ultrasonic emitter drives with a developed radiation surface that most fully satisfy the requirements of the process are needed. A methodology for calculating such drive emitters has been developed and tested [8–12]. Their design represents a multi-disc resonant speaker system (Fig. 4).

Fig. 4. Ultrasonic resonant cavitator with a developed radiation surface.

Figure 5 shows a possible implementation of technological equipment for ultrasonic cavitation fabric cleaning technology [13, 14].

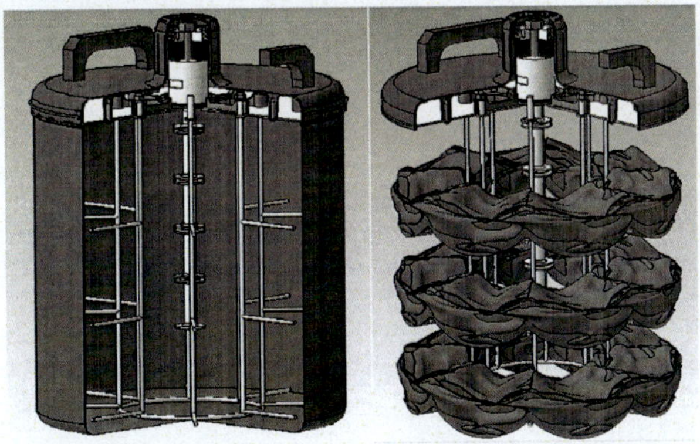

Fig. 5. An example of the implementation of technological equipment for ultrasonic cavitation technology for cleaning elastic surfaces.

3 Conclusion

The expediency of applying the effects accompanying the phenomenon of ultrasonic cavitation for the cleaning of solid and greasy film contaminations of fabric surfaces is experimentally confirmed.

It is shown that cavitation processing of fabric leads to a decrease in its strength. However, this decrease in strength is significantly less than with traditional drum-type cleaning. Cavitation cleaning allows to process drawn thread works made from delicate fabrics. The restoration of the color parameters of faded fabrics is confirmed. The effect of inactivation of harmful and dangerous microorganisms for human health is an important advantage of ultrasonic cavitation fabric cleaning technology.

The achievement of high cleaning efficiency due to the introduction of medium and high intensity ultrasonic vibrations into the liquid is experimentally confirmed. For this purpose, it is proposed to use multi-disc acoustic systems with a developed radiation surface, which allow simultaneously introducing vibrations of different intensity levels into the technological cavitation chamber. A methodology for calculating such emitter drives is proposed.

References

1. Panov, A.P.: Ultrasonic cleaning of precision parts, Mechanical Engineering (1984)
2. Keller, O.K., Kratysh, G.S., Lubyanitskiy, T.D.: Ultrasonic cleaning. Mechanical Engineering, Leningrad. Dep. (1977)
3. Luhovskyi, O.F.: Ultrasonic cavitation in modern technologies/O.F. Luhovskyi, N.V. Chukhraev. - K.: Publishing and Polygraphic Center "Kyiv. un-t" (2007)
4. Antoniak, P., Stryczek, J., Banas, M., Grysnko, I., Zilinskyi, A., Kovalov, V.: Visualiztion research on the influence of an ultrasonic degassing on the operation of a hydraulic gear pump MATEC Web of Conferences VETOMAC XIV (2018). https//doi.org/10.1051/mateccconf/201821103005
5. Yakhno, O.M.: Study of the possibilities of technology of ultrasonic cavitation cleaning of elastic surfaces/O.M. Yakhno, K.O. Luhovska, A.V. Movchanuk//- Reporter of National Technical University of Ukraine "Kyiv Polytechnic Institute": series "Mechanical Engineering", vol. 58, pp. 234–240 (2010)
6. Lugovska, K.O.: Environmental safety technology of ultrasonic cavitational cleaning/Osteuropa Workshop auf dem 22. Kasseler Abfall- und Bioenergieforum. Uni Kassel Transfer. Ost-West-Wissenschaftszentrum, Kassel (2010)
7. Luhovska, K.O.: Improved Process Ultrasonic Cavitation Treatment Elastic Surface/K.O. Luhovska, O. Yahno//- International Scientific Conference UNITECH 2010, p. II, 379–385 Gabrovo (2010)
8. Luhovska, K.O.: Planning a multivariate experiment and parametric optimization of indicators of ultrasonic cavitation cleaning of elastic radiation surfaces//- Reporter of National Technical University of Ukraine "Kyiv Polytechnic Institute": series "Mechanical Engineering", vol. 62, pp. 241–246 (2011)
9. Kumabe, D.: Vibration cutting/Japanese translation L. Maslennikova/Edited by I.I. Portnova, V.V. Belova.- M.: Mechanical Engineering (1985)
10. Luhovska, K.O.: Method for calculating an ultrasonic resonant cavitator with a developed radiation surface//– Reporter of National Technical University of Ukraine "Kyiv Polytechnic Institute": series "Chemical Engineering, Ecology and Resource Saving", vol. 2, no. 4, pp. 13–19 (2009)

11. Movchanuk, A.V.: Features of the calculation of ultrasonic cavitators with a developed radiation surface for various technological purposes/Movchanuk A.V., Luhovska K.O., Fesich V.P.//- Industrial hydraulics and pneumatics, Vinnytsya, vol. 2, no. 32, pp.48–50 (2011)
12. Kolosov, A.E.: Method for calculating an ultrasonic cavitation device with a radiating plate/A.E. Kolosov, V.I. Sivetsky, E.P. Kolosova, K.O. Luhovska//Chemical and oil-gas engineering, vol. 11, pp. 11–17 (2012)
13. Fesich, V.P.: Computer-aided design system for ultrasonic cavitation drives/V.P. Fesich, K.O. Luhovska, S.A. Krivko//- Industrial hydraulics and pneumatics, Vinnytsya, vol. 3, no. 29, pp. 34–40 (2010)
14. Luhovska, K.O.: Ultrasonic machine for cleaning elastic surfaces/K.O. Lugovska, O.M. Yakhno, O.F. Luhovskyi //Utility Model Patent of Ukraine № 53888, Published. 25.10.2010, Rep. № 20 Lugovska K.O. Method for ultrasonic cavitation cleaning of elastic surfaces/K.O. Luhovska, O.M. Yakhno, O.F. Luhovskyi//Utility Model Patent of Ukraine №. 53888, Published. 25.10.2010, Rep № 2

Mobile Equipment for Ultrasonic Cavitation Inactivation of Microorganisms in the Liquid Environment

Oleksandr Luhovskyi[1], Irina Bernyk[2], Ihor Gryshko[1], Darina Abdulina[3], and Andrii Zilinskyi[1](✉)

[1] "Igor Sikorsky Kyiv Polytechnic Institute", National Technical University of Ukraine, Prosp. Peremohy 37, Kiev 03056, Ukraine
{atoll-sonic,griwko}@ukr.net, zilinski.andrew@gmail.com

[2] Vinnitsa National Agrarian University, Sonyachna St. 3, Vinnitsa 21000, Ukraine
iryna_bernyk@i.ua

[3] D.K. Zabolotny Institute of Microbiology and Virology of the NASU, 154 Acad. Zabolotny St. 03143, Kiev, Ukraine
adara@ukr.net

Abstract. The article is devoted to the study of the possibilities of ultrasonic cavitation inactivation of microorganisms. The possibilities of constructing mobile processing equipment for implementing the technology of ultrasonic cavitation inactivation of harmful and dangerous microorganisms for human health are considered. Two possible ways of increasing the intensity of ultrasonic exposure to biological objects were presented due to the concentration of ultrasonic energy and cavitation treatment in a thin layer on the radiation surface. The results of microbiological analysis as well as recommendations to increase the efficiency of cavitation inactivation of microorganisms had given.

Keywords: Ultrasonic cavitation · Inactivation of microorganisms · The intensity of ultrasonic vibrations · Mobile cavitation equipment

1 Introduction

In industrial processes using liquid environment, it is often necessary to decontaminate working fluids, i.e. to inactivate harmful for human health microorganisms.

Known methods of disinfection of liquid environment can be divided into four main groups, which differ by mechanism of microorganisms' inactivation [1, 2]:

- chemical methods associated with the use of strong oxidants reagents. Chlorine, chlorine dioxide, sodium hypochlorite, ozone, iodine, bromine, hydrogen peroxide, manganese-acid potassium are used as oxidizing agents [3, 4];
- thermal methods realized via liquids' boiling or freezing [5, 6];
- water saturation methods using ions of noble and non-ferrous metals with bactericidal effect - silver, copper, etc. [7, 8];

© The Editor(s) (if applicable) and The Author(s), under exclusive license to Springer Nature Switzerland AG 2021
J. Stryczek and U. Warzyńska (Eds.): NSHP 2020, LNME, pp. 272–281, 2021.
https://doi.org/10.1007/978-3-030-59509-8_24

– physical methods, i.e. ultraviolet radiation, ultrasound, radiation [9, 10].

Each of these methods is characterized by mechanism of inactivation of microorganisms, has its advantages and disadvantages.

In many respects, ultrasonic cavitation method of inactivation is preferred, relating to non-reagent inactivation methods [11–14]. The mechanism of its operation is:

– the damaging effect of the cumulative jet stream formed by the collapse of the cavitation bubble in the immediate vicinity of the microorganism;
– local increase of temperature and pressure in case of cavitation bubble collapse resulting in thermal and mechanical destruction of microorganisms;
– a drop in pressure over the microorganism's length in a standing ultrasonic wave, resulting in a breakdown of the structure of the microorganism;
– intensification of chemical oxidation processes in the cavitation environment, leading to acceleration of the process of inactivation of microorganisms by the formation of chemically active free radicals.

Fig. 1. Effect of formation of shielding cavitation layer on the emitting surface of ultrasonic transducer when ultrasound intensity increases from 18 W/cm^2 (a) to 28 W/cm^2 (b).

This method can be applied at any degree of liquid transparency. It is environmentally safe and is least capable of adversely affecting the human body, as it does not require the use of chemical agents.

The purpose of the proposed article is to analyse problems encountered in the creation of ultrasonic cavitation equipment for disinfecting liquid media and to test its effectiveness.

An important point in the implementation of the technology of ultrasonic cavitation inactivation of microorganisms is the intensity of ultrasonic oscillations, which are introduced into liquid. Low intensity (less than 2 W/cm^2) stimulates intensive reproduction and development of microorganisms [11, 12]. In order to achieve high efficiency of the process of inactivation of microorganisms by the method of ultrasonic cavitation treatment it is necessary to provide the level of intensity of ultrasound above 20–30 W/cm^2. At the same time, the width of the spectrum destroyed in the liquid of microorganisms depends significantly on the intensity of the ultrasonic oscillations. Therefore, the developers of ultrasound cavitation technological equipment for disinfecting fluids aim to ensure the maximum intensity of ultrasonic fluctuations. The problem is that ultrasonic fluctuations with low intensity only (Fig. 1a) can be introduced into the liquid with high efficiency [15]. When the intensity of ultrasonic oscillations increases on the radiating surface of the ultrasonic converter, a two-phase cavitation layer is formed (Fig. 1b), which absorbs and dissipates ultrasonic energy, resulting in the efficiency of the introduction of ultrasonic oscillations into the liquid decreases. When creating stationary industrial flow cavitation plants, this problem is solving due to using of the focusing features of the radiating surface [16–18]. The use of a tubular radiating surface, which is activated on the radial mode of the vibration, allows introducing the low-intensity ultrasound fluctuations effectively into the liquid and then focusing them on the longitudinal axis of the vibrator (Fig. 2). As a result, the intensity of about 120 W/cm^2 can be reach and a wide range of microorganisms can be inactivate.

2 Ultrasonic Cavitation Equipment

During the designing of the mobile and home ultrasound cavitation equipment, it is very difficult to use the considered tubular cavitator, triggered on the radial mode of fluctuations, due to mass-size characteristics. In this case, liquid quarter wave transformers of oscillating pressure (Fig. 2) are often used to increase sound pressure [19, 20].

In Fig. 3 was presented a scheme of construction of a step-type hydraulic transformer of vibrational pressure using a semi-wave ultrasonic piezoelectric transducer consisting of sequentially composed and reinforced radiating lining 5, piezoceramic elements 7 and damping lining 6. The diameter of the radiating surface 9 and the diameter of the larger transformer stage are selected significantly less than half the wavelength. This ensures the piston nature of the vibrations of the transmitter surface and a flat acoustic wave in the inner volume of the cavitation chamber. The ultrasonic transducer is fixed at the nodal point of deformation wave 11, which is set along its length. Due to tightness of the ultrasonic transducer mounting in the pressure transformer housing there is a closed air cavity 8, which does not allow the treated liquid to wet the nodal point of the transducer and reduce the efficiency of ultrasonic energy input into the liquid through

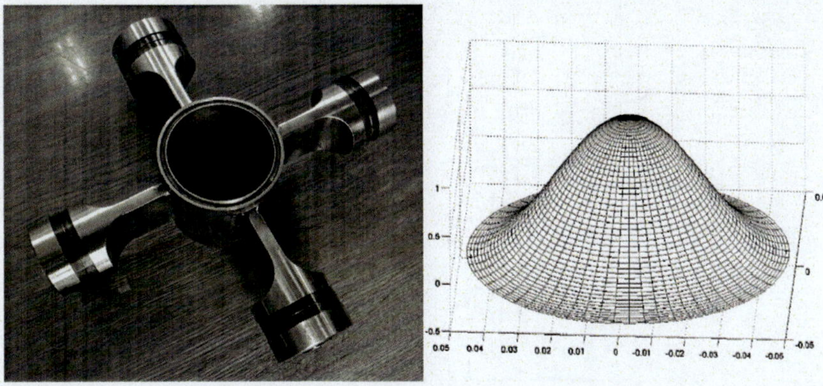

Fig. 2. Ultrasonic flow cavitator with tubular radiating surface and distribution of sound pressure over its cross-section in case of excitation of radial mode of vibration.

the radiating surface. The enlarged area of the radiating surface 9 makes it possible to adjust the load on the piezoelectric transducer and to inject ultrasonic vibrations into the liquid with sufficiently low intensity. This is prevent the formation of cavitation layer 9 on the radiating surface and ensures that ultrasonic vibrations are dispersing into the liquid bulk with maximum efficiency. Since the chambers 1 and 2 of the oscillating pressure transformer have a length divisible by an odd number of quarters of the length of the standing ultrasonic wave in the liquid at the resonant frequency of the piezoelectric transducer, and the stage transition of a larger diameter of the pressure transformer to a smaller diameter stage is organized in the plane of the nodal point of the pressure wave 12, then along the length of the transformer oscillatory pressure is installed standing deformation wave, and in the nodal planes standing deformation wave in the stage of small diameter level of sound pressure increases in proportion to the ratio of areas of the transformer stages. As a result, in a step of a smaller transformer diameter the level of ultrasound intensity sufficient for inactivation of a wide range of dangerous microorganisms for human health will be provided.

In Fig. 4 were shown the scheme of ultrasonic mobile cavitation equipment, which provides cavitation treatment of liquid in thin layer. Ultrasonic wave with high ultrasound intensity is specially demonstrated on the scheme. For this aim using the oscillating speed transformer 3 high amplitude of oscillations of radiating disk 4 at the output end of transformer 3 is provided. Figure 1b shows, that in this case, most of the ultrasonic energy will be absorbed in the cavitation layer with developed cavitation formed on the radiating surface. It is in this layer the most intensive treatment of the liquid with the aim of inactivation of microorganisms will occur. Therefore, the scheme under consideration provides for the organization of narrow gaps 5 and 6 between the radiating disk surface and housing 1, through which all the treated liquid was proceed. In addition, the casing made of non-magnetic material has an output channel of resonant length, in which a standing deformation wave 8 is set, at the nodal points of which there are intensive cavitation zones, providing additional cavitation treatment of the liquid. The output channel of the body 1 is covered with powerful permanent magnets 7, which ensure the structuring of fluid after destructive intensive cavitation treatment and due to alternating

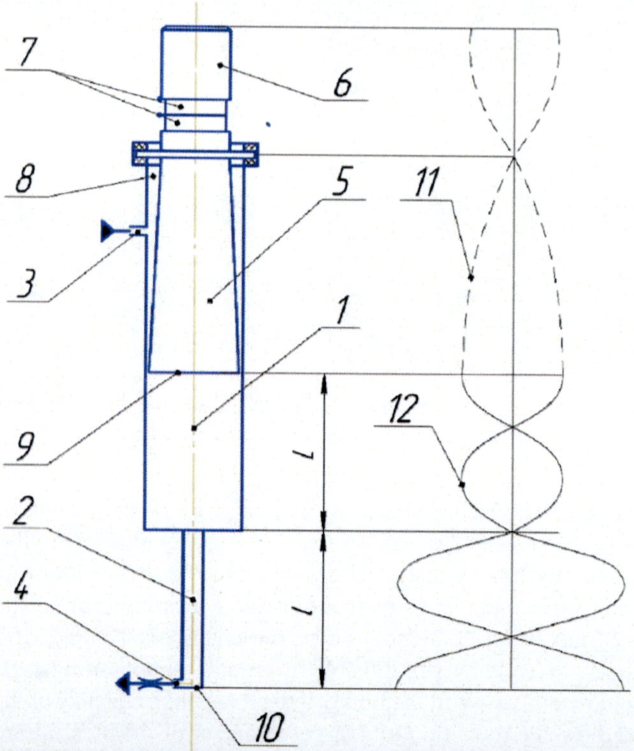

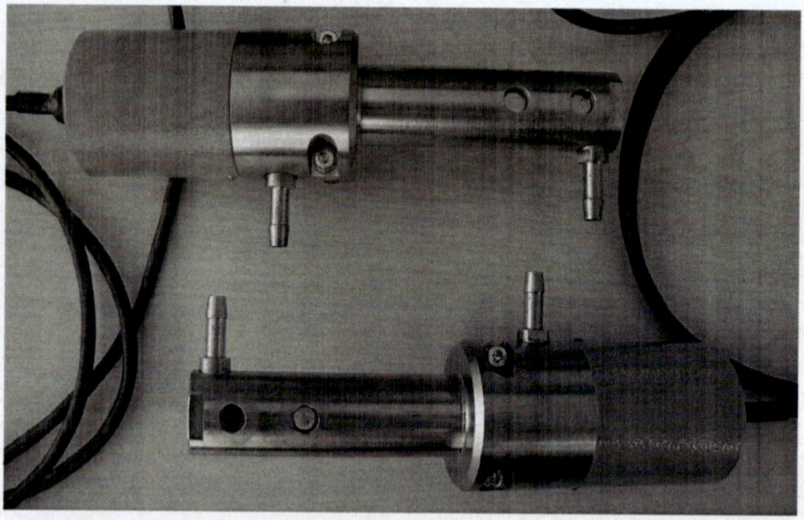

Fig. 3. Scheme and appearance of mobile ultrasonic cavitation equipment with oscillating pressure transformer. (1 - stage of large diameter of pressure transformer; 2 - stage of small diameter of pressure transformer; 3 - inlet connection; 4 - outlet connection; 5 - radiating lining of piezoelectric converter; 6 - damping lining; 7 - piezo-ceramic elements; 8 - air chamber; 9 - radiating surface; 10 - reflecting surface; 11 - deformation wave in piezoelectric converter; 12 - pressure wave in liquid). (Intensity of ultrasonic field 60 W/cm^2; capacity - 2 l/min; power consumption - 40 W).

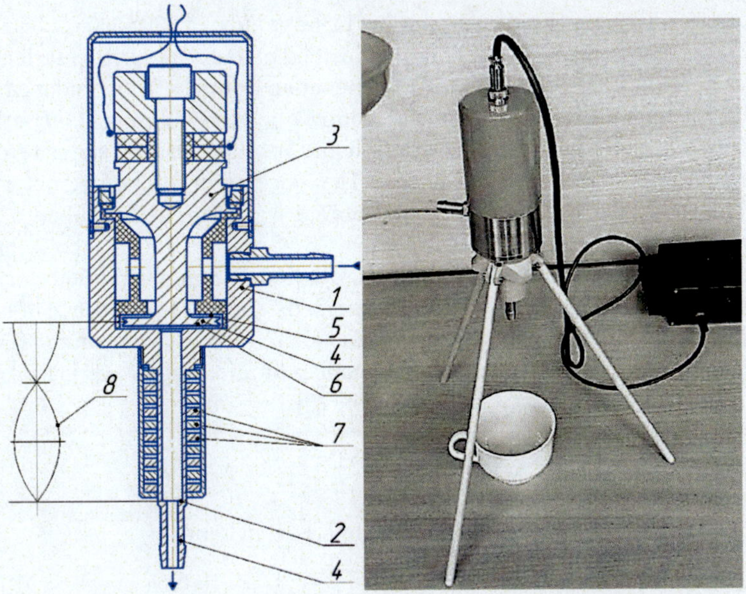

Fig. 4. Scheme and appearance of mobile ultrasonic cavitation equipment with thin layer processing. (1 - case made of non-magnetic material with input and output connectors and resonant output channel; 2 - reflective end of output channel; 3 - piezoelectric ultrasonic radiator with transformer of vibrational speed; 4 - radiating disk of transformer of vibrational speed; 5 and 6 - slot gaps for a flow of processed liquid; 7 - block of permanent magnets; 8 - standing deformation wave in the output channel. (Intensity of ultrasonic field 60… 80 W/cm^2; productivity - 2 l/min; power consumption - 40 W).

magnetic field ensure the removal of heavy metal ions from the liquid, which also adversely affect the human body. Structured water already has healing properties because it has increased permeability through the cell membranes.

3 Microbiological Studies

Microbiological studies were carried out to evaluate the efficiency of water treatment in thin layer in ultrasonic cavitation equipment presented in Fig. 4.

Microorganisms of different species with different structure of the cell walls were selected for the research: gram-negative Escherichia coli, Pseudomonas aeruginosa (non-pathogenic), sulfate-reducing corrosive-relevant bacteria Desulfovibrio sp., Desulfovibrio desulfuricans, Desulfovibrio vulgaris; gram-positive - Bacillus subtilis, Staphylococcus aureus; microscopic fungi Saccharomyces cerevisae. Studies with microorganism's cultures were carried out in the laboratories of the D.K. Zabolotny Institute of Microbiology and Virology NAS of Ukraine.

Sterile tap water with introduction of bacteria in initial titers 106-108 cells/ml was used as a working liquid. The volume of working liquid was 200 ml. After each cycle of liquid spilling through the ultrasonic cavitator 1 ml of sample was taken for analysis and the cavitator was washed with sterile tap water in the volume of 200–400 ml. The number of bacteria was determined by serial dilution method on the appropriate nutrient media: nutrient broth NB (Himmedia, India), YEPD medium, Postgate "B" liquid medium [21, 22]. The determination was carried out as follows: a fixed volume (1 ml) of the test liquid was putted into tubes with the sterile liquid medium and the series of dilutions in liquid media were performed in triplicate. The inoculation was carried out with each dilution, and each dilution to achieve repeatability was sown in 3–5 tubes. The tubes with dilutions were incubated in a thermostat at 28 °C and 37 °C (for E.coli) for 3 to 15 days depending on the microbial growth rate. After incubation, the presence or absence of growth was observed using different indicators: the presence or absence of turbidity, sediment, gas or accumulation of certain metabolic products in the medium. The results were statistically processed; the most likely number of microorganisms per ml of water was calculated using a special McCradey table [23].

4 Results

The results of the tap water treatment are summarized in Table 1.

Table 1. The efficiency of ultrasonic cavitation treatment of the tap water with microbial cultures.

No.	Bacteria culture	Ultrasonic cavitator 40 W, frequency 22 kHz, Treatment length					
		1 cycle			10 cycles		Approximation data. Efficiency 99.99999%
		Initial number of m/o, cells/ml	Residual number of m/o, cells/ml	Efficiency, %	Residual number of m/o, cells/ml	Efficiency, %	Cycle*
1	*Escherichia coli*	$1.0*10^8$	$9.5*10^7$	**5.0**	$1.0*10^6$	**99.0**	**38–40**
2	*Staphylococcus aureus*	$1.5*10^8$	$9.5*10^7$	**36.6**	$1.5*10^6$	**99.0**	**40–42**
3	*Saccharomyces cerevisae*	$1.0*10^8$	$9.5*10^7$	**5.0**	$9.5*10^5$	**99.05**	**35–40**
4	*Pseudomonas aeruginosa*	$1.0*10^8$	$9.5*10^7$	**5.0**	$9.5*10^6$	**90.5**	**55–60**
5	*Bacillus subtilis*	$1.5*10^7$	$1.5*10^7$	**0.0**	$9.5*10^7$	**0.0**	**-****
6	*Desulfovibrio desulfuricans*	$9.5*10^8$	$8.5*10^8$	**10.5**	$4.5*10^6$	**99.0**	**30–36**
6	*Desulfovibrio vulgaris*	$9.5*10^7$	$3.0*10^6$	**96.8**	$1.0*10^6$	**98.9**	**-****
8	*Desulfovibrio* sp.	$1.0*10^8$	$4.5*10^7$	**55.0**	$1.0*10^6$	**99.0**	**60–70**

* Number of cycles required to achieve 99.99999% processing efficiency. (According to drinking water quality indicators), m/o – microorganism.

The last column in Fig. 5 is calculated as prediction based on approximation of the received experimental data for various types of microorganisms after 10 cycles water

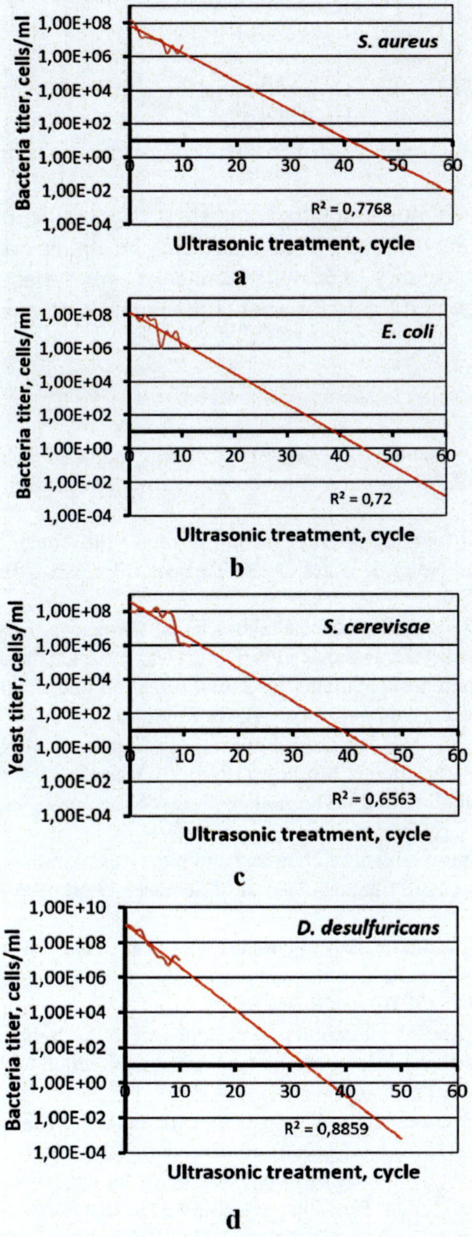

Fig. 5. Data approximation derived from ultrasonic cavitation treatment of water with microorganisms. (a - *Staphylococcus aureus*; b - *Escherichia coli;* c - *Saccharomyces cerevisae*, d - *Desulfovibrio desulfuricans*).

treatment with cavitation. There is shown results for data approximation for the following microorganisms *S. aureus; E. coli; S. cerevisae, D. desulfuricans* (Fig. 5 a, b, c, d).

5 Conclusions

The problems that arise during the introducing the ultrasonic vibrations with increased intensity into a liquid were elucidated. Two methods of increasing the intensity of ultrasonic vibrations to sufficient level for inactivation of the microorganisms in water are considered.

The results of microbiological studies confirm the operability of ultrasonic cavitation equipment. However, they shown that for achieving the almost complete inactivation of microorganisms it is necessary to provide number of cycles manually or automatically due to micro pump a treated liquid for a set of the required number of treatment cycles.

References

1. Luhovskyi, O.F., Movchanuk, A.V., Gryshko, I.A.: Estimation of the water disinfection methods. Bull. Natl. Tech. Univ. Ukraine "Kiev Polytech. Inst." Mach.-Build. Ser. Rev. **52**, 103–111 (2008). (in Russian)
2. Luhovskyi, O.F., Gryshko, I.A., Bernyk, I.M.: Enhancing the efficiency of ultrasonic wastewater disinfection technology. J. Water Chem. Technol. **40**(2), 95–101 (2018). Allerton Press, Inc., ISSN 1063-455X
3. Rachmanin, Y.A., Zholdakova, Z.I., Poliakova, E.E., Kiryanova, L.F., Myasnikov, I.N., Tulskaya, E.A., Artemova, T.Z., Ivanova, L.V., Dmitrieva, R.A., Doskina, T.V.: Joint application of the active chlorine and coagulants for the purification and disinfection of the drinking water. Hyg. Sanitation (1), 449–458 (2004). (In Russian)
4. Pasl, L.L., Karu, Ya.Ya., Melder, H.A., Repin, B.N.: Handbook of Natural and Waste Water Treatment, pp. 36–39. Medicine, Moscow (1994). (in Russian)
5. Kolesov, A.M., Glagolev, L.S.: Thermal method of the waste water disinfection. Hyg. Sanitation (3) (1978). (in Russian)
6. Sashchenko, V.V.: Improvement of chemical and biological parameters of drinking water by the method of programmed freezing. Doctor of Science: 27.00.03. VUO MANEB, Alchevsk (2009). (in Russian)
7. Zolotukhina, E.V., Spiridonov, B.A., Fedianin, V.I., et al.: Water disinfection with nanocomposites on the basis of the aluminum oxide porous and silver compounds. Sorpt. Chromatogr. Processes **10**(1), 78–85 (2010). (in Russian)
8. Shmuter, G.M., Izotova, P.V., Maslenko, A.L., Furman, A.A., Sobolevskaya, T.T.: Hygienic estimation of the electrochemical method of the water disinfection with silver. Hyg. Sanitation (12), 10–11 (1986). (in Russian)
9. Kostiuchenko, S.V.: Ultraviolet irradiation - a modern method of the water disinfection. Water Supply Sanitary Eng. (12), 21–23 (2005). (in Russian), Ch. 1
10. Bernyk, I.M., Luhovskyi, O.F.: Using of physical fields for hydrolysis of protopectin of plant material. Vibr. Technol. Tech. **3**(52), 92–100 (2008). (in Ukrainian). VBAU, Vinnytsia
11. Elpiner, I.E: Ultrasound. Physical-Chemical and Biological Action. Fizmatiz, Moscow (1963). (in Russsian)
12. Feng, H., Barbosa-Cánovas, G.V., Weiss, J. (eds.): Ultrasound Technologies for Food and Bioprocessing. Food Engineering Series. Springer, Heidelberg (2011)

13. Antoniak, P., Stryczek, J., Banas, M., Grysnko, I., Zilinskyi, A., Kovalov, V.: Visualiztion research on the influence of an ultrasonic degassing on the operation of a hydraulic gear pump. In: MATEC Web of Conferences 211, VETOMAC XIV (2018). https://doi.org/10.1051/matecconf/201821103005
14. Luhovskyi, A.F., Gryshko, I.A., Zilinskiy, A.I., Patsola, B.V.: The impact of static pressure on the intensity of ultrasonic cavitation in aqueous media. J. Water Chem. Technol. **40**(3), 143–150 (2018). https://doi.org/10.3103/S1063455X18030050
15. Luhovskyi, O.F, Gryshko, I.A.: Problems of creation of the technological equipment for ultrasonic cavitation water disinfection. Promislova gidravlika i pneumatics **4**(26), 3–6 (2009). (in Russina), Vinnitsa
16. Luhovskyi, O.F., Chukhraev, N.V., Chukhraev, K.V.V.: Ultrasonic cavitation in the modern technologies (2007). (in Russian), K.: MIC "Kyiv. Un-t"
17. Luhovskyi, O.F., Gryshko, I.A., Krivosheev, V.S.: Ultrasonic flow cavitation installation for the technological processes efficiency increase. Visnik NTUU "KPI". Mashinobuduvannya, Visnik NTUU "KPI", vol. 63, pp. 230–232 (2011). (in Russian), Kyiv
18. Luhovskyi, O.F., Gryshko, I.A., Movchanuk, A.V.: Investigation of the ultrasonic tube cavitator operation in the radial vibration mode. Visnik NTUU "KPI". Mashinobuvannya, Rev. **59**, 285–287 (2010). (in Russian), Kyiv
19. Luhovskyi, O.F., Movchanuk, A.V., Zilinskyi, A.I., Gryshko, I.A.: Calculation methods of the rational sizes of the one-dimensional ultrasonic cavitators. Visnik NTUU "KPI". Mashinobuvannya, Rev. **71**, 33–38 (2014). (in Russian), Kyiv
20. Luhovskyi, O.F., Movchanuk, A.V., Luhovskyi, O.O.: Patent UA № 100470; 25.12.2012, Bulletin № 24; Ultrasonic devices for the liquid processing (in Ukrainian)
21. Gromozova, E.N., Voychuk S.I.: Influence of radiofrequency EMF on the yeast Sacharomyces Cerevisiae as model eukaryotic system. In: Biophotonics and Coherent Systems in Biology, pp. 167–175. Springer, Boston (2007). https://doi.org/10.1007/978-0-387-28417-0_13/
22. Iutinskaya, G.A., Purish, L.M., Abdulina, D.R.: Corrosive microbial communities of technogenic ecotopes. M.: LAP Lambert Academic Publishing (2014)
23. Netrusov, A.I., Egorova, M.A., Zakharchuk, L.M. et al.: Workshop on Microbiology: Textbook. Allowance for Students. M.: Publishing Center "Academy" (2005)

Ultrasonic Cavitation Equipment with a Liquid Pressure Transformer

Andrey Movchanyuk, Oleksandr Luhovskyi, Volodymyr Fesich, Iryna Sushko(✉), and Nataliia Lashchevska

National Technical University of Ukraine, «Igor Sikorsky Kyiv Polytechnic Institute», Kyiv, Ukraine
movchanuk@rtf.kpi.ua, {atoll-sonic,f_v_p}@ukr.net, {sushko,ivanyuk}@ros.kpi.ua

Abstract. High-intensity ultrasonic vibrations allow to intensify many physical and chemical processes in liquids. In particular, high-intensity ultrasonic vibrations allow efficiently removing the dissolved gases from a liquid. The ultrasonic equipment consists of ultrasonic generator, surfaces for introducing the ultrasonic vibrations into the liquid, cavitation chamber filled with liquid, in which ultrasonic cavitation occurs in the process fluid under the action of intense ultrasound. As higher the intensity of ultrasound, as higher the intensity of the physical processes accompanying ultrasonic cavitation. The intensity of ultrasound is proportional to the value of sound pressure. As a result, an increase in the sound pressure value intensifies the processes occurring in the process fluid.

This work is devoted to issues related to equipment efficiency improving for introducing the high-intensity ultrasonic vibrations into a liquid. The cavitation layer which shields ultrasonic vibrations is formed on the surface of the emitter with the introduction of high-intensity ultrasound.

The cavitation layer consists of a large number of cavitation bubbles. As a result, a liquid can be considered as a liquid with decreased density. This leads to reduction of the liquid wave resistance.

This significantly reduces the value of sound pressure. The use of a liquid pressure transformer makes it possible to increase the sound pressure at the same magnitude of the oscillation amplitude.

The liquid pressure transformer is a cavitation chamber with a variable cross-sectional area. The length of such transformer should be a multiple of half the length of the sound wave in the liquid. A liquid pressure transformer can be considered as a resonant system.

The paper presents calculations of liquid pressure transformers, modeling of their work and examples of their practical implementation in equipment for biological contaminants removing and degassing.

Keywords: Cavitation chamber · Pressure transformer · Ultrasonic equipment · Liquid processing

© The Editor(s) (if applicable) and The Author(s), under exclusive license to Springer Nature Switzerland AG 2021
J. Stryczek and U. Warzyńska (Eds.): NSHP 2020, LNME, pp. 282–292, 2021.
https://doi.org/10.1007/978-3-030-59509-8_25

1 Introduction

The introduction of high frequency ultrasound into a liquid [1] is accompanied by ultrasonic cavitation [2]. The collapse of cavitation bubbles causes a local increase in temperature and pressure, the occurrence of microcurrents, shock cumulative jets, etc. This allows the use of high-intensity ultrasonic vibrations accompanied by ultrasonic cavitation to intensify biochemical processes [3], degassing process fluids [4], sound chemistry [5], ultrasonic cleaning [6], synthesis of nanomaterials [7], etc. We note that the physical processes accompanying ultrasonic cavitation are similar to those accompanying hydrodynamic cavitation [8]. The main difference lies only in the size of cavitation bubbles.

The cavitation intensity is related to the sound pressure in the fluid and ultrasound intensity. As higher the sound pressure, as higher the cavitation intensity. It is necessary to increase the oscillations amplitude of the ultrasonic transducer working end to increase the sound pressure. In this case, the maximum pressure will be at the working surface of the transducer. This causes the appearance of a cavitation layer with a small acoustic wave impedance compared to the liquid in the pre-cavitation mode [9]. This causes the appearance of a cavitation layer with a small acoustic wave impedance compared to the liquid in the pre-cavitation mode.

This phenomenon prevents the acoustic energy passing into the liquid and significantly reduces the efficiency of the equipment. Liquid treatment often occurs in a closed volume filled with a process fluid, which is called a cavitation chamber. The resonance can be achieved if the chamber dimensions are commensurate with the wavelength of ultrasonic vibrations. In this case, the pressure distribution in the internal volume of the chamber will not be uniform. You can choose the chamber dimensions with less inlet than outlet pressure. In fact, the cavitation chamber will behave like a pressure transformer. In particular, you can find a mode with the less pressure at the transducer working end than the pressure in the internal volume of liquid. This will reduce the influence of the shielding cavitation layer.

The case of a one-dimensional cavitation chamber is of great interest. In this case the transverse dimensions are much smaller than the longitudinal ones. Wherein the front of the acoustic wave can be considered flat. Such chambers, and hence pressure transformers, can be regarded as one-dimensional.

The aim of this research is to develop a mathematical model of a one-dimensional liquid pressure transformer in order to increase the efficiency of ultrasonic technological equipment.

2 The Modeling Approach

We assume that acoustic vibrations propagate along the axis Oz since a one-dimensional problem is considering. It is convenient to use the velocity potential $\tilde{\Phi}$ to simulate the acoustic field in the internal volume of the cavitation chamber [10]. The velocity potential is related to the amplitude of sound pressure P:

$$\mathrm{P} = \rho_0 \frac{\partial \tilde{\Phi}}{\partial t} + const,$$

where ρ_0 is fluid density, t is a time.

Note, that the velocity potential amplitude and the sound pressure amplitude are directly proportional.

For one-dimensional case, the acoustic field in the internal volume of the cavitation chamber can be described using the wave equation for the velocity potential [10]:

$$\frac{\partial^2 \tilde{\Phi}}{\partial z^2} = \frac{1}{c^2} \frac{\partial^2 \tilde{\Phi}}{\partial t^2}, \tag{1}$$

where c is sound velocity in a liquid.

The boundary conditions finding the acoustic field in the internal volume of the cavitator are consider. We will take that one of two idealized cases takes place – absolutely soft walls or absolutely rigid walls. In the case of closed cavitation chamber, we can assume that it has absolutely rigid walls.

In the case of absolutely rigid wall, the amplitude of the oscillating speed of the particles is equal to zero, and therefore the derivative of the velocity potential with respect to the coordinate is equal to zero

$$\frac{\partial \tilde{\Phi}}{\partial z} = 0, \tag{2}$$

The solution to Eq. (1) will be sought by the method of separation of Fourier variables in the form:

$$\tilde{\Phi} = \Phi_z(z) \cdot \Phi_t(t);\ \ \Phi_z = (A \cdot \sin(k \cdot z) + B \cdot \cos(k \cdot z));\ \ \Phi_t = \sin(\omega \cdot t).$$

where $A,\ B$ are constants, k is wavenumber, ω is circular frequency.

Omitting the multiplier harmonic vibrations $\Phi_t = \sin(\omega \cdot t)$, we get:

$$\Phi(z) = A \cdot \sin(k \cdot z) + B \cdot \cos(k \cdot z). \tag{3}$$

We will search for integration constants based on the boundary conditions (2). Closed cavitation chamber with length l is considered (Fig. 1).

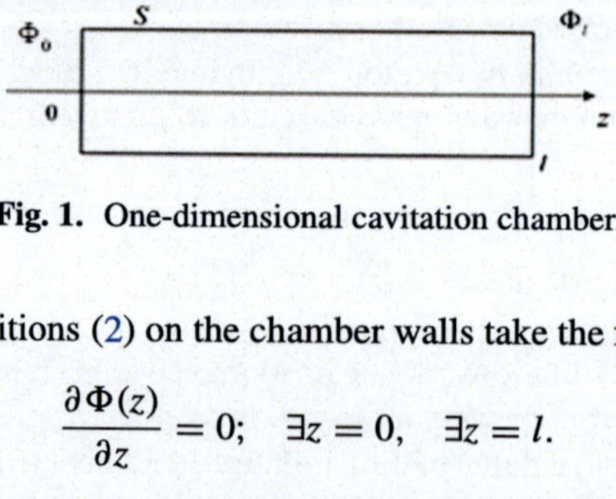

Fig. 1. One-dimensional cavitation chamber.

The boundary conditions (2) on the chamber walls take the following form:

$$\frac{\partial \Phi(z)}{\partial z} = 0;\ \ \exists z = 0,\ \ \exists z = l. \tag{4}$$

We obtain next expressions substituting the boundary conditions in (3):

$$\begin{cases} A \cdot k \cdot \cos(k \cdot 0) - B \cdot k \cdot \sin(k \cdot 0) = 0; \\ A \cdot k \cdot \cos(k \cdot l) - B \cdot k \cdot \sin(k \cdot l) = 0. \end{cases}$$

Solving the system of equations, we get the following:

$$A = 0;$$

$$l = \frac{\pi \cdot n}{k}, \quad n = 0, 1, 2 \ldots$$

Taking into account that $k = \frac{2 \cdot \pi}{\lambda}$, where λ is wavelength in a liquid, it can be assumed:

$$l = \frac{\lambda}{2} \cdot n, \quad n = 0, 1, 2 \ldots$$

The cavitation chamber resonance length must be multiple of an integer number of half-waves.

Solution (3) will take the form

$$\Phi(z) = B \cdot \cos(k \cdot z).$$

It can be seen from obtained solution that the one-dimensional cavitation chamber behaves like a segment of a long line. So it is possible to propose the construction of one-dimensional cavitation chamber, where it is possible significantly to affect the velocity potential due to a change in the camera's cross section along the axis z, in comparison with one-dimensional cavitation chambers with constant cross section. Stepped pressure transformers, consisted of two pipe sections of different diameters, are of particular interest from a technological point of view.

3 The Model and Problem Solution

The cavitation chamber in the form of a stepped pressure transformer (Fig. 2), consisted of two cylindrical chambers with constant cross-section with areas S_1 and S_2 and closed by absolutely rigid covers is considered.

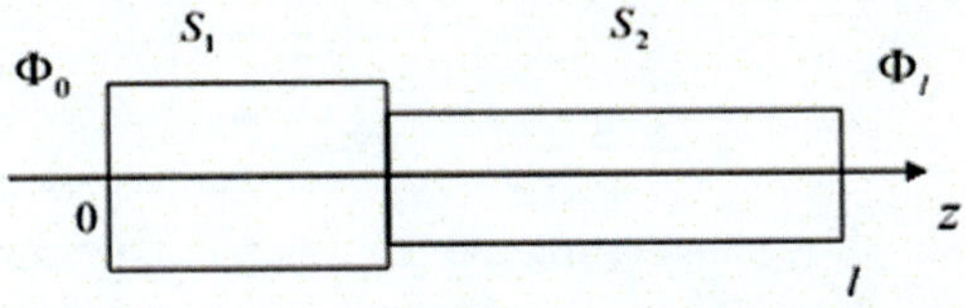

Fig. 2. Cavitation chamber in the form of stepped pressure transformer.

If the total length of the system is equal l, the cylindrical chambers connect at a point with the coordinate $m \cdot l$ $(0 \leq m \leq 1)$.

One more boundary condition at the junction of two cylindrical chambers of different sections is added to already known boundary conditions (4).

The pressure values (velocity potentials) and volume velocities must be equal (the product of the derivative with respect to the coordinate of the velocity potential and the cylinder cross-sectional area) at the junction:

$$\Phi(z+\varepsilon) = \Phi(z-\varepsilon), \tag{5}$$

$$S_1 \frac{\partial \Phi(z-\varepsilon)}{\partial z} = S_2 \frac{\partial \Phi(z-\varepsilon)}{\partial z}. \tag{6}$$

Next notation is introduced for convenience:

$$N = \frac{S_1}{S_2}.$$

The wave equation for the pressure transformer components has the same form as for constant section camera. The solution to the equation will be determined in the following form

$$\begin{cases} \Phi(z) = A \cdot \sin(k \cdot z) + B \cdot \cos(k \cdot z), & 0 \le z \le m \cdot l; \\ \Phi(z) = C \cdot \sin(k \cdot z) + D \cdot \cos(k \cdot z), & m \cdot l \le z \le l, \end{cases} \tag{7}$$

where A, B, C, D are integration constants, found from boundary conditions.

Find the derivatives for the oscillations amplitude by the coordinate have the next form:

$$\begin{cases} \frac{\partial \Phi(z)}{\partial z} = A \cdot k \cdot \cos(k \cdot z) - B \cdot k \cdot \sin(k \cdot z), & 0 \le z \le m \cdot l; \\ \frac{\partial \Phi(z)}{\partial z} = C \cdot k \cdot \cos(k \cdot z) - D \cdot k \cdot \sin(k \cdot z), & m \cdot l \le x \le l. \end{cases} \tag{8}$$

The boundary conditions substituting (4) into the first Eq. (8) leads to the system:

$$\begin{cases} \frac{\partial \Phi(z)}{\partial z} = A \cdot k \cdot \cos(k \cdot z) - B \cdot k \cdot \sin(k \cdot z), & 0 \le z \le m \cdot l; \\ \frac{\partial \Phi(z)}{\partial z} = C \cdot k \cdot \cos(k \cdot z) - D \cdot k \cdot \sin(k \cdot z), & m \cdot l \le x \le l. \end{cases}$$

We obtain the following expressions by solving the system with respect to integration constants.

$$A = 0, \; B = \Phi_0.$$

The system is obtained by substituting the boundary conditions (4) into the second Eq. (8):

$$\begin{cases} \Phi_l = C \cdot \sin(k \cdot l) + D \cdot \cos(k \cdot l), \\ 0 = C \cdot k \cdot \cos(k \cdot l) - D \cdot k \cdot \sin(k \cdot l). \end{cases}$$

Solving the system with respect to integration constants, we obtain:

$$C = \Phi_l \cdot \sin(k \cdot l); \; D = \Phi_l \cdot \cos(k \cdot l).$$

As a result, we obtain the expressions for the distribution of velocity potential amplitude along the axis:

$$\begin{cases} \Phi(z) = \Phi_0 \cdot \cos(k \cdot z), & 0 \le z \le m \cdot l; \\ \Phi(z) = \Phi_l \cdot \sin(k \cdot l) \cdot \sin(k \cdot z) + \Phi_l \cdot \cos(k \cdot l) \cdot \cos(k \cdot z), & m \cdot l \le z \le l. \end{cases} \tag{9}$$

As a result of substituting (5) and (6) in (9) and (8), we obtain a system of two equations:

$$\begin{cases} \Phi_0 \cdot \cos(k \cdot m \cdot l) = \Phi_l \cdot \sin(k \cdot l) \cdot \sin(k \cdot m \cdot l) + \Phi_l \cdot \cos(k \cdot l) \cdot \cos(k \cdot m \cdot l), \\ -\Phi_0 \cdot S_1 \cdot \sin(k \cdot m \cdot l) = S_2 \cdot (\Phi_l \cdot \sin(k \cdot l) \cdot \cos(k \cdot m \cdot l) - \Phi_l \cdot \cos(k \cdot l) \cdot \sin(k \cdot m \cdot l)). \end{cases}$$

Dividing the second equation by the first:

$$-\frac{S_1}{S_2} \cdot tg(k \cdot m \cdot l) = tg((1 - m) \cdot k \cdot l).$$

After substitution we get the following equation:

$$N = -\frac{tg((1 - m) \cdot k \cdot l)}{tg(k \cdot m \cdot l)} \tag{10}$$

The length of pressure transformer can be obtained based on the obtained expression, setting the ratio of the areas N and the asymmetry coefficient m.

As we can see, in general case l and m depend on the ratio N of the transformer components areas.

However, one can find those l and m, for which the next condition is satisfied

$$tg(k \cdot m \cdot l) = tg((1 - m) \cdot k \cdot l) = 0.$$

It is satisfied at $m = 0,5$ and $k \cdot l = \pi$. Such system consists of two pipe sections; length of each is equal to an odd number of $\lambda/4$ (quarter of the wavelength). It should be noted that the total pressure transformer length will be multiple of integer number of $\lambda/2$(half of the wavelength).

In general, the pressure change factor is equal to:

$$M = -N \frac{\sin(k \cdot m \cdot l)}{\sin((1 - m) \cdot k \cdot l)} \tag{11}$$

So in case of symmetrical pressure transformer:

$$M = N.$$

And slightly different expression for the gain coefficient can be received:

$$M = \frac{\cos(k \cdot m \cdot l)}{\cos((1 - m) \cdot k \cdot l)}.$$

4 The Modeling

An analysis of the transcendental equation solution (10) (Table 1) and substitution of the obtained values in (11) (Table 2) for the pressure change factor shows that the maximum pressure increase is observed at $m = 0,5$. It corresponds to a symmetric pressure transformer from two quarter-wave segments.

Table 1. The ratio $\frac{l}{\lambda}$ for stepped pressure transformer.

	N								
m	2	2.5	3	3.5	4	4.5	5	5.5	6
0.1	0.4606	0.4454	0.4324	0.4213	0.4117	0.4032	0.3958	0.3891	0.3832
0.2	0.4463	0.4296	0.4167	0.4063	0.3979	0.3908	0.3849	0.3798	0.3753
0.3	0.4512	0.4376	0.4275	0.4197	0.4135	0.4084	0.4041	0.4005	0.3975
0.4	0.4703	0.4622	0.4563	0.4517	0.4481	0.4452	0.4427	0.4407	0.4389
0.5	0.5	0.5	0.5	0.5	0.5	0.5	0.5	0.5	0.5
0.6	0.5332	0.5436	0.5518	0.5584	0.5638	0.5684	0.5724	0.5758	0.5788
0.7	0.5532	0.5698	0.5828	0.5934	0.6022	0.6097	0.6161	0.6217	0.6266
0.8	0.5474	0.5599	0.5690	0.5758	0.5812	0.5855	0.5890	0.5920	0.5945

Table 2. The gain coefficient $\frac{M}{N} \cdot 100\%$ of pressure transformer

	N								
m	2	2.5	3	3.5	4	4.5	5	5.5	6
0.1	55.774	47.337	41.849	38.013	35.190	33.030	31.328	29.953	28.821
0.2	67.9818	61.800	57.735	54.861	52.723	51.072	49.758	48.688	47.800
0.3	82.075	78.304	75.748	73.899	72.500	71.404	70.521	69.796	69.189
0.4	94.455	93.116	92.169	91.464	90.917	90.481	90.125	89.829	89.579
0.5	100	100	100	100	100	100	100	100	100
0.6	92.970	90.658	88.807	87.287	86.013	84.929	83.994	83.178	82.460
0.7	75.341	67.532	61.369	56.359	52.191	48.660	45.624	42.981	40.655
0.8	59.862	49.674	42.403	36.964	32.748	29.387	26.647	24.371	22.451

Decrease in geometric length is observed with a slight decrease in efficiency at value of asymmetry coefficient $m < 0,5$. General increase in length with decrease in efficiency is observed at $m > 0,5$. Therefore, from a practical point of view, the case $m \leq 0,5$ is most important.

The ratio of the liquid volume in the pressure transformer at $m \leq 0,5$ to the liquid volume at $m = 0,5$ is considered.

$$\frac{V_m}{V_{0,5}} = \frac{(N-1) \cdot m + 1}{0,25 \cdot N + 0,25} \cdot \left(\frac{l_m}{\lambda}\right) \cdot 100\%.$$

An analysis of this expression (Fig. 3a) and normalized one with respect to the maximum pressure increase coefficient at $m = 0,5$ (Fig. 3b), calculated according to

(11), shows that the introduction of asymmetry can be an effective mean of reducing the processed fluid volume.

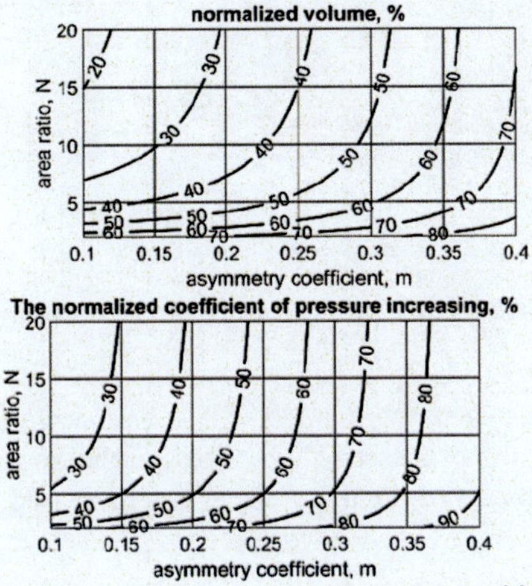

a) b)

Fig. 3. Normalized: a - volume of pressure transformer; b - pressure increase coefficient.

The distribution of the velocity potential, and hence the sound pressure along the axis of the pressure transformer is considered (Fig. 4).

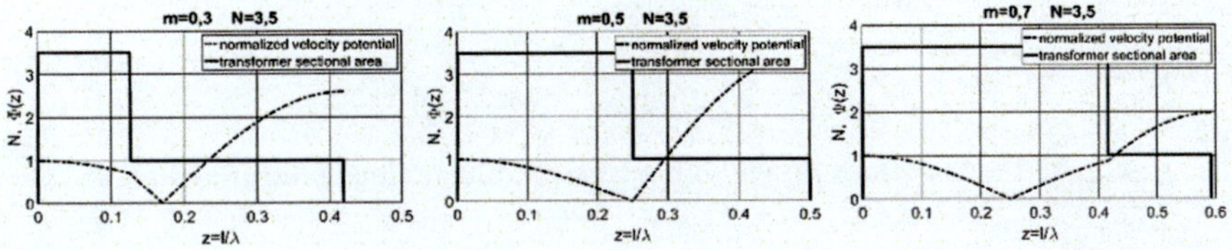

a) b) c)

Fig. 4. The distribution of pressure potential at: a) $m = 0.3$ $N = 3.5$, b) $m = 0.5$ $N = 3.5$, c) $m = 0.7$ $N = 3.5$.

As we can see, for $m < 0,5$ the node of the velocity potential is located "inside" a small stage of the transformer, and for $m > 0,5$ it is "inside" a large stage.

To verify the results, FEM simulation of the pressure distribution in the internal volume of a symmetrical transformer with component diameters of 20 mm and 40 mm is carried out. The lengths of the components are taken equal to a quarter of the wavelength at a frequency of 22 kHz. In the simulation it is assumed that the transformer internal volume is filled with water. The simulation results (Fig. 5) show that the obtained pressure distribution is close to the calculated from (11) values. However, at the junction of components with different diameters, the wave front deviates from the plane one.

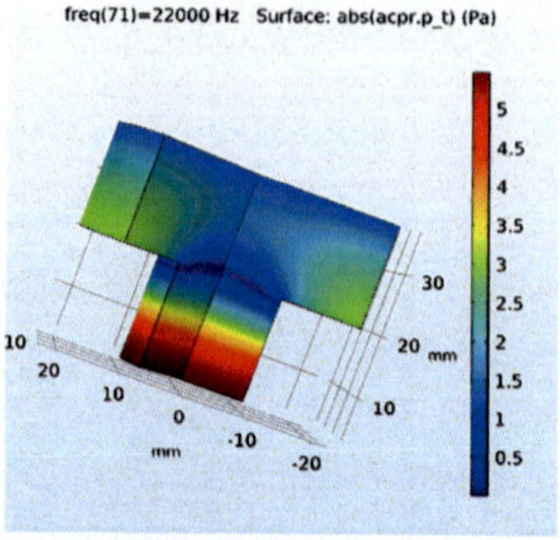

Fig. 5. Pressure distribution in a symmetrical pressure transformer (inlet pressure 1 Pa).

Small-diameter stage is supplemented with a half-wave extension tube (Fig. 6) to increase the volume of the processed fluid. The simulation results show that the high-pressure area in the thin part of the transformer has increased significantly.

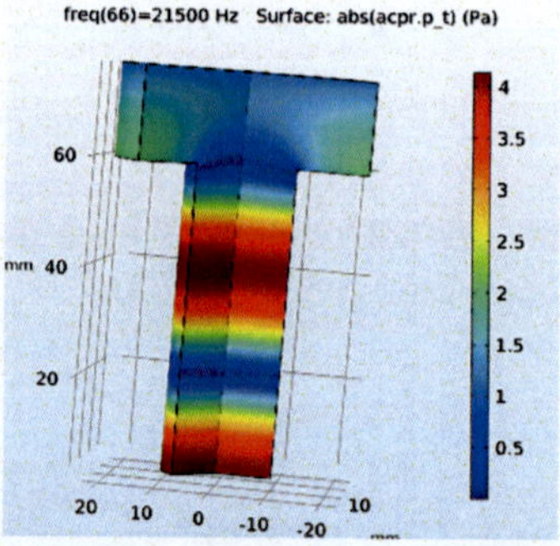

Fig. 6. Pressure transformer with increased volume (inlet pressure 1 Pa).

5 The Practical Implementation

Several types of ultrasonic cavitation chamber construction are developed based on the theoretical studies. Piezoelectric transducers with resonant frequency equal to 22 kHz [11] are used to excite ultrasound. This resonant frequency is calculated on the basis of the chamber sizes (in the form of a pressure transformer) (Fig. 7).

As a result, ultrasonic equipment for processing liquids with high-power ultrasound is developed (Fig. 8). The main difference is in supplying the treated fluid to the internal volume of the cavitation chamber. The developed equipment was used for water treatment in order to inactivate bacteria and degassing.

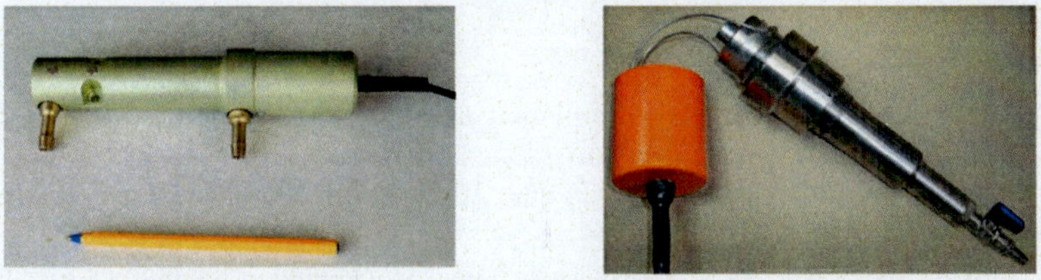

Fig. 7. Prototypes of ultrasonic cavitation chambers with piezoelectric transducers.

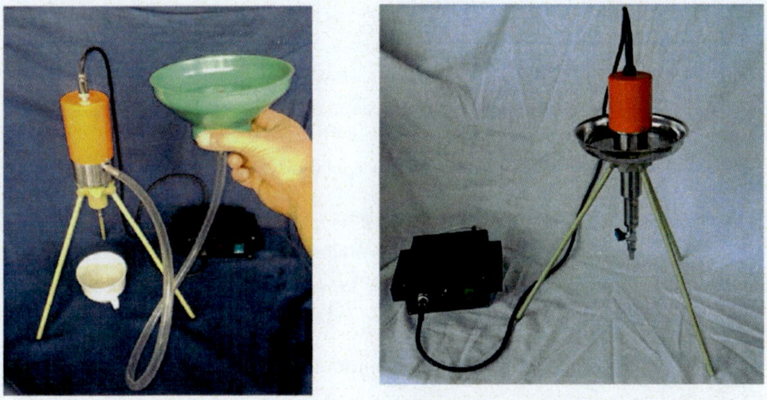

Fig. 8. High power ultrasonic liquid processing equipment.

6 Conclusions

Cavitation chambers in the form of pressure transformers can be the basis for creating new ultrasonic technological equipment for the intensification of processes occurring under the influence of ultrasonic cavitation. The pressure increasing is achieved by changing the cross-sectional area of the cavitation chamber. The dimensions of the chamber must be a multiple of acoustic quarter wave in the liquid. This increasing pressure method reduces the cavitation layer size on the working surface of the ultrasonic transducer. This, in turn, increases the ultrasonic energy introduced into the fluid.

Cavitation chambers as pressure transformers can perform either as symmetrical or asymmetrical ones. A symmetrical transformer consisting of two quarter-wave segments will have the highest coefficient of pressure increasing. The coefficient of pressure increasing will be equal to the ratio of the cross-sectional areas of the constituent parts. The transverse size of the cavitation chamber must be such that the propagation conditions of the plane acoustic wave are maintained. The efficiency of the cavitation chamber in the form of a pressure transformer may be reduced due to distortion of the acoustic wave front.

References

1. Gallego-Juàrez, J.A., Graff, K.F. (eds.): Power Ultrasonics: Applications of High-Intensity Ultrasound. Woodhead Publishing (2015)
2. Kuttruff, H.: Ultrasonics Fundamentals and Applications. Springer, Netherlands (1991). https://doi.org/10.1007/978-94-011-3846-8
3. Feng, H., Barbosa-Cánovas, G.V., Weiss, J. (eds.): Ultrasound Technologies for Food and Bioprocessing. Springer, Heidelberg (2011). https://doi.org/10.1007/978-1-4419-7472-3
4. Antoniak, P., Stryczek, J., Banas, M., Gryshko, I., Zilinskyi, A., Kovalov, V.: Visualization research on the influence of an ultrasonic degassing on the operation of a hydraulic gear pump. In: MATEC Web of Conferences VETOMAC XIV (2018). https://doi.org/10.1051/matecconf/201821103005
5. Ashokkumar, M.: Handbook of Ultrasonics and Sonochemistry. In: Cavalieri, F., Chemat, F., Okitsu, K., Sambandam, A., Yasui, K., Zisu, B. Springer Science + Business Media, Singapore (2016)
6. Ensminger, D., Stulen, F.B. (eds.): Ultrasonics: Data, Equations, and Their Practical Uses. CRC Press/Taylor & Francis Group (2008)
7. Chen, D., Sharma, S.K., Mudhoo, A. (eds.): Handbook on Applications of Ultrasound Sonochemistry for Sustainability, CRC Press/Taylor & Francis Group (2011)
8. Kollek, W., Kudźma, Z., Stosiak, M., Mackiewicz, J.: Possibilities of diagnosing cavitation in hydraulic systems. Arch. Civil Mech. Eng. **7**(1), 61–73 (2007). https://doi.org/10.1016/S1644-9665(12)60005-3
9. Hovem, J.M.: Underwater acoustics: Propagation, devices and systems. J. Electroceram. **19**, 339–347 (2007)
10. Pierce, A.D.: Acoustics an Introduction to Its Physical Principles and Applications, p. 678. Acoustical Society of America. Springer, Heidelberg (1989). https://doi.org/10.1007/978-3-030-11214-1
11. Hamonic, B., Decarpigny, J.N.: Power Sonic and Ultrasonic Transducers Design: Proceedings of the International Workshop, Held in Lille, France, 26–27 May 1987, p. 249. Springer, Heidelberg (1988)

Research of the Influence of Hydraulic Orifice Material on the Hydrodynamic Cavitation Processes Accompanied by Luminescence

Ihor Nochnichenko, Oleksandr Luhovskyi, Dmytro Kostiuk(✉), and Jakhno Oleg

"Igor Sikorsky Kyiv Polytechnic Institute", National Technical University of Ukraine, Kiev, Ukraine
igornoch@gmail.com, {atoll-sonic,kostiukdv}@ukr.net

Abstract. The flow in the throttle channel of the hydrodynamic cavitator was visualized. Previous studies have shown that luminous radiation - luminescence can be observed in the area at the output of the throttle under certain conditions of cavitation. This phenomenon is not acceptable in some hydraulic systems, so it is necessary to study the patterns and conditions of its occurrence. As today there is no solid theory of its origin, this study aims to study the effect of material from which the elements of the throttle are made on the intensity of luminescence. A model of a hydrodynamic cavitator and a series of screws of heterogeneous materials, which are at opposite parts of the triboelectric series, were developed. The model's hydraulic tests showed that the material the throttle elements are made of has some influence on the radiation intensity.

Keywords: Flow visualization · Hydrodynamic cavitation · Luminescence · Hydraulic orifice

1 Introduction

Hydrodynamic cavitation is a complex process that occurs due to a local decrease in the fluid pressure to a certain critical value pcr (in a real fluid, the pcr value is close to the saturation vapor pressure of this liquid at a certain temperature), which can occur in case the fluid velocity increases [1]. As modern hydraulic equipment operates at the high pressure of the working fluid, and the working bodies have a high speed, conditions for the appearance of cavitation in hydraulic systems appear. Such conditions can occur in pumps [2, 3], valves, throttles [4, 5] and other equipment [6–8], and when the bodies are flowing at high velocity [9]. Since small bubbles of vapor or gas are always present in the real fluid, when they move with the flow, they fall into the pressure zone p < pcr and they lose their stability and increase in size. After the transition to the zone of high pressure and the exhaustion of the kinetic energy of the liquid under expansion conditions, the growth of the bubble stops and decreases. If the bubble contains enough gas, after reaching the minimum radius it restores its size and performs several cycles of

© The Editor(s) (if applicable) and The Author(s), under exclusive license to Springer Nature Switzerland AG 2021
J. Stryczek and U. Warzyńska (Eds.): NSHP 2020, LNME, pp. 293–300, 2021.
https://doi.org/10.1007/978-3-030-59509-8_26

damping oscillations, and if there is little gas, the bubble will completely shut down in the first period. So, a rather clearly defined "cavitation zone", filled with moving bubbles, is created near the streamlined body (for example, in a tube with local constriction).

The reduction of the cavitation bubble occurs at high speed and is accompanied by an acoustic impulse (a kind of hydraulic shock), which is stronger when the bubble contains less gas.

The occurrence of cavitation in the equipment may cause noise, vibration and flow ripples. Also, when the hydraulic equipment is made of a transparent material, it can be observed that cavitation is accompanied by luminescence – non-thermal spontaneous light emission by the excited substance. It should be noted that hydrodynamic luminescence is sometimes used as an indicator of advanced cavitation; this phenomenon requires consideration and careful study [7–11].

In this study, the processes observed in the throttle channel, which cross-section size is formed by the hole and the edge of the screw, were considered (Fig. 1).

2 Design of the Experimental Orifice and the Visualization Research Stand

For the study of cavitation processes a model of a hydrodynamic cavitator with a regulated orifice was created. Case of the device is made from polymethyl methacrylate (organic glass) and screws are made from series of different materials to reveal the influence of this factor. The element that controls the cross-section is the screw. The scheme of the tested orifice (Fig. 1) shows the path of the flow – from right to left. The orifice area could be changed by rotating the screw before the experiment and it was the same for all tests. The region where the luminescence occurs is shown by the red circle (L).

A special stand, which can be used to observe cavitation, and luminescence in the model of hydraulic orifice and at certain conditions were designed for the tests of the hydrodynamic cavitation and luminescence in the flow.

For experimental studies of the hydrodynamic cavitation and luminescence in the flow an experimental booth was developed (Fig. 2).

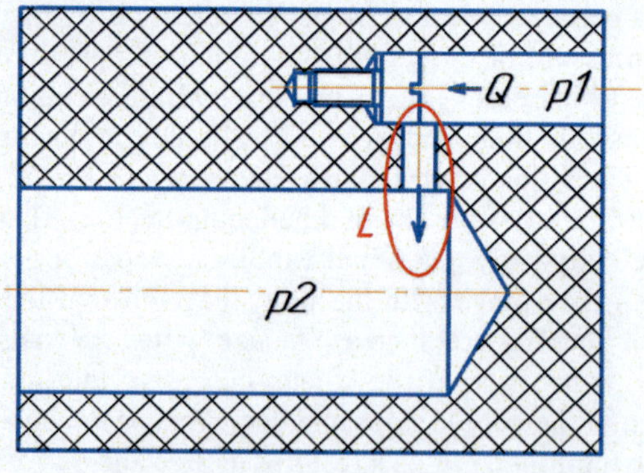

Fig. 1. Determination of the studied area of the cavitator orifice.

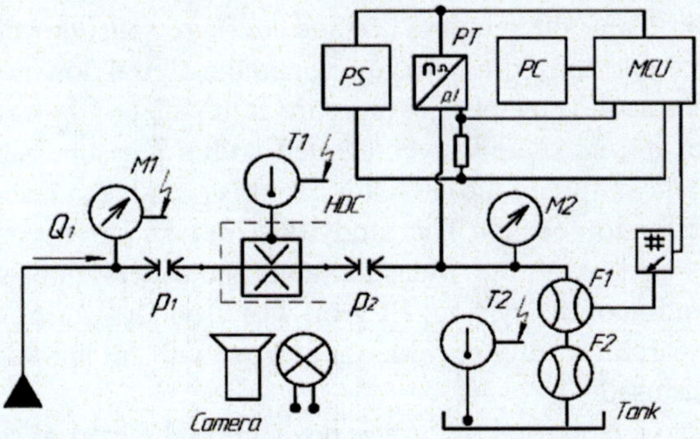

Fig. 2. Circuit diagram of the experimental stand for the study of hydrodynamic cavitation.

The scheme of a stand, based on FESTO workstation, contains a hydraulic power unit that provides flow Q1 and allows setting the pressure p1 at the entrance to the cavitator (HDC). The pressure is controlled by manometers M1 and M2 and pressure transducer.

Using the transducer provides the possibility of recording pressure changes during the operation of the cavitator (at the output) is provided with the help of an analogue pressure transducer (PT). The flow rate was measured by an analogue flow sensor – a turbine flowmeter (F1). To determine the temperature of working fluid a temperature sensor DS18B20 (T1) was used. It was installed at the outlet of the cavitator. All sensors were connected to the signal processing board with the Atmega328 microcontroller (MCU), which was programmed to import data to Excel from the board using PLX-DAQ and save every 1 s on a computer (PC). All devices are connected to the power supply unit (PS).

The appearance of the experimental stand and cavitator with a tested orifice is shown in Fig. 3a and b.

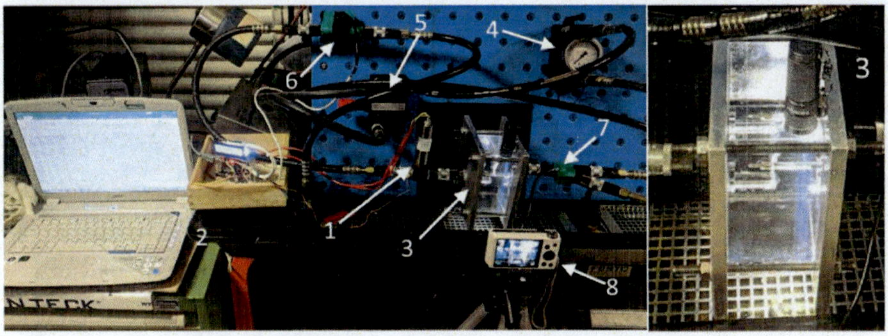

a b

Fig. 3. a - Appearance of the experimental stand 1 - pressure sensor, 2 - PC, 3 - hydrodynamic cavitator, 4 - manometer, 5 - safety valve, 6 - flowmeter, 7 - temperature sensor, 8 - camera; b - detailed view of the cavitator.

Before the experiment, the temperature and pressure sensors were calibrated with a reference thermocouple and a pressure gauge, and the digital flowmeter was calibrated with a reference volume flow meter. The pressure at the inlet of the cavitator was settled with a safety valve, and its value was controlled with a pressure gauge. The pressure sensor, mounted on the output of the cavitator, records possible pulsations of pressure in the system when cavitation occurs. This study did not take into account pressure, flow and temperature changes over time. Measurements and observations were made during the short-term operation of the system. HLP oil was used as working fluid.

For visual research, a digital video camera with a shooting frequency of 120 … 1000 fps was used.

The experiments were carried out at the pressure at the inlet of the cavitator being p1 = 5 MPa, and the fluid flow rate Q = 1 l/min. The temperature of the working fluid is 30°C. The average velocity in the throttle channel is up to 70 m/s.

3 Visualization Research of the Flow Processes in the Cavitator

There are thermal and electrical theories to explain the occurrence of luminescence in the stream [5–11]. The purpose of the study was to consider the emerging luminescence when cavitator elements are made from different materials and to investigate the effect of different parameters on its intensity. There are known so-called triboelectric series, according to which different materials can acquire positive or negative potential during an interaction [12].

Several screws were made from different materials (Fig. 4) which are placed in several parts of the triboelectric row (Table 1). Hydraulic oils have a value of +29 nC/J according to the triboelectric series [13]. These screws were successively inserted into the cavitator, which was tested on the experimental stand. The rising intensity of cavitation and luminescence was observed.

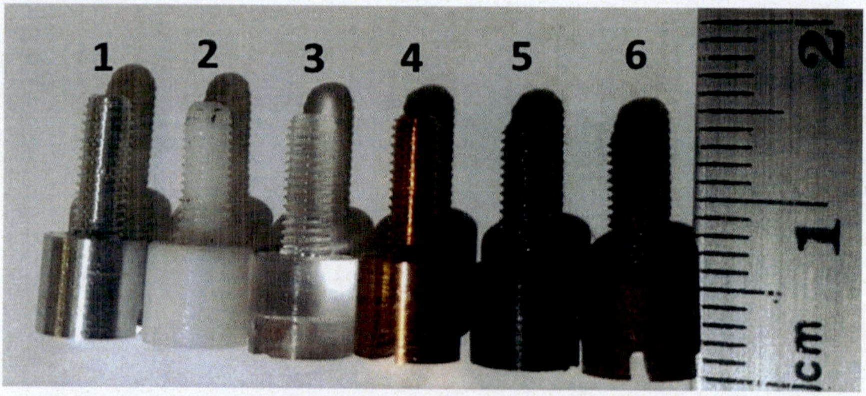

Fig. 4. Screws used in the study. The number corresponds to the material in Table 1.

It should be noted that the luminescence was observed only when the fluid flowed in a certain direction when the cavitator was connected back, luminescence was not

Table 1. Tested screws and their triboelectric values for selected materials.

Material, and position on the Fig. 4	Triboelectric series value, nC/J	Picture
Duralumin (1)	+6	Fig. 6 a, b
Copper (4)	–	Fig. 6 c, d
Brass	0	Fig. 6 e, f
Teflon (2)	−190	Fig. 7 a, b
Polymethyl methacrylate (3)	+10	Fig. 7 c, d
Composite epoxy material (6)	−32	Fig. 7 e, f
Ebonite (5)	−13	Fig. 7 g, h

detected (Fig. 5). The white area in the figures is bubble cavitation. Here, the hydraulic line is filled with microscopic bubbles, which are formed because of the local pressure drop across the throttling element. In Fig. 5a there is a cavitation torch observed.

a b

Fig. 5. Examples of flow in direct and reverse direction: a – direction from right to left – observed luminescence; b – the direction from right to left – no luminescence.

4 Discussion of the Findings

According to the results of the study, several images were acquired and conclusions can be drawn about the effect of the screw material on the intensity of the luminescence appearing downstream the screw in the throttling channel.

The volume of gases in bubbles depends on cavitation coefficient. Critical parameters of the flow depend on pressures value and velocity that indicate the start of stream cavitation can be carried out [1]:

$$\chi = \frac{2(p_1 - p_2)}{\rho \cdot V_1^2},$$

where p_1, V_1 - pressure and velocity of flow, e.g. at the valve inlet, p_2- vapor pressure; ρ- fluid density.

Figures 6 and 7 show pairs of images for different screw materials. In the drawings in one row, the screws are made from the same material, the drawings differ only in the intensity of the external illumination. The images are illuminated to detect the cavern, and to observe the luminescence radiation more clearly without illumination. The images of the caverns are almost identical in appearance to all the materials of the screw, while the differences in the luminescence patterns are noticeable.

Fig. 6. Results of experiments with screws made from conductive materials a, b - Duralumin; c, d - Copper; e, f - Brass.

Thus, observations show that the least intense luminescence was observed when using a screw made from Teflon, which has a value of −190 nC/J in the triboelectric series, while other materials are of an order of smaller magnitude (Table 1). Theoretically, during rubbing of a pair of materials from the triboelectric series the material closer to the positive end of the row will charge positively and the other negatively. Thus, it can be assumed that the luminescence intensity will increase as the positive potential on the screw increases.

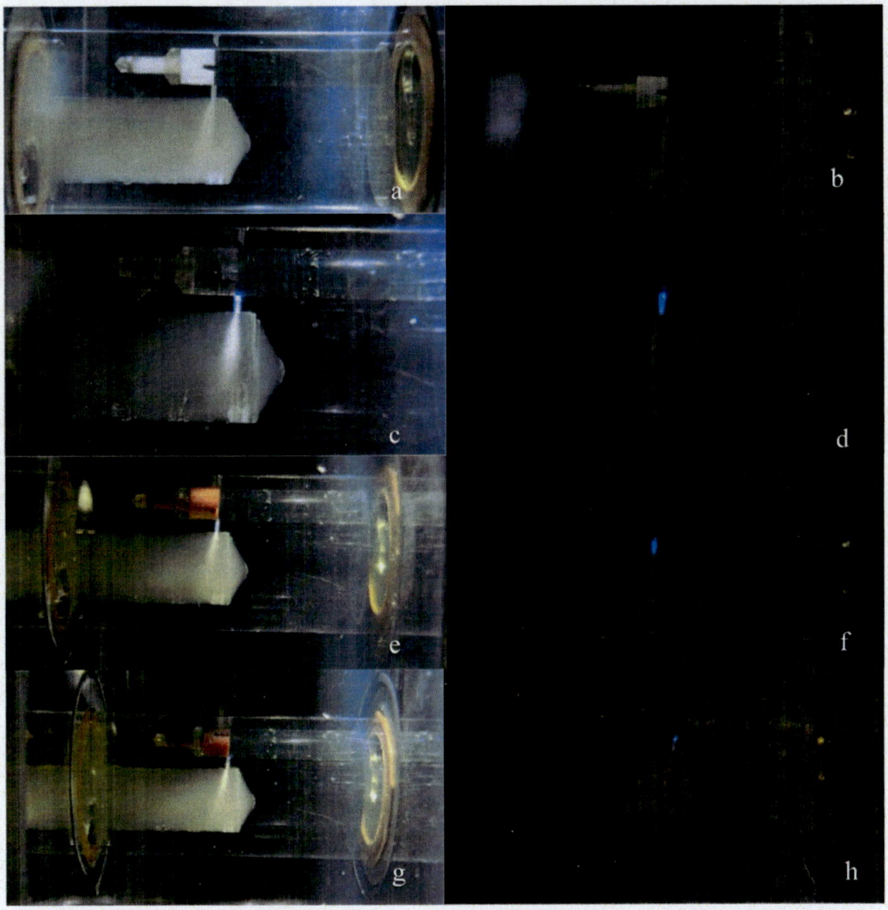

Fig. 7. Results of experiments with screws made from non-conductive materials a, b – Teflon; c, d – Polymethyl methacrylate; e, f – Composite epoxy material; g, h – Ebonite.

5 Conclusion

A research stand and a model of a hydrodynamic cavitator with several screws of heterogeneous materials (Duralumin, Copper, Brass, Teflon, Polymethylmethacrylate, Composite epoxy material, and Ebonite) have been developed according to the triboelectric series to research appearance of luminescence when different materials of the hydraulic orifice are used.

The experiments were conducted at the following conditions: pressure difference on the cavitator $\Delta p = 5$ MPa, fluid flow rate $Q = 1$ l/min, the temperature of the working fluid (HLP oil was used) is 30 °C. The average velocity in the throttle channel is up to 70 m/s.

Studies have made it possible to investigate cavitation in the throttling element and downstream the orifice. Observations have shown that luminescence is observed under certain conditions. Its intensity is determined by many factors, one of which is the material of the throttle element. The study showed a correlation between the luminescence intensity and the location of the material in the triboelectric series, i.e., when non-conductive materials with large negative value at the triboelectric series were used, the intensity of luminescence at the orifice was low. Thus, it is necessary to rationalize the choice of materials during the design of hydraulic systems and devices.

The authors express their gratitude to master students Tomashevskiy A., Sidletskiy V. and Kryvosheiev V. for the help in carrying out experimental and theoretical researches and processing of obtained results.

References

1. Knapp, R.T., Daily, J.W., Hammit, F.G.: Cavitation. McGraw-Hill Book Company, New York (1970)
2. Stryczek, J., Antoniak, P., Jakhno, O., Kostyuk, D., Kryuchkov, A., Belov, G., Rodionov, L.: Visualisation research of the flow processes in the outlet chamber-outlet bridge-inlet chamber zone of the gear pumps. Arch. Civil Mech. Eng. **15**(1), 95–108 (2015)
3. Antoniak, P., Stryczek, J.: Visualization study of the flow processes and phenomena in the external gear pump. Arch. Civ. Mech. Eng. **18**(4), 1103–1115 (2018). https://doi.org/10.1016/j.acme.2018.03.001
4. Banaś, M., Antoniak, P., Marciniak, L., Stryczek, J.: Visualization of flow phenomena in hydraulic throttle valves of plastics. In: VETOMAC XIV MATEC Web Conferences **211**, 19001 (2018)
5. Nochnichenko, I.V., Luhovskyi, A.F., Jakhno, O.M., Kostiuk, D.V., Komada, P., Kozbakova, A.: Experimental research of hydroluminescence in the cavitating flow of mineral oil. Paper presented at the Proceedings of SPIE - the International Society for Optical Engineering (2019)
6. Lockett, R.D., Bonifacio, A.: Hydrodynamic luminescence in a model diesel injector return valve. Int. J. Engine Res. (2019)
7. Margulis, M.A., Pil'gunov, V.N.: The mechanism of the appearance of luminescence and electrification in liquid flows through narrow channels. Russ. J. Phys. Chem. A **83**(10), 1789–1793 (2009)
8. Biryukov, D.A., Gerasimov, D.N.: Spectroscopic diagnostics of hydrodynamic luminescence. J. Molecule Liq. **266**, 75–81 (2018)
9. Farhat, M., Chakravarty, A., Field, J.E.: Luminescence from hydrodynamic cavitation. Proc. R. Soc. A **467**, 591–606 (2018)
10. Nochnichenko, I.V., Luhovskyi, O.F., Kostiuk, D.V.: Study of hydrodynamic luminescence in a cavitation liquid medium. Naukovo-tekhnichnyi zhurnal «Problemy tertia ta znoshuvannia» **3**(84), 57–62 (2019)
11. Luhovskyi, O., Nochnichenko, I., Jakhno, O., Kostiuk, D.: Temperature influence on cavitational mass transfer in the channel of laval nozzle type. J. Tech. Univ. Gabrovo **57** (I), 12–15 (2018)
12. Zou, H., Zhang, Y., Guo, L., et al.: Quantifying the triboelectric series. Nat. Commun. **10**, 1427 (2019)
13. AlphaLab.Inc.: The TriboElectric Series. https://www.alphalabinc.com/triboelectric-series/. Accessed 18 Feb 2020

Check for updates

Comparative Studies of the Dynamic Response of Hydraulic Cylinders with Different Hydraulic Supply Systems Design

Tomasz Siwulski(✉)

Department of Technical Systems Operation and Maintenance, Wrocław University of Science and Technology, Łukasiewicza 5, 50-371 Wrocław, Poland
tomasz.siwulski@pwr.edu.pl

Abstract. As a result of development works, which were carried out by Wrocław University of Science and Technology and by HYDROMAR company and which focused on a new design of hydraulic cylinders characterized by, among others, an increased degree of reliability, it was possible to develop and test a prototype of the new solution. During the implementation of the project, a new hydraulic cylinder testing stand was designed, which allows the identification of dynamic phenomena occurring in the hydraulic power system. Tests of the new solution were carried out in the area of both dynamic and thermal phenomena associated with the exchange of liquid in the cylinder chambers. In terms of dynamics, the research allowed identifying the responses of the cylinder system with supply lines and flow control valves to a variable, external dynamic excitations. The tests also allowed determining the response of the hydraulic cylinders with supply lines and control valves to rapid overloading. The dynamic load was simulated by switching the control valves in a cut-off position during the movement of a working system arm with specific mass parameters. Comparative tests of dynamic response systems were carried out using a classic cylinder and a new design of a hydraulic power supply system. The article presents the results of tests and of the preliminary analysis of the impact that the type of hydraulic power supply system has on the dynamic response of the system. It should be emphasized that the presented results are part of a broad research and implementation project carried out in cooperation between the research and scientific unit and the industrial partner.

Keywords: Dynamic response · Hydraulic cylinder · Boom working system

1 Introduction

Dynamic phenomena occurring in hydraulic systems have been studied in a number of scientific papers, referring to both entire systems and individual elements [1–3]. The dynamic parameters of hydraulic cylinders are one of the basic factors influencing the behavior of working systems during their operation. In practice, determining the dynamic response of cylinder-driven systems is not common. Some research results are

© The Editor(s) (if applicable) and The Author(s), under exclusive license to Springer Nature Switzerland AG 2021
J. Stryczek and U. Warzyńska (Eds.): NSHP 2020, LNME, pp. 301–310, 2021.
https://doi.org/10.1007/978-3-030-59509-8_27

known in the literature, which describe the dynamic parameters of hydraulic supply lines mainly in order to identify pulsation phenomena occurring in liquid ducts [4, 5]. Experimental research is also performed into new designs of hydraulic cylinders with increased efficiency [6, 7]. Meanwhile, users of hydrostatically driven machines report a number of operating problems which increase the costs or affect the work safety [8, 9]. The results of research involving new concepts for hydraulic cylinder control systems are also described in the literature [10–12]. The basis for current research and development works was to identify these problems, and one of the results is a design of a hydraulic cylinder with a new power supply system. The new design has, among other advantages, an improved ability to exchange thermal energy with the system and to remove impurities from the cylinder chambers [13]. It also offers increased operating safety and allows any desired operating configuration without the need to extend the system. One of the important features is the modification of the cylinder's dynamic parameters. These parameters have a significant influence on piston strokes and on the resulting accuracy with which the driven elements are positioned in space. This is of particular importance in the case of machines with boom systems, e.g. drilling rigs or cranes.

2 Theoretical Analysis of the Problem

A hydraulic cylinder which drives mass elements such as members of a working system may be analyzed as a second order dynamic system with damping (Fig. 1). Its basic parameters include:

- loading mass (m) – this parameter depends on the mass and the geometry of the driven working system, and in actual applications it is a function of momentary system geometry,
- stiffness of the elastic element (k) – in the case of hydraulic cylinder, this parameter depends on the parameters of the liquid and on the elastic parameters displayed by the elements of the cylinder and by the hydraulic lines connected to the cylinder,
- system damping parameter (c) – describes the ability of the system to dissipate mechanical energy.

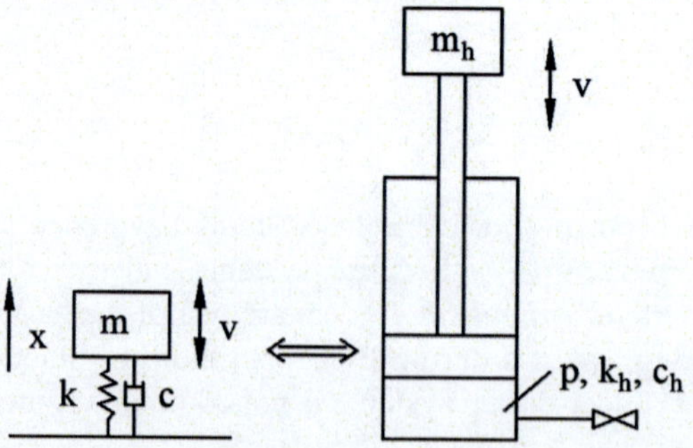

Fig. 1. Physical analogy of a hydraulic cylinder and a second order dynamic system.

In the case of hydraulic cylinders, dynamic analyzes are based on the measurement of pressure changes in time. In the analyzed case, an assumption was made that pressure changes in the piston chamber of a cylinder loaded with external body force and with inertia forces have a vibration frequency equal to the frequency of natural vibrations in the system.

$$x(t) = A \cdot \sin(\omega t + \varphi) <=> p(t) = A_h \cdot \sin(\omega t + \varphi) \tag{1}$$

where:

$x(t)$ – displacement of the mass system [m],
A – vibration amplitude of the mass system [m],
$p(t)$ – hydraulic system pressure [Pa],
A_h – pressure vibration amplitude [Pa],
ω – angular frequency of vibrations [rad/s],
φ – initial phase [rad].

The hydraulic cylinder, however, should not be considered as a undamped system, because it contains energy dissipation. The theoretical dynamic response of a damped system is shown in Fig. 2.

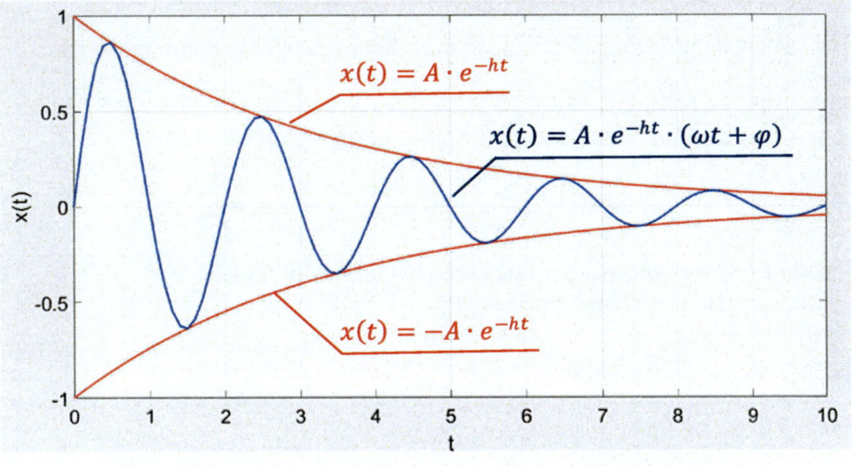

Fig. 2. Theoretical graph of displacement in a damped system.

Analogically, a linear relationship may be assumed between the amplitude of cylinder displacement and the pressure value. This assumption is valid in the case when the analysis concerns a hydraulic system with a specific liquid volume subject to pressure changes and if constant parameters are assumed of liquid compressibility and of system elasticity due to the deformations of cylinder walls and of the hydraulic lines connected to the cylinder.

$$\frac{A}{A_h} = const. \tag{2}$$

The above equations indicate the reasonability behind performing dynamic analyzes of hydraulic cylinders on the basis of pressure values for a constant liquid volume on which the pressure acts. This condition is met if the cylinder is loaded with a mass and given a speed, and if subsequently the control valves are rapidly switched to the closed position. In such case, the kinetic energy of the moving mass will start to act on the hydraulic cylinder and thus on the liquid having constant volume and on the system having constant elastic parameters. The observed pressure changes were assumed suitable for description with a mathematical relationship which describes damped harmonic oscillations.

$$p(t) = A_h \cdot e^{-ht} \cdot (\omega t + \varphi) \tag{3}$$

where:

h – damping measure [1/s],

The system stiffness parameter calculated hydraulically (k_h) can be described with the following relationship:

$$k_h = m_h \cdot \omega^2 = m_h \cdot \left(\frac{2\pi}{T_h}\right)^2 \tag{4}$$

The system damping measure (h), on the other hand, allows defining the attenuation parameter.

$$2h = \frac{c_h}{m} => c_h = 2h \cdot m \tag{5}$$

where:

m_h – loading mass on the system calculated hydraulically [kg],
T_h – periodic time of pressure changes [s],
c_h – system damping parameter calculated hydraulically [$\frac{N \cdot m}{m}$]

The use of the numerical decrement (δ) in relationship to the vibration period of the system and to the damping measure is given by:

$$\delta = ln\frac{A_{hn}}{A_{hn+1}} \tag{6}$$

$$\delta = 2h\frac{\pi}{\omega} \tag{7}$$

where:

$A_{h\,n}$ – vibration amplitude in the nth period [Pa],

After transformations, a relationship is obtained which defines the hydraulically calculated system damping parameter (c_h) referring directly to variables which are measurable and characterize the system response.

$$c_h = \frac{m_h \cdot \omega \cdot \delta}{\pi} \tag{8}$$

3 Parameters of the Compared Cylinders and the Research Objective

Comparative tests of the dynamic response of cylinders were performed for two configuration cases of power supply systems for a single test cylinder. The first was a classic configuration, while the second was a new design proposed by the authors, in which the control valves have been located in blocks mounted directly on the cylinder (Fig. 2). The basic parameters of the tested cylinder are provided in Table 1.

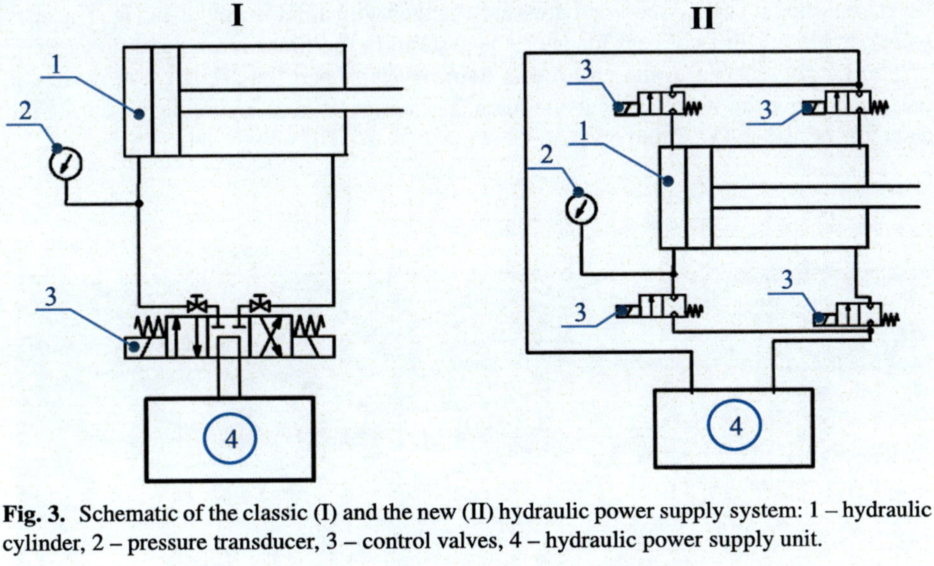

Fig. 3. Schematic of the classic (I) and the new (II) hydraulic power supply system: 1 – hydraulic cylinder, 2 – pressure transducer, 3 – control valves, 4 – hydraulic power supply unit.

Table 1. Parameters of the cylinder.

Cylinder parameter	Value [mm]
Piston diameter, D	80
Piston rod diameter, d	50
Piston stroke, s	495
Supply line length, l	4930

The new architecture of the power supply system limits the influence of hydraulic power supply lines on the compliance of the cylinder when it is stopped. In the analyzed case, the aim of the tests was to identify the dynamic parameters of the new cylinder and compare them to the parameters measured during th tests of the cylinder with the classic power supply system.

4 Test Procedure

The tests were performed on a test stand designed and built in order to conduct a broad spectrum of examinations on hydraulic cylinders (Fig. 4). Details of the test stand configuration are described in a separate publication. In the case of this research, an electronic control system was used, which allows automatic configuration of control valves depending on piston stroke. The system was configured so that the position approximate to the horizontal weighting arm corresponded to the piston rod extension by 20% of the stroke and served as the initial position. During the tests, the weighting arm was first lifted with the use of the tested cylinder to a height corresponding to 80% of the piston rod extension and then dropped and rapidly stopped in the initial position. During the stopping action, the measurement system recorded pressure changes with a frequency of 1 kHz. The tests were performed for two mass loads on the system. In the first case, an additional mass of 500 kg was placed at the end of the boom system, and the second test cycle was performed after the weight was dismounted. The temperature of hydraulic oil during the tests was between 45 °C and 55 °C.

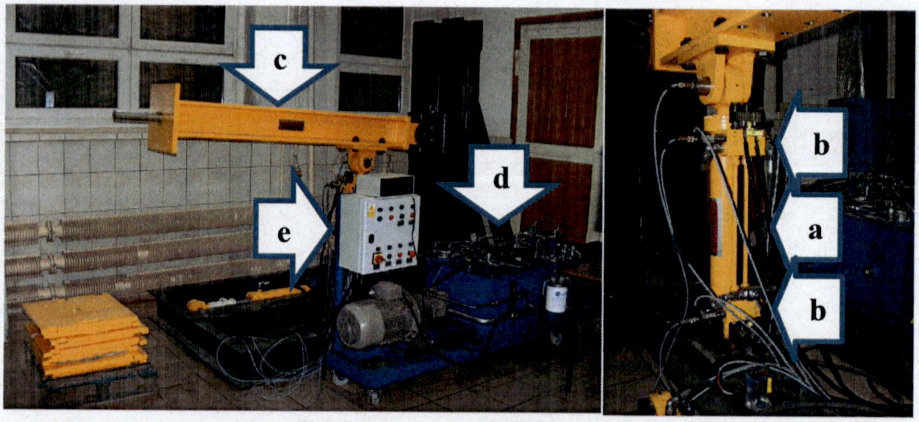

Fig. 4. The test stand and the hydraulic cylinder with the new power supply system. a – tested cylinder, b – valve and measurement block, c – loading arm, d – hydraulic power unit, e – measurement and control system.

5 Results Analysis

As suggested by the previously described theoretical models, the relationship between the hydraulically calculated stiffness parameters and the system damping parameters was assumed as the basis for the comparison of the results. Index "I" was assigned to the classic power supply solution, while index "II" was assigned to the system with the proposed architecture (Figs. 3 and 4). In addition, as the tests were performed in a repeatable manner and the system was always automatically stopped in the same position, the loading mass on the cylinder was assumed to be identical ($m_{h\,I} = m_{h\,II}$). The transformations resulted in the following relationships:

$$\frac{k_{hII}}{k_{hI}} = \left(\frac{T_{hI}}{T_{hII}}\right)^2 \tag{9}$$

$$\frac{c_{hII}}{c_{hI}} = \frac{\omega_{II} \cdot \delta_{II}}{\omega_{I} \cdot \delta_{I}} \tag{10}$$

Figures 5 and 6 show histories of pressure changes during the tests of cylinders with a 500 kg load and with no load on the boom system. In order to enable comparison of results, the time axis on the charts has been shifted in such a way that the value of t = 0 always corresponds to the moment of control valves switch. Negative time values are plotted on the charts to present the values measured before the valves switch.

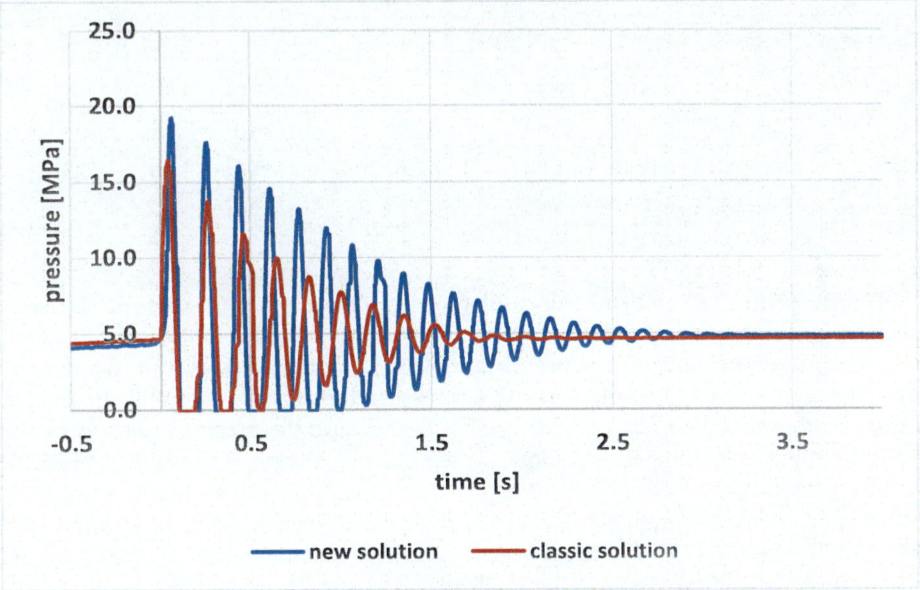

Fig. 5. The course of pressure during stopping the cylinder with additional load.

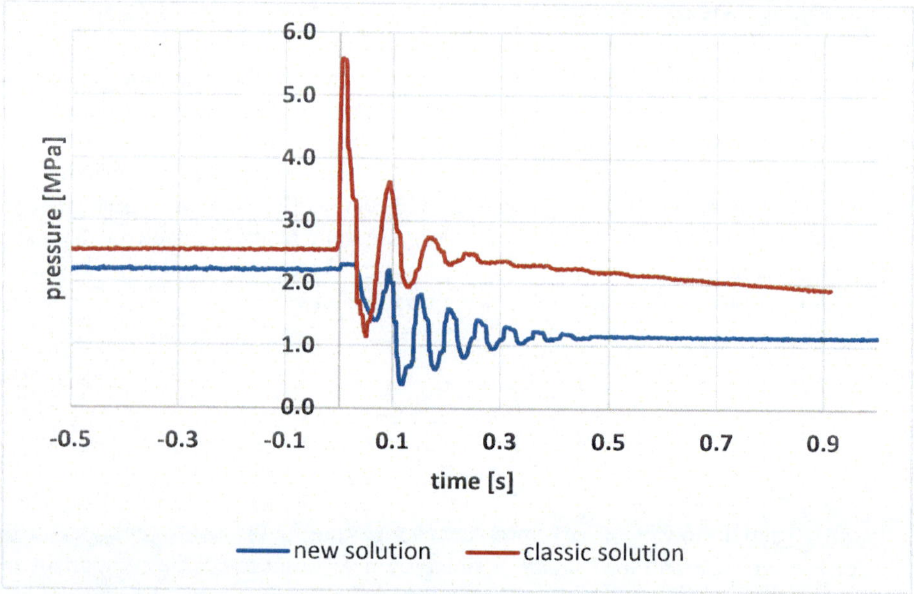

Fig. 6. The course of pressure during stopping the cylinder without additional load.

Table 2. Test results.

Extra load	Solution	T_h [s]	A_{h1} [MPa]	A_{h2} [MPa]	δ [-]	$\frac{k_{hII}}{k_{hI}}$ [-]	$\frac{c_{hII}}{c_{hI}}$ [-]
Yes	New	0.2025	19.24	17.63	$87 \cdot 10^3$	1.15	0.51
	Classic	0.2175	16.44	13.70	$183 \cdot 10^3$		
No	New	0.0600	1.85	1.62	$134 \cdot 10^3$	1.56	0.39
	Classic	0.0750	5.60	3.63	$432 \cdot 10^3$		

The obtained characteristic values of pressure changes are provided in Table 2. Noteworthy is the fact that the time history of pressures recorded in the loaded system is similar to the theoretical response of the dynamic system. However, in the case of zero load on the system, the time history of pressure signal is similar to the theoretical history only in a certain range. Additionally, in this case the maximum values of the recorded pressure are higher than those for the classic solution. The value of pressure in the classic solution lowers considerably with time, after the system is stabilized. These phenomena are probably due to the dynamic influence of the slide valve and the compliant hydraulic power supply line, and will be the topic of future publications. The comparison of stiffnesses in hydraulically calculated systems indicates that with the new power supply, the stiffness of the system increases by approx. 15.4% in the case of the loaded system and by approx. 56.3% in the case of the system not loaded with additional mass. An analogical relationship between the hydraulically calculated system damping parameters indicate that the parameter in the new solution is reduced by

48.9% in the case of the loaded system and by 61.3% in the case of the zero load system. Maximum momentary pressure values recorded for the loaded system were 16.4 MPa in the case of the classic power supply system and 19.2 MPa in the case of the new design. Analogically, for the system with no load the values are 5.6 MPa in the case of the classic power supply system and 2.3 MPa in the case of the new design.

6 Summary

The above test results for a cylinder with two different power supply configurations and for two cases of external loads demonstrate beyond doubt the need for further research into the issue. The identified changes of stiffness and of the damping parameter for a cylinder system loaded with external body force are the basis for designing new hydraulic power supply solutions with improved stiffness parameters and thus, with improved accuracy of movements. The described method for identifying changes in stiffness and dynamic damping of a hydraulic system, based on a relatively simple measurement of pressure changes, can be used in industrial applications. The identified disturbed histories of pressure changes in relation to the model of a second order dynamic system with damping, observed especially in the case of classic cylinders with slide valves, are the basis for the currently conducted research into the influence of individual system elements on the character of cylinder movements. The author believes that the mathematical model presented in this article has been positively verified and can be used as a basis for further research.

References

1. Sochacki, W., Tomski, L.: Free vibration and dynamic stability of a hydraulic cylinder set. Mach. Dyn. Probl. **23**(4), 91–104 (1999)
2. Fiebig, W., Wróbel, J., Cependa P.: Transmission of fluid borne noise in the reservoir. In: Proceedings of the 2018 Bath/ASME Symposium on Fluid Power and Motion Control, FPMC2018, pp. 1–8. ASME, New York (2018)
3. Fiebig, W, Wróbel, J.: System approach in noise reduction in fluid power units. In: Proceedings of the 2018 Bath/ASME Symposium on Fluid Power and Motion Control, FPMC2018, pp. 1–8. ASME, New York (2018)
4. Yang, S., Tao, A., Luo, Y., Zhang, J., Zhou, P., Zhou, L.: Experimental measurements of bulk modulus for two types of hydraulic oil at pressures to 140 MPa and temperatures to 180 °C. In: 10th International Fluid Power Conference, Dresden, pp. 193–204 (2016)
5. Hružík, L., Vašina, M., Bureček, A.: Evaluation of bulk modulus of oil system with hydraulic line. In: EPJ Web Conferences, vol. 45 (2013)
6. Huova, M., Aalto, A., Linjama, M., Huhtala, K., Lantela, T., Pietola, M.: Digital hydraulic multi-pressure actuator – the concept, simulation study and first experimental results. Int. J. Fluid Power **18**(3), 141–152 (2017)
7. Wiens, T.: An efficient, compact and low-cost Dual Cylinder Hydrostatic Actuator (DCHA). Int. J. Fluid Power **19**(2), 80–90 (2018)
8. Garimella, P., Yao, B.: Model based fault detection of an electro-hydraulic cylinder. In: American Control Conference, pp. 484–489 (2005)
9. Shang, L., Ivantysynova, M.: A temperature adaptive piston design for swash plate type axial piston machines. Int. J. Fluid Power **18**(1), 38–48 (2017)

10. Linjama, M., Huova, M., Huhtala, K.: Model-based force and position tracking control of an asymmetric cylinder with a digital hydraulic valve. Int. J. Fluid Power **17**(3), 163–172 (2016)
11. Heybroek, K., Sahlman, M.: A hydraulic hybrid excavator based on multi-chamber cylinders and secondary control – design and experimental validation. Int. J. Fluid Power **19**(2), 91–105 (2018)
12. Destro, M.C., De Negri, V.J.: Method for combining valves with symmetric and asymmetric cylinders for hydraulic systems. Int. J. Fluid Power **19**(3), 126–139 (2018)
13. Warzyńska, U., Siwulski, T.: Numerical analysis of the influence initial position of a piston on fluid exchange process in a hydraulic cylinder. In: Zolotarev, I., Radolf, V. (eds.) Engineering Mechanics 2019: 25th international conference, pp. 403–406. Institute of Thermomechanics of the Czech Academy of Sciences, Prague (2019)

Analysis of the Impact of Vibrations on the Microhydraulic Pressure Relief Valve Taking into Account the Interval Classification of Induction Trees

Michał Stosiak[1], Krzysztof Towarnicki[1(✉)], Marian A. Partyka[2], and Adam Deptuła[3]

[1] Department of Technical Systems Operation and Maintenance, Faculty of Mechanical Engineering, Wroclaw University of Science and Technology, Wroclaw, Poland
{michal.stosiak,krzysztof.towarnicki}@pwr.edu.pl

[2] Department of High Voltage Technology, Institute of Electrical Power Engineering and Renewable Energy, Faculty of Electrical Engineering Automatic Control and Informatics, Opole University of Technology, Opole, Poland
m.partyka@po.edu.pl

[3] Department of Management and Production Engineering, Faculty of Production Engineering and Logistics, Opole University of Technology, Opole, Poland
a.deptula@po.edu.pl

Abstract. The paper indicates some sources of external forces acting on microhydraulic valves. Mechanical vibrations characterized by a wide frequency spectrum were considered as an example of these excitations, with high-frequency vibrations being the most dangerous in microhydraulic components. The impact of external mechanical vibrations on the microhydraulic relief valve results in falling into the vibrations of the valve control element and causes changes in the amplitude-frequency spectrum of pressure pulsation of the system in which the valve works. The resulting pressure pulsation causes further propagation of mechanical vibrations, among others through vibrating microhydraulic pipes. The paper attempts to use the interval classification of induction trees to determine the impact of selected valve design and operating parameters on valve sensitivity to external mechanical vibrations. In the generation of inductive decision trees, a series of tests are carried out in a specific order. The criterion for choosing the attribute used to expand the tree is entropy as a measure of information. For hierarchically ordered classes of measuring intervals in the whole frequency range, induction trees defining the range of the most important measuring points were generated. Then, in order to acquire knowledge from data, it is possible to select the most important individual points and rules determining the relationships between construction parameters.

Keywords: Microhydraulic valve · Vibrations · Induction trees · Optimization

© The Editor(s) (if applicable) and The Author(s), under exclusive license to Springer Nature Switzerland AG 2021
J. Stryczek and U. Warzyńska (Eds.): NSHP 2020, LNME, pp. 311–322, 2021.
https://doi.org/10.1007/978-3-030-59509-8_28

1 Introduction

In microhydraulic systems, a relationship between mechanical vibration and pressure pulsation can be found, and this phenomenon is called coincidence, which can be considered in two ways. On the one hand, the vibrations of the elements of the microhydraulic system result from the pulsating flow of liquids (e.g. vibrations of microhoses, microvalves), and on the other hand, the pressure pulsation is caused by the vibrations of the elements of the microhydraulic system resulting mainly from the excitation of microvalve control elements (e.g. directional control microvalve spool) [1]. The varying pressure of the working medium in the microhydraulic system is a result of the pump output pulsation [2] caused by the kinematics of the displacement elements of the micropump. Ongoing work is continuing to improve the design of pumps to reduce their pulsation of efficiency and noise. An example of innovative testing of pumps is shown in lit. [3]. The nature of the generated pulsation of efficiency takes the shape of a harmonic function. The frequency of pressure pulsations corresponds to the frequency of pulsations of efficiency [4]. Undoubtedly, its effect is the induction of mechanical vibrations, among others microhydraulic hoses. Below is the consequence of the occurrence of pressure pulsations in the form of excitation of pipe vibrations, based on literature measurements [5], in Fig. 1.

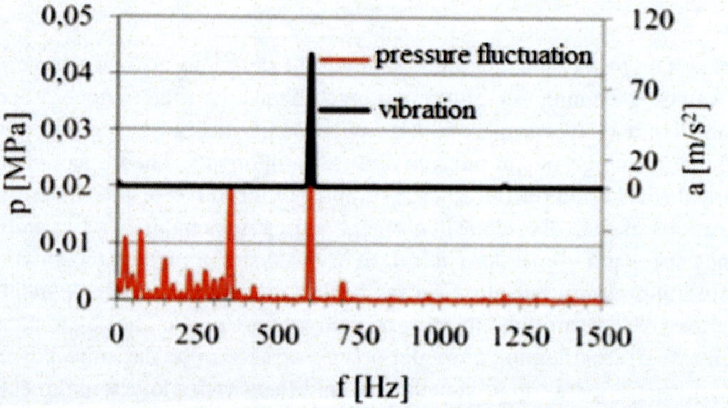

Fig. 1. Amplitude-frequency spectrum external mechanical vibration and pressure fluctuaction, frequency external mechanical vibration f = 600 Hz, $p_{śr}$ = 5 MPa, Q = 1,33 × 10^{-5} m^3/s (0, 8 dm^3/min).

2 Measuring Ring

For testing the impact of external mechanical vibrations on the hydraulic microvalve, a measuring stand was built, which is shown in Fig. 2.

A gear pump with internal gearing was chosen for the construction of the rig due to its advantages, i.e. generation of relatively high pressures, resistance to contamination and affordability. For measuring the average pressure in the microhydraulic system, an

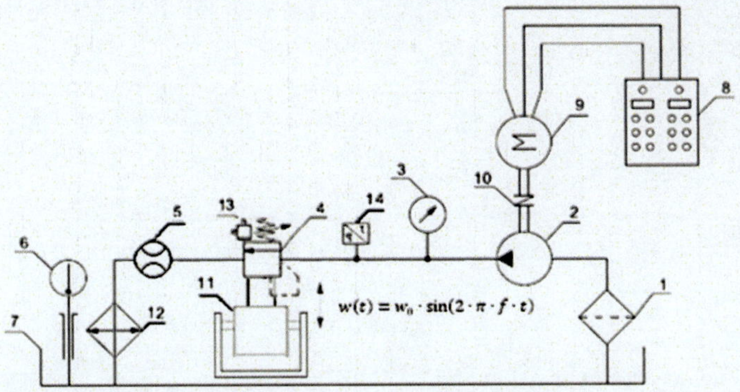

Fig. 2. Diagram of the hydraulic system for testing the impact of external mechanical vibrations on hydraulic microvalves: 1 - oil filter, 2 - gear pump WPH PZ3A1G, 3 - ParkerServiceJunior SCJN-400-01 pressure gauge, 4 - tested microvalve, 5 - Parker KSCVF-002-10 flow meter - 07X, 6 - Elmetron PT-217 thermometer, 7 - oil tank, 8 - control cabinet, 9 - three-phase electric motor, 10 - flexible coupling, 11- electrodynamic mechanical oscillator TiraVib, 12 - oil cooler with controller, 13- Piezotronics 340A16 piezoelectric acceleration sensor PCB, 14-Piezotronics 105C23 piezoelectric pressure sensor PCB.

analog pressure gauge was used on the pressure line located behind the pump pressure port. Subsequently, a piezoelectric pressure sensor was installed behind the pressure gauge to measure pressure pulsations in front of the pressure relief microvalve under test. TiraVib placed on the electrodynamic exciter of mechanical vibrations, allowing to generate vibrations with a given amplitude and frequency. Additionally, the Piezotronics 340A16 PCB acceleration sensor was placed on the housing of the microvalve. Badger Meter BM-OG flow meter was installed next to the tested valve. A cooler is located behind the flow meter, which together with the controller is part of the temperature stabilization system. During tests, Azolla 22 AF oil was used in the hydraulic system ($\rho = 865$ kg/m^3, ν(at 40 °C) $= 22.5$ mm^2/s).

The valves design is based on the classic configuration of the valve with a sharp-edged seat [6] and a closing element based on a conical poppet with an opening angle of 30°. A series of measurements were carried out for a constant vibration amplitude at an excitation frequency in the range of 200–900 Hz in 10 Hz steps. The attached pressure sensor and acceleration sensor were connected to the computer and operated by the PUMA system. The data post processing was performed in the Origin Pro 9.0 and Microsoft Excel environment.

3 Selected Results of Experimental Research

In order to determine the influence of external mechanical vibrations on the pressure relief valve, measurements of flow rates through the valve of 1 dm^3/min were carried out for a fixed opening pressure of the relief valve of 10 MPa. Due to the limited amount of space, selected measurement results are listed below. In Figs. 3 and 4, the results for external frequency of 230 Hz and 660 Hz and flow rate of 1 dm^3/min, respectively were presented.

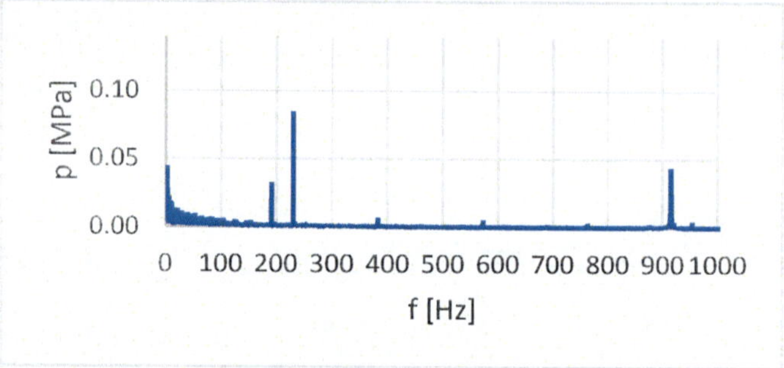

Fig. 3. Amplitude spectrum at a flow rate of 1 dm^3/min, at a pressure of 10 MPa average and external excitation for a frequency of 230 Hz.

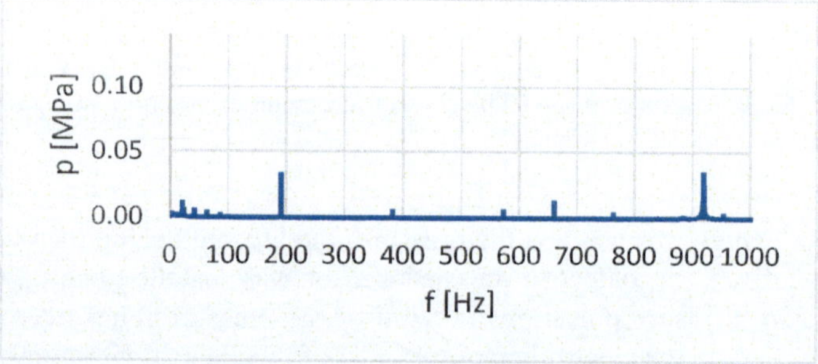

Fig. 4. Amplitude spectrum at a flow rate of 1 dm^3/min, at a pressure of 10 MPa average and external excitation for a frequency of 660 Hz.

4 Interval Classification Using Induction Trees

Classification means dividing any set of elements into groups, which include elements that differ, but are similar, i.e. those that distinguish a given group. A set of elements belonging to one group is called a class, and each element of a class is an object. Class elements may differ, except those properties on which the classification is based. Depending on the type of information available, two issues can be distinguished within the classification:

1. Reference classification; the structure of the category is known, i.e. We have a description of the classes from which the objects originate; this issue is called learning (recognition) with a teacher or under the supervision,
2. Standard classification, known as taxonomy or cluster analysis (clustering), also known as learning or recognition without a teacher.

In general terms, a given set of objects has the following form:

$$\left\{x_i, C^{(i)}\right\} i = 1 \ldots N \tag{4.1}$$

where x is a n-dimensional vector describing the case, while $C^{(i)} \in \{C_1, C_1, \ldots, C_L\}$ is a class label.

4.1 Decision Areas in the Classification

From a geometrical point of view, each x vector from the training sequence can be assigned a point in a suitably defined feature space (observation space), whose dimension is equal to the dimension of the training vector. The classifier divides the feature space into decision areas, i.e. parts of the feature space, such that all points in this area correspond to the same decision (class). The boundaries between these areas are called decision surfaces or decision boundaries. In other words, the boundary of decisions is an area in the space of features for which the relevant criterion functions (functions of belonging to a given class) have the same value.

The optimal classifier selects the class for which the posteriori probability is higher. A given vector x is classified into class C_k when the following relation is met [7, 8]:

$$p(C_k|x) > p(C_j|x)j \neq K \tag{4.2}$$

Where: $p(C_k)$ is the a priori probability of class C_k is the a priori probability of class C_k, while p (x|C_k) is the distribution of samples for class C.

This equation leads to a solution in which the decision boundary between classes lies at the intersection of the class distributions. For example, the best decision limit for separating these two classes (features *X* and *Y*) is the circle (Fig. 5).

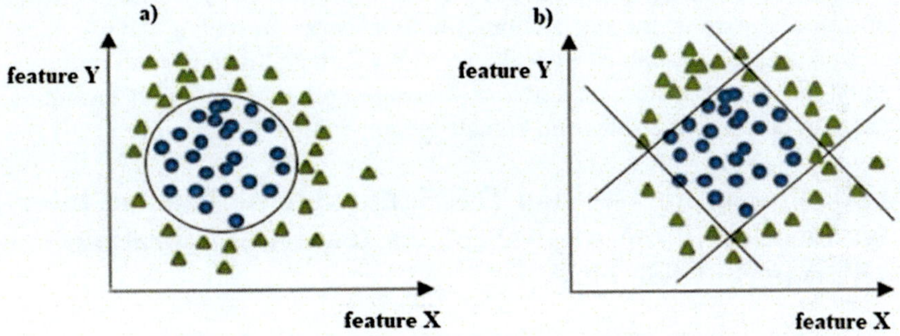

Fig. 5. Optimal a) and not optimal b) decision boundary.

Unfortunately, the shape of decision areas in the analyzed data is usually unknown. Therefore, different models are used for each data and their generalization is examined, and the one whose generalization was the highest is considered the best.

4.2 Machine Learning and Interval Induction

Machine learning is a branch of artificial intelligence dealing with the automatic acquisition of knowledge from data in order to use it in the future [9, 10].

The sources of machine learning include negative experiences with expert systems, in which knowledge is usually acquired through dialogue with an application specialist.

For similar reasons, machine learning is now becoming a popular trend as part of multi- criteria decision support. Machine learning methods are usually embedded in one of two environments:

- unsupervised setting (learning),
- supervised setting (learning).

In an unattended environment, the learning system generates a hypothesis based on the values of the characteristics. The task of the learning system is then to detect and describe certain regularities (regularities, relationships, clusters) in the analyzed data

In this paper, the division of *P (U)* into decision intervals C_k is defined on the set of all U objects. According to the definition, this division meets the following conditions:

- Fullness [11]:

$$\bigcup_{K=1}^{n} C_k = U$$

- separation of decision classes:

$$\forall C_k \in U, \mathrm{C_l} \in U : C_k \in C_l = \emptyset, K, l = \langle 1, n\rangle, k \neq l \tag{4.3}$$

where n is the number of decision classes. The division into decision classes usually results from the field of application and is given a priori (e.g. by an expert in the field of application). There are numerous author/co-author works presenting the use of inductive decision trees, multi-valued logical trees (also with coefficients) and multi-valued logic equations as a decision support tool, discrete optimization and determining the importance of decision variables and knowledge generation, for example [12–14]

5 Application of the Decision Tree Induction Method and Interval Classification in the Analysis of the Impact of Vibration on Themicrohydraulic Valve

Classifier learning is based on minimizing some function, called an error or cost function. In the simplest case, i.e. when all errors are equally unfavorable, this function has the form [15, 16]:

$$E = \sum\nolimits_i Er\left(y(x_i;\ W),\ C^{(i)}\right) \tag{4.4}$$

$$\text{Where function}:\ Er\left(C_i,\ C_j\right) = \begin{cases} 0,\ if\ i = j \\ 1,\ if\ i \neq j \end{cases} \tag{4.5}$$

In general, Er () is a matrix function that determines the cost of incorrectly classifying an object from class Ci to another class, e.g. Cj.

In [17] the analysis was carried out for the range of the frequency of external excitation (200–900 [Hz]) in 10 [Hz] steps for:

- three values of flow rates through the valve: 0.6; 0.8; 1 [dm^3/min],
- spring stiffness c = 7.49 [N/mm].

Measurements were carried out for flow rates through the valve equal to 1, 0.8 and 0.6 [dm^3/min] at an opening pressure of the relief valve 10 [MPa].

The figures (Fig. 6) show example charts with specific intervals for intensity: 0.6 [dm^3/min].

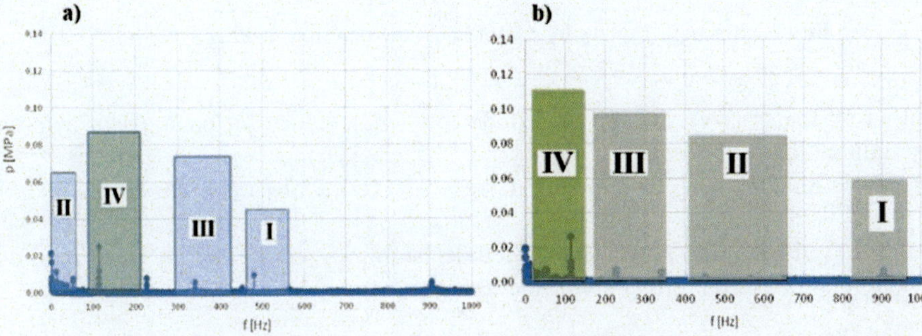

Fig. 6. The most important frame intervals for the amplitude frequency spectrum at a flow rate of 0.6 [dm3/min], pressure 10 [MPa] and external excitation for a frequency of a) 490 [Hz], b) 650 [Hz].

In [17], the effect of intensity on the appearance of the amplitude-frequency spectrum of pulsation of the component pressure with the frequency corresponding to these vibrations and the dominant amplitude was analyzed. The influence of intensity Q plays a decisive role for its second and third range of variability [16].

Currently the work presents a preliminary analysis of the impact of changes in the frequency ranges (200–900 [Hz]) in 10 [Hz] steps for three values of flow rates through the valve: 0.6; 0.8; 1 [dm^3/min] taking into account the frequency spectrum range. Were considered only intervals in which the flow rate plays the most important role for the entire amplitude frequency spectrum were considered.

The concept of entropy was used in the analysis. The information in the set of teaching examples is equal to (Fig. 7):

$$I(E) = -\sum_{i=1}^{|E|} \frac{|E_i|}{|E|} \cdot \log_2\left(\frac{|E_i|}{|E|}\right) \tag{4.6}$$

where:

E - a collection of teaching examples, -$|E_i|$ the number of examples of the i-th object, $|E|$ - the number of examples in the training set E.

In the analysis we take as:

- input (we) attributes of the external excitation frequency (200–900 [Hz]): **[we_1, we_2, we_3, …,we_i] [we_{200}, we_{210}, we_{220}, …, we_{900}]** for amplitude frequency spectra **[we_{200}, ($p_{(f200)}$), we_{210} ($p_{(f210)}$), we_{220} ($p_{(f220)}$), …, we_{900} ($p_{(f900)}$)]**..

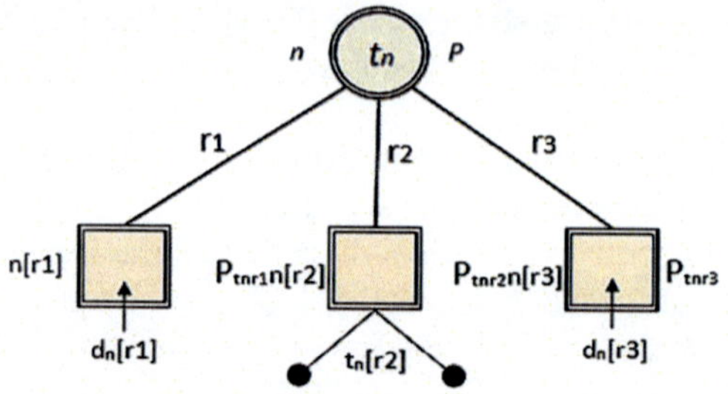

Fig. 7. A sample "query by example" to the decision making system.

– output attributes (**wy**): are the amplitude frequency values for each set of external excitation (200–900 [Hz]):
[h_1 ($f_1 = 200$ Hz), $h_2(f_2 = 210$ Hz), $h_3(f_2 = 210$ Hz),…, $h_3(f_2 = 900$ Hz)].

The analysis uses the Aitech integrated decision program by the DeTreex module. The computational structure is presented in Fig. 8.

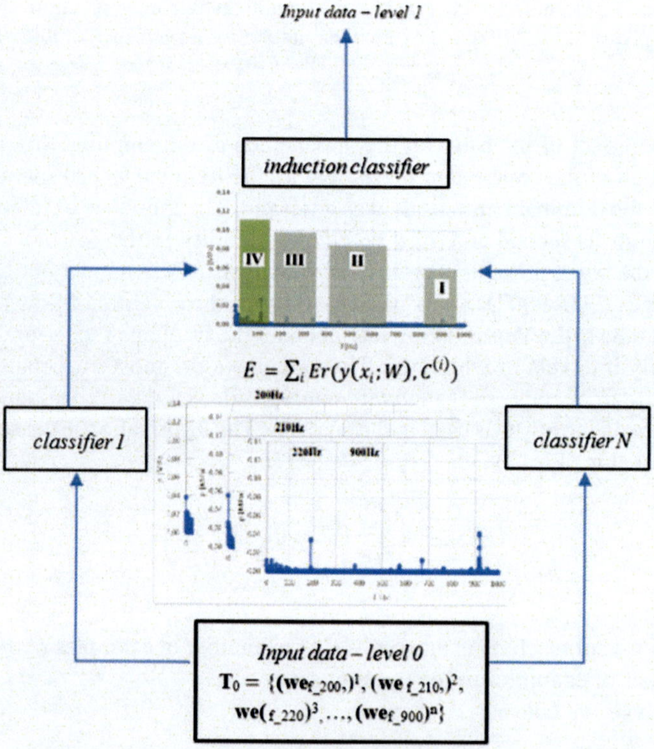

Fig. 8. Integrated decision system.

The induction tree shown in Fig. 9 was generated for the flow rate $Q = 1$ [dm^3/min] and the entire frequency range *f*: 200–900 [Hz] with the division into attributes: [h_1 (f_1 = 200 Hz), h_2 (f_2 = 210 Hz), h_3 (f_3 = 210 Hz), ..., h_k (f_k = 900 Hz)]. An instance is classified by starting at the root node of the tree, testing the attribute specified by this node, then moving down the tree branch corresponding to the value of the attribute as shown in the Fig. 9. This process is then repeated for the subtree rooted at the new node. The designations h_1, h_1,... h_m are respectively the hypotheses represented by the subtrees T_1, T_2, ... T_m. Each of the finite number of possible test results corresponds to a branch leading from the node to the subtree. If a node contains a test, then the result it i_t= { i_1, i_2, i_3, ...i_m}, and the corresponding branches lead to subtrees T_1, T_2, ... T_m, then the hypothesis represented by this node can be given for each example $x \in X$ write as follows:

$$h(x) = \begin{cases} h_1(x)\ \mathit{if}\ t(x) = i_1 \\ h_2(x)\ \mathit{if}\ t(x) = i_2 \\ \cdots \\ h_m(x)\ \mathit{if}\ t(x) = i_1 \end{cases} \quad (4.7)$$

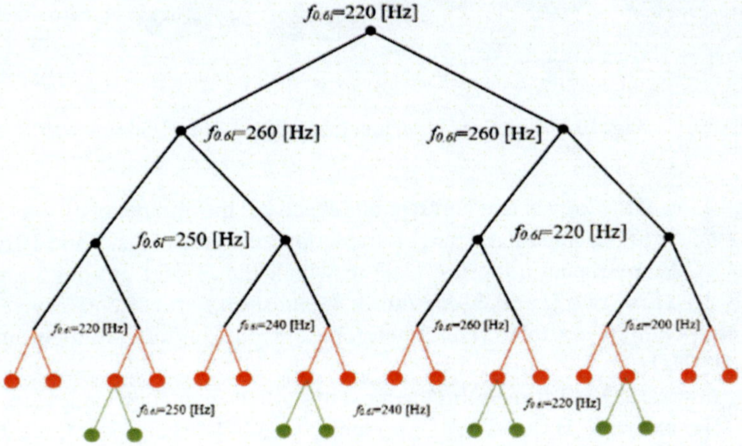

Fig. 9. Induction tree for the learning file from Fig. 8- for Q = 0.6 [dm^3/min].

The same analysis is presented in Fig. 10 for the flow rate $Q = 1$ [dm^3/min]

The tree in Fig. 9 is, however, symbolically represented using a proprietary graphical interface. However, the tree shown in Fig. 10 is presented as a graphical interface directly from the DeTreex program.

In the event that each operating parameter, assuming numerical values from a specific range, will be marked with a logical variable of two or multi- value, then it is possible to discretization such numerical ranges. The set of all numerical combinations forms an induction tree. The number of sub-ranges of a given interval means a set of branches in one twig bundle (possibility of implementation), and the number of all paths from the

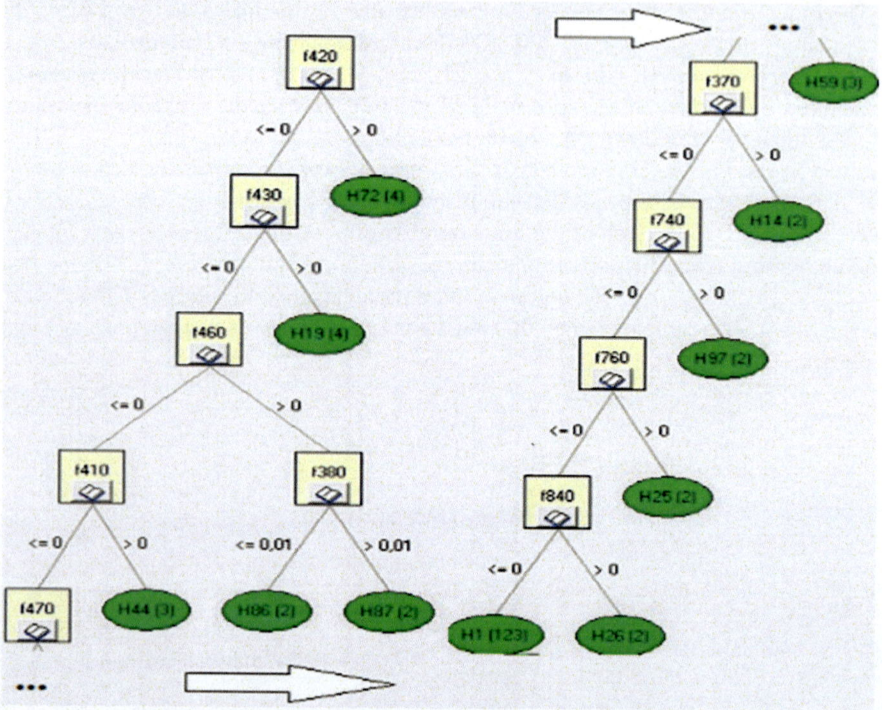

Fig. 10. Minimized induction tree for $Q = 1[\mathrm{dm}^3/\mathrm{min}]$.

bottom to the top of the logical tree corresponds exactly to the number of all combinations of discrete values of the considered ranges. Induction trees from Figs. 9 and 10 designate the ranges of the frequency ranges f [Hz] in which the impact of sound pressure is greatest. For the flow rate $Q = 0.6$ [dm^3/min], the most important frequency values are: $f = 220$ [Hz], 250 [Hz] and 260 [Hz]. Therefore, the most important range can be for f = {220–260 [Hz]}.

For the flow rate $Q = 1$ [dm^3/min] the most important frequency values are: $f = 420$ [Hz], 430 [Hz] and 460 [Hz] and less important: $f = 370$ [Hz] and 380 [Hz]. Thus, the most important range can be considered for: $f = 370$–460 [Hz].

These are quantities describing operating parameters. In subsequent publications, the analysis will concern the determination of the importance rank of construction parameters, such as spring constant, valve diameter, shape of the valve plug. The analysis will be carried out in the currently most important designated compartments using induction trees and taking into account the analysis using neural networks presented in [17].

6 Conclusions

When analysing the impact of frequency changes [(f_1 = 200 Hz), (f_2 = 210 Hz), (f_3 = 210 Hz), …, (f_4 = 900 Hz)] on pressure pulsations, it can be stated that its greatest impact is for the ranges f = 220–260 [Hz] - induction tree in Fig. 8.

The analysis should be carried out for three flow rates through the valve: 0.6; 0.8; 1 [dm^3/min]. In the case of analysis for intensity 1 [dm^3/min], the importance of the impact of frequency in its higher values increases. On the induction tree of Fig. 10, the largest impact is in the range of 420–460 [Hz]. Further analysis should also take into account the flow rate of 0.8 [dm^3/min]. The present work is another one regarding the impact of design parameters (such as spring stiffness, closing member weight, seat diameter, etc.) and operational (valve opening pressure, valve flow rate, etc.) affecting the vibrations of hydraulic and noise.

Ultimately, the analysis should be carried out for:

- various changes in the construction parameters: analysis of springs with different stiffness and/or change of the shape of the microvalve,
- various pressure changes as an operating parameter.

References

1. Stosiak, M., Towarnicki, K., Lubecki, M.: The influence of external mechanical vibrations on hydraulic microvalves. Test stand design and preliminary tests. Interdiscip. J. Eng. Sci. **6**(1), 48–52 (2018)
2. Kollek, W.: Basics of Design, Modelling Operation of Microhydraulic Components And Systems. Publishing house PWR, Wrocław (2011)
3. Śliwiński, P.: New satellite pumps. Key Eng. Mater. **490**, 195–205 (2011)
4. Kudźma, Z., Kułakowski, K., Stosiak, M.: Selected problems in the operation of microhydraulic systems. Drives Control **16**(4), 78–83 (2014)
5. Stosiak, M.: Identification of Vibration Impact and Methods of their Reduction in Selected Hydraulic Valves. Publishing house PWR, Wrocław (2015)
6. Kudźma, Z., Stosiak, M.: Micro-max valve. Patent PL221214B1 (2016)
7. Michalski, R.S., Carbonell, J.G., Mitchell, T.M.: Machine Learning: An Artificial Intelligence Approach, vol. III. Morgan Kaufmann Publishers Inc., San Francisco (1983)
8. Weiss, S., Kulikowski, C.A.: Computer Systems that Learn: Classification and Prediction Methods From Statistics, Neural Nets, Machine Learning and Expert Systems. Morgan Kaufmann Publishers Inc., San Francisco (1991)
9. Michalski, R.S., Bratko, I., Kubat, M.: Machine Learning and Data Mining. Methods and Applications. Wiley, Hoboken (1997)
10. Chung-Min, W., Ching-Lin, H., Kuei-Lun, Ch.: A Hybrid multiple criteria decision making model for supplier selection. Math. Prob. Eng. 8 (2013)
11. Kotsiantis, S.B.: Artificial intelligence review. Decis. Trees: Recent overv. **39**(4), 261–283 (2013)
12. Deptuła, A., Partyka, M.A.: Application of dependence graphs and game trees for decision decomposition for machine systems. J. Autom. Mob. Robot. Intell. Syst. **5**(3), 17–26 (2011)
13. Deptuła, A., Macek, W., Partyka, M.A.: Analysis of loading history influence on fatigue and fracture surface parameters using the method of induction trees. In: MATEC Web of Conferences, vol. 252, no. 08003, pp. 1–6, Lublin (2018)
14. Deptuła, A., Osiński, P., Radziwanowska, U.: Decision support system for identifying technical condition of combustion engine. Arch. Acoust. **41**(3), 449–460 (2016)
15. Hall, J.G.: Engineering knowledge engineering. Expert Syst. **29**(5), 427–525 (2012)

16. Brunsson, N.: The Consequences of Decision-making. Oxford University Press, New York (2007)
17. Stosiak, M., Towarnicki, K., Partyka, M., Deptuła, A.: Analysis of the impact of vibrations on the microhydraulic valve, including neural networks and induction trees. In: XXIII Scientific Conference "Problems of Development of Working Machines", pp. 52–52. Department of Mechanics and Fundamentals of Machine Design at the Opole University of Technology, Szklarska Poręba (2020)

Experimental Research into the Influence of Operational Parameters on the Characteristics of Pressure Pulsation Dampers

Urszula Warzyńska(✉)

Department of Technical Systems Operation and Maintenance, Wroclaw University of Science and Technology, Lukasiewicza 5, 50-371 Wroclaw, Polska
urszula.warzynska@pwr.edu.pl

Abstract. Reciprocating compressors operated in gas transmission systems are equipped with passive pressure pulsation dampers protecting the system against the negative effects of excessive pulsation and the related dynamic phenomena. The transmission loss characteristics are determined by means of numerical simulations or measurements at test stands. Most often, however, both the simulations and the bench tests cover either only some assumed simplified operating conditions, e.g. experimental tests of acoustic characteristics, without taking into account the flow of medium, or numerical simulations without taking into account the geometry of the entire installation (the anechoic boundary condition). The article presents a proposed methodology for determining the transmission loss characteristics of pulsation dampers with allowance for variable operating parameters and with the use of a specially designed test stand equipped with a piston compressor and a variable volume damper. In addition, the influence of operational parameters (flow rate and pressure) on pressure pulsation attenuation was tested. The research results presented in the article may form the basis for the development of new improved mathematical models simulating the characteristics of pulsation dampers with allowance for the operational parameters of a system.

Keywords:: Piston compressors · Pressure pulsation · Pulsation damper

1 Introduction

Piston compressors working in process and gas transmission installations are equipped with passive (reactive) dampers protecting the compressor and the pipeline system against the negative effects of excessive pressure pulsation. Uncontrolled pressure pulsation in gas installations carries the risk of both damage to system components and serious failures leading to system downtime. In addition, pressure pulsation can reduce the efficiency of the system and cause indication errors of the measuring equipment. Gas pressure pulsation results from the cyclic flow of gas through a reciprocating compressor. The movement of the piston in the compressor cylinder generates pulsations at

© The Editor(s) (if applicable) and The Author(s), under exclusive license to Springer Nature Switzerland AG 2021
J. Stryczek and U. Warzyńska (Eds.): NSHP 2020, LNME, pp. 323–332, 2021.
https://doi.org/10.1007/978-3-030-59509-8_29

a fundamental frequency corresponding to the crankshaft rotational speed. Valves and other compressor components cause subsequent harmonic pulsations with frequencies that are a multiple of the fundamental frequency. After passing through a damper (anti-pulsation bottles), a pressure pulsation propagates in a gas as a wave and can interact with the system by undergoing reflection or interference. The amplitude of individual harmonics of pressure pulsations can be significantly increased as a result of resonance – if the frequency of excitations from the compressor coincides with the frequency of natural vibrations of the piping system [1, 2]. For this reason, anti-pulsation bottles of piston compressors must be optimally selected and arranged, and this procedure should result from the analysis of the operating conditions and the geometrical parameters of the installation. The correct selection of dampers should allow increasing the reliability of the reciprocating compressor and the entire installation as well as minimizing the initial investment and operating costs. The selection of the appropriate damper type is based on analytical calculations (Helmholtz model or recommendations of API 618 [3]), numerical simulations [4, 5] or experimental tests. Two methods have been described so far in the literature for the experimental determination of the transmission loss characteristics of passive pulsation dampers. The first of these is the two-microphone method, in which the pulsation damper is excited by means of a sound source with a wide frequency spectrum, and a free gas flow is ensured at the output of the damper (anechoic boundary condition). Two microphones are mounted in a pipe at the inlet and at the outlet of the damper in order to capture the traveling and reflected waves [6]. Experimental studies of damping characteristics, including mean flow of medium, were also undertaken. In this case, the damper was excited by a centrifugal compressor operating at different rotational speeds and a free flow of air was provided at the outlet of the damper. Attenuation characteristics were determined using two microphones mounted before and after the damper [7]. The two-microphone method has also been adopted to determine the four-field matrix coefficients in order to perform 3D numerical calculations of dampers [8]. Acoustic measurement methods are also widely used in other systems, e.g. for measuring noise resulting from positive displacement machines operation and system vibrations [9]. In the second method of determining the transmission loss characteristics, pressure transducers are used to measure the dynamic pressure component during the flow of a working medium. In publications [10, 11] the source of pulsations was a piston compressor, and the system was terminated with an air tank. The research concerned a system with a volume-choke-volume filter and with a Helmholtz resonator.

2 Experimental Tests

For the purpose of experimental research on the impact of operating parameters on the characteristics of pressure pulsation dampers, a test stand was built. It comprised a piston compressor together with an electric motor and an inverter to control the rotational speed, as well as a pulsation damper with adjustable volume and a pipe system with fittings. Dynamically variable pressure was measured with a system consisting of two Trafag NAH industrial pressure transducers along with an HBM signal processing and recording system. The system also measured the compressor suction and discharge temperatures (Fig. 1).

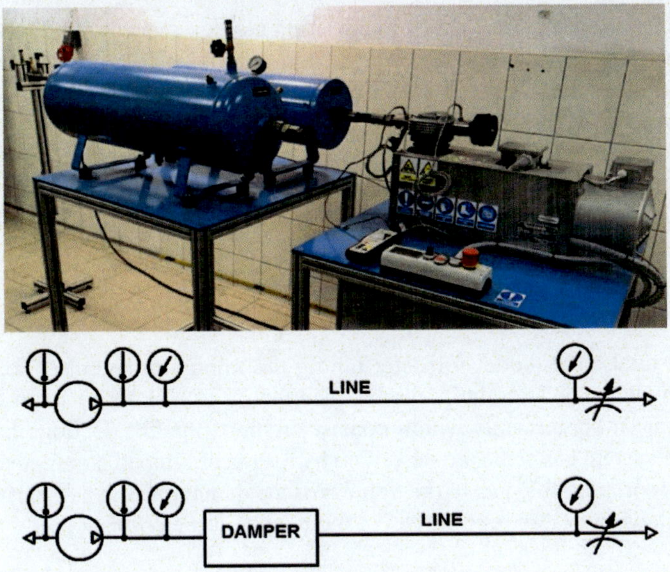

Fig. 1. Test stand for determining the transmission loss characteristics of gas pressure pulsation dampers.

The piston compressor has the ability to work in the range of 900–2500 rpm at a nominal pressure of 8–10 bar. The volumetric flow at 8 bar pressure and 1500 rpm is 125 l/min. Air was used as the working medium in the system. Experimental studies of pressure pulsation measurement were carried out on two installations:

- a system consisting of a straight steel pipe with a diameter of 28 mm and a length (including valves at the end of the line) equal to 2100 mm, ended with an open throttle valve (free outflow of air),
- a system with a pulsation damper with adjustable volume, internal diameter 200 mm and internal chamber length set to 650 mm, 290 mm and 160 mm, ended with a pipe section with valves and an open throttle valve at the outlet.

Table 1. Operating parameters during the tests.

	Pipe system		System with a damper	
Rotational speed [rpm]	Suction pressure [bar (g)]	Discharge pressure [bar (g)]	Suction pressure [bar (g)]	Discharge pressure [bar (g)]
1000	1.0	1.4	1.0	1.3
1500	1.0	2.1	1.0	2.0
2000	1.0	2.7	1.0	2.6

The test cases included gas pressure pulsation measurement at 1000 rpm, 1500 rpm and 2000 rpm. The static pressure in the installation depended on the flow resistance and rotational speed, but the fixed operating pressure was not determined for each measurement case (throttle valve open – free outflow of gas). The air temperature on the suction and discharge sides during the measurements was maintained at 23 °C at the start of each measuremet in order to ensure a constant sound propagation velocity in the medium, which was 348 m/s. The operating parameters are presented in Table 1.

3 Pulsation Damper Selection

The simplest analytical model for determining the minimum chamber volume of the pulsation damper is the Helmholtz model. The theory of Helmholtz resonators can be applied to system components which consist of short pipeline sections having small volumes. The system elements are described by means of lumped parameters – acoustic compliance – compressible gas in the volume of the damper acting as a spring, acoustic inertance – incompressible plug of gas in the neck volume corresponding to vibrating mass.

The natural frequency of a chamber damper f_0 is given by:

$$f_o = \frac{c}{2\pi}\sqrt{\frac{S}{LV}} \tag{1}$$

where: c – speed of sound, S – section area of outlet line, L – outlet line length, V – damper volume.

The natural frequency of a damper f_0 must be lower than the first harmonic of pressure pulsation, and the cut-off frequency f above which the low-pass filter will suppress the signal is given by the formula:

$$f = \sqrt{2}f_o \tag{2}$$

The minimum volume of a chamber damper V can therefore be determined by transforming formulas (1) and (2):

$$V = \frac{S \cdot c^2}{2\pi^2 \cdot f^2 \cdot L} \tag{3}$$

Based on the formula (3), a volume damper with an internal diameter of 200 mm and a length of 650 mm was selected for the analyzed installation in order to ensure the occurrence of a resonant frequency below the excitation frequency at the lowest rotational speed of the compressor. It should be emphasized that the analytical formula based on the Helmholtz resonators theory serves only for preliminary calculations of the required volume and generally leads to a significant oversizing of the dampers. In addition, the use of a model with lumped parameters is limited by small dimensions of the damper and outlet line in relation to the length of the highest harmonic of the analyzed wave excitation. Obtaining an adequately accurate solution is conditioned by the length of the longest damper element not exceeding ¼ of the highest harmonic wavelength.

If this condition is not met, standing waves (acoustic resonance) are generated in the system, which may cause additional resonances in the range of frequencies determined as attenuated. For this reason, the final selection of a damper for installation should be confirmed by numerical simulations or measurements of pressure pulsations in the system.

4 Results Analysis

As a result of the measurements, time history of pressure pulsations was obtained in the system with a straight pipe section and in the system with a chamber damper, depending on the compressor rotational speed and the operating pressure in the installation. The graphs (Fig. 2 and Fig. 3) show one period of a pressure pulsation wave based on signals from pressure transducers located directly at the outlet of the piston compressor and at the end of the installation before the throttle valve at 1000 rpm, 1500 rpm and 2000 rpm.

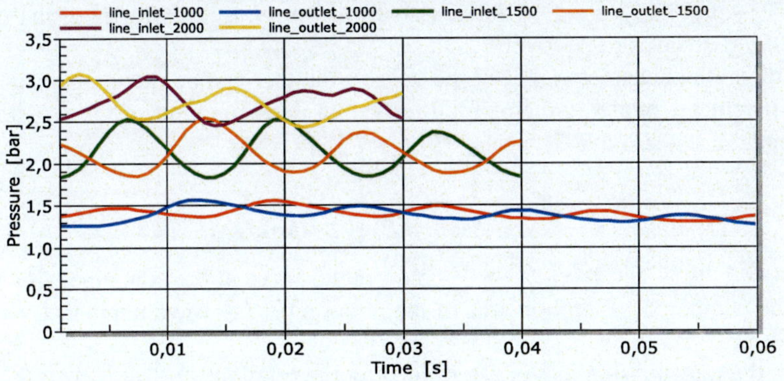

Fig. 2. Pressure pulsation in a system with a straight pipe section at 1000 rpm, 1500 rpm and 2000 rpm compressor speeds.

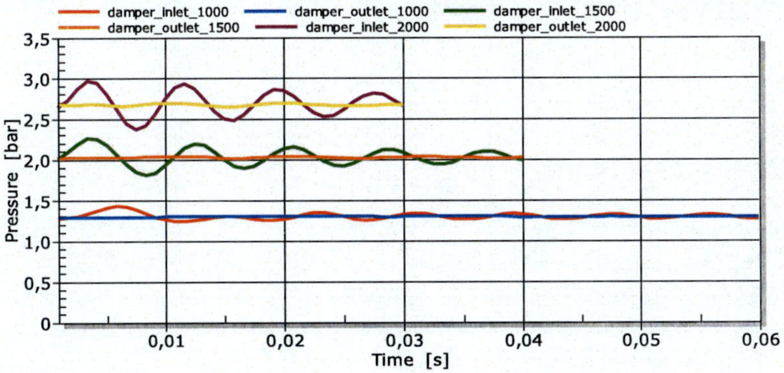

Fig. 3. Pressure pulsation in a system with a damper at 1000 rpm, 1500 rpm and 2000 rpm compressor speeds.

In the case of a system without a damper, the pressure pulsation wave is almost not damped along the pipeline. Only the recorded pressure pulsation signal at the end of the pipeline is shifted in phase relative to the pressure pulsation wave at the pipe inlet. The use of a chamber damper significantly reduces the pressure pulsation amplitude at each analyzed compressor rotation speed.

Further analysis of the test results was based on the determination of the transmission loss characteristic, which describes the decrease of the acoustic signal power between the inlet and outlet of the pulsation damper and is most often determined from the ratio of pressure pulsation amplitudes at the damper inlet and outlet for a given harmonic of a signal. For the purpose of determining the transmission loss function, Fourier transforms were made from pressure time signals measured with pressure transducers in order to determine the amplitude of pulsations in the frequency domain. The damping function can then be calculated from the formula:

$$TL = 20log\frac{FFT_1}{FFT_2} \tag{4}$$

where: FFT_1 and FFT_2 are the Fourier transforms of the time histories of the incident and transmitted waves respectively.

The theoretical characteristics of the volume damper can be calculated using a distributed parameter model, which simplifies to the determination of the transmittance coefficient T_t:

$$T_t = \frac{4}{A^2cos^2kL + B^2sin^2kL}$$

where: $A = 1 + \frac{S_C}{S_A}$ and $B = \frac{S_C}{S_B} + \frac{S_B}{S_A}$, S_A – section area of the inlet pipe, S_B – section area of the damper, S_C – section area of the outlet pipe, k – wave number, L – damper length.

Then, the transmission loss is determined by the relation:

$$TL = -10logT_t$$

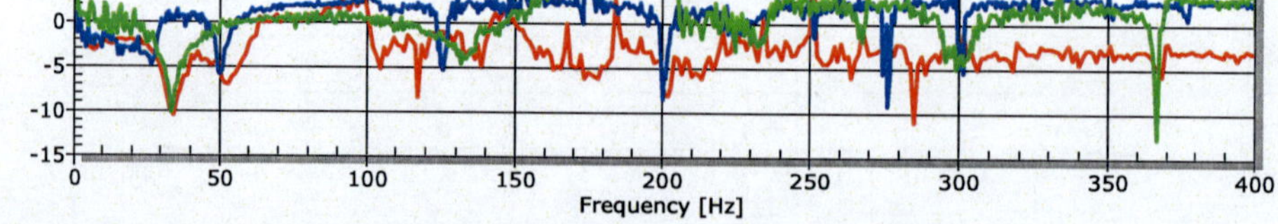

Fig. 4. Transmission loss characteristics for the system with a straight pipe at 1000 rpm, 1500 rpm and 2000 rpm.

The diagram (Fig. 4) shows the transmission loss function depending on the frequency in a system consisting of a straight pipe at compressor speeds of 1000 rpm, 1500 rpm and 2000 rpm. If the amplitude of the pulsation at the outlet is much higher than the amplitude of the pulsation at the inlet, then the transmission loss function reaches the minimum, which means resonance, i.e. amplification of the pulsation at a given frequency. The analysis of the transmission loss characteristics in a system with a straight pipe section indicates that the pressure pulsation is not damped in the system. For frequencies 33 Hz, 117 Hz, 200 Hz, 275 Hz, 283 Hz and 367 Hz, the excitation frequencies coincide with the natural frequencies of the system, causing resonance. The maximum attenuation function determines the highest attenuation at a given frequency. At 2000 rpm, pulsation suppression can be seen at 200 Hz.

The graph (Fig. 5) shows the transmission loss characteristics depending on the frequency in the installation with the tested chamber damper at compressor rotational speeds of 1000 rpm, 1500 rpm and 2000 rpm.

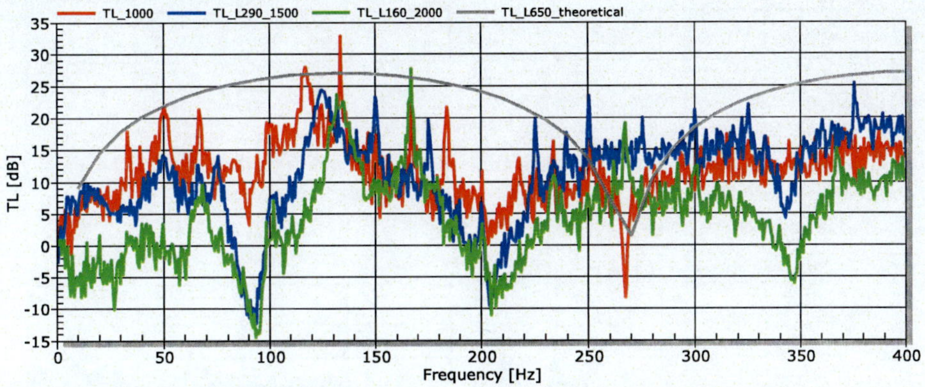

Fig. 5. Transmission loss characteristics for the system with a damper at 1000 rpm, 1500 rpm, 2000 rpm and theoretical transmission loss characteristic of the damper installed in the system.

The analysis of the damping function curve for a rotational speed of 1000 rpm indicates that it corresponds approximately to the theoretical characteristic. The analytically determined maximum attenuation is 27 dB at 134 Hz, while the actual maximum attenuation reaches 32 dB at the same frequency. The theoretical and actual resonant frequencies of the damper are 267 Hz. Introducing to the system a chamber damper with a volume larger than required should improve the attenuation of pressure pulsations due to increased acoustic compliance. The diagram of the transmission loss characteristics (Fig. 5) shows that, in fact, the oversized dimensions of the damper do not significantly affect the reduction of the pressure pulsation amplitudes in the system.

It should be emphasized that the selection of dimensions of the pressure pulsation damper for a compressor with a variable speed is carried out in relation to the lowest frequency of excitations (in this case the frequency of the first harmonic is 16.7 Hz for 1000 rpm). The attenuation range of the damper should include at least ten harmonics, in this case up to 167 Hz. In the analyzed case, the selected pulsation damper meets the above requirements. However, if the compressor operates at a higher speed, then, based on Helmholtz resonator theory, the dimensions of the damper can be much smaller. Assuming a constant diameter of the muffler ($D_W = 200$ mm), for a rotational speed of 1500 rpm the required length of the damper is 290 mm, and for a rotational speed of 2000 rpm – only 160 mm.

The diagrams in Figs. 6, 7and 8 show the actual transmission loss characteristics of dampers installed in the system and the theoretical transmission loss characteristics for compressor speeds of 1000 rpm, 1500 rpm and 2000 rpm.

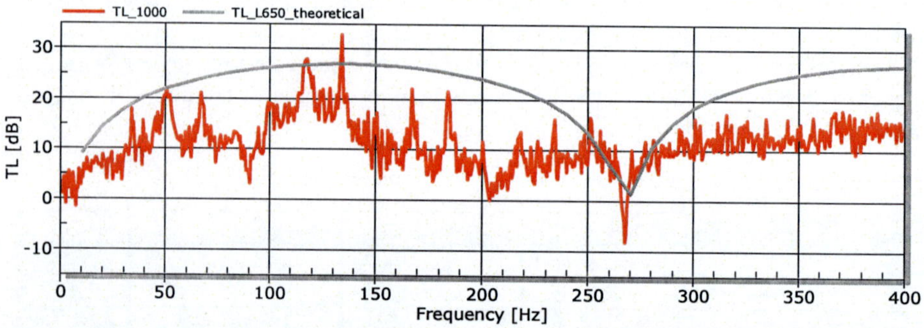

Fig. 6. Theoretical and measured transmission loss characteristics of a damper with the length of 650 mm selected on the basis of compressor rotational speed of 1000 rpm.

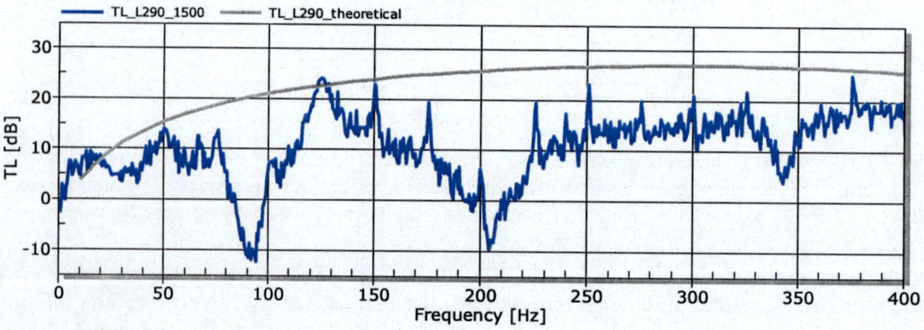

Fig. 7. Theoretical and measured transmission loss characteristics of a damper with the length of 290 mm selected on the basis of compressor rotational speed of 1500 rpm.

The analysis of the results presented in the graph (Figs. 7 and 8) suggests that the reduction in the volume of the damper for a higher compressor speed does not provide attenuation of pulsations in the required frequency range. The measured characteristics of dampers selected according to the compressor rotational speed of 1500 rpm and 2000 rpm show additional resonance frequencies at 93 Hz, 204 Hz and 345 Hz not resulting directly from the compressor excitation frequency, but caused by vibrations of other components of the installation. In addition, the smallest volume of the damper selected for the rotation speed of 2000 rpm does not ensure the suppression of pulsations

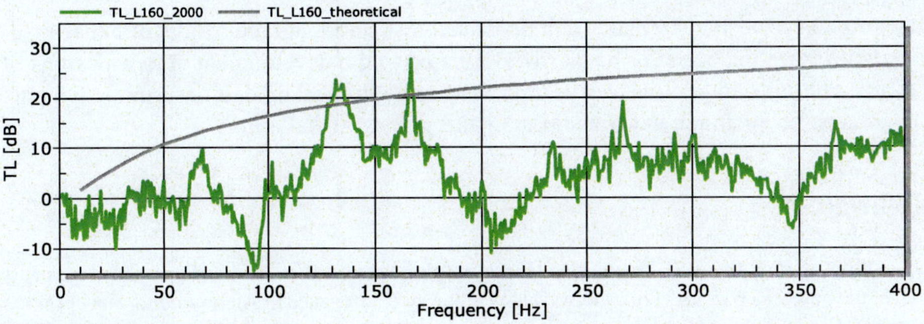

Fig. 8. Theoretical and measured transmission loss characteristics of a damper with the length of 160 mm selected on the basis of compressor rotational speeds of 2000 rpm.

in the range of 0–60 Hz, in which the fundamental frequency of compressor excitation is located (about 33 Hz), but strengthens them, which is also visible in the spectral analysis of the signal presented in Fig. 9.

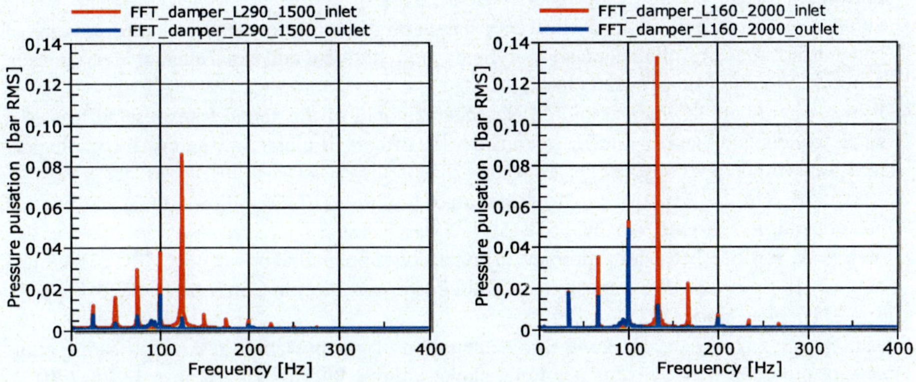

Fig. 9. Amplitude-frequency spectra of pressure pulsation in a system with a damper length of L = 290 mm for a rotational speed of 1500 rpm and in a system with a damper length of L = 160 mm for a rotational speed of 2000 rpm.

5 Conclusion

The selection of pulsation dampers only on the basis of theoretical calculations or experimental research without taking into account the fluid flow and actual operating parameters may lead to incorrect decisions regarding the required dimensions of the damper, and therefore does not ensure the required damping of pulsation amplitudes. Experimental research presented in the article directly indicates that analytical formulas and basic mathematical models may not allow the exact prediction of resonances resulting from system operational parameters. The presented methodology for determining the transmission loss characteristics at a dedicated test stand and the results of experimental studies form the basis for further work related to the development of new designs of variable volume dampers, as well as improved mathematical models describing dynamic phenomena occurring in gas transmission and process installations.

References

1. Atkins, K.E., Pyle, A.S., Tison, J.D.: Understanding the pulsation & vibration control concepts in the new API 618 fifth edition. In: Gas Machinery Conference in Albuquerque, New Mexico (2004)
2. Soedel, W.: Sound and Vibrations of Positive Displacement Compressors. CRC Press (2007)
3. API STD 618 Reciprocating Compressors for Petroleum, Chemical, and Gas Industry Services, Fifth Edition, Errata 1 2009, Errata 2 2010. American Petroleum Institute (2007)
4. Kollek, W., Osiński, P., Warzyńska, U.: Comparison of methods for modeling pressure pulsation dampers (in polish). In: Hydraulics and Pneumatics 2016: International Scientific-Technical Conference, pp. 59–65. SIMP, Wroclaw (2016)
5. Warzyńska, U., Kollek, W.: Modelling of transmission loss characteristics of reactive pulsation dampers. In: Steinrück, H. (ed.) 11th EFRC Conference 2018, pp. 198–205. EFRC, European Forum for Reciprocating Compressors, Wien (2018)
6. Selamet, A., Radavich, P.M.: The effect of length on the acoustic attenuation performance of concentric expansion chambers: an analytical, computational and experimental investigation. J. Sound Vib. **201**(4), 407–426 (1997)
7. Byrne, K., Skeen, M., Kessissoglou, N.: Measurement of the sound transmission loss of a small expansion chamber muffler to consider the effects of mean flow and wall compliance. In: Proceedings of Acoustics, pp. 257–263 (2006)
8. Kadam, P., Kim, J.: Experimental formulation of four poles of three-dimensional cavities and its application. J. Sound Vib. **307**, 578–590 (2007)
9. Fiebig, W, Wróbel, J.: System approach in noise reduction in fluid power units. In: Proceedings of the 2018 Bath/ASME Symposium on Fluid Power and Motion Control, FPMC2018, pp. 1–8. ASME, New York (2018)
10. Liu, B., Feng, J., Wang, Z., Peng, X.: Attenuation of gas pulsation in a reciprocating compressor piping system by using a volume-choke-volume filter. J. Vib. Acoust. **134**(5) (2012). https://doi.org/10.1115/1.4006234
11. Jia, X., Liu, B., Feng, J., Peng, X.: Attenuation of gas pulsation in the valve chamber of a reciprocating compressor using the Helmholtz resonator. J. Vib. Acoust. **136**, 5 (2014)

Influence of Pressure Inside a Hydraulic Line on Its Natural Frequencies and Mode Shapes

Jakub Wróbel[1](✉) and Jędrzej Blaut[2]

[1] Wrocław University of Science and Technology, Łukasiewicza 7/9, 50-370 Wrocław, Poland
jakub.wrobel@pwr.edu.pl

[2] AGH University of Science and Technology, aleja Adama Mickiewicza 30, 30-059 Kraków, Poland
blaut@agh.edu.pl

Abstract. Vibrational behaviour of hydraulic power system elements is crucial in aspects of noise problems. Hydraulic lines used in engineering systems to transfer hydraulic energy are also a structural and fluid coupling between vibrating elements of the entire system. Excitation generated in different places of the hydraulic systems due to external loads, control elements or flow pulsation are all transferred through hydraulic lines on which resonances may occur. Experimental investigations were undertaken in order to identify the influence of hydraulic pressure on the natural frequencies and mode shapes of a basic hydraulic line model. Modal analysis was conducted on a hydraulic line at several pressure increments to identify the change in natural frequencies and mode shapes.

Keywords: Hydraulic system · Vibration · Natural frequencies · Hydraulic line.

1 Introduction

Hydraulic systems are commonly used in various branches of industry due to several advantages over electrical, mechanical or pneumatic systems. High energy density, flexibility in system shaping, reliability, ease of automation - some of many advantages that contribute to the popularity of hydraulic systems. On the other hand, there are some drawbacks of hydraulic power systems such as equipment maintenance, medium conditioning, heat generation, and last but not least noise. In many cases the participation of hydraulic equipment in production site noise emission is significant especially if the noise exposure time effect is taken into account [1].

Main focus of hydraulic equipment manufacturers and researchers is scoped on the better reliability and higher operational pressures of hydrostatic pumps [2–7] and control elements [8–12] and acting elements [13–17]. Some researchers address the problem of noise and vibration of hydraulic pumps, conducting geometry optimization with the criterion of lowest noise levels and allowable pump housing strength [11]. Additional work regarding the dynamics of hydraulic system elements such as pumps and valves is presented in the following literature [17–20].

© The Editor(s) (if applicable) and The Author(s), under exclusive license
to Springer Nature Switzerland AG 2021
J. Stryczek and U. Warzyńska (Eds.): NSHP 2020, LNME, pp. 333–343, 2021.
https://doi.org/10.1007/978-3-030-59509-8_30

Noise generated by hydraulic systems can be categorized into structure-borne noise, fluid-borne noise and air-borne noise [21–29]. Air-borne noise mainly connected to the operation of electric motor cooling fans and air heat exchanges commonly used to reduce hydraulic fluid temperature. Structural vibrations are coupled with fluid-borne noise due to the hydraulic system specific. System elements vibrations caused by normal operation or external loads are transmitted through the system structure as well as through the hydraulic medium which fills the system. Approaching the phenomenon from the other side, pressure ripple or pressure peaks are transmitted through oil to the structure of the system where noise can be emitted [21, 22, 26].

Research indicates the hydraulic pump as the main source of hydraulic system internal vibration [24, 27] both structure and fluid borne. Reduction of noise at the hydraulic power pack was investigated in previous own work [28], however structural vibrations as well as pressure ripple that do not cause noise at the power pack can be transmitted to valves or acting elements via hydraulic fluid and flow channels.

Hydraulic lines, either rigid steel pipelines or flexible hoses, transfer the hydraulic energy as well as structure and fluid borne noise between system components. It was observed in laboratory experiments that the resonance vibrations of a hydraulic line can be transferred through mounting elements and induce several times larger structures to emit high levels of noise. Several researchers undertaken the investigations of hydraulic hose and pipeline vibrations. The research focuses mainly on experimental investigations of steel or fibre braided flexible hoses vibration transfer function [29]. Additional research on micro hydraulic flexible hoses was present in [30] where the effect of flow rate and pressure on the hose natural frequency was shown. Another literature position concerns with analysis and vibration modelling of simply supported flexible and rigid pipelines [31]. Other researchers indicated that the external loading exerted on large technical object can influence its dynamic characteristics [32].

FEM modal analysis of structural elements is commonly used in case of mechanical systems resonance problems. The same principle could be applied to hydraulic lines although the influence of oil and pressure should be investigated and eventually implemented into FEM simulations. The point of the research presented in this paper is to investigate the influence of oil and static pressure value in a commercial rigid hydraulic line on its natural frequencies and mode shapes.

2 Experimental Investigations

Two commercially available St 37.4 steel hydraulic lines with outer diameters of 16 (mm) and 10 (mm) and wall thickens respectively of 2 (mm) for the first one and 1.5 (mm) in case of the smaller one were chosen for experimental investigations. Each 1500 (mm) long line was firmly attached at its end to a massive foundation. One side of the line was caped and the other one was connected to a laboratory hydraulic hand pump used to introduce hydraulic pressure. Fixed attachment conditions shown in Fig. 1 were chosen for experiments due to the resemblance to actual hydraulic line mounting procedure. It is clear that in case of industrial setups additional support clamps are recommended by manufacturers and the spacing between them depends from the pipe diameter. In case of a 10 (mm) and 16 (mm) lines the recommended support spacing is respectively about 1(m) and 1.5(m) for larger diameters. During all experiments The supports were set to 1.5 (m) distance in order to allow for a more direct result comparison.

First measurements were carried out without the presence of oil in the system in order to instigate the frequencies and mode shapes of the line structure. Next set of measures was conducted after the line was filled with oil, air bladed and sealed. Experimental modal analysis was carried out at static pressure of 5, 10 15, 20, 25 and 30 (MPa) for each specimen.

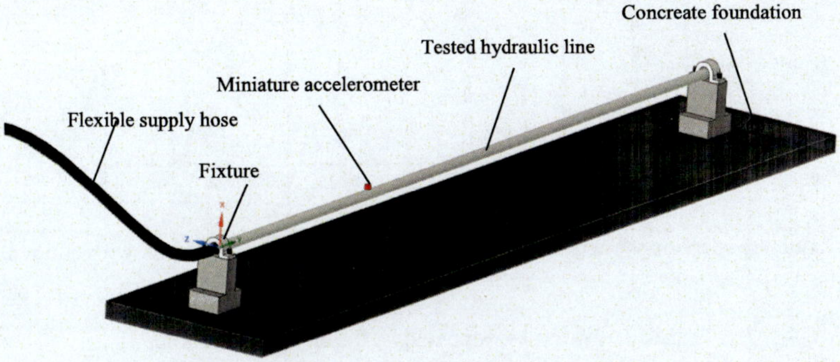

Fig. 1. Experimental investigation test setup.

Impact hammer (PCB 086C01) excitation was chosen for the modal analysis with eight acceleration measurement points at each tested hydraulic line. The excitation point was stationary while a three axial accelerometer (PCB J356A03) was moved with each measurement. Impact force was controlled by use of modal hammer impacting rig which guided the hammer and allowed similar excitation forces between measurements. Data acquisition system was set to 500 (Hz) bandwidth with 3200 spectral lines and three averages at each measurement points. A simple line geometry model was prepared in Spectral Dynamics Star Modal software for the purpose of mode shape representation and investigations of its eventual change due to introduction of oil and pressure. Signal post processing was used to compare the summarized FRFs– Mode Peaks for each test case.

Figure 2 presents the summarized FRFs acquired for the 10 mm hydraulic line without oil and after the introduction of oil. Each peak indicates a natural frequency of the hydraulic line with corresponding mode shape overlaid on the plot. As expected in case of a beam like element, each next mode shape is a representation of sine function, where higher modes are evolving to more complex sine shapes.

The introduction of hydraulic fluid did not change any of the mode shapes. It can be seen that the natural frequencies are shifted towards lower values when oil is introduced into the line. The change in frequencies presented in Table 1 is especially visible at frequencies above 100 (Hz). Based on literature as well as own experience and observations in case mechanical systems with low damping even small change in excitation frequency can lead to resonance or out of it. At higher frequencies the change in natural frequency value can significantly influence the vibrations of the hydraulic line. It is also worth mentioning that the presence of oil increased the frequency response function values at peaks.

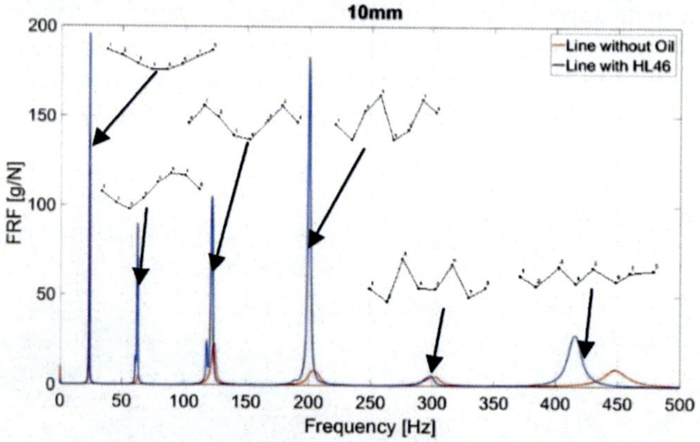

Fig. 2. Frequency response functions (FRF) with mode shapes–10 mm line with and without oil.

Table 1. Natural frequencies–10 mm hydraulic line.

	1st. mode (Hz)	2nd. mode (Hz)	3rd. mode (Hz)	4th. mode (Hz)	5th. mode (Hz)	6th. mode (Hz)
Empty line	23.75	63.59	123.9	203.9	302.3	447.7
Oil filled	23.28	62.66	122	199.7	299.1	415.5
Difference	−0.47	−0.93	−1.9	−4.2	−3.2	−32.2

Results acquired for the 16 (mm) line are presented Fig. 3. where summarized FRFs without oil and after the introduction of oil were compared. Similarly, to previous case the introduction of oil did not change any of the mode shapes. It can be also observed

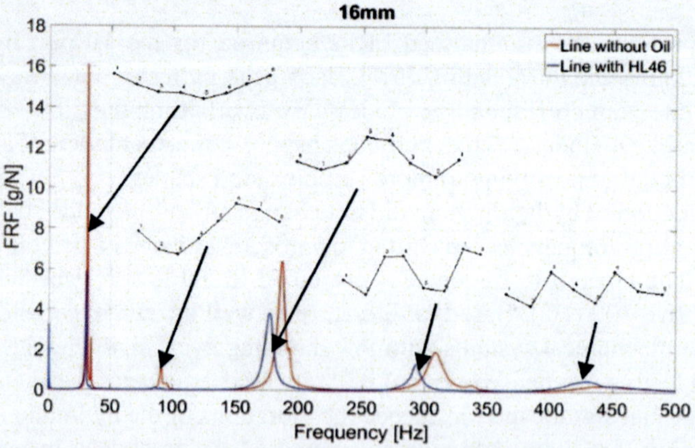

Fig. 3. Frequency response functions with mode shapes – 16 mm line with and without oil.

that the natural frequencies shift towards lower frequency values once again, especially in the higher frequency range.

Table 2. Natural frequencies – 16 mm hydraulic line.

	1st. mode (Hz)	2nd. mode (Hz)	3rd. mode (Hz)	4th. mode (Hz)	5th. mode (Hz)
Empty line	31.56	91.09	186.6	308.9	446
Oil filled	30.63	86.72	176.9	293	428.6
Difference	−0.93	−4.37	−9.7	−15.9	−17.4

It can be seen in Table 2 that in case of the 16 (mm) line the shift of frequencies is greater than in case of the 10 (mm). Differences are significant starting from the second natural frequency. Contrary to the 10 (mm) line presence of oil decreased the response function values at peak and in case of the 2nd mode strong reduction in value is visible.

Further analysis deals with the influence of static pressure on the natural frequencies. Both specimen ware loaded with static hydraulic pressure increments to investigate eventual change in frequency response functions. All presented pressure values are relative pressures. Calculated mode shapes did not differ from ones presented in Fig. 2 for the 10 (mm) and in Fig. 3 for the 16 (mm) specimen and remained unchanged throughout the incremental pressure increase.

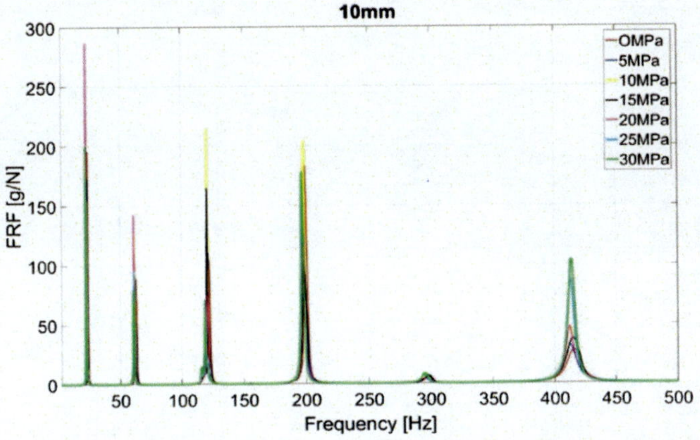

Fig. 4. Frequency response functions 10 (mm) – Incremental pressure increase.

At first glance no visible changes in frequency signals are visible in Fig. 4 thus additional plots were prepared in Fig. 5, Fig. 6, Fig. 7 which present frequency ranges accordingly to present modes. It is visible that the increase of static pressure shifts the peak FRF values towards lower frequencies. This phenomenon can be seen in case of first five mods. In case of other higher mods, the tendency to shift towards lower frequencies

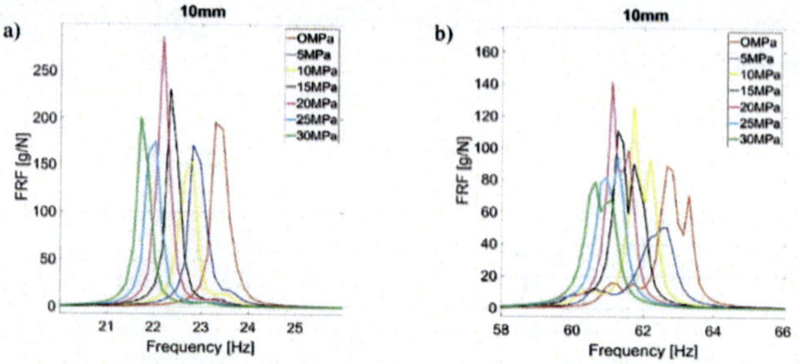

Fig. 5. Frequency response functions 10 (mm) - narrow frequency range: a) 1st Mode, b) 2nd mode.

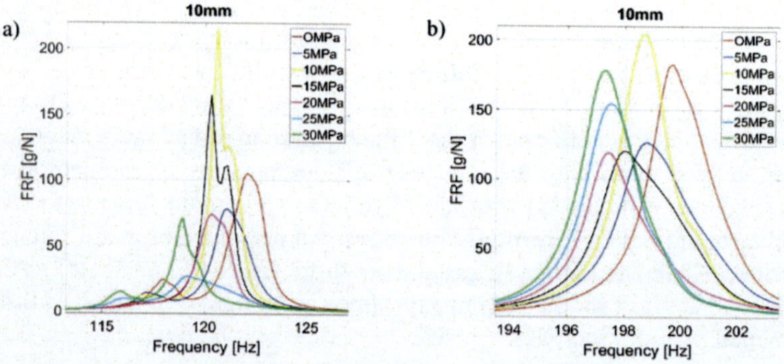

Fig. 6. Frequency response functions 10 (mm) - narrow frequency range: a) 3rd Mode, b) 4th Mode.

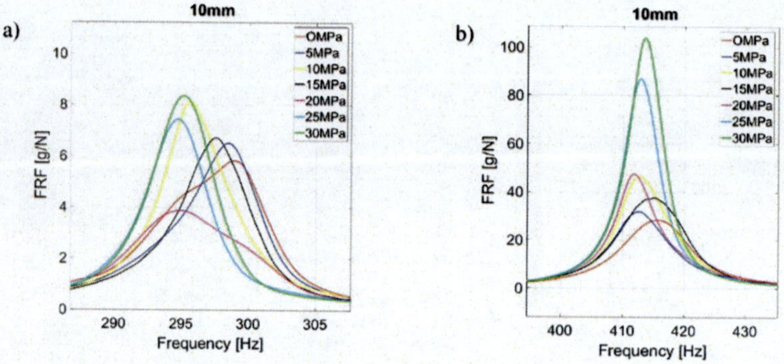

Fig. 7. Frequency response functions 10 (mm) - narrow frequency range: a) 5th Mode, b) 6th mode.

with increasing pressure is noticeable, although the not as clear as in case of the three first mods. Last sixth mode does present some changes in frequency due to pressure but contrary to previous mods the decrease in frequency value is not always increasing with pressure increment.

Second set of measurements was carried out for the 16 mm hydraulic line. As in case of previous measurements the broadband FRF representation does not indicate to any noticeable changes in frequencies at peak (Fig. 8).

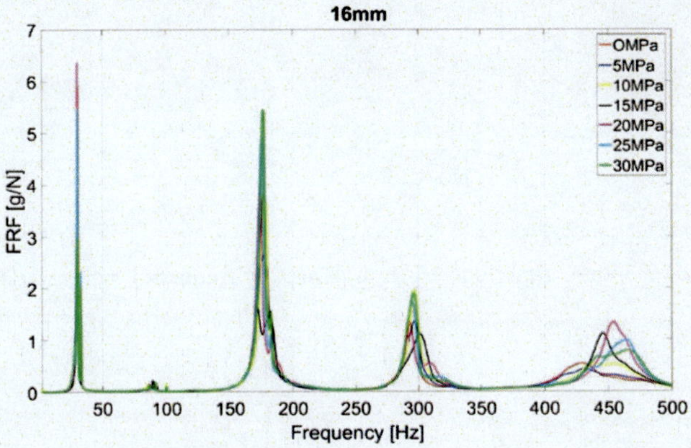

Fig. 8. Frequency response functions 16 (mm) – Incremental pressure increase.

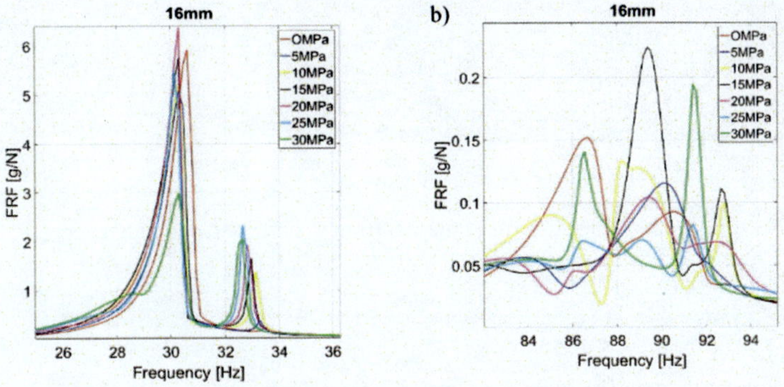

Fig. 9. Frequency response functions 16 (mm) - narrow frequency range: a) 1st mode, b) 2nd mode.

Additional plots were prepared for mode narrow frequency bands. In case of the first mode the changes in frequencies are visible but no trend can be observed. Second mode shows similar behaviour, the frequency shift tends towards increase and is stronger in case of mid pressures of 10 and 15 (MPa).

Błąd! Nie można odnaleźć źródła odwołania. a) presents the FRFs at third mode. Similarly, as in case of previous two modes the change in frequency under load is present although no trends can be observed. In case of the fourth mode presented in **Błąd! Nie można odnaleźć źródła odwołania.**b) the increase of pressure is showing upshift of frequencies at peak although strongest change is taking place at 15 and 10 (MPa). Finally, the fifth mode is showing frequency shift towards higher values at peak and the change is increasing with pressure load.

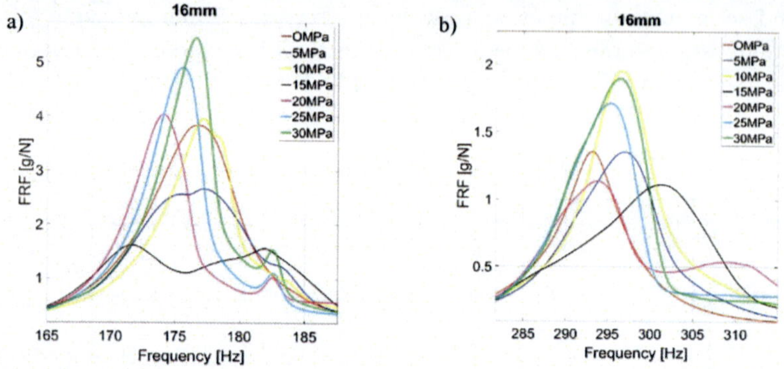

Fig. 10. Frequency response functions 16 (mm) - narrow frequency range: a) 3rd mode, b) 4th mode.

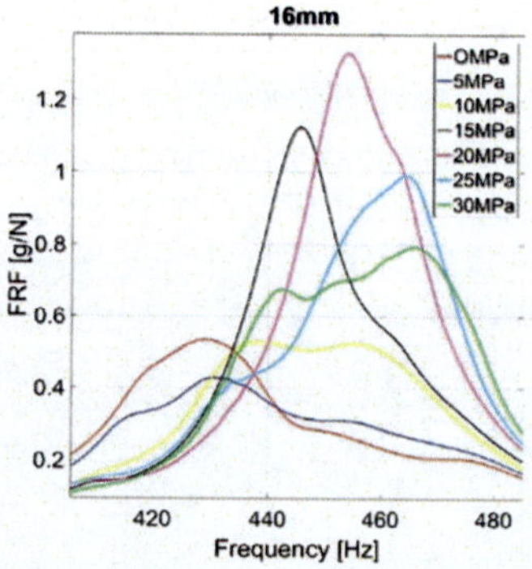

Fig. 11. Frequency response functions 16 (mm) - narrow frequency range: 5th mode

3 Discussion

Conducted experiments indicate that introduction of oil into hydraulic lines will change the natural frequencies of the system. The additional weight of oil should contribute to lowering of natural frequencies [33–36] of mechanical systems and this was also observed in case of investigated hydraulic lines. This statement, however obvious and self-evident shows that even small amount of oil can change the dynamic properties of hydraulic lines and at least the additional mass effect should be included into FEM simulations.

Additional investigations are needed in order to identify the eventual hydraulic line structural stiffness to oil mass ratio and the fluid structure interactions caused by oil presence. The stiffness of the 10 (mm) hydraulic line significantly lower than the 16 (mm) line on the other hand oil mass was equal to 150 (g) in case of the 16 (mm) and 50 (g) in case of 10 (mm) lines.

Results presenting hydraulic pressure induced natural frequency change should be treated as a preliminary investigation of this phenomenon. Additional tests are needed on larger population of specimen in order to provide more detailed explanation. In case of the

10 (mm) hydraulic line, incremental hydraulic pressure increase lead to a corresponding mode peak frequency downshift within 0–5 (Hz). The change in frequencies was weaker in case of last mode placed at higher frequencies.

Contrary to the observations stated above, the measurements carried out on the 16 (mm) hydraulic line did not indicate any consistent influence of the increased hydraulic pressure. Furthermore, at frequencies higher than 250 (Hz) the mode peak frequencies shifted upwards within 0–35 (Hz).

Measurement errors and uncertainties concerning experimental modal analysis were reduced by implementation of an impact hammer excitation rig which allow to apply the excitation force at repeatable location and at comparable force levels although the excitation force value should not influence the FRFs as long as excitation energy is sufficient and the investigated object is linear. Experimental results presented above may indicate that steel hydraulic lines filled with oil, at certain configuration of pipe diameter and wall thickens exhibit nonlinearities which should be investigated in future research.

Further investigations should be conducted with use of electrodynamic shaker and at even higher frequency resolutions in order to capture the eventual frequency changes in detail. A broader population of specimen is needed e.g. various internal and external diameter hydraulic lines in order dismiss uncertainties and define eventual nonlinearities.

References

1. Korbiel, T., Stepien, B., Batko, W., Baranski, R., Mleczko, D., Pawlik, P., Blaut, J.: Recognition of the 24-hour noise exposure of a human, archives of acoustics, vol. 42, pp. 601–607, Warsaw (2017)
2. Mucchi, E., Dalpiaz, G., Rivola, A.: Dynamic behaviour of gear pumps: effect of variations in operational and design parameters. Meccanica **46**, 1191–1212 (2011)
3. Kollek, W., Osiński, P.: Modelling and Design of Gear Pumps. Oficyna Wydawnicza Politechniki Wrocławskiej, Wrocław (2009)
4. Kollek, W., Osiński, P., Warzyńska, U.: The influence of gear micro pump body asymmetry on stress distribution. Pol. Marit. Res. **24**(93), 60–65 (2017)
5. Bosch Rexroth Corporation: Whisper pumps. VPV vane pumps – Catalog, Bethlehem (2014)
6. Bosch Rexroth Corporation, Close-Coupled Motor Pump Groups, Bethlehem (2004)
7. Voith Turbo GmbH & Co. KG, Motor/pump hybrid system EPAI for high and medium-pressure applications, Heidenheim (2008)
8. Pobędza, J., Sobczyk, A.: Properties of high-pressure water hydraulic components with modern coatings. In: Materials Engineering and Technology: selected, Peer Reviewed Papers from the 2013 International Conference on Advances and Trends in Engineering Materials and their Applications (ATEMA 2013), pp. 100–107. Trans Tech Publications Ltd. (2013)
9. Smid, G., Drew, J., Qatu, M., Dougherty, M.: Optimizing the power steering components to attenuate noise and vibrations. In: European Conference on Vehicle Noise and Vibration, London (1998)
10. Stosiak, M.: The modelling of hydraulic distributor slide-sleeve interaction. Arch. Civ. Mech. Eng. **12**(2), 192–197 (2012)
11. Bittner, U.: Structural optimization of axial piston pumps- modelling, validation and topology optimisation (in German), Karlsruhe technical papers – Institute of Engineering Mechanics, Scientific Publishing (2013)

12. Siwulski, T.: Study on the friction forces in a spool - sleeve pair of hydraulic directional control valve. In: Engineering Mechanics 2018: 24th International Conference: Extended Abstracts, 14–17 May 2018, pp. 761–764. Czech Republic, Institute of Theoretical and Applied Mechanics of the Czech Academy of Sciences, Svratka (2018)
13. Warzyńska, U., Siwulski, T.: Hydraulic cylinder with chamber fluid exchange support. In: 30th Conference on Construction machinery design challenges (in Polish), Zakopane, 24–27 January 2017, pp. 1–10, Institute of Mechanised Construction and Rock Mining, Warsaw (2017)
14. Siwulski, T., Warzyńska, U.: Numerical simulations of the liquid exchange process in a chamber of a new hydraulic cylinder solution in the aspect of the angular nozzle arrangement. Cuprum. **1**, 21–31 (2019)
15. Warzyńska, U., Siwulski, T.: Numerical analysis of the influence initial position of a piston on fluid exchange process in a hydraulic cylinder. In: Engineering Mechanics 2019: 25th international Conference: Book of Full Text, 13–16 May 2019, pp. 403–406. Institute of Thermomechanics of the Czech Academy of Sciences, Svratka (2019)
16. Mucchi, E., Dalpiaz, G.: Numerical vibro-acoustic analysis of gear pumps for automotive applications. In: Proceedings of ISMA2012-USD (2012)
17. Stosiak, M.: Vibration insulation of hydraulic system control components. Arch. Civ. Mech. Eng. **11**(1), 237–248 (2011)
18. Stosiak, M.: Ways of reducing the impact of mechanical vibrations on hydraulic. Arch. Civ. Mech. Eng. **15**(2), 392–400 (2015)
19. Skaistis, S.: Noise Control in Hydraulic Machinery. Marcel Dekker Inc., New York (1988)
20. Crocker, M.J.: Handbook of Noise and Vibration Control. Wiley, Hoboken (2007)
21. Li, B., Moore, S.: Noise control for fluid power systems. In: Internoise (2014)
22. Fiebig, W.: Noise reduction in hydraulic systems (in German). In: VDI Progress reports, no. 11, p. 337 (2008)
23. Edge, K.A., Johnston, D.N.: New method for evaluating the fluid borne noise of hydraulic components. In: 7th International Fluid Power Symposium, Bath (1986)
24. Fiebig, W.: Location of noise sources in fluid power machines. Int. J. Occup. Saf. Ergon. (JOSE) **13**(4), 441–450 (2007)
25. Fiebig, W.: Noise control of fluid power units. In: 23rd International Congress on Sound and Vibration, 10–14 July, Athens (2016)
26. Johnston, D.N.: Generation and transmission of fluid borne noise. In: IMechE Seminar on Environmental Impact of Fluid Power Systems, Bath (1999)
27. Chenxiao, N., Xushe, Z.: Study on vibration and noise for the hydraulic system of hydraulic hoist. In: Proceedings of 2012 International Conference on Mechanical Engineering and Material Science (MEMS) (2012)
28. Fiebig, W., Wróbel J.: System approach in noise reduction in fluid power units. In: Proceedings of the 2018 Bath/ASME Symposium on Fluid Power and Motion Control, FPMC2018, 16–19 October 2018, pp. 1–8. ASME, Bath, New York (2018)
29. Fiebig, W., Harla, K.: Vibration transsmission in hydaraulic hoses. In: Construction Machinery Design Challenges –23rd Conference, Zakopane, 25–28 January 2010, pp. 1–14, Lodz University of Technology Publishing , Łódź (2010)
30. Bocian, M., Lubecki, M., Stosiak, M.: Experimental investigations of the vibration of hydraulic microhoses. In: Engineering Mechanics 2018: 24th International Conference, 14–17 May 2018, pp. 97–100. Institute of Theoretical and Applied Mechanics of the Czech Academy of Sciences, Svratka (2018)
31. Czerwiński, A., Łuczko, J.: Vibrations of steel pipes and flexible hoses induced by periodically variable fluid flow. Mech. Control **31**(2), 63–71 (2012)

32. Pietrusiak, D., Smolnicki, T., Stańco, M.: The influence of superstructure vibrations on operational loads in the undercarriage of bulk material handling machine. Arch. Civ. Mech. Eng. **17**(4), 855–862 (2017)
33. Norton, M.P., Karczub, D.G.: Fundamentals of Noise and Vibration Analysis for Engineers, 2nd edn. Cambridge University Press, New York (2003)
34. Everest, A.F.: Master Handbook of Acoustics. McGraw-Hill (2001)
35. Ewins, D. J.: Modal Testing: Theory and Practice, Research Studios Press Ltd. (1984)
36. Avitabile, P.: Experimental Modal Analysis, Sound and Vibration, Lowell (2001)

Check for updates

Study of the Structural Materials Cavitation Strength in Ultrasonic Technological Equipment

Andrii Zilinskyi(✉), Oleksandr Luhovskyi, Ihor Gryshko, and Vasyl Kovalev

National Technical University of Ukraine, "Igor Sikorsky Kyiv Polytechnic Institute", Prosp., Peremohy 37, Kiev 03056, Ukraine
zilinski.andrew@gmail.com, {atoll-sonic,griwko}@ukr.net, vaskovalev@gmail.com

Abstract. The article considers the problems of ensuring the durability of technological equipment for ultrasonic cavitation. The results of planning and conducting an experimental investigation of the structural materials erosive destruction from which ultrasonic emitters and parts of cavitation chambers are presented. The study allowed us to determine the most preferred materials in terms of cavitational strength in the presence of fatigue stresses.

Keywords: Ultrasonic vibrations · Ultrasonic cavitation · Experiment planning · Cavitational strength of materials · Cavitation chambers

1 Introduction

In the fields of mechanical engineering, chemistry, food industry, agriculture and medicine etc., technologies widely use the effects that accompany the phenomenon of ultrasonic cavitation in a technological fluid. Such technologies include dispersion, cleaning of hard and elastic surfaces, fine spraying, filtering, mixing at the molecular level, degassing, activating liquids, extraction, disinfection, etc. [1–3].

These technologies use the effects that occur when an ultrasonic wave is introduced into a liquid. In the rarefaction phase of the ultrasonic wave, cavitation bubbles are formed from cavitation nuclei present in the liquid, which, having made one or more oscillations, collapse in the compression phase. Moreover, if the bubble collapses in an unlimited volume of liquid, then spherical shock waves diverge from it. If the collapse occurs near some surface, for example, near the cavitation chamber, the surface of a microorganism or an adjacent bubble, due to the curvature of the velocity field and are capable to destroy these surfaces then energy-intensive cumulative jets are formed.

In addition, during the collapse of cavitation bubbles, a local increase in temperature up to 1000 °C and pressure up to 1000 MPa is observed, and oxidative processes in the cavitation area are also activated [4]. Oscillating and collapsing cavitation bubbles create intense microflows that can mix components at the molecular level [5].

These accompanying ultrasonic cavitation effects can improve the efficiency and quality of technological processes. However, when creating ultrasonic cavitation technological equipment, problems arise associated with the durability of the ultrasonic

© The Editor(s) (if applicable) and The Author(s), under exclusive license to Springer Nature Switzerland AG 2021
J. Stryczek and U. Warzyńska (Eds.): NSHP 2020, LNME, pp. 344–352, 2021.
https://doi.org/10.1007/978-3-030-59509-8_31

transducers emitting surfaces, as well as with the durability of the cavitation chambers surfaces themselves.

In some cases, ingress of wear products on the surfaces of technological equipment into the process fluid is unacceptable. This applies, for example, to the technological process of extracting important medicinal components from plant materials in the pharmaceutical industry. When creating filters with the effect of self-regeneration, cavitational erosion of a filter element made of a thin mesh reduces the life of equipment [6–8].

The purpose of this article is to determine experimentally the resistance of various structural materials to ultrasonic cavitation to ensure the necessary durability of technological equipment.

2 Development of Procedure for Conducting Multivariate Experiment and Assessment of Measurement Accuracy

Preliminary experimental studies have shown that the relative consumption of structural material is a function of following three main parameters:

$$\Delta V = f(N, t, E)\ [\%] \tag{1}$$

where: N - power of cavitation equipment, W; t is the duration of cavitation exposure, min; E is Young's modulus, GPa.

Taking in account the large volume of experiments in a one-factor study, it is advisable to study the influence of these factors on the relative volume consumption of structural materials in the form of a multivariate experiment to obtain the regression equation for the response function — the relative volume consumption of structural materials using a multivariate experiment of type 23 using the Box-Wilson method [9, 10].

The choice of variation ranges of the function (1) factors was carried out in such way that any combination of them, in the ranges provided for by the experimental plan, could be realized and would not lead to contradictions. So, search experiments were carried out, which made it possible to determine the area in which the necessary combinations of factor levels would be stably implemented.

All factors that enter to function (1) are values having different dimensions and different orders. Therefore, to obtain the response surface of this function, the factor coding operation performed, which is a linear transformation of the factor space [9] according to the formula:

$$x_j = \frac{\left(\tilde{X_j} - \tilde{X_{j0}}\right)}{I_j} \tag{2}$$

As a result of search experiments, the following values were determined for each factor: X_{j0} - the main level of the factor; X_{jmax}, X_{jmin} - the upper and lower levels of the factor; αX_{jmax}, αX_{jmin} - maximum and minimum factor levels; - starry shoulder; L_j is the interval of variation.

Table 1. Levels of factors and variation ranges.

Factors	Factor levels					Variation interval
	−1.681	−1	0	+1	+1.681	
x_1 – power of cavitation equipment, W	100	160.8	250	339.2	400	89.2
x_2 – duration of cavitation exposure, min	0	48.6	120	191.4	240	71.4
x_3 – Young's modulus, GPa	55.2	87.6	135.1	182.6	215	47.5

The following values of factor levels were established on a conditional scale: minimum –1, average 0, maximum +1 and stellar values –1.681, +1.681. The true values of the factors determined during the search experiments are presented in Table 1.

Preliminary experiments showed that the dependence of the relative volume flow of structural materials on parameters of ultrasonic cavitation process cannot be adequately described by either a linear or quadratic regression model. Therefore, there was decided to present a regression model from exposure factors with fractional degrees. The values of fractional degrees were results of preliminary search experiments. Thus, the mathematical model can represent as:

$$\Delta V = f\left(N^{0,7306}, t^{2,282}, E^{1,155}\right)[\%] \tag{3}$$

To select the values of such an impact factor as Young's modulus, structural materials with close values of this parameter were selected (Table 2).

Table 2. Young's modulus of structural materials

Structural materials	TEFLON 7	Class30B	C37000	AISI 321	30HGSA
E – Young's modulus, GPa	55.2	100	105	196	215

The required number of experiments for the case of quadratic regression determine by the formula:

$$N = 2^k + 2k + N_0(k) \tag{4}$$

where: k is number of factors; 2 - variation levels number; N_0 is the number of experiments in the plan center, N_0 (k) = N_0 (3) = 6.

For our case, there were $N = 2^3 + 2 \cdot 3 + 6 = 20$ experiments.

To carry out rotatable central compositional planning (RCSC) of the second order on the basis of a full-factor experiment of type 2^3, an experimental planning matrix was compiled.

It was planned to obtain the following quadratic regression model with first-order interaction effects:

$$y = b_0 + b_1 x_1 + \ldots + b_3 x_3 + b_{12} x_1 x_2 + \ldots + b_{23} x_2 x_3 + b_{11} x_1^2 + \ldots + b_{33} x_3^2 \quad (5)$$

where: y is the objective function (relative consumption of structural materials volume); b_0, b_1, ..., b_3, b_{13}, ..., b_{23}, b_{11}, ..., b_{33} are the regression coefficients.

To determine the current estimates b_0, b_1, ..., b_3, b_{13}, ..., b_{23}, b_{11}, ..., b_{33}, the least squares method was used [10]:

$$B = Y\Phi^{-1} \quad (6)$$

where: $B = \begin{bmatrix} b_0 \\ b_1 \\ \ldots \\ b_k \end{bmatrix}$ – matrix that contains regression coefficients;

$\Phi = F^T F$ – Fisher information matrix; $Y = \begin{bmatrix} y_1 \\ y_2 \\ \ldots \\ y_N \end{bmatrix}$ – matrix that contains the results of experiments according to the experiments planning matrix; $F = \begin{bmatrix} 1 & x_{1,1} & \ldots & x_{1,k} \\ 1 & x_{2,1} & \ldots & x_{2,k} \\ \ldots & \ldots & \ldots & \ldots \\ 1 & x_{N,1} & \ldots & x_{N,k} \end{bmatrix}$ - a matrix that contains the factors values $x_{i,j}$ (where i is the experiment number according to the planning matrix, j is the factor number); k is the factors number; N is the experiments number according to the experiment planning matrix.

The adequacy of the regression model tests by the Fisher criterion [11]:

$$F = \frac{S_{ad}^2}{S_{rep}^2} \leq [F(f_1, f_2)], \quad (7)$$

where: S_{ad}^2 - dispesion of adequacy; S_{rep}^2 - reproducibility dispersion; $[F(f_1, f_2)]$ is the critical value of the Fisher criterion, which is equal to the value of the Fisher distribution; $f_1 = N - d$ is the number of freedom degrees of the dispersion adequacy; $f_2 = N_{0-1}$ is the number of degrees of reproducibility freedom dispersion; d is the number of significant regression coefficients (5).

The calculated value of the criterion F compare with the critical one and for $F > [F(f_1, f_2)]$ the regression model was considered inadequate.

The adequacy variance determine by the formula [11]:

$$S_{ad}^2 = \frac{1}{f_i} \Sigma_{i=0}^{N} \left(y_i - \widetilde{y_i}\right)^2 \quad (8)$$

where: y_i is the result of i-th experiment conducted according to the planning matrix; $\widetilde{y}_i$ - the result of i-th value of the experiment provided by the regression model (5).

The significance of the regression coefficients determine by the t-Student criterion [11]:

$$t_i = \frac{|b_i|}{S_{rep}\sqrt{c_{i,i}}} > [t(f_2)], \tag{9}$$

where: $[t(f_2)]$ is the critical value of t-Student test, which is equal to the Student distribution value; $c_{i,\,i}$ is the corresponding element of the matrix Φ^{-1}.

The calculated value of the criterion ti compares with the critical one and for $|t| \leq [t(f_2)]$ the second regression coefficient consider insignificant.

The number of repeated experiments at each point of the experimental design determine by the formula [11]:

$$n \geq \frac{1 + \gamma + 2n_{mes}}{1 - \gamma}, \tag{10}$$

where: γ is the confidence probability that the measurement error is within acceptable limits; n_{mes} - the number of measurements that are discarded.

According to the authors [12, 13] recommendations, the confidence probability when normalizing the quantile estimate of the resulting and random errors of the measuring technique select in the range (0.8 … 0.9). Then for $n_{mes} = 0$ we have:

$$n \geq \frac{1 + (0.8 \ldots 0.9)}{1 - (0.8 \ldots 0.9)} = 9 \ldots 19. \tag{11}$$

3 The Results of Multivariate Experiment

For the actual factor values, the regression equation after dropping the factors with insignificant regression coefficients for the response function - the relative consumption of structural materials volume according to the multivariate experiment has the form:

$$\Delta V = 0.0506N^{0,7306} - 0.00805E^{1,155} + 1.716 \cdot 10^{-7}N^{0,7306}t^{2,282} - 4.762 \cdot 10^{-5}N^{0,7306}E^{1,155} - 1.639 \cdot 10^{-8}t^{2,282}E^{1,155} - 3.658 \cdot 10^{-4}N^{1,4612} - 1.6 \cdot 10^{-11}t^{4,564} + 1.523 \cdot 10^{-5}E^{2,31}[\%] \tag{12}$$

A negative result obtained using the regression model (12) should be taken as zero, in which the structural material destruction under the influence of ultrasonic cavitation erosion does not occur. According to Student's criterion, all factors except the free term b_0 and x_2, first-order interaction effects, and quadratic effects turned out to be significant.

So, $S^2_{rep} = 0.0002202$; $S^2_{ad} = 0.0003654$; $F = 1.659 < [F] = 4.784$, that is, according to the Fisher criterion, the hypothesis of regression model adequacy (12) can be considered correct with 95% confidence. The correlation coefficient was $R = 0.951$, which indicates a high accuracy of the results.

From the according to Student's criterion among the studied parameters, the power of the cavitation device affects the relative of structural materials consumption most of all, and the duration of cavitation least of all.

In Fig. 1 it is shown the response surfaces of the objective function — the relative consumption of structural materials volume and their two-dimensional sections in the planes of influence parameters, which allow one to clearly illustrate the dependence of this function on individual influence parameters.

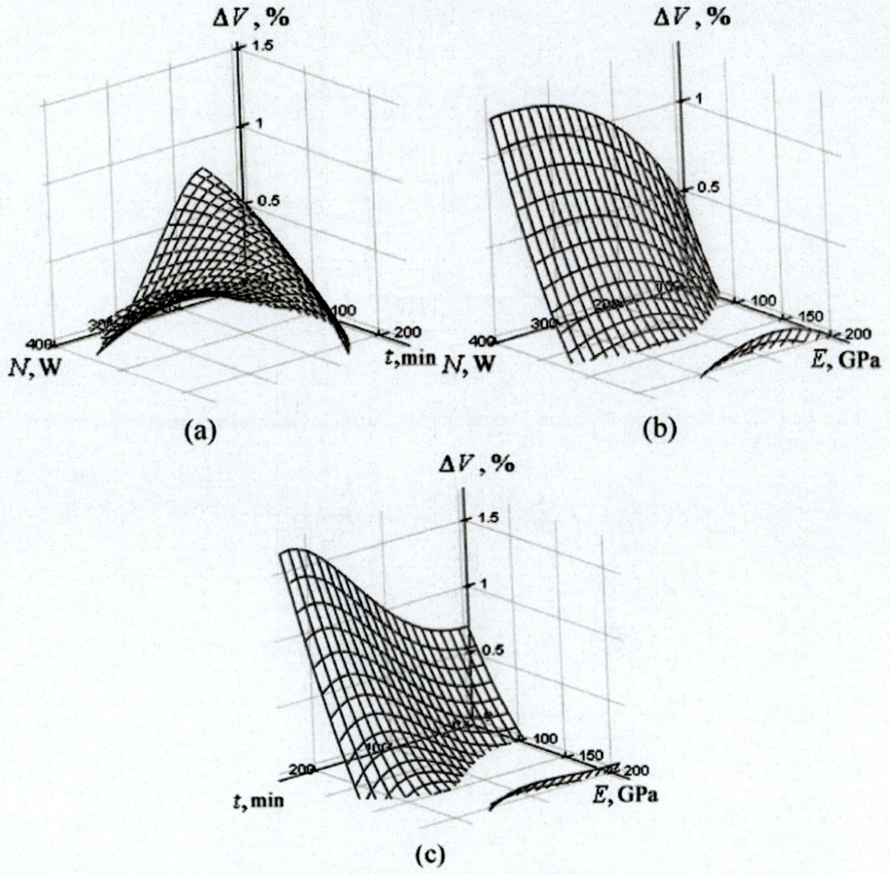

Fig. 1. The response surface of objective function is the relative consumption of structural materials volume and their two-dimensional sections in the planes of influence parameters: a)$\Delta V = f(N, t)$; b) $\Delta V = f(N, E)$; c) $\Delta V = f(t, E)$.

4 Results of a Full-Scale Experiment

Full-scale experiments carried out on a special bench (Fig. 2), which is a cavitation cylindrical flow chamber, on the outer forming surface of which piezoelectric ultrasonic linear displacement drives that excite the cylindrical chamber in a radial mode of oscillation were rigidly mounted. Such an excitation regime made it possible, due to the focusing properties of the chamber emitting inner surface, to create along the longitudinal axis of the chamber a region of developed cavitation generated by high-intensity ultrasound. The intensity controlls by the supply voltage of the ultrasonic transducers.

To increase the results reliability, a constant fluid flow provides in the cavitation chamber, which ensured a constant number of cavitation nuclei in the chamber (Fig. 3). The experiments carried out at a static pressure of 0.4 MPa and a constant temperature. In order to determine the material consumption after each experiment, the test sample weighed on a high-precision analytical balance of the AXIS A250 series. The results are

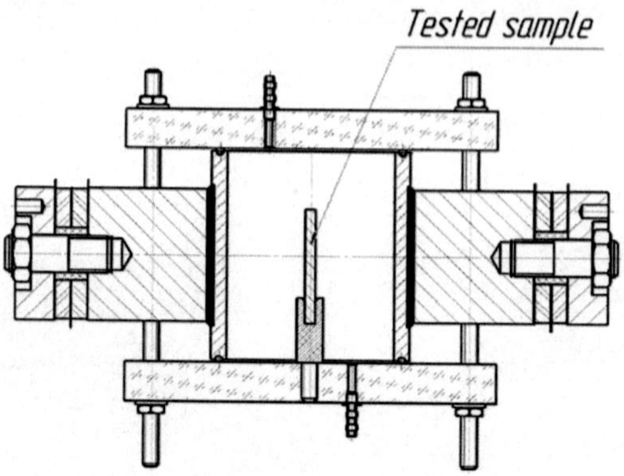

Fig. 2. Ultrasonic bench for experiments on cavitation erosion of structural materials.

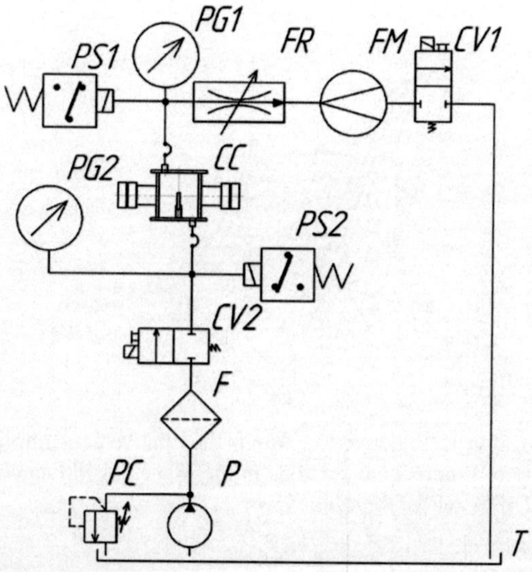

Fig. 3. Hydraulic diagram of the experimental bench. (C - cavitation chamber; CV1 - control valve 1; CV2 - control valve 2; F - filter; FM - flow meter; FR - flow control; P - pump; PC - pressure control; PG1 - pressure gauge 1 PG2 - pressure gauge 2; PS1 - pressure switch 1; PS2 - pressure switch 2; T – tank).

presented graphically in Fig. 4. According to the Student's criterion it was found, the power of the cavitation device most affected by the relative loss of structural materials volume. Cavitation duration is most not affected the relative loss of structural materials volume.

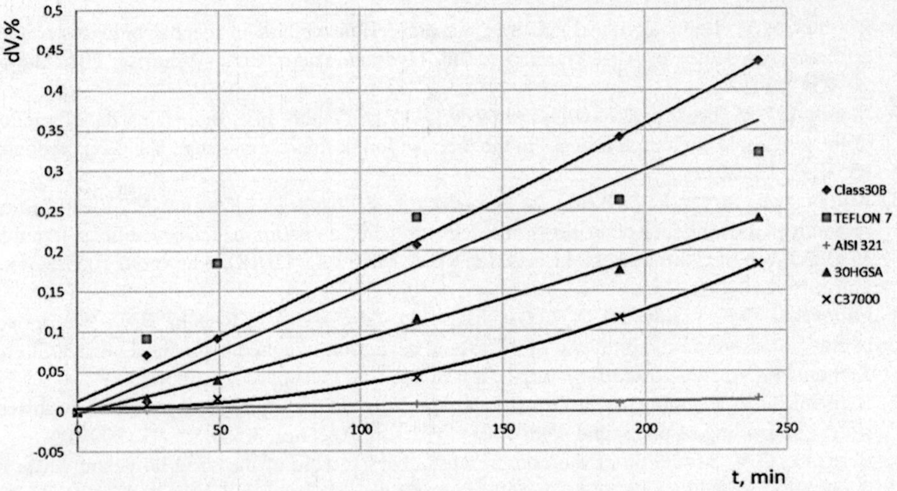

Fig. 4. The mass flow of structural materials under the action of ultrasonic cavitation with ultrasonic power emitters, 250 W.

5 Conclusions

The results obtained make it possible to evaluate the cavitation resistance of structural materials and make an informed choice of materials for use in ultrasonic technological equipment. To increase the durability and reliability of ultrasonic cavitation equipment, it is necessary to expand the number of tested materials types and increase the test duration, since the obtained results on structural materials consumption under the influence of cavitation erosion indicate that the influence of fatigue strength is characteristic for individual materials, even cavitation-resistant, upon reaching which there is an intensive material destruction.

References

1. Panov, A.P.: Ultrasonic Cleaning of High-Precision Parts. Mashinostroenie, Moscow (1984)
2. Luhovskyi, O.F., Chuhraev, N.V.: Ultrasonic cavitation in modern technologies [In Russian]. Kyiv, VPC "Kiev University", 244 (2007)
3. Yakhno, O.M., Lugovska, K.O., Movchaniuk, A.V.: Research of the possibilities of the technology of ultrasonic cavitation cleaning of elastic surfaces, Vestnik Nacional'nogo tehnicheskogo universiteta "Kievskij politehnicheskij institute". Serija mashinostroenie **58**, 234–240 (2010)
4. Fedotkin, I.M., Gulyy, I.S.: Cavitation, cavitation technology and technology, their use in industry (theoretical basis of production, calculations and design of cavitation heat generators) Part I. Kiev: JSC "OKO" (2000)
5. Bernik, I.N., Bernik, I.M., Lugovskiy, A.F., Movchanyuk, A.V., Laszok, A.V.: Ultrasound field parameters research in the process of hydrolysis-cavitation extraction pectin. In: Proceedings of the National Technical University of Ukraine "Kyiv Polytechnic Institute". A series of engineering. no (57), pp. 82–87 (2009)

6. Liashok, A.V., Luhovskyi, A.F.: Ultrasonic dispersion of fluid in mechatronic systems of artificial microclimate. All-Ukrainian J. Ind. Hydraul. Pneumatics - Vinnitsa, 2011 **34**(4), 20–25 (2011)
7. Marchuk, L.V., Prokopenko, O.F., Luhovskyi, O.F., Gryshko, I.A.: Increasing the efficiency of the ultrasonic cavitation process in the inactivation of microorganisms. Vib. Eng. Technol. **63**, 108–113 (2011)
8. Antoniak, P., Stryczek, J., Banas, M., Gryshko, I., Zilinskyi, A., Kovalov, V.: Visualization research on the influence of an ultrasonic degassing on the operation of a hydraulic gear pump, MATEC Web of Conferences 211. – 2018, VETOMAC XIV (2018). doi.org/10.1051/matecconf/201821103005
9. Luhovskyi, O.F., Luhovskyi, A.F., Gryshko, I.A., Zilinskiy, A.I., Patsola, B.V.: The impact of static pressure on the intensity of ultrasonic cavitation in aqueous media. J. Water Chem. Technol. **40**(3), 143–150 (2018). https://doi.org/10.3103/S1063455X18030050
10. Bereziuk, O.V.: Planning of a multifactorial experiment for the study of vibration hydropower-water compacting of municipal solid waste. Vib. Eng. Technol. **3**(55), 92–97 (2009)
11. Bereziuk, O.V.: Modeling of the compression characteristic of the solid domestic waste in the waste disposal vehicle on the base of the computer program "PlanExp". Bull. Vinnytsia Politechnic Inst. **6**, 23–28 (2016)
12. Levshina, E.S., Novitskiy, P.V.: Electric measurements of physical values: (Measuring converters). Energoatomizdat (1983)
13. Novitskiy, P.V., Zograf, I.A.: Assessment of the errors of the measurements results. Energoatomizdat (1991)

New Materials and Special Solutions in Fluid Power Technology

Volumetric Efficiency of a Hydraulic Pump with Plastic Gears Working with Tap Water

Michał Banaś(✉)

Wrocław University of Science and Technology,
ul. Łukasiewicza 5, 50-370 Wrocław, Poland
michal.banas@pwr.edu.pl
http://fprg.pwr.wroc.pl/

Abstract. Water is a natural working fluid, which is increasingly used in hydraulic systems due to environmental requirements. Making plastic parts allows us to enjoy the benefits of this liquid without introducing expensive materials or coatings. In the hydraulic pump, gears are a key component, therefore the type of material used to make them and the process of their production affect the operating parameters of the pump. The article presents the results of tests of volumetric efficiency of a water-supplied hydraulic pump with gears from a couple of materials (*PPS+GF40* and *PEEK*) made by two methods: injection moulding and machining. The results were compared with the plots for the pump with gears from *POM* when working with hydraulic oil. The influence of used materials and manufacturing methods on pump efficiency was discussed.

Keywords: Polymers · Composites · Gear pump · Injection moulding · Machining

1 Introduction

Different fluids are used in hydraulic systems. The most common are mineral oils from the HL group, but in some cases, non-flammable liquids are also used, e.g. emulsions (HFA, HFB), glycol-based liquids (HFC) and other high water-based fluids (HWBF). Water was a natural working fluid at the beginning of the development of hydraulic systems, but the benefits of other liquids, primarily mineral oils, limited its use. Currently, we observe the return to tap water as a working medium. Main reasons are widespread accessibility and environmental neutrality. The use of water, which is an aggressive agent for currently used materials, requires to introduce appropriate protective coatings [1] or other materials, e.g. stainless steel [2], which increase costs and are a barrier for the development of this direction.

One way to overcome these disadvantages is to use plastics. Development works on plastic components are carried out in many R&D centers. Research

© The Editor(s) (if applicable) and The Author(s), under exclusive license to Springer Nature Switzerland AG 2021
J. Stryczek and U. Warzyńska (Eds.): NSHP 2020, LNME, pp. 355–366, 2021.
https://doi.org/10.1007/978-3-030-59509-8_32

includes basic components such as pumps [3,4], actuators [5,6] and valves [7,8], but also others, for example pipelines [9]. In this paper, an attempt was made to answer the question, whether a hydraulic pump with plastic gears designed for working with oil can be used directly with tap water and what operating parameters can be expected due to the lower viscosity.

2 Hydraulic Systems from Plastics

The use of plastics in hydraulic systems has many advantages, which authors divide into designing, technological and economical. At the same time, they draw attention to limitations related to strength and stiffness, variability of properties and waste management [10]. Difficulties in modelling the impact of external factors on the material mean that research directions include both the use of basic polymers and the development of composites. Researchers modify polymers to achieve specific properties using chemical methods such as chain length change [9] and physical - by choosing the composition of additives [11]. They are also looking for structural composites using carbon and glass fibres in various forms [5] and using designing measures [12].

Three methods of manufacturing hydraulic elements from plastics can be pointed out: machining processing [10], injection moulding [8] and additive methods [7]. Machining ensures the best geometrical parameters of the manufactured elements, which affects the high hydraulic parameters of the systems. However, it is expensive and not cost-effective with industrial application. The most advantageous method of producing plastic components is injection moulding, which requires many components to get a low price. Geometric accuracy depends on the selection of proper process parameters. The choice in between are additive methods that allow us to get different shapes [13] while keeping a low price for a smaller number of made elements. However, there are problems with the roughness (surface finish) of such components, which often limit their use in hydraulic drives.

One of the key elements necessary to generate energy in the hydraulic system is a hydraulic pump. For a plastic gear pump, gears [12] and bodies [14] and their deformations are very important. Water, in addition to numerous environmental benefits over hydraulic oils, has the major drawback of being low viscosity (Table 1). The difference in density does not play a significant role, as it only increases the mass of liquid in the system. In contrast, the viscosity of water is over 150 times lower than oil at $\theta = 25$ °C. When changing oil to water, one should take into account the increase of leakages and decrease of volumetric efficiency. Lower water viscosity also limits the formation of a lubricating film in rotary joints and can increase friction.

Table 1. Selected parameters of oil and water at temperature $\theta = 25$ °C [15].

Fluid	Hydraulic oil	Water
Designation	HLP-68	Tap water
Viscosity (mm^2/s)	150	0.89
Density (kg/m^3)	870	997

3 Object of the Study

A hydraulic pump with cycloidal gears made of plastics, which is shown in Fig. 1, is discussed in this paper. A set of gears is formed by an external 4 and an internal 5 toothed gear. Gear 4 is located on the input shaft 6 and fixed with three keys 7, and wheel 5 is placed in a ring in the aluminium body 2. The pump does not include leakage compensation, therefore the discharge space in body 2 is limited directly by two covers: the front cover 1 and the rear cover 3. Gears 4 and 5 rotate and pump the liquid between the ports made in the rear cover 3 – from the input port (*IN*) at the pressure of p_1 to the output port (*OUT*) at the pressure of p_2.

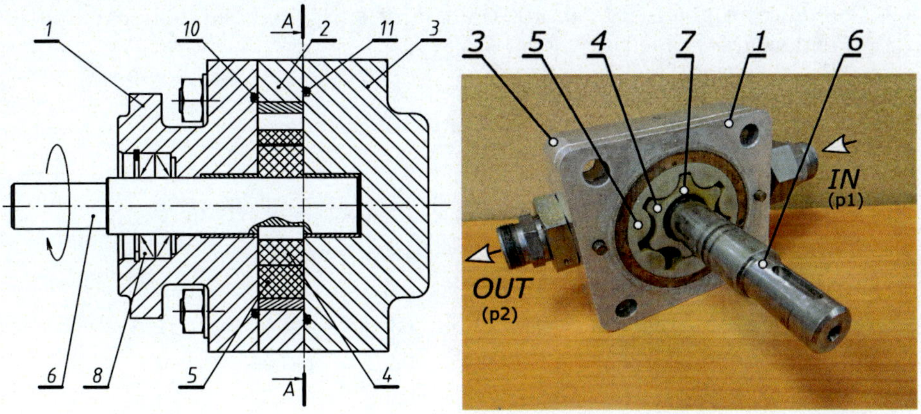

Fig. 1. Tested gear pump: a) cross-section [16,17], b) view of the pump with an example gear set; 1 – front cover, 2 – body, 3 – rear cover, 4 – external toothed gear, 5 – internal toothed gear, 6 – input shaft, 7 – keys, 8 – shaft seal, 9, 10 – static seal

External leakages are limited by static seals 9 and 10 between the body and the covers and a set of lip seals on the drive shaft 8. An internal channel is made in the pump body to drain leaks to the input channel (*IN*).

The gear sets made of plastics were installed in the tested pump. All sets consist of a 6-tooth external gear and 7-teeth internal gear (Fig. 2). They were made of two materials: gears (a) are made of a composite based on polyphenylene

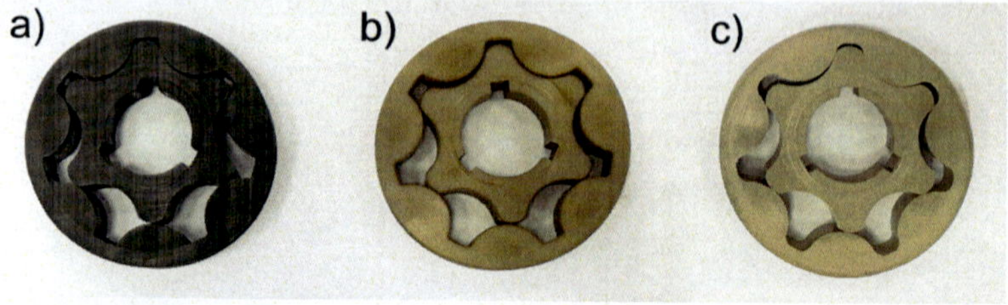

Fig. 2. Tested gear sets; injected: a) PPS+GF40, b) PEEK, machined: c) PEEK

sulphide with 40% glass fibre (*PPS+GF40*), gears (b) and (c) of poly(ether)-etherketone (*PEEK*). Two methods were used to manufacture the gears: (a) and (b) were made using *injection moulding*, and set (c) was made by *machining*.

Materials for the gears were selected based on the analysis of [16] taking into account the basic criterion, which is the use in systems with hydraulic oil. In addition to the required parameters R_e and stiffness E, the absorption of oil, water from oil and the friction coefficient are important as well (Table 2[1]). The density and water absorption of all materials are similar. The addition of glass fibre means that *PPS+GF40* has the best mechanical parameters (R_e and E). In terms of stress, *PEEK* is not much weaker. However, it should be remembered that the share of glass fibre limits the use of such modified polymer due to friction and wear.

Table 2. Selected parameters of materials of tested gears (average values according to the manufacturer's catalogue [18]).

Parameter (unit)	POM-C	PPS+GF40	PEEK
Density (kg/m^3)	1.41	1.65	1.32
Modulus of tensile elasticity E (GPa)	3.0	14	4.0
Tensile Strength, Yield Re (MPa)	68	185	110
Water absorption (%)	0.2	0.0	0.2
Friction coefficient (dynamic)	0.46	–	0.44

4 Test Stand

Figure 3 shows the test stand. The pump 1, driven by the electric motor 2, supplies water through the flow control valve 3 to the tank 5. The valve 3 is used to change the fluid flow resistance in the pump delivery line and to control

[1] Density was determined according to DIN 53479, modulus of elasticity and stress at the yield limit according to DIN EN ISO 527, moisture absorption at saturation at 23 °C, 50 % relative humidity according to DIN EN ISO 62 [18].

the flow rate Q_p. The change in flow rate Q_p results in a change in pressure p_2 loading the pump. The pressure relief valve 4 works as a safety valve. The flow rate Q_p was measured with the flow transducer 7 and the pressure p_2 in the output line was measured with the pressure transducer 6. The input pressure p_1 was not measured and was assumed to be $p_1 = 0\,\text{MPa}$. The rotational speed of the pump n_p was changed in the control system 9 of motor 2 in a certain range. The operating fluid temperature θ was measured by a temperature transducer 8 and maintained by an additional temperature control system. Figure 3b shows additional equipment such as the filter 10 and the cooling and heating system 11 for controlling the temperature of the fluid.

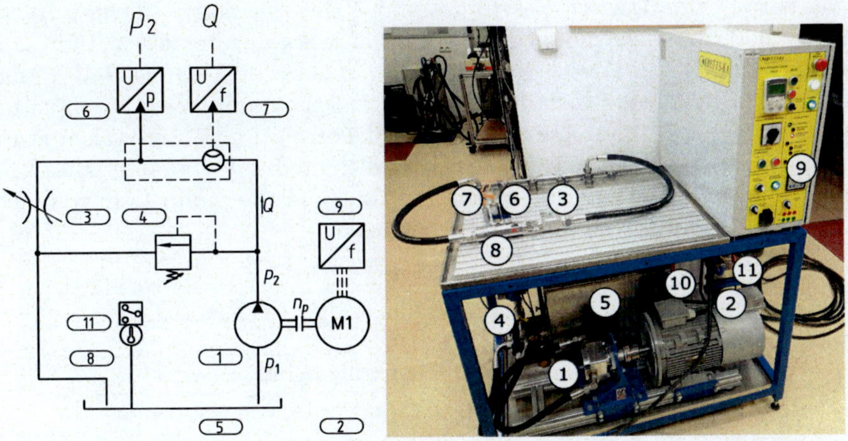

Fig. 3. Test stand: a) diagram of the hydraulic system, b) view of the stand; 1 – pump, 2 – electric motor, 3 – flow control valve, 4 – pressure relief valve, 5 – tank with additional equipment, 6 – pressure transducer, 7 – flow rate transducer, 8 – temperature transducer, 9 – electric motor control system, 10 – filter, 11 – temperature control system

Standard for testing hydraulic pumps [19] requires measurement of the input p_1 and output p_2 pressure at a certain distance from the pump ports along the flow line. The conducted tests omitted the measurement of the input pressure p_1 and the distance from the pump to the measuring point of pressure p_2 was increased due to the lower viscosity of the fluid (Table 1) and the resulting lower pressure drop. It was clearly stated that the characteristics were determined in relation to the delivery pressure p_2, and not the pressure difference $\Delta p = p_2 - p_1$. During the tests, due to the structure of the pump body (Fig. 1), no internal leakage Q_{int} was measured. It was determined using the indirect method after tests. Parameters of accessible transducers, that can work with tap water, are shown in Table 3.

Table 3. Parameters of transducers.

Quantity	Symbol	Accuracy	Full scale
Temperature	θ	±2 K	125 °C
Pressure	p_2	±0.05 MPa	10 MPa
Flow rate	Q_p	2%	60 lpm
Rotational speed	n_p	±2 rpm	6000 rpm

5 Results and Analysis

The purpose of the research was to determine the volumetric efficiency η_v and indirectly determine the flow rate of internal leaks Q_{int} as a function of the outlet pressure p_2. The tests were carried out according to the full factorial plan (Table 4) by changing the pressure p_2 and the pump speed n_p. The temperature was kept in the range of (24–27) °C. Three samples were taken at each measuring point and the mean value was calculated. All characteristics relate to the output pressure p_2. Due to the instability of the pump (low volumetric efficiency), some measuring points were omitted.

Table 4. Measuring points according to the full factorial plan.

Parameter	Symbol	Measuring points
Pressure (MPa)	p_2	0.05, 0.1, 0.15, 0.2, 0.3, 0.4, 0.6, 0.8
Rotational speed (rpm)	n_p	750, 1000, 1500

Based on measurements Q_p and p_2, the volumetric efficiency η_v was determined from the relationship

$$\eta_v = \frac{Q_p}{Q_{th}}$$

with the theoretical efficiency Q_{th}

$$Q_{th} = q_{th} \cdot n_p \tag{1}$$

taking into account the theoretically determined value of the derived capacity q_{th}.

Internal leakage flow rate Q_{int} was calculated as a difference between the theoretical flow rate Q_{th} calculated from the formula (1) and the actual flow rate Q_p

$$Q_{int} = Q_{th} - Q_p$$

In this work, in terms of comparison of different gear sets (with different dimensional deviations from nominal values) the common value $q_{th} = 10\ \text{cm}^3/\text{rev}$ indicated in the publications [14, 20] has been used. Plots also include the results of the research presented in [21] for gears made of *POM* (Table 2) and working with oil HLP-68 (Table 1) within a range up to 1 MPa. The authors used the recommendations of the standard [22] to determine the theoretical derived capacity q_{th}. Due to the different methodology for determining the volumetric efficiency η_v when presenting plots, the efficiency values were recalculated. It resulted in lower volumetric efficiency of the pump with gears with *POM* than the authors presented in [21].

During the tests, any external leakage was observed, but some problems with the sealing of the input shaft occurred. An increase in temperature was observed in the area of cooperation of the shaft 6 (Fig. 1) with the sealing ring 8. The shaft made of steel had lower temperature in the place of contact with water (working medium) than in the area of cooperation with the rotary shaft sealing ring and in the place of coupling attachment. The temperature of this joint was not measured, but observed only and compared to other pump components.

Figure 4 shows the volumetric efficiency of the pump η_p with gears made of different materials as a function of the outlet pressure p_2 for the rotational speed $n_p = 1500$ rpm. The volumetric efficiency of the pump η_v when working with water is less than for oil (at $p_2 = 0.4$ MPa $\eta_v = 0.827$ for oil and $\eta_v < 0.704$ for water). The differences are observed in the whole range of operating ($p_2 \leq 1.0$ MPa), which is related to the lower water viscosity ν. The volumetric efficiency for *PPS+GF40* decreases from $\eta_v = 0.464$ to $\eta_v = 0.104$ at $p_2 = 0.4$ MPa. For gears with *PEEK-mach* it is the range 0.802–0.543 at $p_2 = 0.8$ MPa, and for gears with *PEEK* it is $0.617 - 0.236$ at $p_2 = 0.6$ MPa.

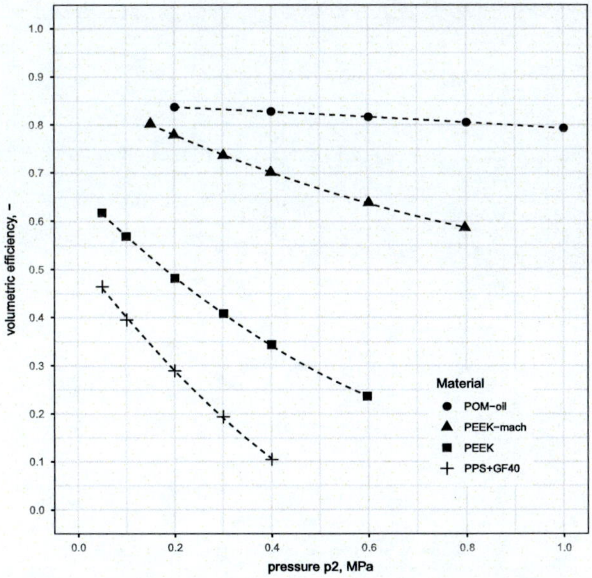

Fig. 4. Pump volumetric efficiency η_p as a function of outlet pressure p_2 for gears made of different materials for constant speed ($n_p = 1500$ rpm)

Comparing the gears from *PEEK* and *PPS+GF40* we observe that the strength of the materials does not play here an important role. The latter are more durable ($R_e = 185$ MPa) relative to the gears from *PEEK* ($R_e = 110$ MPa) (Table 2), but the efficiency is lower. This may be due to shrinkage, which demonstrated to be larger and resulted in greater clearances, despite the use of glass fibre. Comparison of the gears from *PEEK* shows that the method of manufacturing is more important for dimensions than material. Machining ensures better stability of axial and radial dimensions of gears.

Similar conclusions result from the analysis of Fig. 5, which shows the dependence of the internal leaks of the pump Q_{int} as a function of the outlet pressure p_2 for different materials at pump speeds $n_p = 1000$ rpm and $n_p = 1500$ rpm. When comparing leaks for water, it is noticed that they are many times larger than for oil. With gears from *PEEK-mach* the difference is small – approx. 35% ($Q_{int} = 2.46$ lpm for *POM* relative to $Q_{int} = 3.32$ lpm for *PEEK-mach* at $p_2 = 0.2$ MPa), but with gears from *PEEK* and *PPS+GF40* it exceeds 100%, and is even many times larger ($Q_{int} = 10.7$ lpm for *PPS+GF40*). Considering the difference in viscosity of liquids, however, it should be noted that the difference in internal leaks is smaller than that resulting from the viscosity ratio of the liquids themselves. This may be related to the method of determining the theoretical derived capacity of the pump q_{th}, which has been overestimated or by large gaps in the gear sets at which viscosity no longer plays such a role. Due to the course of the graphs, it can be concluded that leaks are independent of the strength and stiffness of the material.

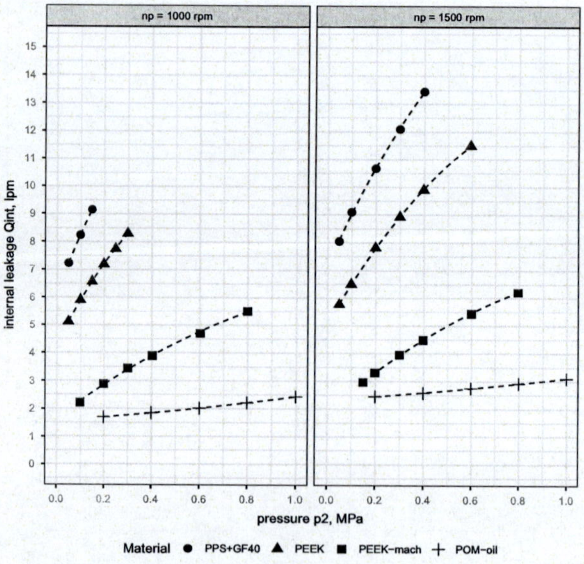

Fig. 5. Flow rate of the pump internal leakage Q_{int} as a function of the outlet pressure p_2 for several speeds ($n_p = 1000$ rpm, $n_p = 1500$ rpm) for gears from different materials

The effect of the utilised material on volumetric efficiency can be seen in Fig. 6. In the case of *POM*, the difference in volumetric efficiency η_v is clearly smaller than for other materials. The volumetric efficiency is $\eta_v = 0.836$ at $p_2 = 0.2$ MPa. At $p_2 = 1.0$ MPa it is in the range of 0.723–0.793 depending on speed. For gears from *PEEK-mach*, the initial value is similar to gears from *POM* ($\eta_v = 0.766$ at $p_2 = 0.1$ MPa). The increase in pressure causes a significant decrease in efficiency, which with increasing pressure is strongly dependent on the rotational speed from $\eta_v = 0.340$ to $\eta_v = 0.586$ at $p_2 = 0.8$ MPa. The biggest differences are observed with the injected gears from *PEEK* and *PPS+GF40*. In addition to lower initial efficiency ($\eta_v < 0.617$ for *PEEK* and $\eta_v < 0.464$ for *PPS+GF40*, we observe a significant decrease in efficiency as pressure increases p_2. At a speed of $n_p = 750$ rpm, the pump cannot operate at pressure $p_2 > 0.2$ MPa ($\eta_v < 0.1$). Also this plot indicates no significant differences between materials cooperating with water (*PPS+GF40*, *PEEK*). The biggest difference is related to the hydraulic fluid (*POM*) and to the use of gears made by machining (*PEEK-mach*).

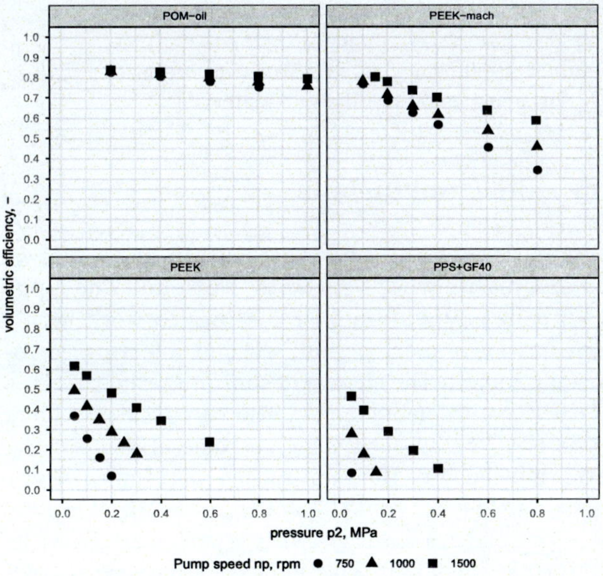

Fig. 6. Pump volumetric efficiency η_p as a function of outlet pressure p_2 for gears made of different materials for several speeds ($n_p = var$)

In Fig. 7 has been presented the volumetric efficiency of the pump η_p as a function of the outlet pressure p_2 at different speeds $n_p = var$ for gears made of different materials. The pump with gears from *PPS+GF40* and *PEEK* works in a limited range ($p_2 < 0.2$ MPa) at a speed of $n_p = 750$ rpm. Only the increase in speed to $n_p = 1500$ rpm allows to reach the working pressure $p_2 > 0.4$ MPa. With machined gears from *PEEK-mach*, the pump works stably up to $p_2 = 0.8$ MPa.

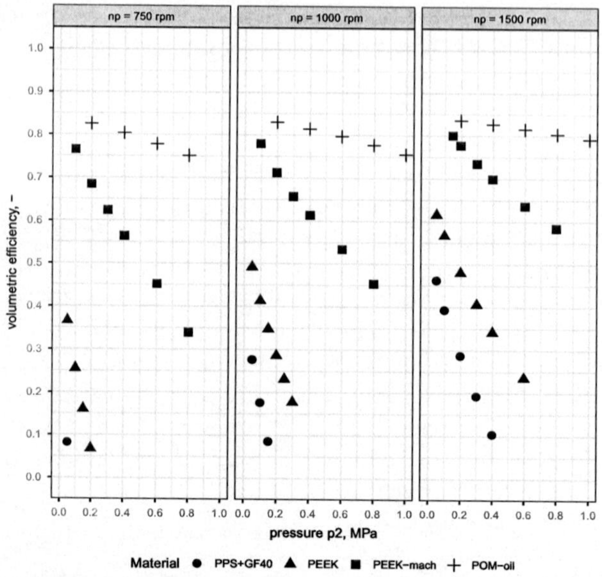

Fig. 7. Pump volumetric efficiency η_p as a function of outlet pressure p_2 at different speeds ($n_p = var$) for gears made of different materials

6 Summary and Conclusions

The hydraulic pump designed to work with oil can also work with water. The aluminium body and plastic wheels are water-resistant. The input shaft is made of carbon steel without any additions or coatings to work with water, but the tests were carried out for a short period, without affecting corrosion processes. Further works should include the change of shaft material to water-resistant, e.g. stainless steel. Other requirements should also be considered for work with fluid with non-oil properties (e.g. viscosity, aggressiveness).

The volumetric efficiency of the pump when working with water is less than when working with oil (e.g. $\eta_v = 0.827$ and $\eta_v = 0.710$ at $n_p = 1500$ rpm and $p_2 = 0.4$ MPa). Water has a lower viscosity, which increases internal leakage and reduces volumetric efficiency. This indicates the need to reduce radial and axial gaps by means of design (compensation of axial clearness) or technological measures (improving the quality of gears).

When choosing gear material, we need to pay attention to the parameter determining the water absorption, which is important for tap water, but also for oil in which minimum amounts of water may occur (up to 0.2%). In tests, lower efficiency of the pump with gears from *PPS+GF40* with the lowest absorption coefficient was observed, but the impact of the manufacturing method on leaks could not be separated.

Changing the material and increasing strength does not significantly affect the volumetric efficiency of the pump (e.g. $\eta_v = 0.342$ for *PEEK* and $\eta_v = 0.104$

for *PPS+GF40* at $n_p = 1500$ rpm and $p_2 = 0.4$ MPa). At low operating pressure, the strength and stiffness of the gears, depending on the material, do not play a significant role. The injection moulding of gears is marked by a significant difference in radial and axial dimensions. This is related to the dependence of the final dimensions on the established injection parameters, on after-injection shrinkage (up to 24 h), and on secondary shrinkage as well.

Machining the gears, resulting in smaller gaps, significantly reduces leaks and increases the pump volumetric efficiency (e.g. $\eta_v = 0.177$ for gears made by *injection* and $\eta_v = 0.658$ for gears made by *machining* with $n_p = 1000$ rpm and $p_2 = 0.3$ MPa).

Striving to increase the accuracy of *injected* gears by optimising process parameters, improving the design solution, and technology of gear with adaptation to the injection process is recommended. The introduced changes will allow to obtain a better shape and surface and limit the impact of the manufacturing method, e.g. shrinkage, on the size of axial and radial clearances. The use of liquids with a lower viscosity than oil requires greater accuracy.

Unfavourable cooperation of the rubber seal (NBR) with the steel shaft in the rotary joint was observed. Water has weaker lubricating properties (e.g. viscosity), which increases the requirements for seals. It is worth to consider changing the seal to not requiring the formation of a lubricating film, as with hydraulic oil, by introducing a different design solution or by using other materials to enable the seals to work with water.

References

1. Krawczyk, J., Sobczyk, A., Stryczek, J., Walczak, P.: Tests of new methods of manufacturing elements for water hydraulics. Mater. Res. Proc. **5**, 200–205 (2018). https://doi.org/10.21741/9781945291814-35
2. Water Hydraulics in Water Power Stations and Water Level Control Systems. DKCFN.PB.000.P1.02/521B0787. Danfoss Industri Service, Denmark (2006)
3. Rodionov, L., Rekadze, P.: Experimental vibroacoustic research of a gear pump made of different materials. Procedia Eng. **176**(204), 636–644 (2017). https://doi.org/10.1016/j.proeng.2017.02.307
4. Gamez-Montero, P.J., et al.: Magnet-sleeve-sealed mini trochoidal-gear pump prototype with polymer composite gear. Energies **10**(10), 1458–1475 (2017). https://doi.org/10.3390/en10101458
5. Lubecki, M., Stosiak, M., Leśniewski, T.: Comparative studies of tribological properties of selected polymer resins for use in hydraulic systems. Bimonthly Tribologia **6**(288), 31–37 (2019). https://doi.org/10.5604/01.3001.0013.7764
6. Stryczek P., Przystupa F., Banaś, M.: Research on series of hydraulic cylinders made of plastics. In: 2018 Global Fluid Power Society PhD Symposium (GFPS) Samara, pp. 1–7. IEEE (2018) https://doi.org/10.1109/GFPS.2018.8472385
7. Banaś M., Antoniak P., Marciniak L., Stryczek J.: Visualization of flow phenomena in hydraulic throttle valves of plastics. In: The 14th International Conference on Vibration Engineering and Technology of Machinery (VETOMAC XIV), MATEC Web of Conferences, Lisbon, Portugal, pp. 1–6 (2018) https://doi.org/10.1051/matecconf/201821119001

8. Ing, A., Bonanno, A., Paoluzzi, R.: Use of fiber reinforced plastics in cartridge valves manifold. In: The 8th International Fluid Power Conference, Dresden, Germany (2012)
9. Urbanowicz, K., Duan, H.-F., Bergant, A.: Transient flow of liquid in plastic pipes. Strojniški vestnik - J. Mech. Eng. **2**(66), 77–90 (2020). https://doi.org/10.5545/sv-jme.2019.6324
10. Stryczek, J., Banaś, M., Krawczyk, J., Marciniak, L., Stryczek, P.: The Fluid Power Elements and Systems Made of Plastics. Procedia Eng. **176**(204), 600–609 (2017). https://doi.org/10.1016/j.proeng.2017.02.303
11. Capanidis, D., Kowalewski, P., Krauze, M.: Study of resistance to abrasive wear of multicomponent polyoxymethylene composites. Bimonthly Tribologia **1**(265), 7–19 (2016). https://doi.org/10.5604/01.3001.0010.7575
12. Biernacki, K.: Analysis of the material and design modifications influence on strength of the cycloidal gear system. Int. J. Precis. Eng. Manuf. **3**(16), 537–546 (2015). https://doi.org/10.1007/s12541-015-0073-6
13. Huang, Y., Leu, M.C., Mazumder, J., Donmez, A.: Additive manufacturing: current state, future potential, gaps and needs, and recommendations. J. Manuf. Sci. Eng. **137**(1), 1458–1469 (2015). https://doi.org/10.1115/1.4028725
14. Krawczyk, J. Stryczek, J.: Design and experimental research of a plastic gerotor pump. In: 2018 Global Fluid Power Society PhD Symposium (GFPS) Samara , pp. 1–8. IEEE (2018) https://doi.org/10.1109/GFPS.2018.8472370
15. Lide, D.R.: CRC Handbook of Chemistry and Physics, 85th edn. CRC Press LLC, Florida, USA (2004)
16. Stryczek, J., Bednarczyk, S., Biernacki, K.: Strength analysis of the polyoxymethylene cycloidal gears of the gerotor pump. Arch. Civ. Mech. Eng. **4**(14), 647–660 (2014). https://doi.org/10.1016/j.acme.2013.12.005
17. Antoniak, P.: Modeling the flow phenomena in gerotor pumps. In: Proceedings of the 3rd FPNI-PhD Symposium on Fluid Power, Terrassa, Spain, pp. 1–6 (2004)
18. Plastics, http://www.plastics.pl. Accessed 12 Feb 2020
19. ISO 4409. Hydraulic fluid power – Positive-displacement pumps, motors and integral transmissions – Methods of testing and presenting basic steady state performance (2019)
20. Krawczyk, J., Stryczek, J.: Badania pompy gerotorowej z kołami zebatymi wykonanymi z PPS (Experimental research of gerotor pump with gears made of PPS). Hydraulika i Pneumatyka **34**(1), 15–18 (2014)
21. Bednarczyk, S., Biernacki, K., Stryczek, J.: Application of plastics in manufacture of the gerotor pump. In: The Twelfth Scandinavian International Conference on Fluid Power SICFP'11, Tampere University of Technology, Tampere, Finland, pp. 369–383 (2011)
22. ISO 8426. Hydraulic fluid power–Positive displacement pumps and motors–Determination of derived capacity (2008)

Torque Transmitted by Multi-plate Wet Clutches in Relation to Number of Friction Plates and Their Dimensions

Marcin Bąk(✉), Piotr Patrosz, and Paweł Śliwiński

Gdańsk University of Technology, Narutowicza 11/12, 80-233 Gdańsk, Poland
marcin.bak@pg.edu.pl

Abstract. In the paper results of experimental tests on multi-plate wet clutch torque capacity for various diameters and numbers of friction plates are presented. Construction of an apparatus for laboratory tests of wet clutch engagement, drag torque and maximum torque is shown. Methodology of maximum torque capacity experiment and hydraulic circuit diagram, which supplies a motor and an actuator for the experiment are described. Analytically calculated values of torque capacity are compared to results obtained with laboratory tests. Additionally, the article includes FEM analysis, that was used to obtain pressure distribution on every friction surface in relation to force applied to the assembly and to determine differences in these distributions. The analysis has helped to determine a mean radius of sliding interface on each surface and to compare their values to other plates and to result calculated analytically. A diagram is shown in the article, that demonstrates pressure concentration as a function of number of friction surfaces.

Keywords: Multi-plate clutch · Friction · Experiments

1 Introduction

As a result of increased use of automatic transmissions in heavy machinery like tractors or telescopic handlers in recent years, it is needed to gain knowledge about effects occurring in a wet clutch in relation to number and dimensions of friction plates and their influence on torque transmitted by the assembly. It would facilitate design of adequate assembly, that fulfils crucial requirements such as short time of engagement and disengagement, sufficient torque capacity, low inertia and drag torque, durability i.e.

Multi-plate wet clutch is a friction clutch with multiple friction plates. Growth of number of friction plates allows to transmit higher torque compare to clutch with only one friction plate. Typical multi-plate construction is shown in the Fig. 1. Working principle is based on applying clamping force into friction and separator plates. It causes engagement of the clutch and transmitting torque between two elements due to friction occurrence. Problems with multi-plate clutches relate to deformation of friction and separator plates, as well as blocking and pressure plates [2–4]. Additionally friction

© The Editor(s) (if applicable) and The Author(s), under exclusive license
to Springer Nature Switzerland AG 2021
J. Stryczek and U. Warzyńska (Eds.): NSHP 2020, LNME, pp. 367–376, 2021.
https://doi.org/10.1007/978-3-030-59509-8_33

appearing on connections between shaft and friction plates and separator plates and hub reduce clamping axial force [5, 6].

In this paper study of wet clutch performance, capacity etc. in function of various dimensions of friction plates and their number is presented. Additionally, numerical approach is presented. FEM analysis was used to obtain value of mean radius and to compare it with result obtained by analytical equation.

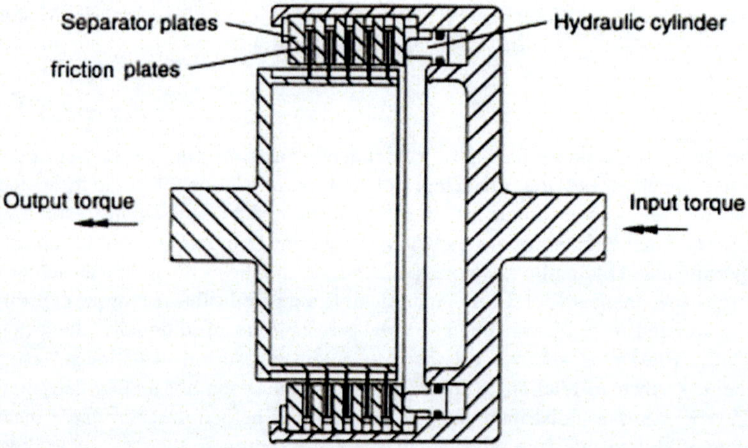

Fig. 1. Multi-plate wet clutch [1].

2 Experiments

Many series of experiments were conducted on designed wet clutch test stand. They were performed in order to collect all the crucial data such as angular velocity, torque, displacement of a hydraulic piston and pressure in its chamber, temperature of lubricating oil. It is essential to measure physical quantities mentioned in previous section as a function of time, to accurately determine relation between them [7].

2.1 Test Stand

The designed test stand allows to study drag torque, time of engagement or disengagement, maximum torque capacity and durability of the clutch. Versatility of the device is its main advantage. The designed apparatus enables experiments with parameters presented in the Table 1. It is possible to use two different hydraulic motors, GMS-100 and GMT-250 [8]. It enhances the experimental ability, due to increased rotational velocity possible to achieve by GMS-100, while GMT-250 allows higher torque, because of its higher displacement. Typical diameter ratio between internal friction diameter D_{i}- and outer friction diameter D_o, shown in the Fig. 1, in cars and heavy machinery is between 0.6 and 0.8. While, in relation to the data presented in the Table 1, it is easy to notice

Table 1. Test bench range of parameters.

	Range	Unit
Drive shaft angular velocity	0–900 (GMS-100) 0–500 (GMT-250)	rpm
Torque	0–250 (GMS-100) 0–400 (GMT-250)	Nm Nm
Normal force	0–31000	kN
Maximum outer diameter	134	mm
Minimum internal diameter	36	mm

that the test rig allows to conduct experiments with ratio even less than 0.3. Also, test stand enables to examine various friction lining materials.

In the Fig. 3 test stand is presented. The torque transmitted by the clutch is measured by torque transducer, that also couples hydraulic motor shaft with the apparatus shaft. The apparatus works as a multi-plate brake, where the hub is stationary, connected to foundation, while the shaft rotates with friction plates. Separator plates are mounted in the hub, alternately with friction plates (Fig. 2). The test rig allows conducting various experiments such as:

- measurement of engagement or disengagement time,
- measurement of maximum torque transmitted by the clutch assembly,
- drag torque measurement,

with combinations of different number and dimension of friction and separator plates. The torque transmitted by the clutch depends on axial force generated by the hydraulic actuator, number of discs and their dimensions. Axial force is applied by the piston to the pressure plate. If the normal force occurs, it clamps friction and separator plates, hence the clutch is engaged.

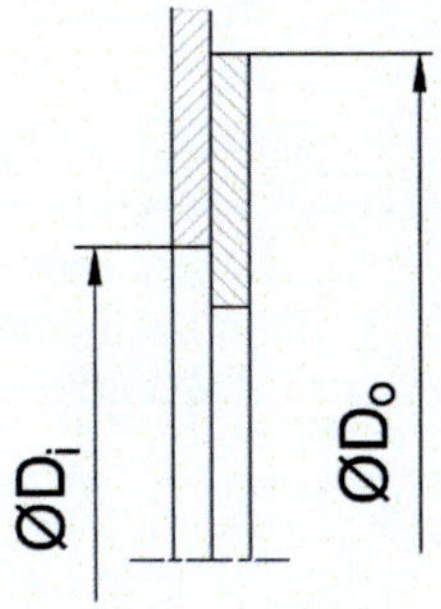

Fig. 2. Internal and outer friction diameter.

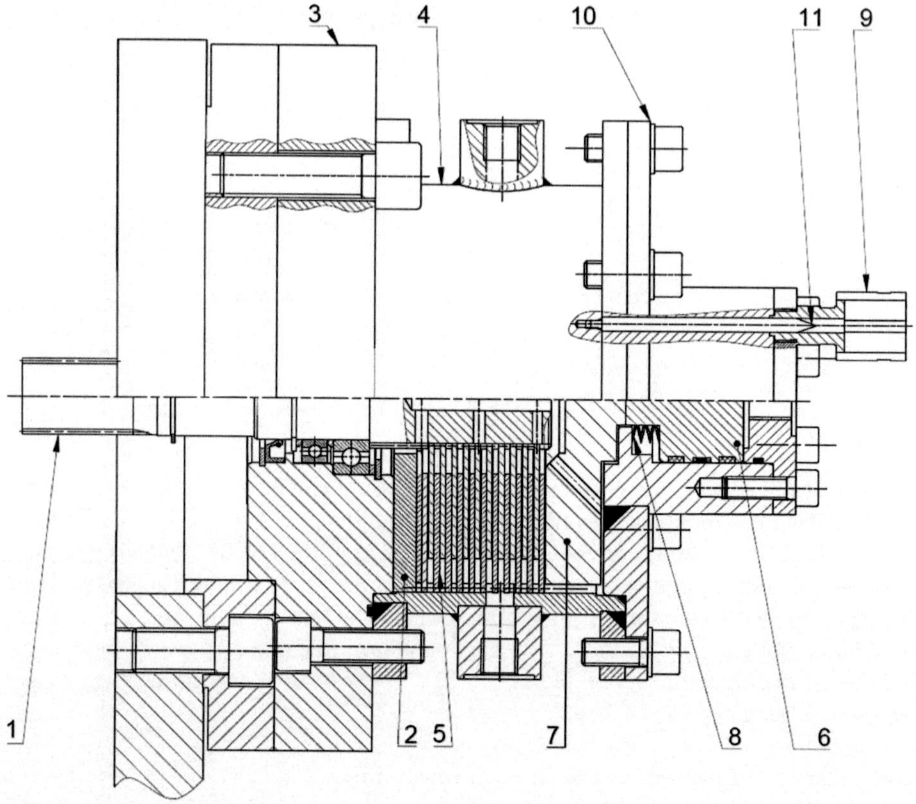

Fig. 3. Test rig: 1-shaft, 2-blocking plate, 3-cover, 4-hub, housing, 5-friction plates, 6-piston, 7-pressure plate, 8-package of disc springs, 9-position sensor holder, 10-cover, 11-position sensor tappet.

Hydraulic circuit that supplies hydraulic actuator and motor is shown in the Fig. 4. During maximum torque capacity experimental tests following procedure was valid. First, pressure p_1 was adjusted. Then, set-up of SV_2 valve was slowly increasing, till occurrence of slippage, which means that torque capacity of the clutch was exceeded. After the described cycle, the process continued further but with increased adjustment of SV_1 valve.

Axial force applied to the plates package is determined by pressure in actuator chamber and varies as a function of set-up of pressure relief valve SV_1. Angular velocity of hydraulic motor is set by two port adjustable flow control valve. Maximum torque appearing on hydraulic motor shaft depends on adjustment of pressure p_4. All the signals from the sensors were recorded by Hydac HMG 4000.

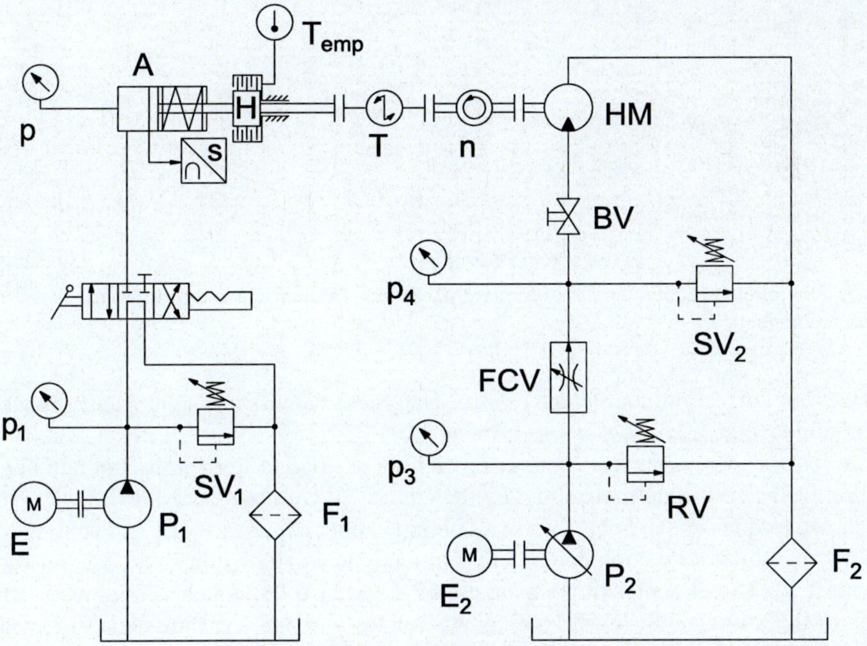

Fig. 4. Hydraulic circuit: P-pumps, SV-safety valves, RV- relief valve, BV-ball valve, FCV-flow control valve, p-pressure transducers, F-filters, HM-hydraulic motor, A-actuator, T_{emp}-temperature transducer, H-clutch, T-torque transducer, n-angular velocity sensor, E-electric motor.

2.2 Experimental Results

In relation to construction of a test rig it is obvious, that normal force applied by the piston to the pressure plate depends on pressure in the cylinder and force generated by deflected package spring of disc springs, hence piston position. For this reason it is necessary to gain a knowledge about springs characteristic, as well as axial load as a function of pressure. Both springs package and axial force tests were carried three times to receive reliable, accurate data. Those information allows to accurately calculate real clamping force. The obtained results are presented in Fig. 5a) and 5b). Package of the springs were tested on compression testing machine. The operation mentioned above is presented in Eq. (1):

$$F_a(p, s) = F_p(p) - F_s(s) \tag{1}$$

where $F_a(p,\ s)$ is axial load applied to the pressure plate, $F_p(p)$ is normal force generated by pressure acting on piston area as a function of pressure p and $F_s(s)$ is force generated by spring as a function of their deflection s.

As presented in literature [6], clutch assembled by increased number of friction and separator plates transmits lower torque in relation to initial value than it could be as calculated with analytical equation, with proportionally higher number of friction

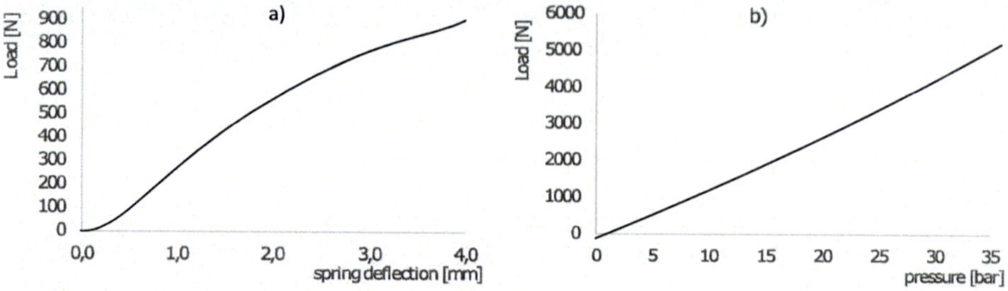

Fig. 5. Diagrams: a) characteristic of a spring package, b) axial force as a function of pressure in actuator chamber.

surfaces. A correction coefficient used in analytical calculation significantly reduces maximum torque.

Series of experiments were conducted on the test stand to study impact of number of friction plates on torque transmitted by the clutch. To compare, in following diagrams are shown torque capacity characteristics obtained both experimentally and analytically without including correction coefficient. For analytical calculation static friction coefficient of steel-steel connection was assumed as equal 0.05, which is minimum value given in literature [6, 9]. In Fig. 6 diagrams relates to diameters ratio equal 0.7, while in Fig. 7 applies to ratio equal 0.8. For both situations internal diameters are the same. In both Figs. 5 and 6 in three out of four diagrams experimental maximum torque is higher than analytical result, which means that friction coefficient have higher value than established in calculations.

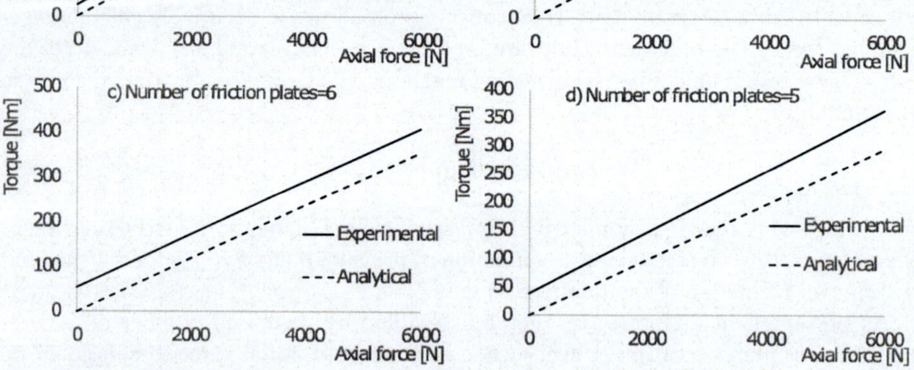

Fig. 6. Torque transmitted by a wet clutch, plates with diameter ratio equal 0.7.

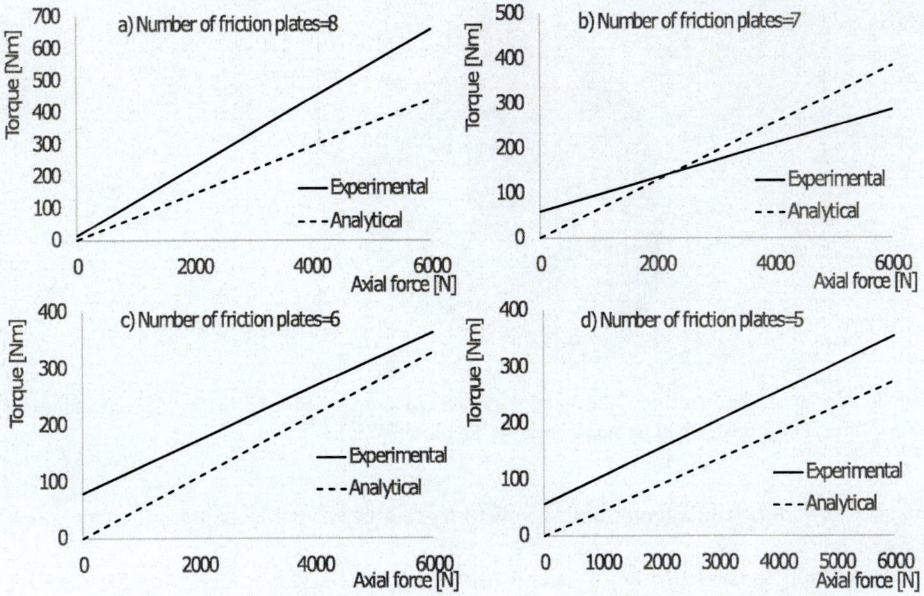

Fig. 7. Torque transmitted by a wet clutch, plates with diameter ratio equal 0.8.

3 Finite Element Method Analysis

Due to crucial influence of pressure distribution on contact surfaces on clutch performance and durability, it is essential to have an insight into the distribution as a function of number of friction and separator plates [2, 10].

3.1 Model

In the Fig. 8 model of the clutch is presented with supports and applied load. The presented model consists of 331739 elements, with number depending on number of friction and separator plates used in simulations.

3.2 FEM Results

Results of FEM analyses shows pressure concentration appearing on relatively small diameter equal to internal diameter of friction annulus area. Additionally, maximum values differ for every friction surface. It confirms both analytical and numerical analyses conducted in the past by various authors. For better understanding, the effect is shown in the Fig. 9, with comparison to friction plate used in tests. It is observable on the real friction plate that area with intensified wear match to area of pressure concentration estimated by numerical analysis. The effect of unequal pressure distribution is

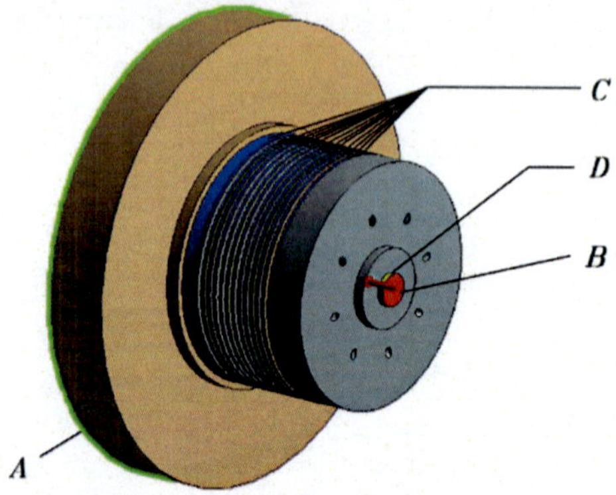

Fig. 8. Model of the clutch for FEM analysis: A-fixed support, B-axial force, C-elastic support, D-displacement (constrained perpendicular to the clutch axis).

considerably disadvantageous, and significantly reduces possible axial force that clamps the assembly.

In relation to the FEM analysis shown in the Fig. 10, it is noticeable, that number of friction surfaces greatly influence on maximum value of pressure on contact surface. With increased number of friction plates the pressure concentration obtain lower value, which corresponds to stiffness of the clutch package. A reason for the effect is that low number of plates in more influenced by pressure plate deformation. Package with more plates involved, more efficiently compensate impact of the deformation on pressure distributions. Effect of unequally distributed pressure is also found in relation to different thicknesses of plates. Thicker plate sustain less deformation, hence pressure on its surfaces is more equal.

Fig. 9. Comparison of maximum contact pressure: left side-real friction clutch after maximum torque capacity experiment, right side-numerical result of contact pressure distribution on friction surface.

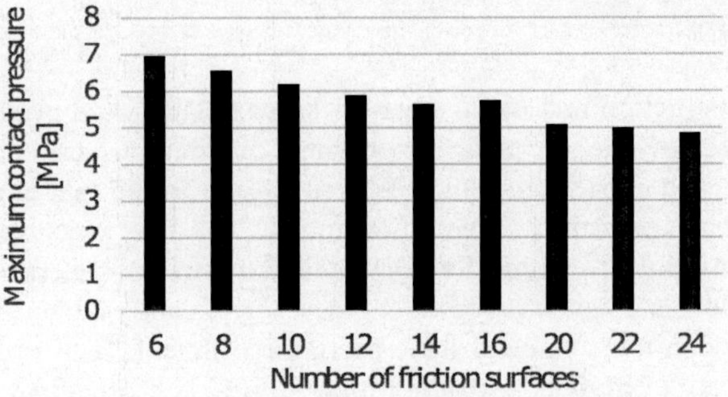

Fig. 10. Maximum contact pressure as a function of number of friction surfaces.

3.3 Mean Radius Comparison

Mean radius by definition is equal to torque transmitted by the friction on particular friction surface, divided by multiplication of friction coefficient and axial force applied to the clutch assembly.

Numerical analysis enables to calculate mean radius and compare its value to an effective mean radius, when it is assumed that pressure is equally distributed on friction surfaces. The two dimensions are compared in the Fig. 11, as a function of no. of friction surface, where friction surface number 1 is the closest to the pressure plate. It is observable, that analytical value is higher than numerically obtained. Disadvantage of the result is torque capacity for specified friction coefficient is lower than established with calculation. Also, for the first three friction surfaces mean radius obtained numerically decreases, but for further surfaces its value consequently increases.

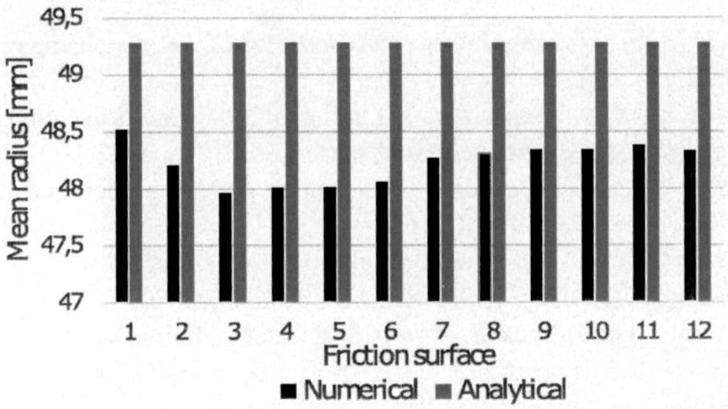

Fig. 11. Comparison of numerically and analytically obtained mean radius.

4 Conclusions

In the article construction and parameters of the test stand were presented. Influence of varying number of friction plate on pressure distribution and concentration was described. Conducted experiments and FEM analyses showed that number of friction plates greatly impacts pressure concentration and mean radius, which value is within the range 1–1.6 mm lower than analytically calculated. Experimental test revealed that torque capacity of a multi-plate clutch is for most cases, significantly higher than analytically calculated for the minimum value of static friction coefficient. The results demonstrates considerable impact of number of friction plates on clutch behaviour, what will be a subject of future research carried by authors.

Conducted numerical simulations has achieved satisfactory agreement with experimental results, which proves that established boundary conditions are correct. Furthermore, it is reasonable to use FEM in future research.

Further studies will be aimed to determine universal mathematical expression describing pressure distribution and mean radius as a function of number of plates. Additionally, other laboratory tests will be performed to investigate multi-plate clutch performance and effects occurring in the assembly.

Acknowledgments. The article includes the research conducted in project funded by The National Centre for Research and Development within the framework of program LIDER:

Project no.: LIDER/22/0130/L-8/16/NCBR/2017.

Project title: Hydro-mechanical automatic gearbox for agricultural vehicles and heavy machinery.

Project no.: LIDER/22/0130/L-8/16/NCBR/2017.

References

1. Holgerson, M.: Friction and temperature characteristics of wet clutch engagement. Licentiate thesis. Lulea (1997)
2. Yu, L., et al.: Numerical and experimental studies of a wet multidisc clutch on temperature and stress fields excited by the concentrated load. Tribol. Trans. **62**(1), 8–21 (2019)
3. Yang, L.: Pressure distribution of a multidisc clutch suffering frictionally induced thermal load. Tribol. Trans. **59**(6), 983–992 (2016)
4. Yu, L., et al.: Investigation on the thermodynamic characteristics of the deformed separate plate in a multi-disc clutch. Eng. Fail. Anal. **110**, 104385 (2020)
5. Abdullah, O., et al.: The influence of frictional facing thickness on the contact pressure distribution of multi-disc dry clutches. FME Trans. **46**, 33–38 (2020)
6. Osiński, Z.: Clutches and brakes (in Polish). PWN (2000)
7. Bąk, M.: Influence analysis of design factors on torque transmitted by a wet clutch. Master thesis, Gdańsk (2019)
8. Hydraulik Nord: catalogue GMS, GMSS, GMSW, GMT, GMTS, GMTW, RE 14 060/03.96. www.stok.khadamathydraulic.com. Accessed Nov 2018
9. Orlinghaus: Technical Information. Edition 02 (2004)
10. Zagrodzki, P., Truncone, S.: Generation of hot spots in a wet multidisk clutch during short-term engagement. Wear **254**, 474–491 (2003)

The Configuration of Circulating Unsteady Flows in the Spacecraft Spherical Tank

Vasyl Kovalev(✉)

National Technical University of Ukraine "Igor Sikorsky Kyiv Polytechnic Institute", Prosp. Peremohy 37, Kiev 03056, Ukraine
vaskovalev@gmail.com

Abstract. The proposed material presents the results of experimental and theoretical studies of the hydrodynamic mechanism for the development of secondary circulating flows in spherical geometry tanks as applied to the tasks of ensuring flight stability and effective controllability of a spacecraft at near-Earth orbit. It has been established that secondary flows arising in the reservoir meridional planes have the form of circulations (one or several) and are capable to influence the distribution of velocity vector components—azimuthal, radial, and meridional.

In turn, the hydrodynamic picture of the three-dimensional flow significantly affects the distribution of force effects from the liquid to the tank walls and the internal guiding baffles, which provides the basis for their more accurate calculation and accounting in the onboard software package of the spacecraft. Noticeable differences between the experimental results and theoretical data obtained by other authors are established.

The velocity profiles constructed from the measurement results give a complete three-dimensional picture of the hydrodynamic mechanism for the circulations development and their effect on the circular torque or pressure distribution on the radial damping baffles.

Keywords: Spacecraft · Spherical tank · Velocity · Circulation · Damping baffles · Polynomial approximation

1 Introduction

The technological development of the leading countries of the world offers new global challenges for the wider exploration of outer space, not only near the Earth, but also deep space. Promising designs of spacecraft that implement the tasks set are higher requirements for reliability, controllability, flight duration, as well as environmental safety.

Currently, oxygen, hydrogen, as well as kerosene, heptyl and others are the widespread fuel types for spacecraft. During the operation of spacecraft propulsion systems, one of instability sources is the uncontrolled fuel flows in tanks. Forceful effects on the spacecraft design from the fluid moving in tanks under the influence of inertia forces can significantly affect the flight trajectory and regime [1].

© The Editor(s) (if applicable) and The Author(s), under exclusive license to Springer Nature Switzerland AG 2021
J. Stryczek and U. Warzyńska (Eds.): NSHP 2020, LNME, pp. 377–385, 2021.
https://doi.org/10.1007/978-3-030-59509-8_34

Compensation of such disturbances is carried out by a special orientation and stabilization system of spacecraft, during the operation of which fuel is consumed, whose reserves in flight are strictly limited. Therefore, efficient and reliable control of the spacecraft with the rational fuel use will allow solving several technically complex problems: to predict possible types of instabilities during the flight, reduce fuel flowrate, increase flight reliability and controllability of the spacecraft, as well as increase their environmental safety.

One type of fluid flow on board is its rotation around the axis and associated viscous friction on the tank walls and the dynamic pressure on the internal damping baffles. Therefore, studies of such flows seem to be very relevant.

2 Formulation of the Problem

For laboratory studies of the hydrodynamic processes in the spacecraft fuel tanks at various stages of flight, the creation of special experimental stands and procedures for flight conditions modeling is required. This allows, taking into account the well-known and special similarity criteria, to study the features of liquid fuel flows and to develop effective methods for the influence compensation of these destabilizing factors [2].

To ensure a rational layout of the spacecraft propulsion system, the spherical shape of the reservoir is widely used in the structures. The sphere and its modifications (spheroid, lentil), schematically presented in Fig. 1, allow the most efficient use of the space inside the spacecraft.

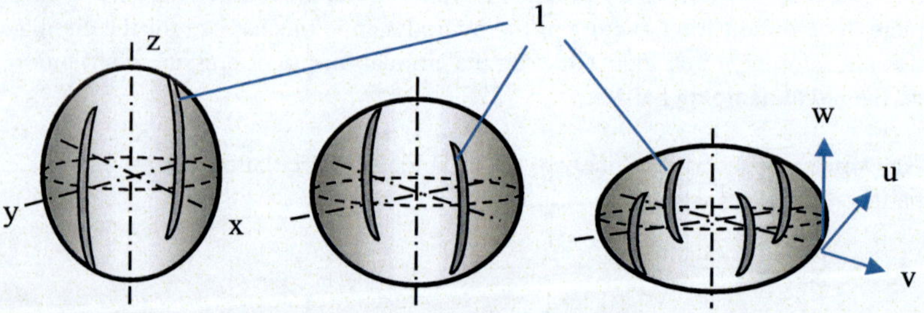

Fig. 1. Options for the performance of spacecraft fuel tanks: a - spheroidal, b - spherical, c – lentil; 1 – crescent near wall baffles; velocity vector components: u – azimuthal; v – radial; w – meridional

The construction of a physical model of the liquid fuel flows in a spherical tank was implemented on the basis of experimental centrifuge-type stand in the hydrodynamics laboratory of NTUU "Igor Sikorsky KPI", the general view and structural diagram of which are shown in Fig. 2. The stand main objective is to provide a wide range of angular velocities of the model tank around the vertical axis and their sharp changes, for example, start or braking of rotation.

A rotary Table (2) is mounted on the power bed (1) of the stand with the possibility of rotation around the vertical axis, on which the frame (3) is mounted. In the frame there is a model tank (4) with liquid, on which velocity sensors (5) are installed, which

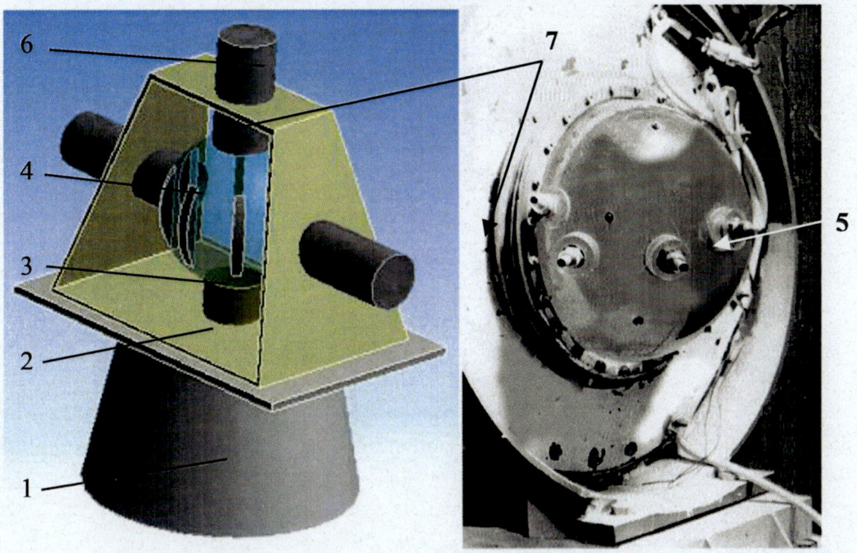

Fig. 2. An experimental "centrifuge" type stand for rotating flows modeling: a) a schematic diagram; b) model tank in a power frame: 1 - power bed; 2 - rotary table; 3 - frame; 4 - tank model; 5 - precision clamps of velocity sensors; 6- rotating contact device; 7 - torque measuring device.

are included in the electronic measurement and recording system associated with the stand with using a rotating contact device (6).

The schematic diagram of hydraulic drive is shown in Fig. 3. It contains an axial piston pump (1) with a flow control system (2), a distributor (3) with electromagnetic control, a hydraulic motor (4), the shaft (5) of which is rigidly connected to the rotary table of the stand. The working pressure in the hydraulic system sets by pressure regulator (6).

To expand the possibilities of experimental measurements, the design of the distributor (3) provides reverse rotation of the hydraulic motor shaft (3) in the range of angular velocities from 0.05 to 0.5 1/s. In addition, due to the principle of flow stepwise regulation in the hydraulic system, the drive control panel provides a fairly quick change in angular velocities within 0.1–0.3 s and good repeatability. The kinematic scheme of the stand and the pump supply allows you to get the rotation speed of the turntable in the range of ± 5.0 r/s with an error of the order of 1.5–2.7% [2, 3].

The experimental procedure is as follows. After acceleration of the model fluid to solid-state rotation with the tank (this is indicated by zero readings of the hot-wire anemometer velocity sensors), a command is sent from the control panel to brake the rotary table while turning on the measuring and recording equipment. Measurements by velocity sensors at various points of the flow allow you to build a velocity field in a tank specific section and establish the dependence of the velocity components on the flow time.

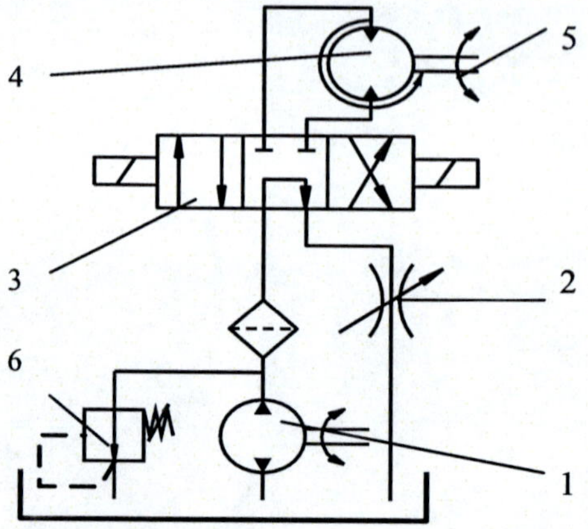

Fig. 3. Hydraulic circuit of the rotary table rotation drive: 1 – pump; 2 – adjustable throttle; 3 – distributor; 4 – hydraulic motor; 5– stand drive shaft; 6 - pressure regulator.

3 Analysis of Flows Modeling Results

The force effects on the walls and internal damping devices of model were determined using device 7 (Fig. 2). To eliminate the influence of friction forces in the upper and lower bearings, the principle of unloading the rotation axes was used.

To simulate flows in the laboratory, the Navier-Stokes equations and criteria dimensionless complexes were used, for example, Reynolds numbers (2), Rossby numbers (1), and others. The weight contribution of the equation terms characterizing the rotational motion is determined as follows

$$Ro = \frac{F_{inertia}}{F_{coriolis}} = \frac{|u\nabla u|}{2|\Omega \times u|} = \frac{u}{2\Omega R \sin\theta}; \quad (1)$$

$$\mathrm{Re} = \frac{F_{inertia}}{F_{visco}} = \frac{2|\Omega \times u|}{|\nu \nabla^2 u|} = \frac{\Omega R^2 Sin\,\theta}{\nu}, \quad (2)$$

where: R – sphere radial coordinate, m; Ω - sphere angular velocity, 1/sec; θ – angular spherical coordinate; T – flow time, sec.

According to well-known expressions, the fluid viscous friction on the wall is proportional to

$$F_{visco} \approx \frac{\rho \nu u R^2}{\delta}, \quad (3)$$

where δ is the thickness of the boundary layer (in our case, a wall inhibited flow).

The Fig. 4a shows the results of measuring the peripheral component u of the velocity vector of the inertial fluid flow in a model sphere with a diameter of 0.3 meters, which shows the exponential nature of velocity attenuation in time of the transient flow, which can be approximated by the expression:

$$u = A \cdot e^{-Bt}, \quad (4)$$

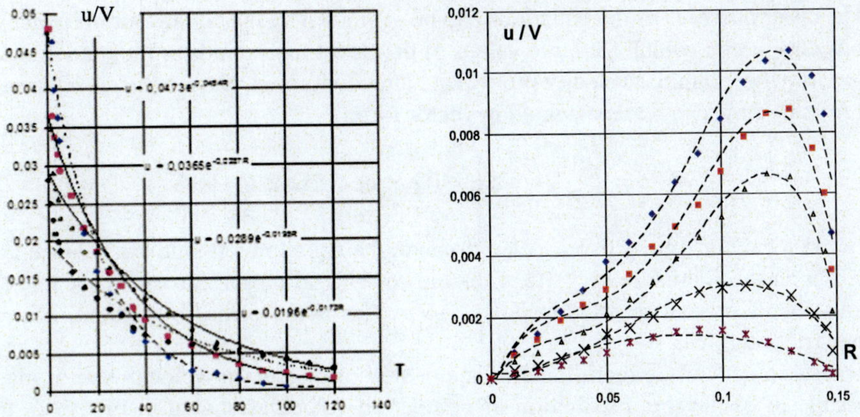

(a) (b)

Fig. 4. The measurement results of the velocity vector azimuthal component u/V: a) depending on the transition process time T at various distances to the rotation axis; b) velocity profiles in the sphere equatorial plane for various decay time T.

where the proportionality coefficients A and B vary respectively in the range of A = 0.0196–0.0473, B = 0.0173–0, 0454.

Figure 4b shows the velocity profiles depending on the distance R from rotation axis. The curves extremes at R = 0.12–0.14 m indicate the existence of a certain boundary for the interaction of the near-wall flow, where the influence of viscous diffusion R = 0.13–0.15 m, as well as a quasi-solid flow in the region R = 0–0.13 m, where the inertial properties of the main flow [4]:

$$\mathrm{Ro} = \mathrm{a}_0 + \mathrm{a}_1 \cdot \mathrm{R} + \mathrm{a}_2 \cdot \mathrm{R}^2 + \mathrm{a}_3 \cdot \mathrm{R}^3 + \mathrm{a}_4 \cdot \mathrm{R}^4, \tag{5}$$

where: a_i are the proportionality coefficients, which vary within the limits a_0 = 0,0002–0,0004, a_1 = 0,111–0,185, a_2 = –3,3–4,97, a_3 = 47,7–66,1, a_4 = –199–268. In this case, estimating the weight contribution of the equation terms, the equation term a_0 can be neglected.

Determination of the convergence interval of a power series reduces to the Leibniz criterion application for alternating polynomials, which provides for a decrease in partial sums with increasing degree of the polynomial. In this case, these values increase, which indicates the limited convergence interval. As the time of the transition process increases, the polynomial estimation of the peripheral velocity u diagrams is somewhat simplified and reduces to the equations of the third and then second order:

$$\mathrm{Ro} = 0,0002 - 0,01 \cdot \mathrm{R} + 0,926 \cdot \mathrm{R}^2 - 5,49 \cdot \mathrm{R}^3, \tag{6}$$

$$\mathrm{Ro} = -0,0004 + 0,042 \cdot \mathrm{R} - 0,254 \cdot \mathrm{R}^2, \tag{7}$$

moreover, the free terms a_0 = 0.0002 and 0.0004 can also be neglected due to their smallness.

The order decrease of the equations can be explained by the significant influence of viscous dissipation, which for large values of the flow time extends to the predominant space part, where laminar velocity profiles are already formed. Navier-Stokes equations for a rotating flow can be represented in vector form:

$$\frac{\partial u}{\partial t} + u\nabla u = -\frac{1}{\rho}\nabla p - \Omega \times (\Omega \times r) - 2\Omega \times u_R + \nu\nabla^2 u, \quad (8)$$

where the local and convective accelerations on the equation left side are balanced by the terms on the right side $\Omega \times (\Omega \times r)$- the centrifugal inertia forces, $2\Omega \times u_R$- the Coriolis forces present in rotating flows, $\frac{1}{\rho}\nabla p$ - the radial pressure gradient, $\nu\nabla^2 u$ - viscous friction forces.

Measurements of the meridional component w of the current velocity vector made it possible to obtain the distribution of circulation velocities shown in Fig. 5. In the measurements, the coordinate angle $\theta = \pi/6$, $\pi/3$ and $\pi/2$, as well as the distance R to the sphere center, were changed.

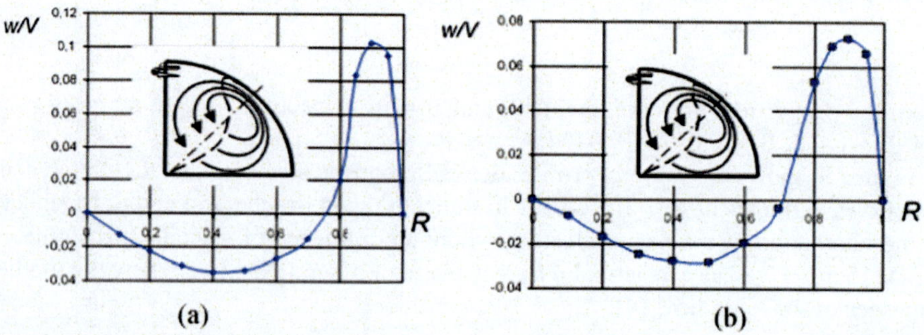

Fig. 5. Meridional velocity w/V profiles when one circulation occurs at $Re_0 = 1200$: a – T = 0.42; b – T = 5.4.

A curves (Figs. 4 and 5) analysis make it possible to determine the characteristic sections (zones) of the meridional velocity w and set the coordinates of the circulation center (Fig. 5). For example, the velocity w diagram at $\theta = \pi/3$ contains two asymmetric zones, the left of which in the near-axis region of the sphere denotes the descending flow from the sphere top to its center, and the right, in the wall region, the upward flow [4]. As can be seen from the figure, the circulation center is located at a distance of R = 0.75 and a coordinate angle $\theta = 40^0$–47^0. In this case, the maximum values of the descending and ascending branches of the secondary current differ 3.2 times, indicating a fairly intense flow of the near-wall region and a slow, $w_i/\Omega_0 R_0 = 0.029$, in the axial.

The approximation of a similar curve with two extrema increases the polynomial order to the sixth and makes up an equation of the form

$$w_i/\Omega_0 \cdot R_0 = A \cdot R^6 + B \cdot R^5 + C \cdot R^4 + D \cdot R^3 + E \cdot R^2 + F \cdot R + G, \quad (9)$$

where A = −34.8, B = 93.5, C = −92.8, D = 42.0, E = −8.34, F = 0.45 and G = −0.0014. Due to the smallness of the coefficients F and G, we neglect them. Thus, we obtain such equation

$$w_i/\Omega_0 \cdot R_0 = R^2\left(A \cdot R^4 + B \cdot R^3 + C \cdot R^2 + D \cdot R + E\right). \quad (10)$$

With an increase in the dimensionless decay time T = 10.4 of the azimuthal velocity u, a shift of the circulation center to the rotation axis (to R = 0.69) and a decrease in the peak values of the meridional velocity w are observed. Approximating the obtained curve, we use a sixth-order polynomial with the coefficients: A = −13.2, B = 32.3, C = −29.1, D = 12.2, E = −2.27, F = 0.063 and G = −0.0001. As in the previous case, the last two terms of the equation can be neglected.

The effect of viscous diffusion at large flow time values also affects the meridional flow development, reducing velocities, smoothing their peak values and even more shifting the circulation center to the axis of sphere rotation. For example, at T = 11.3, the meridional velocity plot will take the form shown in Fig. 5b. This curve can be represented by a fifth-order polynomial

$$w_i/\Omega_0 \cdot R_0 = B \cdot R^5 + C \cdot R^4 + D \cdot R^3 + E \cdot R^2 + F \cdot R + G, \quad (11)$$

where B = −2.51, C = 4.74, D = −2.82, E = 0.71, F = −0.12 and G = 0.0002. As in the above cases, the last two members of the polynomial can be neglected.

4 Theoretical Background of the Circulations Study

According to the theoretical hypotheses of H. Greenspan [5], meridional flows in reservoirs can comprise not only single-circulation structures, but also multi-circulation ones. Similar features of secondary flows, whose characteristic structures are presented in Fig. 6, were experimentally installed using a system of precision hot-wire anemometric transducers [4].

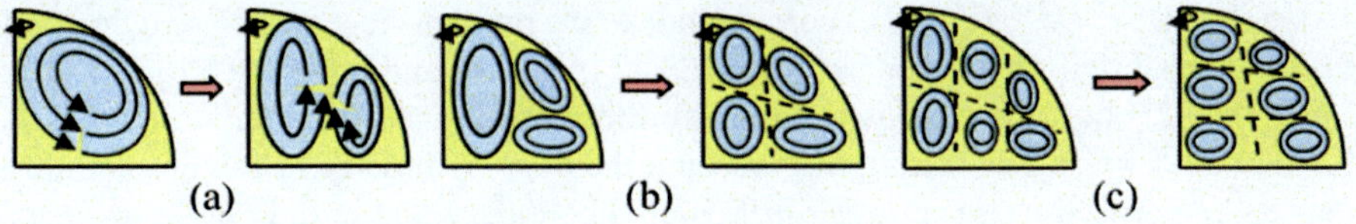

Fig. 6. The evolution of meridional circulations in a spherical tank during liquid inertial rotation: a) Re = 1600–2140; b) Re = 2730–3840; c) Re = 4500–5180.

Moreover, an increase in the initial Reynolds numbers (Re_0 = 1600–2140) led to a fragmentation of the circulation - from one to two, then at Re_0 = 2730–3840 - from three to four, at Re_0 = 4500–580 five to six circulation, etc. An interesting feature of such formations is that, with a decrease in the instantaneous Reynolds numbers Re_i, the circulation intervals shift toward a decrease in Reynolds numbers. For example, the transition from a 3-circulation structure to a 2-circulation structure occurred at Re_i = 1520–2040, that is, ΔRe_i = 80–100. In a numerical study of flows in a cylinder, Xinjun C. [6] established an important property of circulation migration in the meridional plane.

In our opinion, an important feature of the sphere circulations is their ability to significantly affect the three-dimensional distribution of the azimuthal component u, deforming the velocity profiles and reorienting the vectors of torque of viscous fluid friction on the walls. Accurate determination of the correct direction of the power compensation effect from the spacecraft can reduce the duration and increase the efficiency of its work, to clarify the magnitude and exact direction of the circular compensation moments.

The picture of azimuthal velocity distribution in the sphere equatorial plane should be supplemented by the results of a numerical calculation of the liquid particles trajectories in the tank meridional plane, which qualitatively correspond to the hypothetical graphs of H.Greenspan's [5] and are shown in Fig. 7.

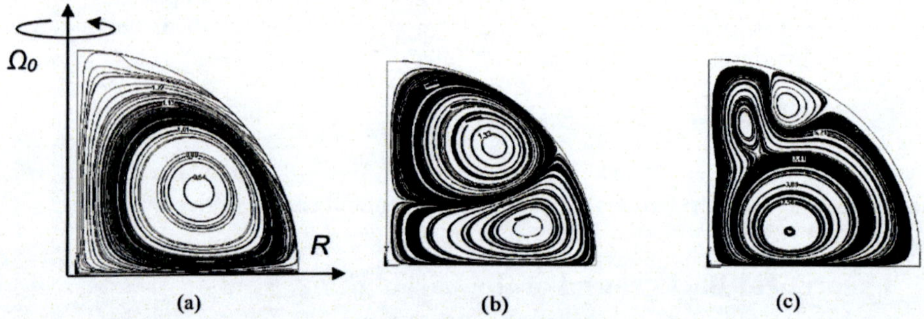

Fig. 7. The trajectory of liquid particles during the attenuation of circulation in the sphere: a – 1 circulation, $Re_0 = 970$; b – 2, $Re_0 = 1150$; c – 3, $Re_0 = 1420$.

A comparison of the circulation structure, as a result of numerical calculations with known data, allows to conclude the mathematical model we adopted is correct in the range of initial Reynolds numbers, $Re_0 = 700$–12500. For example, the presence of one circulation at $Re_0 = 970$ (Fig. 7 a) corresponds to the predicted secondary structure, and the formation of two circulations at $Re_0 = 1150$ (Fig. 7 b) and three - at $Re_0 = 1420$ (Figure 7 c) confirms the possibility of further crushing into smaller structures [4].

When modeling flows taking into account the axial symmetry of the problem, it is necessary to verify the correctness of the model boundary and initial conditions based on the Navier-Stokes equations. To do this, we examined the test problems of flow past baffles in tanks with an infinitely large radius of surface curvature, the calculation results of which are shown in Fig. 8.

For example, when flowing around a 6-element (Fig. 8) structure, where the baffles are located at distances $l = l_0/R_0 = 0.73$–1.0, the dimensions of the tangled wake behind the first baffle significantly exceed the distance at which the second baffle is separated from the first

$$l = l + \Delta l = 1,25\ldots 1,42 \cdot l_0, \tag{12}$$

therefore, it can perceive only some part of the flow dynamic pressure [3]. Thus, the effectiveness of a baffle series is reduced, and to increase its influence, the baffle should

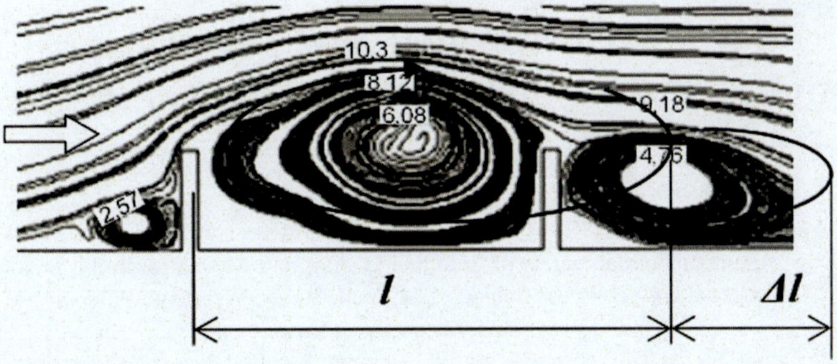

Fig. 8. Contour images of circulations in the wake of continuous adjacent baffles at $Re_0 = 920$ [3].

be located outside the wake. The distance between the baffles should be set equal to l_1, which is proportional to the circulation in the wake.

5 Conclusions

The measurement results of the velocity vector components indicate the exponential nature of the velocity fields attenuation and strongly nonlinear features of the velocity distribution over the tank radius. This is due to the rather strong influence of secondary circulating structures, which are capable of crushing and reorienting the torque effects from the liquid side in the tank space.

In addition, as it was established as a modeling result, the circulations can significantly affect the distribution of force effects on the tank walls and internal guiding baffles, which in turn affect the flow in the free space of the tank that is not covered by baffles.

References

1. Belyayev, N., Belik, N., Uvarov, E.: Reactive spacecraft control systems (in Russian). Machine-building, Moscow (1979)
2. Kovalev, V., Yakhno, O.: Hydrodynamic features of circulating flows in closed volumes. Monography, (in Russian) Knowledge, Donetsk (2014)
3. Kovalev, V.: The structure of inertial flows in a fuel tank with guide baffles. Mech. Adv. Technol. **79**, 71–75 (2017)
4. Kovalev, V.: The evolution of circulations in inertial flows of an incompressible fluid in a spherical reservoir. Mech. Adv. Technol. 81 (2017)
5. Greenspan, H.: The theory of rotating fluids. (in Russian) - Gydrometeoizdat, Moscow (1975)
6. Xinjun, C.: A numerical study of the recirculation zones during spin-up and spin-down for confined rotating flows. J. Theoret. Comput. Fluid Dyn. **1**, 31–496 (2003)

Check for updates

Functional Verification and Performance Studies of the Gerotor Pump Made of Plastics

Justyna Krawczyk(✉)

Department of Fundamentals of Machine Design and Mechatronic Systems, Wroclaw University of Science and Technology, Wroclaw, Poland
justyna.krawczyk@pwr.wroc.pl

Abstract. Paper presents the prototype of the gerotor pump with body and gear set made of plastics, its functional tests (checking measurements, run-in process) and experimental research of the efficiency characteristics. The positive results of previous researches of plastics gear sets allowed for the next step in the design of the entire plastic body of the gerotor pump. Experimental research was aimed at checking the proper functioning of the pump with a plastic body and verification of operational factors, impact of speed and pressure on the pump characteristics. Functional tests and experimental studies confirmed the correctness of the structural solution of the pump body. The pump with a plastic body was working correctly as a low pressure unit at low range of rotational speed. Experimental research of the plastic gerotor pump confirm that plastics as a construction material can be used in hydraulics.

Keywords: Gerotor pump · Cycloidal gears · Plastics

1 Introduction

The basic energy generator in hydraulic systems are gear pumps. Three types of gear pumps can be distinguished: external gear pumps, internal gear pumps and gerotor pumps.

Gear set is an essential element influencing the operation of the gear pump. Thanks to this element, gerotor pumps gain an advantage over external and internal gear pumps. They have a very compact design, smaller dimensions and lower weight. With low and light units they can obtain high efficiency thanks to the intertooth displacement chambers which have a large volume. This types of pumps are reliable and durable. In order to take the next step in the development of gerotor pumps, current market trends should be followed.

In the construction of hydraulic machines and components, as well as in other areas, the reduction of the weight and dimensions of manufactured elements, as well as reduction of the cost of their manufacture is increasingly being sought. However, to make this possible, it is necessary to search for new construction materials and new manufacturing methods. A new direction in the field of construction of hydraulic components and systems is the use of plastics as a construction material [1–4]. The use of plastics is advantageous for structural, technological, operating [5] and economic reasons. Due to current trends, the Fluid Power Research Group (FPRG) from the

© The Editor(s) (if applicable) and The Author(s), under exclusive license
to Springer Nature Switzerland AG 2021
J. Stryczek and U. Warzyńska (Eds.): NSHP 2020, LNME, pp. 386–395, 2021.
https://doi.org/10.1007/978-3-030-59509-8_35

Department of Fundamentals of Machine Design and Mechatronic Systems from the Wroclaw University of Science and Technology, began work to create a complete hydraulic system, in which basic elements such as a pump, cylinder [6] or valves [7, 8], or at least their main parts will be made of plastic [9, 10].

Work on the hydraulic system with plastic elements started with the design and production of a gerotor pump. Initially, only the epicycloidal gear set, was made of plastic and the rest of the pump: body, shaft and bearings, were made of metal. The papers [9, 12] presents the research of a gerotor gear set made of plastics and it was proved that gerotor sets made of various materials can work in gerotor pumps with an aluminum body. Theoretical and experimental research of the gerotor pump with gears made of plastic confirm the possibility of using plastics as a construction material in hydraulics.

These positive results allowed for the next step in the design of the entire plastic body of the gerotor pump. The work [17] presents a design solution of a gerotor pump with a plastic body, while [14] presents the design principles of a gerotor pump with plastic gears and body. Therefore, the goal of the paper was established to presents the prototype of the complete plastic gerotor pump and its functional tests (checking measurements, run-in process) and experimental research of the efficiency characteristics.

2 Plastics as a Construction Material

The analysis and selection of plastic is the first and key stage in the design of hydraulic components. From a wide set of parameters characterizing the plastic, a narrower set of parameters is chosen, which are important from the point of view of the pump body design. Those are:

- yield strength *Re* and Young's modulus *E* - characterizing the strength of the material,
- shrinkage *S* and coefficient of linear expansion *W* - characterizing the dimensional stability of the material during body manufacture,
- working temperature *T* and coefficient of water absorption *A* - characterizing the dimensional stability of the material during operation,
- price and availability of plastic on the market.

In gerotor pumps so far designed and tested by the FPRG, plastics POM, PPS and PEEK were used to make gear sets [11–13]. Experimental research on plastic gears began with the basic POM material [15, 16]. This material is easily available on the market, cheap and easy to machine processing. It also has good strength properties (E = 3000 MPa, Re = 60 MPa), good dimensional stability during manufacture (S = 0,0285 cm/cm, W = 0,0001 cm/°K) and good dimensional stability during operation (A = 0,7%, T = 100 °C). In subsequent phases of experimental research, materials with better properties were used, i.e. PPS [11, 12] and finally PEEK [9]. It was decided to use the same material selection cycle to make the entire pump body. Therefore, POM was used to make the first prototype of the pump body [14, 17]. Experimental research is aimed at checking the proper functioning of the pump with a plastic body, therefore, to eliminate the effect of deformation of the gear set on the pump characteristics, it was decided to choose a better, more stable PEEK material for the gerotor gear set. PEEK has higher than POM strength properties (E = 12700 MPa, Re = 230 MPa), higher dimensional stability during manufacture (S = 0,005 cm/cm, W = 0,00004 cm/°K) and higher dimensional stability during operation (A = 0,4%, T = 240 °C).

3 The Plastic Pump Prototype

Using the modification rules presented in [17, 18] a new modified shape of the gerotor pump body was developed. Figure 1 shows parts of a gerotor pump with a plastic body. The front (Fig. 1a), middle (Fig. 1b) and back (Fig. 1c) bodies are made of POM plastic, while the gear set (Fig. 1d) was made of PEEK plastic. The driving shaft is made of steel and the slide bearing (Fig. 1b) in which the gear set rotates is made of bronze. Due to the fact that we are dealing with a unit - prototype product, in order to eliminate the costs associated with the implementation of the injection mold, the pump prototype was made by machining processing.

(a) (b) (c) (d)

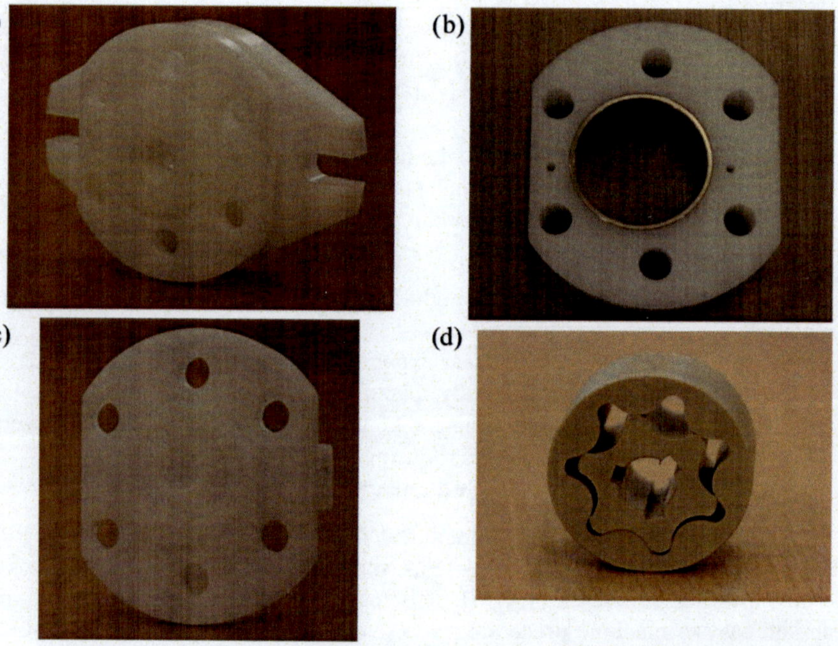

Fig. 1. Gerotor pump body parts made of plastics: a) front body; b) middle body with slide bearing; c) back body; d) epicycloidal gear set

The produced elements were subjected to checking measurements. The surfaces that have a significant impact on the assembly and proper functioning of the pump have been measured, i.e. the diameters of the bearings, driving shaft and gear set, as well as the width of the wheel assembly and central body. Real, measured dimensions were consistent with drawing dimensions, the accuracy of the elements was within the dimensional tolerances adopted in the technical documentation. All surfaces were smooth and clean.

After completing the measurements, checking and verifying surfaces, pump was assembled. The assembly proceeded without any disturbances, which was favored by the precise production of the body parts. All necessary elements, i.e. bearings, dowels, seals, gear set and driving shaft were embedded in the body, then the complete pump was tightened with six bolts and a torque $M_s = 30$ Nm. After tightening, it was possible to rotate the driving shaft, and thus a gear set, without jamming and the need to apply a lot of force, which indicates the correct assembly of the pump.

The assembled pump was placed on the test stand. The pump was attached to the console by means of a specially formed mounting flange in the pump's front body and two fastening screws tightened with a torque $M_s = 30$ Nm. Appropriate dimensions and fits made it possible to maintain concentricity between the pump shaft and the drive shaft of the electric motor. The connection of both shafts was ensured by two bellows couplings between which a torque meter was mounted. After attaching the pump to the console at the stand, supply lines were tightened to it. Figure 2 shows the view of the pump on the test stand.

Fig. 2. Pump mounted on the test stand

4 Experimental Research of the Pump Model

Figure 3a shows a diagram of a test stand for conducting basic hydraulic research of a gerotor pump with a plastic body. The test stand enables measurements of hydraulic parameters, i.e. flow (*Q [l/min]*) and the pressure at the inlet and outlet of the pump

(a)

(b)

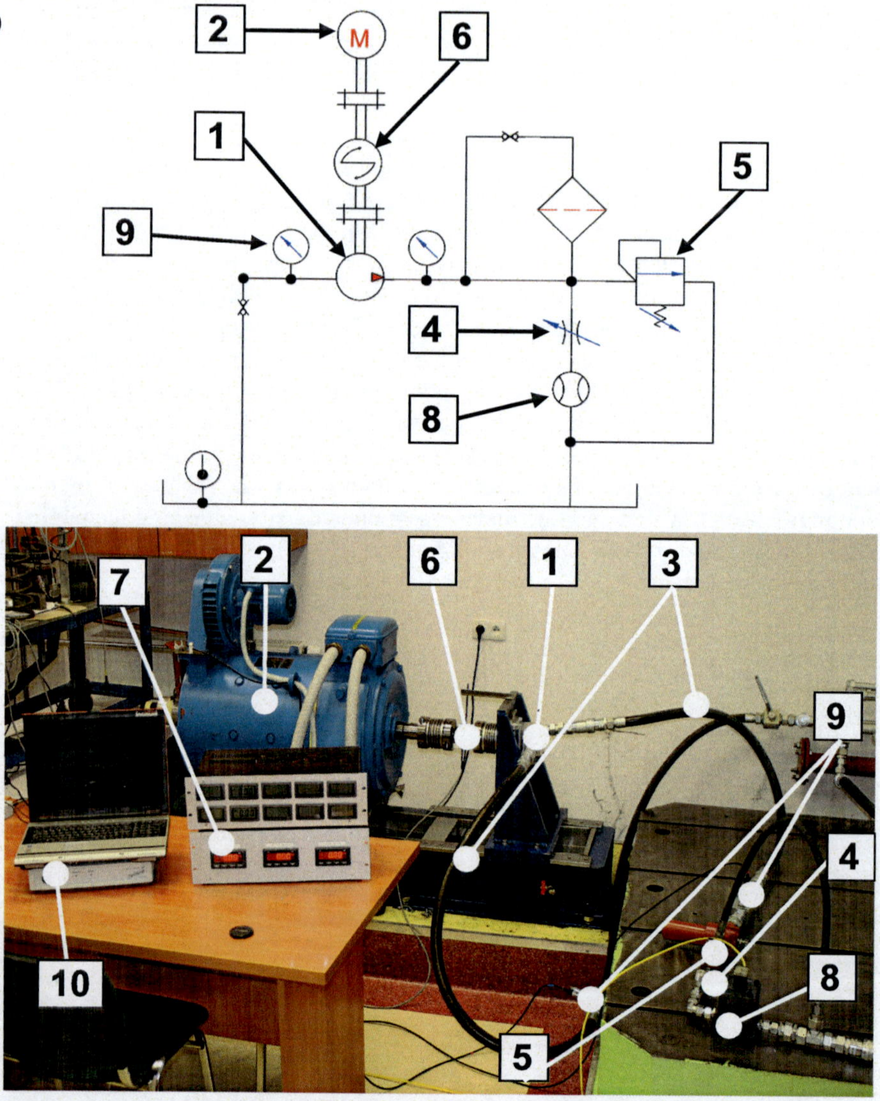

Fig. 3. Test stand: a) diagram of a test stand; b) view of the stand at the FPRG Laboratory: 1 – tested pump, 2 – electric motor, 3 – input and output lines, 4 – throttle valve, 5 – overflow valve, 6 – torque meter, 7 – tachometer, 8 – flow meter, 9 – pressure sensors, 10 - signal amplifier

(*p [bar]*), and measurements of mechanical parameters, i.e. torque (*M [Nm]*) and rotational speed (*n [rpm]*).

The view of the actual stand is shown in Fig. 3b. The tested pump (1) is driven by an electric motor (2). Connection of the pump with the hydraulic system is ensured by supply lines (3). A throttle valve (4) is used to load the pump. The overflow valve (5) was used to protect the pump and electric motor against overload. The drive torque of the pump was measured with a torque meter (6), and the rotational speed with a tachometer (7). The flow rate was measured with a flow meter (8) and the pressure in the input and output lines with pressure sensors (9). Signals from measuring instruments will be sent to a computer using a laboratory signal amplifier (10). The CatmanEasy program was used to archive and analyze measurements on a computer.

The basic hydraulic research program of the gerotor pump included:

- checking the correct operation of the pump with a plastic body,
- determination of volumetric η_v and total η efficiency of the pump depending on various values of rotational speed,
- determination of volumetric η_v and total η efficiency of the pump depending on various values of working pressure.

5 Results and Analysis of Basic Hydraulic Research

5.1 Run-in Process

Before commencing the detailed hydraulic characteristic tests, the pump was started and run-in was performed. It has been found that the pump works correctly, performs its function continuously and without interference. There was no unsealing of the body and no external leaks.

After run-in, the pump was dismantled and its parts checked in order to determine the effects of parts' collaboration and their possible wear. During the inspection of the pump parts, the effects of lapping plastic were noticed. The lapping effects are shown in Fig. 4. Figure 4a shows the material overflow in the form of a circle. It was caused by friction of the rotating gear set on the faces of front and back body. As a result, the material was "stuffed" into the free space between the gear set and the bearing in which the gears were rotating. Figure 4b shows a blocked leakage channel that drains hydraulic oil from under the shaft sealing ring. This is the effect of lapping, excess material has been "pushed" into the channel hole, which caused it to clog. The lapping effects have been removed, and no such effects appeared on subsequent tests.

(a)

(b)

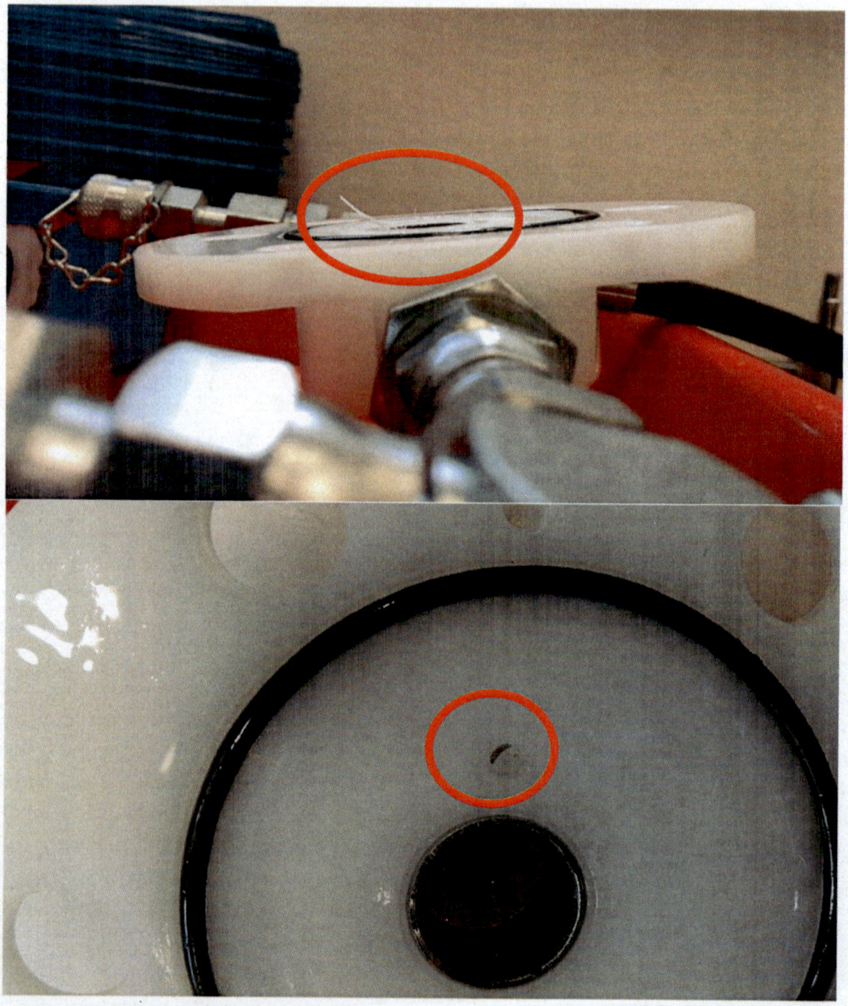

Fig. 4. Plastic lapping products: a) overflow of material in the form of a circle; b) blocked leakage channel that drains hydraulic oil from under the shaft sealing ring

5.2 Basic Hydraulic Characteristics

After checking the parts and removing the lapping effects, the pump was reassembled on the test stand and basic hydraulic characteristics were tested. Figure 5 shows the characteristics of volumetric η_v and total η efficiency depending on the output pressure, for different shaft rotational speeds. The drawing shows that the pump was operating in the working pressure range $p = 0\text{–}20$ bar and rotational speed $n = 500\text{–}1000$ rpm. The volumetric efficiency η_v varied in the range $\eta_v = 89\text{–}70\%$ while the total efficiency η in the range $\eta = 73\text{–}50\%$. Relatively low efficiency prevented further loading of the pump, which could lead to seizing. The figure shows that the volumetric η_v and total

η efficiency decrease with increasing output pressure p and increasing rotational speed n of pump shaft.

This is explained by the fact that as the pressure and speed increase, the body deformations increase [19], which cause the formation of uncontrolled internal gaps, which in turn cause the formation of increased leaks in the pump which causes a decrease in volumetric η_v and total η efficiency.

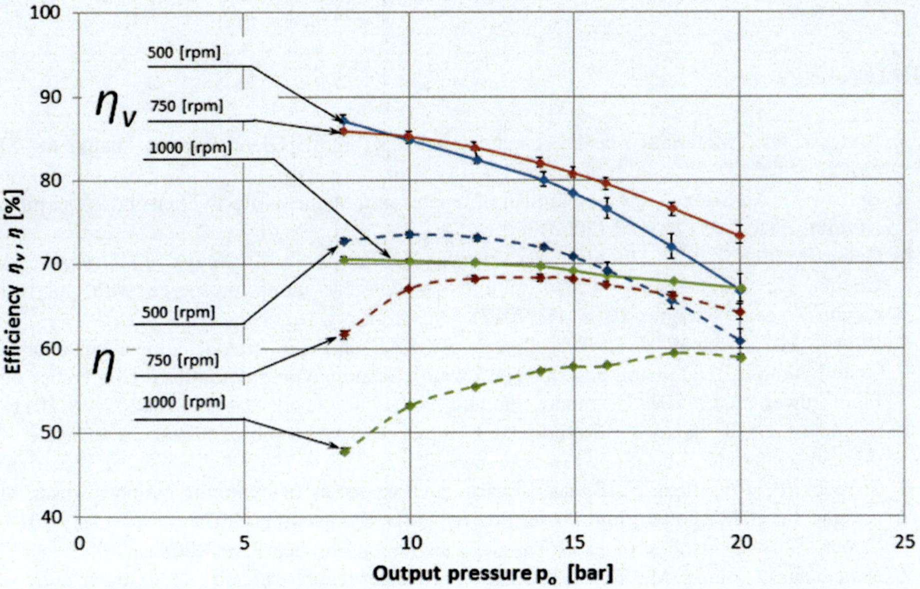

Fig. 5. Characteristics of volumetric and total efficiency depending on the output pressure η_v, $\eta = f(p_o)$, for different pump shaft rotational speeds n = 500 ÷ 1000 obr/min

6 Summary and Conclusions

Experimental research of the plastic gerotor pump confirm that plastics as a construction material can be used in hydraulics.

Experimental studies confirmed the correctness of the structural solution of the pump body. The pump with a plastic body was working correctly as a low pressure unit, in the pressure range $p = 0 \div 20$ bar and rotational speed in the range $n = 500 \div 1000$ rpm, obtaining volumetric efficiency $\eta_v = 89\text{–}70\%$ and total efficiency $\eta = 73\text{–}50\%$. However, the selection of POM material for the pump body proved to be not completely accurate. This material is not suitable for working with higher technical parameters $p > 20$ bar and $n > 1000$ rpm. Materials with better parameters should be looked.

At the same time, the need to test the deformation of the pump body is visible, to check the influence of operational factors such as pressure, rotational speed and time of their operation on the values of axial and radial deformations of the gerotor pump body.

The pump body made of plastic should work in the elastic range so that the body deformations in critical places take values lower than the permissible deformations [20], so that there are no internal leaks that reduce the technical parameters of the pump.

It is anticipated that as a result of further improvement of the entire construction and technological process it will be possible to obtain a pump with higher technical parameters, working at higher working pressures of the range $p = 6$ MPa, with volumetric efficiency $\eta_v = 70\%$ and total efficiency $\eta = 60\%$.

References

1. Bonanno, A.: Matarialli plastici, caratte ristiche a limitti. Oleodynamica Pneum 68–73 (2008)
2. Bełzowski, A., Stróżyk, P.: Assessment of repair reinforcement of polyester-glass fibre pipe. KOMPOZYTY **2**, 179–184 (2008)
3. Gamez-Montero, P.J., Antoniak, P., Castilla, R., Freire, J., Krawczyk, J., Stryczek, J., Codina, E.: Magnet-sleeve-sealed mini trochoidal-gear pump prototype with polymer composite gear. Energies **10**, 1–18 (2017). https://doi.org/10.3390/en10101458
4. Rokala, M., Kaskinen, K.T.: The effect of PV-RAE on PEEK-MADE slipper behavior in water hydraulic axial piston unit. In: The Twelfth Scandinavian International Conference on Fluid Power SICFP 2011, Tampere, Finland, vol. 1, no. 4, pp. 89–101, 18–20 May 2011
5. Rodionov, L., Rekadze, P.: Experimental vibroacoustic research of a gear pump made of different materials. Proc. Eng. **176**, 636–644 (2017)
6. Stryczek, P., Przystupa, F., Banas, M.: Research on series of hydraulic cylinders made of plastics. In: 2018 Global Fluid Power Society Ph.D. Symposium (GFPS), 18–20 July 2018, Samara, Russia. Institute of Electrical and Electronics Engineers Inc. (2018)
7. Marciniak, L., Banaś, M., Stryczek, J.: The design and theoretical and experimental study of the plastic hydraulic valves. In: 9th FPNI Ph.D. Symposium on Fluid Power, Florianópolis, SC, Brazil, 26–28 October 2016 (2016)
8. Banaś, M., Antoniak, P., Marciniak, L., Stryczek, J.: Visualization of flow phenomena in hydraulic throttle valves of plastics. MATEC Web Conf. **211**, 19001 (2018). https://doi.org/10.1051/matecconf/201821119001
9. Stryczek, J., Banaś, M., Krawczyk, J., Marciniak, L., Stryczek, P.: The fluid power elements and systems made of plastics. Proc. Eng. **176**, 600–609 (2017)
10. Krawczyk, J., Stryczek, J.: Układ hydrauliczny z elementami wykonanymi z tworzyw sztucznych; Hydraulic system with parts made of plastics. Górnictwo Odkryw **54**(3), 52–57 (2013)
11. Krawczyk, J., Stryczek, J.: Badania pompy gerotorowej z kołami zębatymi wykonanymi z PPS. Hydraul. i Pneumatyka **34**, 15–18 (2014)
12. Krawczyk, J., Stryczek, J.: Construction and experimental research on plastic cycloidal gears used in gerotor pumps. In: 8th FPNI Ph.D. Symposium on Fluid Power, FPNI 2014, Lappeenranta, Finland, 11–13 June (2014)
13. Krawczyk, J., Stryczek, J.: Pompa gerotorowa z tworzyw sztucznych. In: Napędy i sterowania hydrauliczne i pneumatyczne 2015, Szklarska Poręba-Jakuszyce, 18–20 marca 2015. Wrocław : Ośrodek Doskonalenia Kadr SIMP, pp. 164–170 (2015)
14. Krawczyk, J., Stryczek, J.: Design and experimental research of a plastic gerotor pump. In: 2018 Glob. Fluid Power Soc. Ph.D. Symp. (GFPS), 18–20 July 2018, Samara, Russ. 1–8 (2018). https://doi.org/10.1109/GFPS.2018.8472370

15. Stryczek, J., Bednarczyk, S., Biernacki, K.: Gerotor pump with POM gears: design, production technology, research. Arch. Civil Mech. Eng. **14**(3), 391–397 (2013). https://doi.org/10.1016/j.acme.2013.12.008
16. Biernacki, K., Stryczek, J.: Analysis of stress and deformation in plastic gears used in gerotor pumps. J. Strain Anal. Eng. Des. **45**, 465–479 (2010). https://doi.org/10.1243/03093247JSA630
17. Krawczyk, J., Stryczek, J.: Designing of the gerotor pump body made of plastics. In: 9th FPNI Ph.D. Symposium on Fluid Power, Proceedings of ASME, Florianópolis, SC, Brazil, 26–28 October 2016, pp. 1–6 (2016)
18. Krawczyk, J.: Metodyka kształtowania korpusów maszyn hydraulicznych z tworzyw sztucznych na przykładzie pompy gerotorowej. Wroclaw University of Science and Technology, doctorate thesis (2017)
19. Kujawa, M., Kowalewski, P., Wieleba, W.: The influence of deformation under tension on some mechanical and tribological properties of high-density polyethylene. Polymers (Basel) **11**, 1429 (2019). https://doi.org/10.3390/polym11091429
20. Stefan, S.: Napęd hydrostatyczny: Podręcznik. Wrocław, WNT (1984)

Check for updates

Numerical and Experimental Analysis of the Base of a Composite Hydraulic Cylinder Made of PET

Marek Lubecki[1(✉)], Michał Stosiak[1], and Małgorzata Gazińska[2]

[1] Faculty of Mechanical Engineering, Department of Technical Systems Operation and Maintenance, Wroclaw University of Science and Technology, Wroclaw, Poland
{marek.lubecki,michal.stosiak}@pwr.edu.pl

[2] Faculty of Chemistry, Department of Polymer Engineering and Technology, Wroclaw University of Science and Technology, Wroclaw, Poland
malgorzata.gazinska@pwr.edu.pl

Abstract. The paper presents the process of designing a base made of PET that will be used as an element of a hydraulic cylinder made of composite materials. The requirements for the designed element include the possibility of working with mechanical and thermal loads comparable to conventional cylinders. The choice of material based on the set criteria and assigned weights, experimental research of its real properties, numerical calculations of the element as well as experimental verification on the real object were presented. The material's non-linearity and the change of its parameters as a function of working temperature were taken into account.

In order to obtain real values of tensile strength and Young's modulus, static tensile tests were carried out in accordance with ISO 527–2. The obtained results allowed to include non-linear material properties in the numerical calculations. To know the glass transition temperature and at the same time make sure that the elements made of PET material can be used at temperatures up to 80stC, measurements were carried out using the DMA (Dynamic Mechanical Analysis) method in the temperature range from −120 °C to 250 °C.

Numerical calculations were made using Ansys Mechanical 19.0 software. After the design stage was completed, a base prototype was made that was subjected to static tests in the pressure range up to 25 MPa. Element strains were recorded using strain gauges, which allowed validation of the numerical model.

Keywords: Composite hydraulic cylinder · PET · FEM · DMA

1 Introduction

In recent years, there has been a large development of hydraulic elements and systems, with particular emphasis on the downsizing trend of reducing the dimensions or weight of components while maintaining existing operating parameters (nominal pressures,

© The Editor(s) (if applicable) and The Author(s), under exclusive license to Springer Nature Switzerland AG 2021
J. Stryczek and U. Warzyńska (Eds.): NSHP 2020, LNME, pp. 396–405, 2021.
https://doi.org/10.1007/978-3-030-59509-8_36

efficiency, resistance to operating conditions). The development trend aimed at minimizing dimensions has been named microhydraulics [1]. Another trend in the design of hydraulic elements is the use of new materials, e.g. for pumps, valves or hydraulic cylinders. Research work is being carried out on the use of plastics such as polyoxymethylene (POM) in the construction of the above-mentioned of elements [2].

In hydraulics, composite materials (mostly in the matrix of epoxy resins, reinforced with continuous carbon fibers) are currently used for the manufacturing of accumulators. The gain resulting from the weight reduction is significant in this case, and the design methodology and manufacturing technology show many analogies to well-known composite pressure vessels [3, 4]. The second area of hydraulics where composite materials are used is the manufacture of cylinders and piston rods hydraulic actuators [5].

The rest of the parts of the hydraulic actuator - the end caps and the piston - have a much more complicated geometry. To make them from composite materials, it would probably be necessary to give up continuous fiber reinforcement in favor of staple fibers or powder reinforcement. One could then make such details by methods of injection or 3D printing. However, the easiest way to reduce the weight of these components is to make them out of polymeric materials. In addition to reducing the weight of the element, the elimination of metallic materials brings additional benefits in the form of increased resistance to the work environment, which is particularly unfavorable for hydraulic drive components [6].

This paper presents the process of designing the base of composite hydraulic cylinder. The base itself was made of PET (polyethylene terephthalate).

2 Material and Object of the Tests

2.1 PET Material

The tested material is polyethylene terephthalate (PET) with the trade name TECAPET White manufactured by Ensinger Plastics. It is a thermoplastic plastic from the group of polyesters widely used in various industries for the production of fibers (in the textile industry), bottles and other food containers, as well as machine parts and as a matrix of composites. The basic material properties declared by the manufacturer are presented in Table 1.

Table 1. Basic properties of TECAPET material, declared by the manufacturer.

Parameter	Value
Young's modulus [MPa]	3100
Tensile strength [MPa]	79
Bulk modulus [MPa]	2700
Glass transition temperature [°C]	81

2.2 Tested Base

The tested element was the bottom of the hydraulic cylinder. Its task is to seal one of the cylinder ends (opposite to the piston rod) and to supply and drain the working fluid from the cylinder. The technical drawing of the designed end cap is shown in Fig. 1.

The bottom will be connected to the opposite gland by means of four screws (pins), thus preventing these elements from moving along the cylinder axis. The connection to the hydraulic system will take place through a steel connector located coaxially with the cylinder.

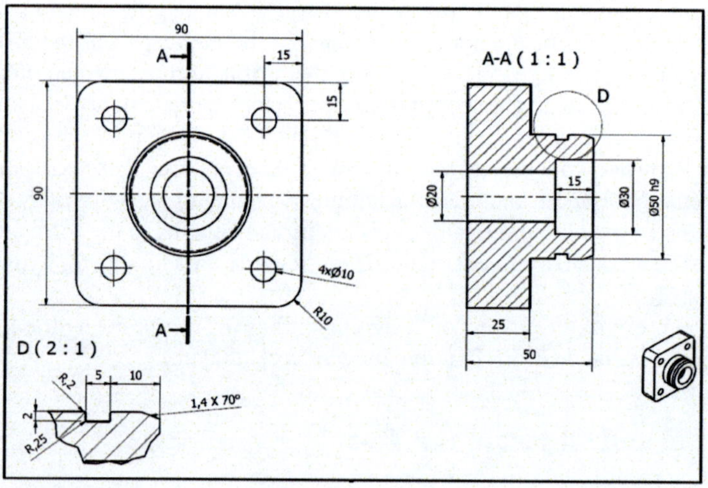

Fig. 1. Technical drawing of a tested base of a hydraulic cylinder.

3 Methods

3.1 DMA Measurements

The DMA method was used to study viscoelastic behavior of PET, to determine the complex modulus and to locate glass transition temperature. A sinusoidal load with a given frequency and force amplitude was applied to the sample and material deformation was measured. The change in sample temperature over time has allowed the identification of the glass transition temperature.

To determine the glass transition temperature, shear modulus and flexural modulus, DMA tests were carried out in two modes - shear and three point bending. The tests were performed on a METTLER TOLEDO DMA/SDTA1 + dynamic mechanical analyzer, having a force range from 1 mN to 40 N and a frequency range from 0.001 Hz to 1000 Hz.

The diagram of tests in shear mode is shown in Fig. 2 b). They consist of placing two disks of known geometry, made of test material (2), between the fixed parts (3) and the movable parts (1) of the holder and pressing them together, which prevents slipping.

Then the movable part (1) of the handle is set in a vertical oscillating motion with a given value of frequency and force amplitude. At the same time, the ambient temperature is changed. Registration of the displacement allows to determine the value of the shear modulus as a function of temperature and to determine the glass transition temperature of polymers characterized by a rapid decrease in the storage modulus value. The tests were carried out at a frequency of 1 Hz and in a temperature range from −120 °C to 200 °C, with heating rate of 3 °C/min.

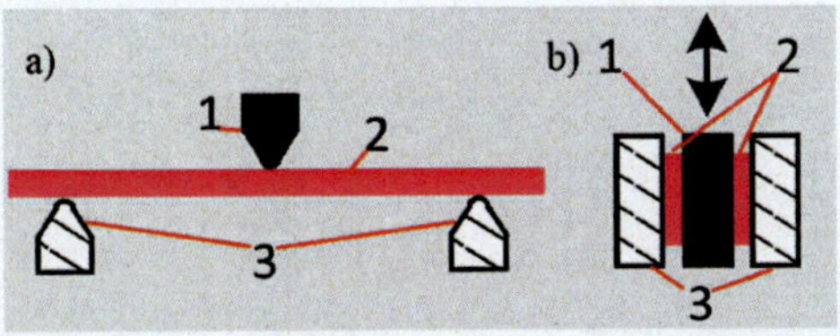

Fig. 2. Scheme of a DMA measurement in a) three-point bending mode, b) shear mode.

Diagram of tests in bending mode is shown in Fig. 2 a). They rely on cyclic, three-point bending of a beam with known geometry. The sample (2) is placed on fixed supports (3), after which it is preloaded with a pin (1). Then the pin (1) performs oscillating vertical movements with a given force amplitude and frequency and the measuring system reads the displacement value, which allows calculating the flexural modulus (which in case of isotropic material is equal to Young's modulus). The key is to set the correct offset force, so that the pin will be in constant contact with the sample during the test, while the deformation of the sample will remain in the linear range. The tests were carried out at ambient temperature.

3.2 Tensile Tests

The static tensile test of plastics was carried out in accordance with ISO 527–2 [7]. In the case of the described tests, type 1B samples in the number of 6 were used. The tests were carried out on a Tinus Olsen H25KT universal testing machine and the speed was 1 mm/min.

3.3 Strain Gauges Measurements

To validate the FEM models, measurements of the base's deformations were made using strain gauges. For this purpose, TENMEX TF-5 strain gauges were used, the basic parameters of which are given in Table 2. The HBM QuantumX MX440B DAC interface connected to a PC and Catman AP software were also used.

Places where sensors are located on the base are shown in Fig. 3. To glue the strain gauges in accordance with the manufacturer's recommendations, ThreeBond 1731 cyanoacrylate adhesive was used. The tested elements were placed on a test stand intended to simulate real working conditions and were loaded with quasistatic internal pressure in the range of 0–25 MPa.

Table 2. Basic parameters of strain gauges

Parameter	Value
Gauge constant [−]	2.19 ± 2%
Resistance [Ω]	120 ± 2%

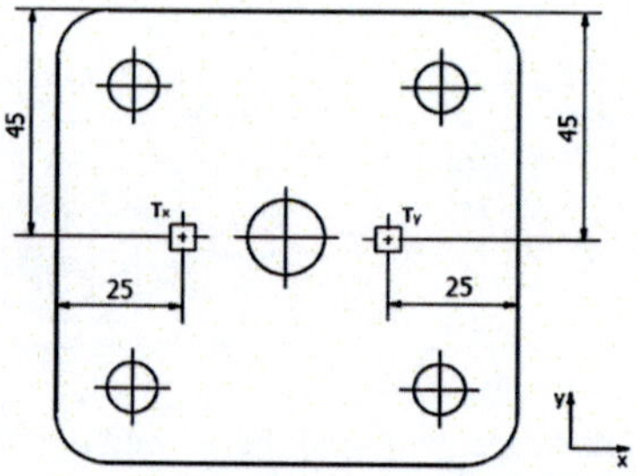

Fig. 3. Diagram of the location of strain gauges on the base.

3.4 FEM Modelling

FEM calculations were carried out using ANSYS Workbench 19.0 software. In the SpaceClaim module, a model of a ¼ of the base was created (Fig. 4). The hydraulic connector and washer for the nut were also modeled.

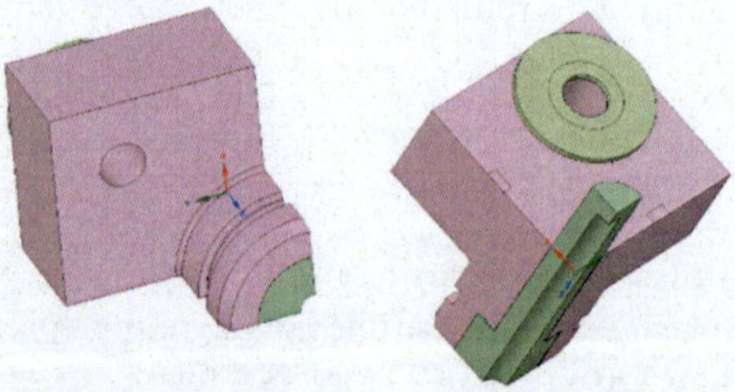

Fig. 4. Model of ¼ of the base geometry.

Model components have been given symmetry constraints to the XZ and YZ planes to simulate the behavior of the entire element. Frictional contacts with a friction coefficient of 0.2 were defined between the washer and the base as well as the hydraulic connector and the base.

The model was meshed with higher-order tetra finite elements 1 mm in size, which gave a total of 102599 elements and 159640 nodes.

The model was supported by taking away all degrees of freedom of the upper part of the washer (a) by simulating the operation of the screw connection and by taking away the possibility of translation along the X and Y axes in the screw hole (b) as well as in one of the cylindrical parts cooperating with the cylinder (c) (Fig. 5). The element was loaded with 20 MPa pressure simulating the actual pressure acting on the element during operation (d).

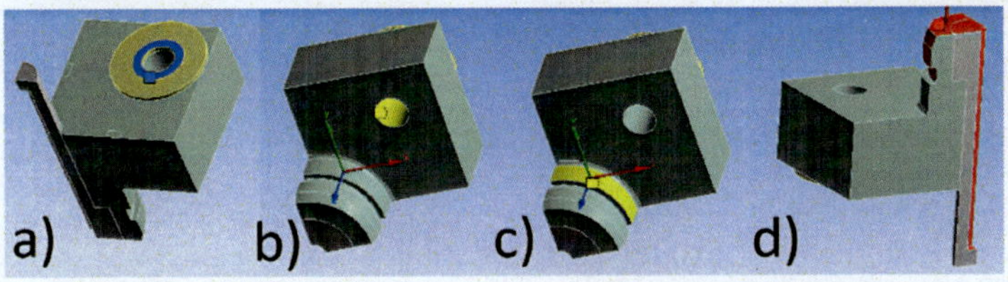

Fig. 5. Supports (a, b, c) and load (d) of the FEM model.

4 Results

4.1 DMA Results

The first part of DMA testing in bending mode was the determination of the offset force required for correct measurements. Figure 6 presents the results of such preliminary tests. The measurements were carried out at a constant frequency of 1 Hz. The offset force was changed over time and the value of the complex Young's modulus was recorded. The highest stable value of the E* module was adopted as the criterion for choosing the right offset force. It was obtained for an offset force of 12 N.

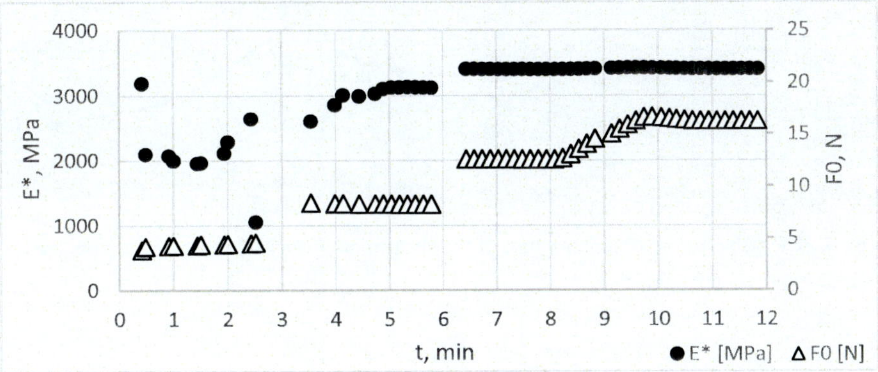

Fig. 6. The value of complex Young's modulus and offset force in time.

Then proceeded to the main part of the measurements. The sample was loaded with an initial force of 12 N and then the pin made oscillatory movements with increasing displacement amplitude. The force registered by the device allowed calculating the complex Young's modulus E* and its real part E' and the imaginary part E". Figure 7 shows the change in module E* value and amplitude of pin displacement during the test.

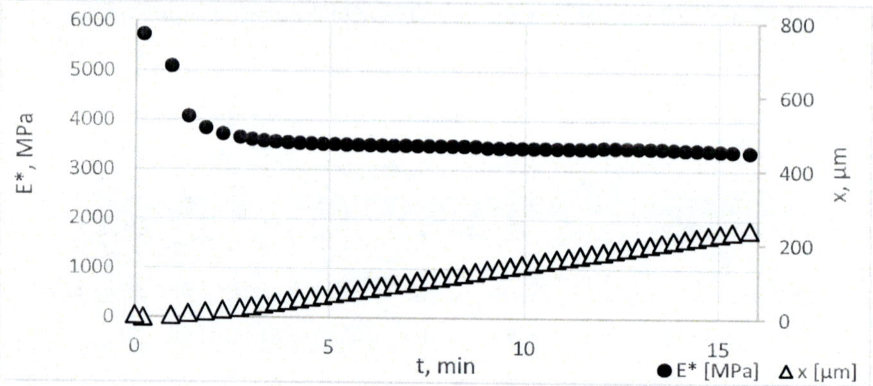

Fig. 7. The value of complex Young's modulus and displacement during the test.

One can see that as the amplitude of displacement increases, the E* module decreases, after which it stabilizes around 3400 MPa.

In shear mode, the shear modulus value was determined depending on the sample temperature. This allowed the material glass transition temperature to be determined. The glass transition temperature was determined by several methods. In this study, a method of finding the temperature at which the maximum peak of the imaginary part of the shear modulus occurs was used. The glass transition temperature determined in this way was 86 °C (Fig. 8).

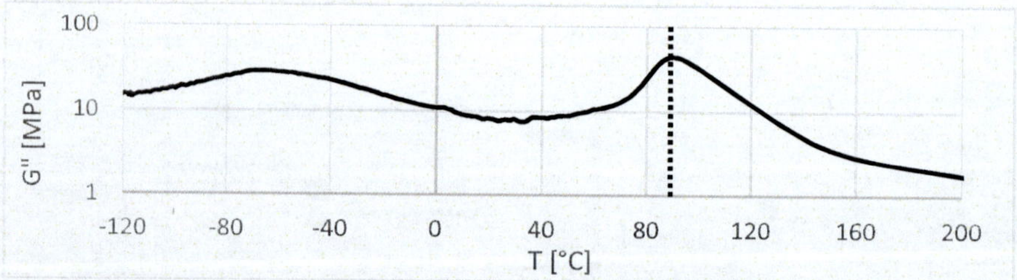

Fig. 8. The value of the imaginary part of the shear modulus as a function of temperature.

4.2 Tensile Test Results

Table 3 shows the material parameters obtained from the static tensile test. It can be seen that they differ (especially the value of Young's modulus) from those given by the manufacturer. Figure 9 presents an example of a stress strain diagram.

Table 3. Material constants obtained from the static tensile tests

Parameter	Value
Young's modulus [MPa]	3652 ± 240
Tensile strength [MPa]	73.3 ± 3.2

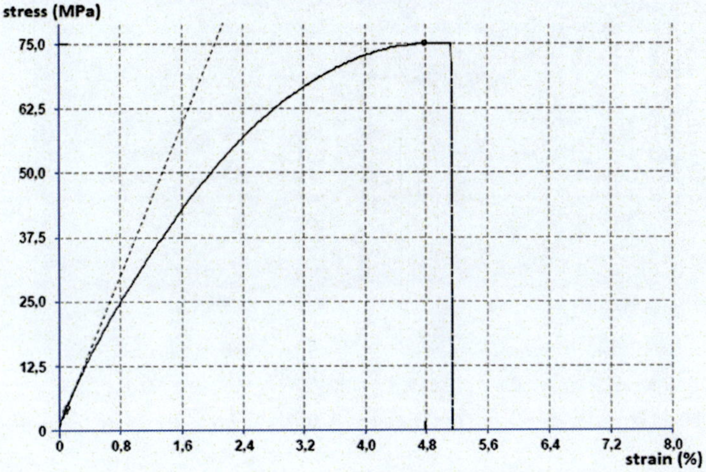

Fig. 9. Stress strain diagram for TECAPET White material obtained from static tensile test.

4.3 FEM Results

Material data obtained from the static tensile test (including material nonlinearity) was implemented for FEM calculations. Figure 10 presents strain results of strain gauges respectively in the X (left) and Y (right) axes obtained by FEM simulation. The average strains were 450 μm/m in the X axis and 2016 μm/m in the Y axis.

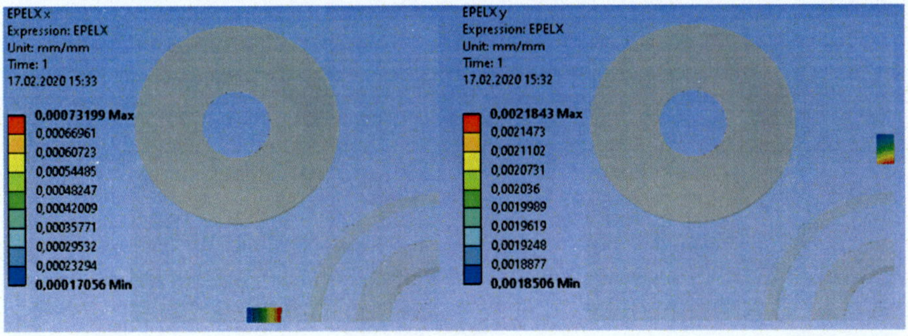

Fig. 10. Strain gauges strains in the X axis (left) and in the Y axis (right) resulting from FEM calculations.

Figure 11 shows the reduced stress according to the Huber-von Mises hypothesis. The maximum stress value is 45.96 MPa, which is much lower than the tensile strength of the material used.

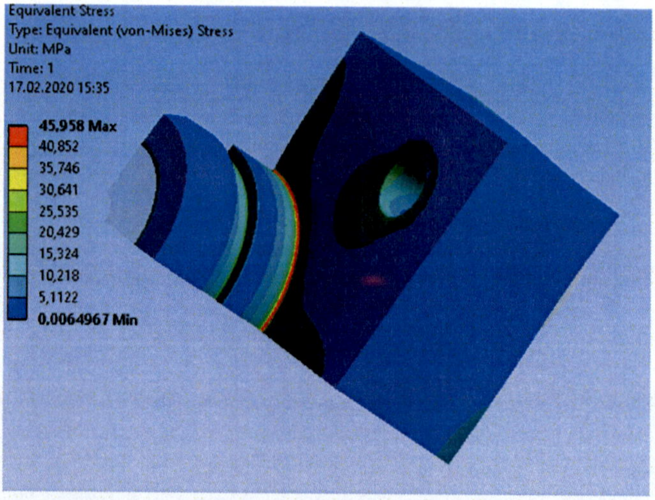

Fig. 11. Reduced stress according to the Huber-von Mises hypothesis occurring in the tested base

4.4 Strain Gauge Tests Results

Figure 12 shows the relationship between the pressure inside the cylinder and the strains recorded by the strain gauge in the Y axis. Similarly, Fig. 13 shows the results for the strain gauge in the X axis.

For a pressure of 20 MPa, strain in the X axis was 576 μm/m, and in the Y axis 5450 μm/m.

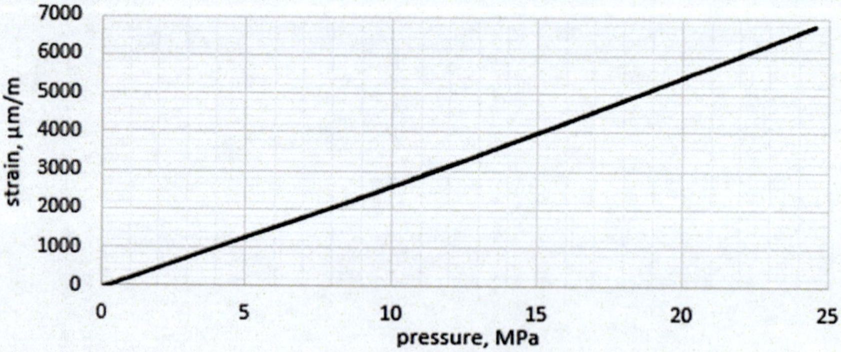

Fig. 12. Strain in the Y axis recorded by strain gauges

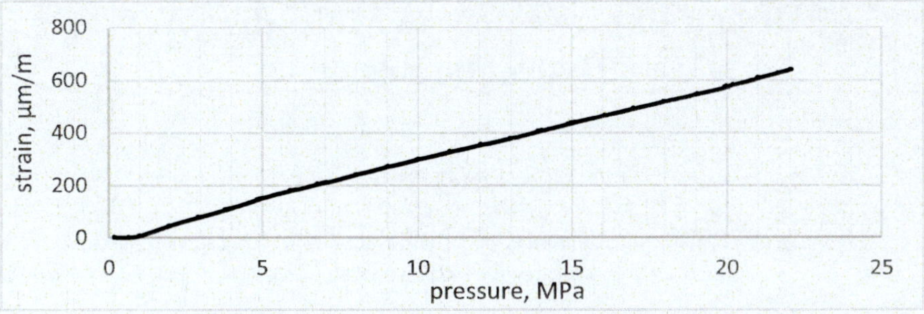

Fig. 13. Strain in the X axis recorded by strain gauges

5 Conclusion

The paper presents the results of PET plastic material tests, FEM simulation of the hydraulic cylinder base and their experimental validation. It has been shown that such an element made of polymeric material can successfully work in a hydraulic system where the working pressures reach 20 MPa and the oil temperature does not exceed 80 °C (which is a common recommendation when designing hydraulic systems).

Acknowledgments. Calculations have been carried out using resources provided by Wroclaw Centre for Networking and Supercomputing (http://wcss.pl), grant "Obliczenia wytrzymałościowe wysokociśnieniowych elementów maszyn z tworzyw sztucznych i materiałów kompozytowych.".

References

1. Kollek, W.: Microhydraulic components and systems – fundamentals of design, modelling and operation (in Polish). Oficyna Wydawnicza Politechniki Wrocławskiej, Wrocław (2011)
2. Stryczek, J., Banaś, M., Krawczyk, J., Marciniak, L., Stryczek, P.: The fluid power elements and systems made of plastics. Procedia Eng. **176**, 600–609 (2017)
3. Błażejewski, W.: Composite pressure vessels reinforced with fibers according to mosaic patterns (in Polish). Oficyna Wydawnicza Politechniki Wrocławskiej, Wrocław (2013)
4. Błażejewski, W., Czulak, A., Gąsior, P., Kaleta, J., Mech, R.: SMART composite high pressure vessels with integrated optical fiber sensors. In: Masayoshi, T., (eds.). Sensor and Smart Structures Technologies for Civil, Mechanical and Aerospace Systems 2010, Proc. SPIE, vol. 7647 (2010)
5. Solazzi, L.: Design and experimental tests on hydraulic actuator made of composite material. Compos. Struct. **232**, 111544 (2020)
6. Śliwiński, P.: The influence of water and mineral oil on volumetric losses in the displacement pump for offshore and marine applications. Pol. Marit. Res. **2**(102), 173–182 (2019)
7. International Standards Office: ISO 527–2:2012 Plastics — Determination of tensile properties — Part 2: Test conditions for moulding and extrusion plastics (2012)

Design Improvement of Multi-disc Wet Hydraulic Brake

Milos Matejic(✉)

Faculty of Engineering, University of Kragujevac, 34000 Kragujevac, Serbia
mmatejic@kg.ac.rs

Abstract. Nowadays the most commonly used brakes are dry ones. However in special situations, harsh and aggressive environments, there was a need for more durable and more powerful brakes. The biggest advantage of wet multi-disc brakes is that they can be used in big variety of conditions, from extremely cold to the extremely hot, from very clean to very dusty environment and so on. Because of their very responsible part in heavy machinery design, wet multi-disc brakes presents a very attractive and hard task as for mechanical engineer as for researchers all around the world. This paper presents a practical approach to wet multi-disc brake design improvement. Namely, the most of present wet multi-disc brakes uses a radial placed compression springs for brake activation. This solution leads to lots of difficulties during brakes manufacturing and brakes maintenance. At this paper is presented concept of compression springs replacement with belleville springs. Concept is proven by hand and FEM calculations. At the end of the paper conclusion and further research directions are given.

Keywords: Multi-disc wet hydraulic brake · Compression spring · Belleville spring

1 Introduction

Brakes are devices most commonly used for stopping or decelerating various types of machines such as: vehicles, aircrafts, machinery, machinery moving parts and etc. According to the way of activating brakes can be divided into mechanic, hydraulic or electromagnetic. Related to coolant usage brakes can be divided to dry brakes, which are not use coolant and wet brakes which are use coolant in their operation mode. Nowadays dry brakes have much more spread appliance related to wet ones. Dry brakes are used in cars, trucks, trailers, agricultural machinery and etc. Dry brakes are mostly used where the lower braking moment is required. Dry brakes have one two very big advantages related to wet ones: dry brakes are much less expensive and they are much simpler for replacement. When the torque needs to be effectively stopped the wet brakes are in advantage. This situation occurs at heavy machinery and equipment, especially in industries where the extreme conditions are occasional. Wet brakes are commonly used in: mining industry, food processing, transportation systems, special types of vehicles and

© The Editor(s) (if applicable) and The Author(s), under exclusive license
to Springer Nature Switzerland AG 2021
J. Stryczek and U. Warzyńska (Eds.): NSHP 2020, LNME, pp. 406–415, 2021.
https://doi.org/10.1007/978-3-030-59509-8_37

etc. Wet brakes main advantages related to dry brakes are: longer life-cycle, much bigger reliability, operation in very harsh conditions. The wet brakes is also called oil cooled disc brakes because they are using oil as coolant during their operation and also very common is oil usage for their activation.A dry brake are made with a single drum system, which causes heating up when they are put under pressure. Dry brake heating causing undue strain to the whole brake system. This phenomenon can contribute difficulties in dry brake operation and even the failure of the brake. The better solution is to use a wet brakes which are closed system. A wet brake has multiple discs, which it uses so that the instantly moving stopping. The multiple discs engagement is commonly done by hydraulics, while the oil between multiple discs prevents wet brake from overheating. Because of their interesting design, various processes happening, big advantages they have in operation in extreme conditions, the wet brakes are very attractive task for improving and researching among many researcher and companies all around the world.

One of the first papers in this research field was introduced by Zagrodzki [1], by exanimating of thermo mechanical phenomena in multidisc clutches and brakes. He used numerical method in order to explain phenomena appearing on various friction surfaces. Jullien et al. [2] did a research about behavior of wet clutches operating under continuous running conditions with a new carbon based material. This research group made comparisons between experimental and theoretical (hydrodynamic) results which indicated that the lubricant does not provide complete separation of the disc faces. In the research about wet brakes and clutches engagement characteristics Holgerson [3], has developed and apparatus of those characteristic measurement. In his paper he presented apparatus which can vary conditions for wet brake engagement: sliding velocity, inertia, force rate, and lubrication. He covered this research both theoretically and experimentally. Zhanling, et al. [4], has performed elastoplastic finite element analysis for wet multidisc brake during lasting braking. Their research was covered by mathematical model and numerical confirmation of presented model. Naunheimer et al. [5] in their book Fahrzeuggetriebegave the latest wet brakes and clutches designs and concepts data. Höhn et al. [6] presented how special additives in oil influence on wet brake discs. Liang et al. [7] conducted a numerical and experimental studies of a wet multidisc clutch on temperature and stress fields. Zhao et al. [8] presented the tribological characteristics of Cu-based friction pairs in a wet multidisc clutch under non-uniform contact. Jianzhong et al. [9] developed mathematical model in order to investigate the steady-state and axisymmetric flow conditions in deformed discs. They also run an experiment based on their model and get very similar results, which validates their mathematical model. Very interested way to obtain optimal parameters of multidisc wet brake is shown in paper [10]. The authors used heuristic optimization methods to solve various problems, among them multidisc magnetic brake parameters as well. Based on that paper, same method can be used in optimizing multidisc wet brakes. The idea for multidisc wet brake design improvement is born by reading paper [11] about unusual spring system. In that paper are shown ortho-planarsprings used for linear motion.

In this paper a practical approach to wet multi-disc brake design improvement is given. The most of present multi-disc wet brakes solution uses a radial placed compression springs for brake engagement. Solution with radial placed compression springs leads to lots of difficulties during brakes manufacturing, brake assembling and maintenance.

This paper presents concept with compression springs replaced bybellevillesprings. The concept feasibility is proven by analytical and FEM calculations. At the end of the paper conclusion and further research directions are given.

2 Old Design of Wet Brake

The old design of wet brake has a radially placed compression springs. Those springs by one end goes into 1st housing, and by second end they are mounted into wet brake piston. In Fig. 1 is shown old design of wet brake.

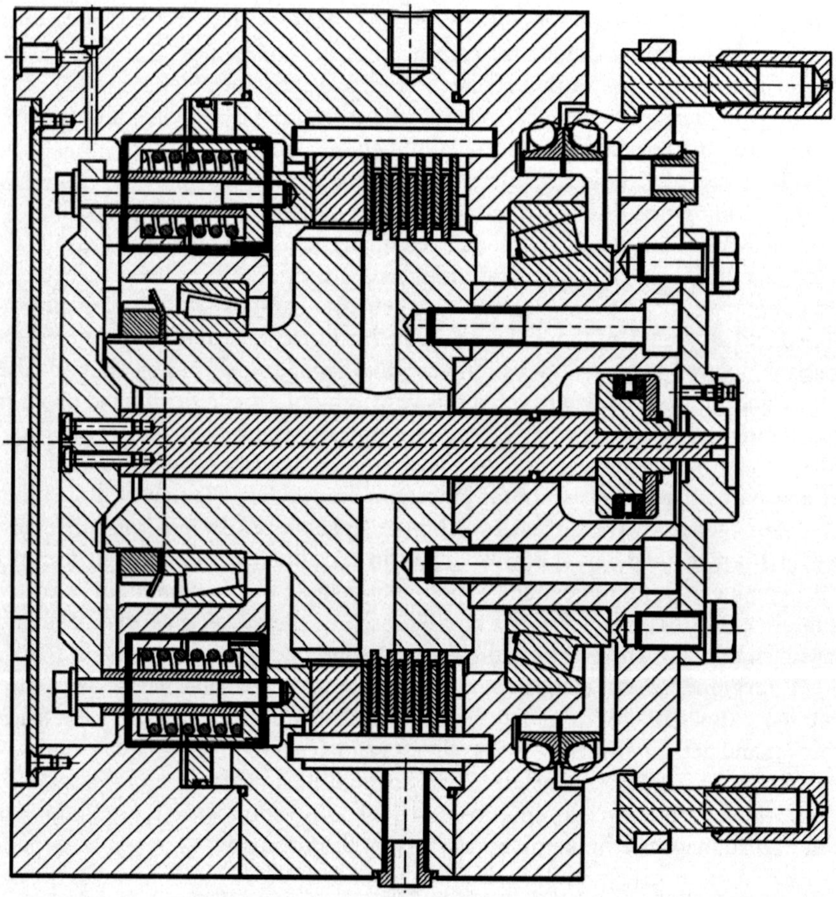

Fig. 1. Old design of wet brake cross-section.

In Fig. 1 compression springs are marked with black rectangulars. There are 8 compression springs in old design of wet brake and they are placed by 45° related to each other. Compression springs are pre-loaded with hex-head bolts thru each of spring. There are a couple of problems with this wet brake conception:

- First problem is production of housing and piston because of centric holes for compression springs placement;
- Second problem is assembling of this wet brake type, it takes a lot of time and lot of precision;
- Third problem is related for operation, if one or more springs fails, brake must go to general maintenance.

Spring failure can be caused by various causes. For example, very common cause can be initial crack which can be caused with pitting (when the spring is compressed fully), or material errors. In Fig. 2 is shown failed spring caused by initial crack [12].

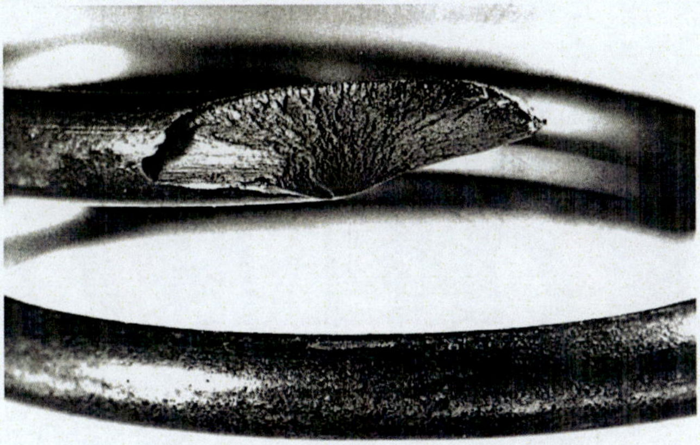

Fig. 2. Failed spring by initial crack.

By modern production technologies first and second problem can be partially overridden. The problem override methods requests expensive and time taking process. Consideration of third problem is taken into account after series of FEM analyses. FEM analysis has been performed in Autodesk software, firstly in Autodesk Inventor, and secondly in Autodesk Simulation Mechanical. In this paper is only given the results from Autodesk Inventor due to partnership restrictions on this project. This research is result of author's collaboration in industry practice project. FEM analyses were for three cases: with all compression springs operative, without one spring and without two springs. The results of FEM analysis are given in Fig. 3.

In Fig. 3a is shown normally operation mode with engagement force. It that case piston moving is by default 2 units. In Fig. 3b is shown that deflection is concentrated in one side if one spring is excluded from simulation. In practice this case should be analog to 1 spring fail. The piston movement should not be completely possible and this case could cause serious wet brake damage. In the Fig. 3c is shown FEM analysis result if 2 compression springs fails. In that case wet brake would be much damaged, and almost impossible for maintenance. Because of the problems described here, wet brake redesign process has been conducted in order to eliminate or partially override described problems.

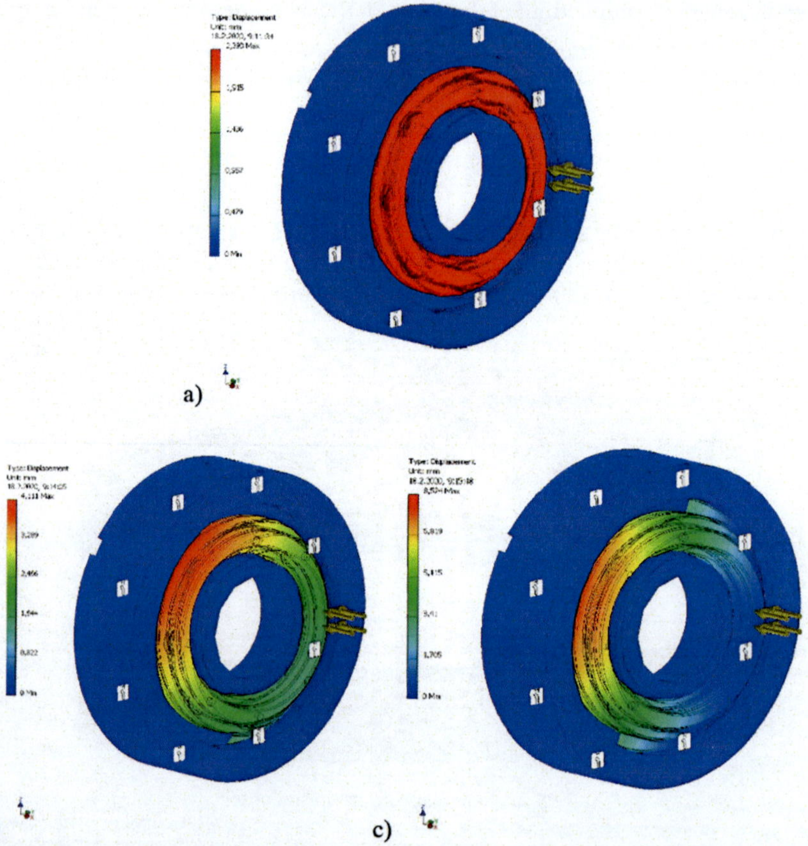

Fig. 3. FEM analysis results: a) with all springs; b) without 1 spring; c) without 2 springs.

3 Wet Brake Redesign Process and Calculations

Before redesign process the axial engagement forces needed to be determined. Axial force for brake activation is calculated by Eq. (1).

$$F_a = \frac{S_\mu \cdot K_a \cdot T_w}{i \cdot \mu \cdot r_\mu} \tag{1}$$

where: F_a – axial engagement force, K_a – working conditions factor, T_w – brake requested torque, i – number of contact surfaces, μ – friction coefficient and r_μ – friction radius.

Friction radius in this case was constraint factor. Friction radius, r_μ, is calculated by Eq. (2).

$$r_\mu = \frac{2}{3} \cdot \frac{r_{out}^3 - r_{in}^3}{r_{out}^2 - r_{in}^2} \tag{2}$$

where: r_{out} – outside radius of friction surface and r_{in} – inside radius of friction surface.

Working condition factor is used for mining industry as it was in original design solution. Friction coefficient is adopted friction materials with lubrication. for Based on old design parameters (requested torque $T_w = 13{,}52$ kNm)wet brake axial engagement force parameters are calculated. The requested torque came by request from author partners from industry, described in Sect. 2. The calculated parameters pare given in Table 1.

Table 1. Contact surfaces and axial engagement force.

Number of friction plates	Number of contact surfaces, i	Engagement force, N
5 friction plates	$i = 10$	231679
6 friction plates	$i = 12$	178066
7 friction plates	$i = 14$	152628
8 friction plates	$i = 16$	118711
9 friction plates	$i = 18$	106839
10 friction plates	$i = 20$	89033

Because of easier result insight data from Table 1 is given on diagram showed in Fig. 4.

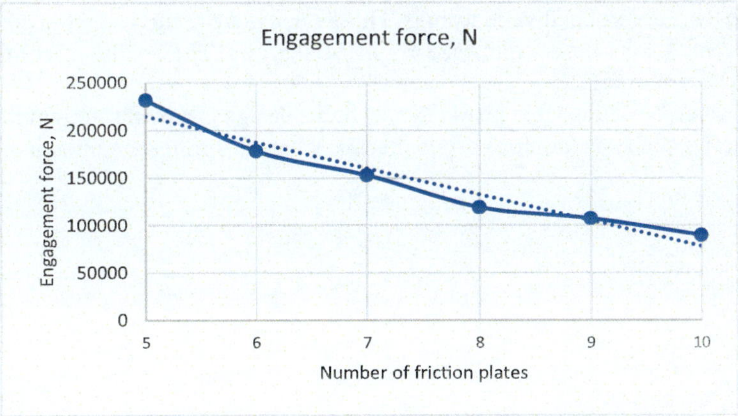

Fig. 4. Diagram of axial engagement forces related to number of friction plates.

Based on the presented data, and geometric limitations it is adopted to have 9 friction plates instead of 6 in the old design. By adding 3 more friction plates axial engagement force is lowered by 40% percent.

Based on Mubea catalog [13] the material for belleville spring is adopted, alloyed steel 50CrV4. For proper selection of belleville springs number and size the characteristic spring stress needed to be determined. Characteristic points for stress determination are shown in Fig. 5.

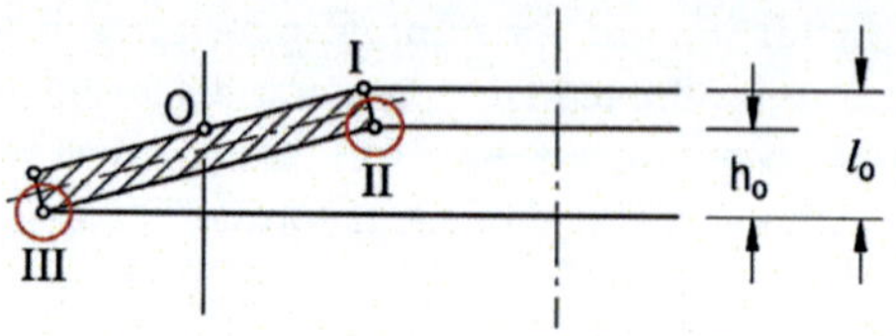

Fig. 5. Characteristic stress points.

where : I, II and III force engagement points, l_0 spring segment outer height, h_0, spring segment inner height.

Characteristic stress in shown stress points is calculated according to Almen-Laszlo simplified calculation method. By that method the most influential factors of calculation are spring geometry, spring travel and spring material.

4 New Improved Design of Wet Brake

The new design of wet brake have 3 more friction plates related to old design. This is introduced in order to engage the wet brake by lower axial force. The hex-head bolts which are used for brake assembling are placed around new set of springs. The same number of bolts is used, and only difference is that the bolts are placed in slightly bigger circle and as well radially placed. In Fig. 6 is shown new improved design of wet brake.

In Fig. 6 the new set of bellivelle springs are highlighted by black rectangulars. The new design has 2 pairs of bellivelle springs. Those springs are set in X position. Regularity of the presented solution is confirmed by FEM analysis. The results of confirmation analysis is shown in Fig. 7.

In Fig. 7 can be noticed that piston in new brake design is moving straight translator, which is very good and similar to original brake when all springs are operational.

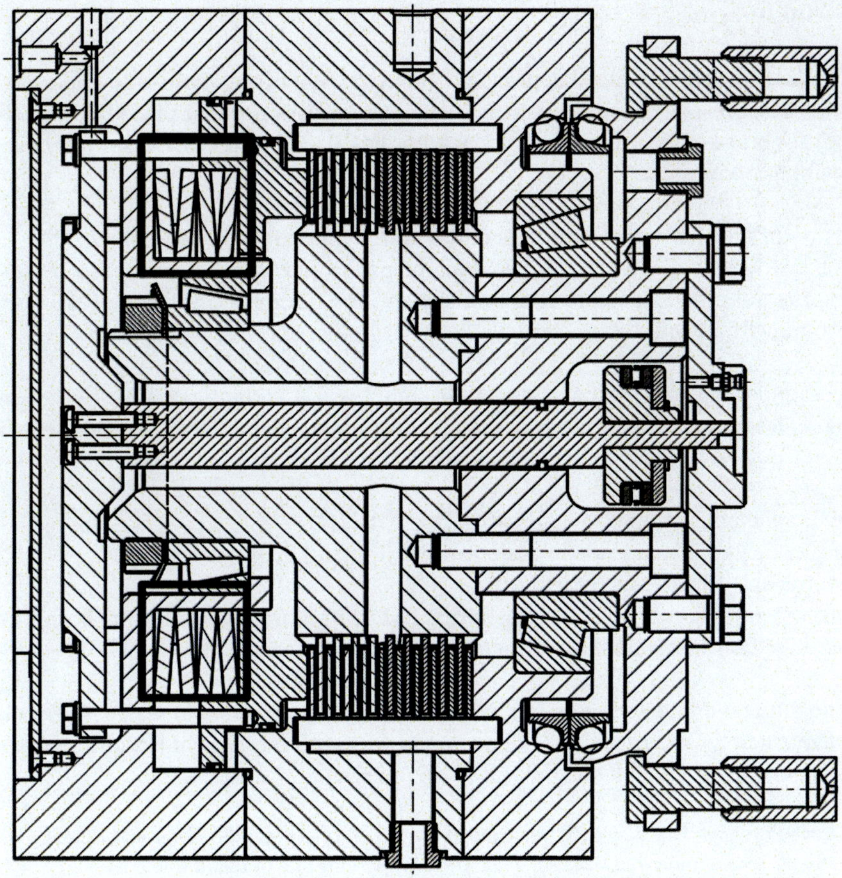

Fig. 6. New improved design of wet brake.

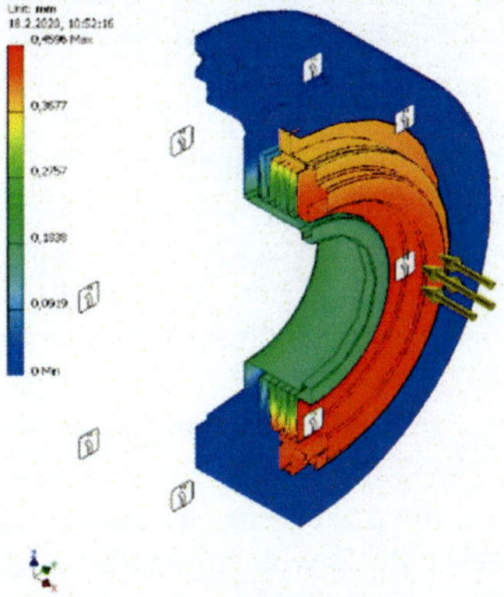

Fig. 7. FEM analysis of improved wet brake design.

5 Conclusions

This paper presents a practical approach to improvement of wet multi-disc brake design. The most of present multi-disc wet brakes solution uses a radial placed compression springs for brake engagement. Solution with radial placed compression springs have the following problems:

- lots of difficulties during brakes manufacturing which increase wet brake costs,
- problems with brake assembling and maintenance and
- brake is not able to operate if one of the springs fail, or even if one of the springs become with unequal characteristics related to other springs.

The improved design concept is with compression springs replaced by belleville springs. The concept feasibility is proven by analytical and FEM calculations. The new, improved design have following advantages:

- lower axial engagement force by 40%,
- simpler design which leads to easier assembling and maintenance,
- lower costs of brake manufacturing,
- brake operation is not related to one spring failure because piston pressure is uniformly distributed on whole contact surface of bellivelle spring.

The future work of author on this problem is focusing to do small scale brakes, which can have much wider appliance. The main problem with this type of brake is the overall manufacturing cost, which can be as well an interesting research topic.

Acknowledgment. This research was funded by the Ministry of Education and Science of the Republic of Serbia under the contract TR33015. We would like to thank to the Ministry of Education, Science and Technological Development of Republic of Serbia for their financial support during these investigations.

References

1. Zagrodzki, P.: Analysis of thermomechanical phenomena in multidisc clutches and brakes. Wear **140**(2), 291–308 (1990)
2. Jullien, A., et al.: Behaviour of wet clutches operating under continuous running conditions with a new carbon based material. Tribol. Ser. **18**, 303–312 (1991)
3. Holgerson, M.: Apparatus for measurement of engagement characteristics of a wet clutch. Wear **213**(1–2), 140–147 (1997)
4. Zhanling, J., et al.: Elastoplastic finite element analysis for wet multidisc brake during lasting braking. Therm. Sci. **19**(6), 2205–2217 (2015)
5. Naunheimer, B.BH., Ryborz, J.: Fahrzeuggetriebe. Springer Vieweg (2019)
6. Höhn, B.R., Pflaum, H., Layher, M.: Schmierstoffadditivierung zur Beeinflussung des Reibungsverhaltens nasslaufender in Reibschaltelemente. Tribologie-und Schmierstoffkongress GETLUB proceedings, pp. 259–268 (2010)

7. Liang, Y., et al.: Numerical and experimental studies of a wet multidisc clutch on temperature and stress fields excited by the concentrated load. Tribol. Trans. **62**(1), 8–21 (2019)
8. Zhao, E., Ma, B., Li, H.: The tribological characteristics of cu-based friction pairs in a wet multidisk clutch under nonuniform contact. J. Tribol. **140**(1) (2018)
9. Jianzhong, C., et al.: Investigation of flow between deformed disks in hydro-viscous drive. Tribol. Int. **121**, 287–301 (2018)
10. Kamboj, V.K., et al.: An intensify Harris Hawks optimizer for numerical and engineering optimization problems. Appl. Soft Comput. **89**, 106018 (2020)
11. Parise, J.J., Howell, L.L., Magleby, S.P.: Ortho-planar linear-motion springs. Mech. Mach. Theory **36**(11–12), 1281–1321 (2001)
12. Prawoto, Y., Ikeda, M., Manville, S.K., Nishikawa, A.: Design and failure modes of automotive suspension springs. Eng. Fail. Anal. **15**(8), 1155–1174 (2008)
13. https://mubea-discsprings.com/disc-spring-catalog/. Accessed Jan 2020

Check for updates

Measurements of the Hydraulic Fluids Compressibility

Leszek Osiecki(✉)

Gdańsk University of Technology, Narutowicza St. 11/12, 80-933 Gdańsk, Poland
losiecki@pg.edu.pl

Abstract. Performance of the hydrostatic high-pressure drive systems is affected by the changes of working fluid's volume. Presence of air bubbles in the fluid cause this problem to be even more serious. To study this phenomenon precise measurements of fluid's bulk modulus are necessary. Different measurement methods are applied, but they are either limited to low pressure range or give inaccurate results.

To solve the problem the new measurement device was built. It allows to directly measure volume changes of both pure fluid or fluid-air mixture within the wide pressure and temperature range. Both adiabatic and isothermal module may be determined, the tested fluid may be pure or aerated.

Measurements of different fluids were made including vegetable oils (rapeseed, sunflower and linseed), mineral oil and synthetic oil.

Keywords: Bulk modulus · Vegetable oil · Aerated oil

1 Introduction

Compressibility of hydraulic fluids causes a number of harmful phenomena [1]. In pumps and motors volume changes of the fluid filling so called dead volume reduce the volume efficiency of those machines [2, 3]. The dead space volume in modern high pressure piston pumps is in the range of 25 ÷ 50% of their stroke volume. This value however refers to their full displacement setting only. Reduction of pump displacement causes simultaneous growth of the dead volume and fall of stroke volume. As a result the dead-to-stroke volume ratio rises quickly, and may easily exceed 500% [3]. At an operating pressure of 40 MPa and a maximum displacement setting, this corresponds to a drop in efficiency of approx. 1% for fluid with a bulk modulus of 1800 MPa. At low displacement setting this drop can even exceed 20% [4]. Another undesirable phenomena caused by the compressibility of the working fluid are: reduction in the rigidity of the entire hydraulic system, which may result in vibration or reduced positioning accuracy and increased noise emission [1, 2, 4].

The degree of volumetric deformation of a given liquid under the influence of pressure changes determines the bulk modulus K, whereas depending on the conditions of its measurement, one can distinguish the isothermal module K_I corresponding to slow pressure changes and the adiabatic (isentropic) module K_A when the pressure changes are

© The Editor(s) (if applicable) and The Author(s), under exclusive license to Springer Nature Switzerland AG 2021
J. Stryczek and U. Warzyńska (Eds.): NSHP 2020, LNME, pp. 416–426, 2021.
https://doi.org/10.1007/978-3-030-59509-8_38

rapid [5, 6]. For a given liquid and given measurement parameters (pressure, temperature, speed of pressure changes) the value of the modulus K is a characteristic, invariable parameter of this liquid. As pressure increases, the speed of the volume changes decrease, which corresponds to the non-linear relationship shown in Fig. 1.

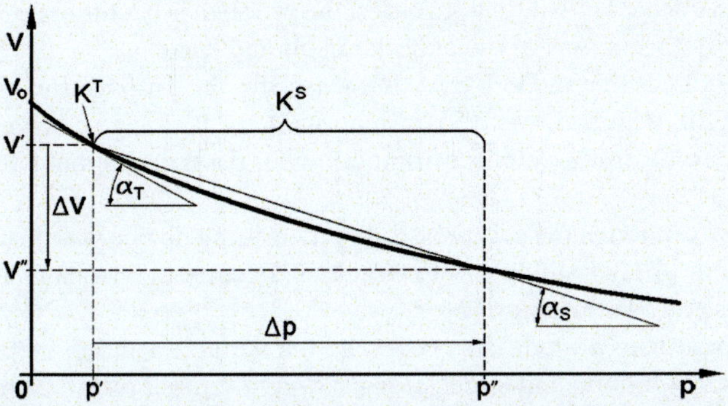

Fig. 1. Interpretation of tangent bulk modulus K^T and secant K^S.

Figure 1 shows two possible interpretations of the modulus of elasticity: K^T - tangent to the curve V(p) (temporary at a given point) and K^S - secant passing through two points of this curve (average for a given interval):

$$K^T = -V'\left(\frac{\partial p}{\partial V}\right) = \frac{-V'}{\tan \alpha_T} \tag{1}$$

$$K^S = -V'\left(\frac{\Delta p}{\Delta V}\right) = \frac{-V'}{\tan \alpha_S} \tag{2}$$

The measurement results included in the chapter 3 present values of the secant isothermal module K_I^S in the pressure range from 0 to the measured value.

2 Measuring Method

There are two direct methods of compressibility measurement: variable volume method and constant volume method [7]. In the first one certain amount of measured fluid fills

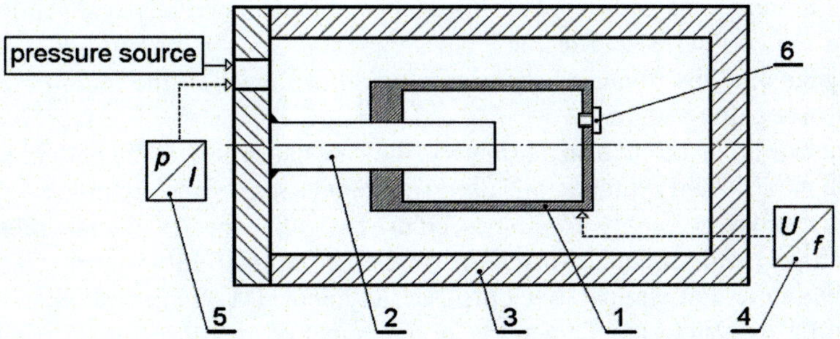

Fig. 2. New fluid compressibility meter [8].

closed cylinder equipped with a tightly fitted, movable piston. The force acting on the piston is the source of pressure in the cylinder and the displacement of the piston is a measure of the change in volume of the liquid filling the cylinder. A variation of this method is a tightly closed flexible bellows filled with the tested liquid. The constant volume method measures the relationship between the pressure in a cylinder and the volume of liquid injected into it. In both cases, the problem is to eliminate errors resulting from the elastic deformations of the cylinder itself and leaks.

The third, indirect method is based on measuring the propagation of sound waves in the tested liquid, which allows determining its density changes. With this method it is only possible to determine the isentropic module, the measurement accuracy is also limited.

An improved version of the compressibility measuring device was proposed and built in the Department of Hydraulics and Pneumatics, Faculty of Mechanical Engineering, Gdańsk University of Technology [2].

The instrument uses a modified version of the variable volume method. The measuring chamber 1 is mobile and slides along the tightly fitted piston 2. After filling the chamber 1 with tested fluid the socket 6 is tightly plugged and the whole set in put into the pressure tank 3 filled also with fluid (may be different from the tested one). The tank 3 is connected to the high-pressure pump set acting as a pressure source. Because chamber 1 may freely slide on the piston 2 pressure in the tank 3 is equal to the pressure inside the chamber 1. Movements of the chamber 1 along the piston 2 reflect changes in the volume of the tested liquid subjected to variable pressure. Simultaneous measurement of pressure with transducer 5 and displacement with transducer 4 allows to determine the course of changes in the liquid compressibility module. Because the pressures inside and outside the chamber 1 are equal, this eliminates the elastic deformations of the chamber and errors resulting from leaks. The design of the device is the subject of a patent claim [8]. The accuracy class of pressure and displacement transducers used is 0.25%. The device allows the measurement of both isothermal and isentropic bulk modulus. Tank 3 strength and the parameters of the available pump set allowed testing in the pressure range $0 \div 70$ MPa.

3 Measurements of Compressibility of Vegetable Oils [9]

In addition to the high bulk modulus, the working fluids in hydrostatic drive systems are subject to numerous additional requirements: chemical and physical stability, good lubricating and cooling properties, low environmental damage, working in a wide temperature range and low price. As no known liquid can meet all the requirements, new alternatives are being sought.

Vegetable oils can be such an alternative, they are characterized by better biodegradability than mineral and synthetic oils used so far and a high viscosity index. In addition to the known and increasingly used rapeseed oil, two other species of vegetable oil were tested: sunflower and linseed. Information on their compressibility was not found in the available literature. For comparative purposes, a typical HM 46 mineral oil and 5 W–40 synthetic oil were also tested. Due to the large number of measurements carried out, the article presents only the most interesting, selected results.

Figures 3 and 4 show that values of the bulk modulus for all three vegetable oils at a temperature of 60 °C exceed the parameters of mineral oil by approx. 3 ÷ 15%, while the difference decreases with increasing pressure. Synthetic oil gets even worse results, in extreme cases the difference reaches 30%. The temperature drop to −10 °C increased the bulk module by an average of 35 ÷ 40% for all oils, however, for mineral oil this increase was about 45%, thanks to which it exceeded vegetable oils.

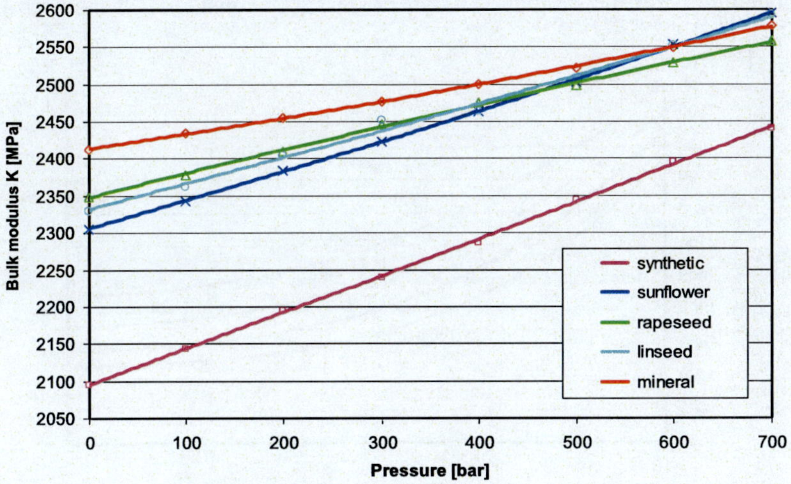

Fig. 3. Changes in bulk modulus at −10 °C.

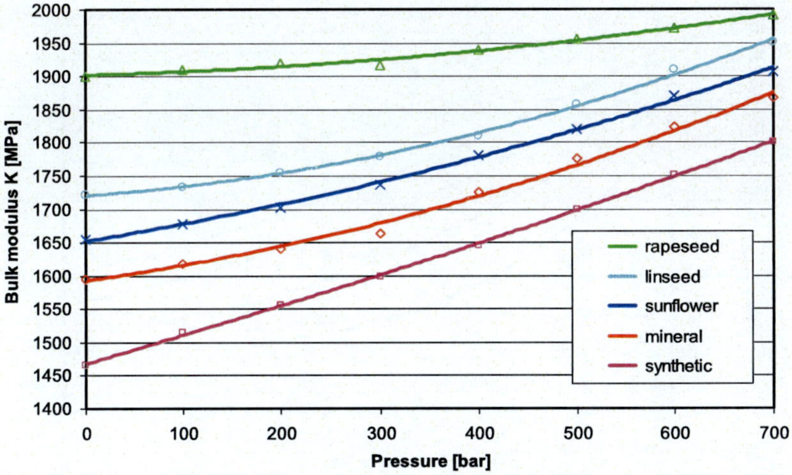

Fig. 4. Changes in bulk modulus at 60 °C.

From the point of view of the operation of hydrostatic systems, the results obtained at elevated temperatures are more important, which corresponds to the typical operating conditions of these systems in steady state. Under these conditions, the advantage of all vegetable oils (especially rapeseed one) over mineral and synthetic oil is visible.

Under conditions of temperature reduced to −10 °C, mineral oil is comparable to vegetable oils. However, these conditions only apply during start-up of the machine operating in low ambient temperature.

Comparing Figs. 5, 6, 7 and 8 it can be stated that vegetable oils in the whole range of pressure changes show a weaker dependence of the bulk modulus on temperature changes than mineral and synthetic oil, which should be considered as their advantage.

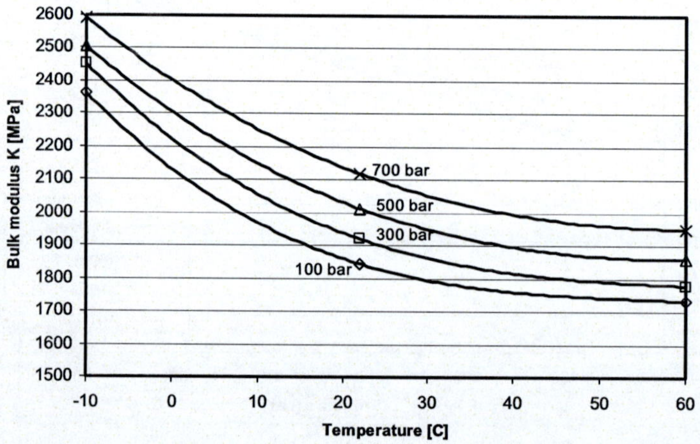

Fig. 5. Changes in linseed oil bulk modulus as a function of temperature.

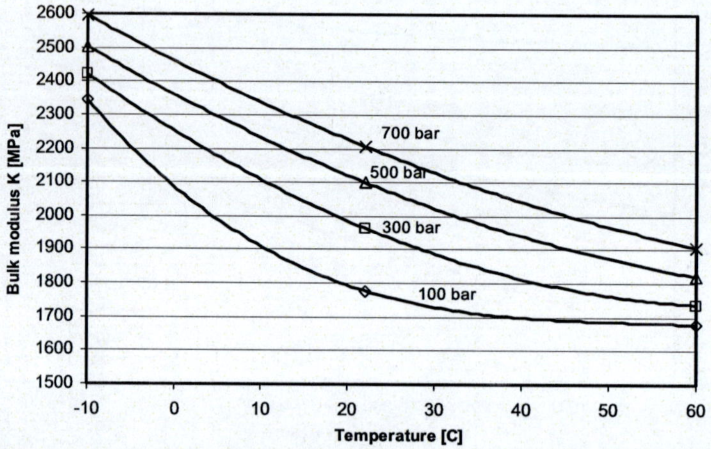

Fig. 6. Changes in sunflower oil bulk modulus as a function of temperature.

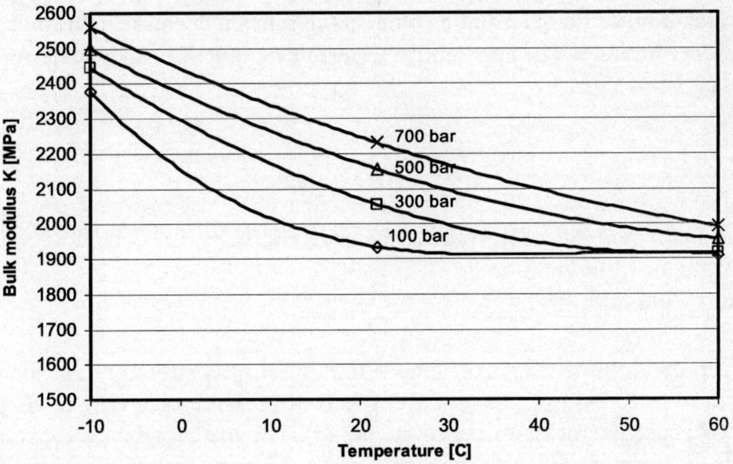

Fig. 7. Changes in rapeseed oil bulk modulus as a function of temperature.

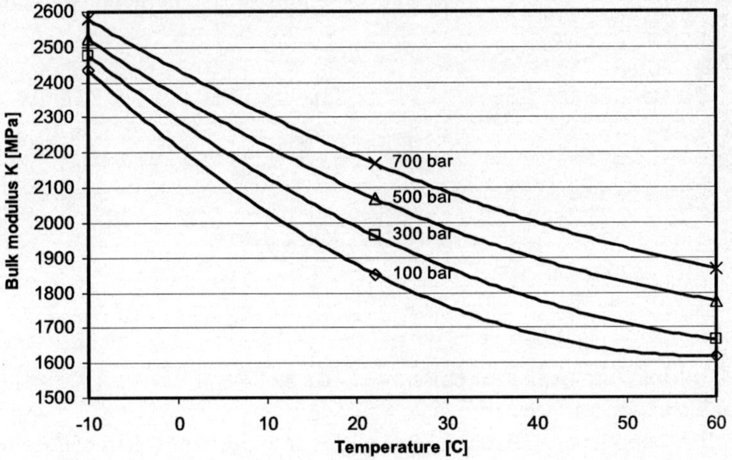

Fig. 8. Changes in mineral oil bulk modulus as a function of temperature.

4 Bulk Modulus of the Aerated Liquid

Under real conditions, the working fluid in hydrostatic drive systems often occurs in an aerated form. The most harmful form of air pollution are tiny air bubbles with a diameter in the order of tenths of a millimeter. Since the viscosity of hydraulic fluids is usually from a dozen to over 100 mm^2/s, complete separation of liquid and air in the tank is not possible, because it requires too much time.

The effect of oil aeration is, among others, a significant decrease in the bulk modulus of the liquid-air mixture. The apparent value of this module K_1 can be determined from the formula [10]:

$$\frac{1}{K_1} = \frac{1}{K_p} + \frac{\varepsilon_p}{p_t} \tag{3}$$

where: K_p – bulk modulus of the pure (deaerated) fluid
ε_p - volumetric air-to-liquid ratio
p_t – temporary pressure

However, the volumetric air content in the liquid ε_p is strongly dependent on different parameters (pressure, temperature, speed of pressure changes), it also changes over time (because the air dissolves in the liquid). The mathematical description of all phenomena occurring in a mixture of liquids and gases during pressure changes requires numerous simplifying assumptions and nevertheless the obtained models are complicated. Depending on the assumptions made, a number of models can be found in the literature: The simplest models Merritt and Cho ignore the phenomenon of air dissolving in liquids [8, 11–14]:

$$K^S_{Mer} = \frac{K^S_{Ap}}{1 + \varepsilon_p \left(\frac{K^S_{Ap}}{\kappa \cdot p_t} - 1 \right)} \tag{4}$$

$$K^T_{Cho} = \frac{(1 - \varepsilon_p) + \left(\frac{p_n}{p_t}\right)^{\frac{1}{\kappa}} \cdot \varepsilon_p}{\frac{\varepsilon_p}{\kappa \cdot p_t} \cdot \left(\frac{p_n}{p_t}\right)^{\frac{1}{\kappa}} + \frac{1 - \varepsilon_p}{K^S_{Ap}}} \tag{5}$$

where: p_n – normal (atmospheric) pressure = 101 325 Pa,
κ - polytrophic exponent
Feldmann, IFAS and Ruan-Barton models taking into account the transition of air into solution are an extension of the above models:

$$K^T_{Feld} = \frac{e^{\left(\frac{p_n - p_t}{K}\right)} + \frac{\varepsilon_p}{1 - \varepsilon_p} \cdot \left(\frac{p_n}{p_t}\right)^{\frac{1}{\kappa}}}{\frac{e^{\left(\frac{p_n - p_t}{K}\right)}}{K} + \frac{\varepsilon_p}{\kappa \cdot p_n \cdot (1 - \varepsilon_p)} \cdot \left(\frac{p_n}{p_t}\right)^{\frac{1+\kappa}{\kappa}}} \tag{6}$$

$$K^T_{IFAS} = \frac{(1 - \varepsilon_p) \cdot \left(1 + \frac{m \cdot (p_t - p_n)}{K_0}\right)^{-\frac{1}{m}} + \varepsilon_p \cdot \left(\frac{p_n}{p_t}\right)^{\frac{1}{\kappa}}}{\frac{(1 - \varepsilon_p)}{K_0} \cdot \left(1 + \frac{m \cdot (p_t - p_n)}{K_o}\right)^{-\frac{m+1}{m}} + \frac{\varepsilon_p}{\kappa \cdot p_n} \cdot \left(\frac{p_n}{p_t}\right)^{\frac{1+\kappa}{\kappa}}} \tag{7}$$

where: m – factor taking into account the effect of pressure on the deaerated oil K module:

$$K(p_t) = K_0 + m \cdot p_t \tag{8}$$

$$K_{R-B}^T = \frac{1 + \left(\frac{p_n}{p_t}\right)^{\frac{1}{\kappa}} \cdot \frac{\varepsilon_p}{1-\varepsilon_p} \cdot \frac{p_c - p_t}{p_c - p_n}}{\frac{1}{K_A} + \left[\left(\frac{p_n}{p_t}\right)^{\frac{1}{\kappa}} \cdot \frac{\varepsilon_p}{1-\varepsilon_p} \cdot \frac{1}{p_c - p_n}\right] \cdot \left(\frac{p_c - p_t}{\kappa \cdot p} + 1\right)} \tag{9}$$

The Ruan-Burton model is only valid for pressures below the critical pressure p_c. Above this value, which may vary in function of additional parameters, full air dissolution in oil is assumed, so the bulk modulus has a value of K_A^T corresponding to the deaerated liquid. However, such discontinuity and uncertain value of the critical pressure are disadvantages of this model.

Results obtained using different models differ significantly (Fig. 9). To clarify doubts, it is therefore necessary to carry out the most accurate measurements possible. Precise measurements made using the new instrument may clarify some doubts and help in finding the new model more relevant to the results of the experiment. For this purpose preliminary measurements were made using mineral oil HM 46 with certain degree of aeration. Results were compared with the IFAS and Feldmann models, regarded as the best ones (Figs. 10 and 11).

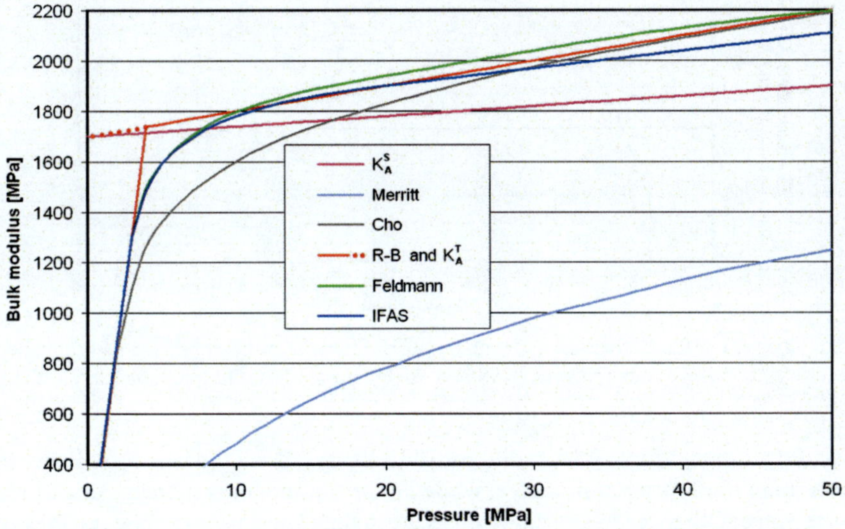

Fig. 9. Comparison of changes in the bulk modulus of aerated oil according to the models available in the literature [6].

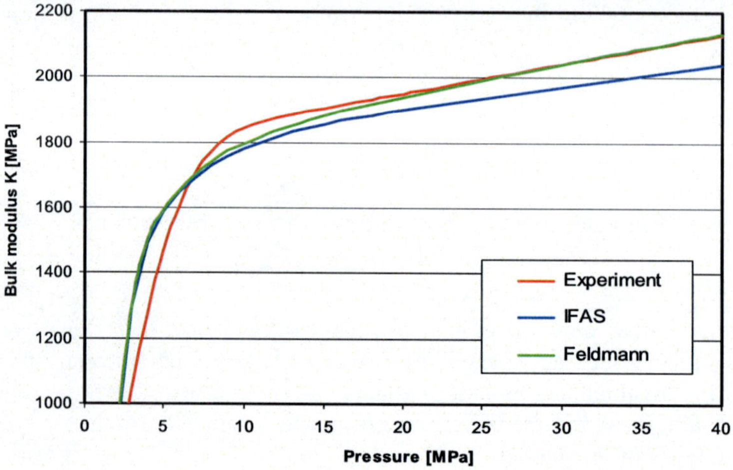

Fig. 10. Changes in the isothermal, tangential bulk modulus of mineral oil HM 46 with a degree of aeration $\varepsilon_p = 2\%$ compared with IFAS and Feldmann models. Oil temperature: 25 °C [2].

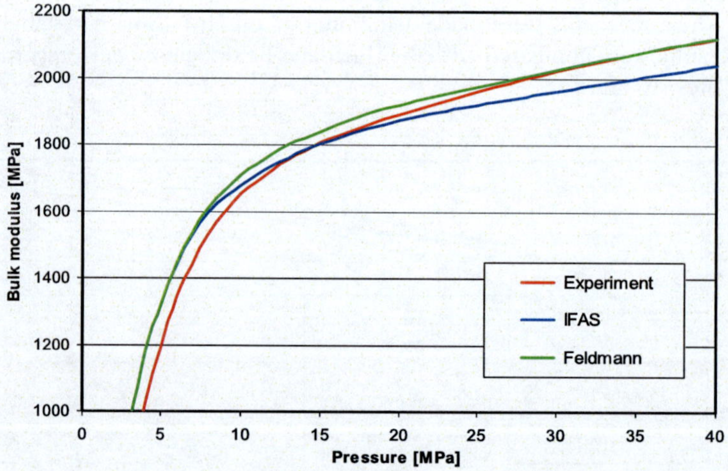

Fig. 11. Changes in the isothermal, tangential bulk modulus of mineral oil HM 46 with a degree of aeration $\varepsilon_p = 4{,}8\%$ compared with IFAS and Feldmann models. Oil temperature: 25 °C [2].

For the given measurement conditions, IFAS model shows the best compliance with the experiment in low pressure range, while Feldmann model is more accurate in high pressure range. Other models strongly differ from the experiment results. As a result of his investigation Załuski proposed a modification of the Feldmann model [2]:

$$K^T_{Feldmod} = \frac{\left[1 + \frac{m(p_t - p_n)}{K}\right]^{-\frac{1}{m}} \cdot e^{\left(\frac{p_n - p_t}{K}\right)} + \frac{\varepsilon_p}{1-\varepsilon_p} \cdot \left(\frac{p_n}{p_t}\right)^{\frac{1}{\kappa}}}{\left[1 + \frac{m(p_t - p_n)}{K}\right]^{-\frac{m-1}{m}} \cdot \frac{e^{\left(\frac{p_n - p_t}{K}\right)}}{K} + \frac{\varepsilon_p}{\kappa \cdot p_n \cdot (1-\varepsilon_p)} \cdot \left(\frac{p_n}{p_t}\right)^{\frac{1+\kappa}{\kappa}}} \tag{10}$$

To confirm the correctness of the proposed model, further tests are necessary, which, however, require the construction of a new, more accurate measuring instrument adapted to tests in a wider pressure range.

5 Conclusions

The test results show favorable values of the compressibility modulus obtained by vegetable oils. In the whole range of pressure and temperature changes, the module for these oils significantly exceeded the parameters of synthetic oil. Mineral oil exceeded vegetable oils at negative temperatures, corresponding to the moment of machine start-up, but was clearly worse at steady-state temperatures. It should be emphasized here that the tested vegetable oils did not contain refining additives found in the compared mineral and synthetic oils, so they cannot be treated as full-fledged hydraulic liquids. Nevertheless, the results obtained should be considered promising and worth further, more detailed research.

Tests carried out using the new device fully confirmed the usefulness of the new concept. Bulk modulus values obtained for mineral oil fully coincide with the results available in the literature, which positively indicates the accuracy of the measurements carried out. The tests however indicated also the direction of further construction work. In an existing device, the transducer 5 measures the pressure in the tank 3 (see Fig. 2). In principle, it should be equal to the pressure in measuring chamber 1. Because of a sealing ring, installed between the chamber 1 and the piston 2 to eliminate leaks, friction that appears in this node causes however a pressure difference of up to 2 bar between the two spaces, which in practice doubles the error of the pressure transducer used. Measuring pressure directly in the measuring chamber should eliminate this problem. The range of pressures achieved in the device is limited by the maximum operating pressure of the available pump set. This can be bypassed by using a pressure multiplication system. Another version of the compressibility meter, taking into account the above observations, is currently being built and should enable more accurate measurements of bulk modulus changes at pressure exceeding 100 MPa.

References

1. Jing, L., Gu, L., Sun Y.: Research on the influence of oil bulk modulus on performance of the hydraulic system coupling. In: Applied Mechanics and Materials, vol. 543–547 (2014)
2. Załuski, P.: Influence of the position of the swash plate rotation axis on the volumetric efficiency of the axial piston pumps (in Polish). Ph.D. thesis, Gdańsk University of Technology, Poland (2017)
3. Osiecki, L.: Volume efficiency loss in axial pump caused by fluid compressibility. Hydraulika a Pneumatika nr 1/2002, Slovakia (2002)
4. Osiecki, L.: Commutation Units of Hydraulic Axial Piston Machines. Dissertation, Gdańsk University of Technology, Poland (2006). (in Polish)
5. Hayward, A.: How to estimate the bulk modulus of hydraulic fluids. Hydraul. Pneumatic Power **16**, 28–40 (1970)
6. Patrosz, P.: Compensation of pressure peaks in variable displacement piston pump with cam driven commutation. Ph.D. thesis, Gdańsk University of Technology, Poland (2017). (in Polish)

7. Kim, S., Murrenhoff, H.: Measurement of effective bulk modulus for hydraulic oil at low pressure. J. Fluids Eng. **134** (2012)
8. Załuski, P., Patrosz, P.: Device for measuring compressibility of liquids or gases. Patent claim nr P 421 783, Poland (2017)
9. Zadrożna, D.: Measurements of the volume deformation module of hydraulic fluids. MSc Thesis, Gdańsk University of Technology, Poland (2019). (in Polish)
10. Szydelski, Z.: Automotive vehicles. Hydraulic Drive and Control. WKiŁ Warszawa, p. 566 (1993). (in Polish)
11. Yu, J., Chen, Z., Lu, Y.: Variation of an oil effective bulk modulus with pressure in hydraulic systems. J. Dyn. Syst. Meas. Control **116**, 146–150 (1994)
12. Ruan, J., Burton, R.: Bulk modulus of air content oil in a hydraulic cylinder. In: Proceedings of 2006 ASME International Mechanical Engineering Congress and Exposition. IMECE2006 – Fluid Power Systems and Technology Division (2006)
13. Bair, S., Michael, P.: Modelling the pressure and temperature dependence of viscosity and volume for hydraulic fluids. Int. J. Fluid Power **11**, 37–42 (2010)
14. Gholizadeh, H., Burton, R., Schoenau, G.: Fluid bulk modulus: comparison of low pressure models. Int. J. Fluid Power **13**, 7–16 (2012)

Modern Materials and Surface Modification Methods Used in the Manufacture of Hydraulic Actuators

Justyna Skowrońska(✉), Jakub Zaczyński, Andrzej Kosucki, and Łukasz Stawiński

Instytut Obrabiarek i Technologii Budowy Maszyn, Zakład Obrabiarek, Politechnika Łódzka, ul. Stefanowskiego 1/15, 90-924 Łódź, Poland
{justyna.skowronska,jakub.zaczynski}@dokt.p.lodz.pl,
{andrzej.kosucki,lukasz.stawinski}@p.lodz.pl

Abstract. The article presents an overview of various materials of which pistons, piston rods and cylinders of hydraulic actuators are currently made. Surface modification issue, aimed at improving both strength and operational properties of the components is also discussed.

The materials, of which the basic parts of the actuators are made, will be discussed in detail. In the case of cylinders, it is usually quality steel. Pistons are made of ductile iron or structural steel. Sometimes aluminium alloys are also used for reducing the weight of the element while maintaining adequate mechanical strength. Piston rods are currently one of the most complex in machining parts. They are made of surface-hardened or quenched and tempered steel (possibly stainless steel) and are most often coated with chromium. Recently, many alternatives to chromium have been created due to its harmfulness for people and environment. These are, for example, tungsten carbide coatings and amorphous carbon coatings. Usage of composite materials and plastics is also listed among modern technologies for making components of actuators.

Keywords: Structural materials · Coatings · Power hydraulics · Hydraulic actuators

1 Introduction

Hydraulic actuators, in the simplest terms, are engines that are designed to convert the fluid energy into mechanical energy. The most common hydraulic actuator consists of a cylinder, piston, piston rod, head and set of seals (Fig. 1).

The piston and the piston rod perform reciprocating motion, and the distance between the piston end positions is called the actuator stroke [1]. In this article, three main parts of a regular actuator will be discussed in more detail: cylinder, piston and piston rod.

Components of actuators should meet a lot of requirements such as corrosion resistance, wear resistance and have appropriate tribological parameters. High temperature resistance is also an important aspect. Hydraulic actuators operate at temperatures up

© The Editor(s) (if applicable) and The Author(s), under exclusive license to Springer Nature Switzerland AG 2021
J. Stryczek and U. Warzyńska (Eds.): NSHP 2020, LNME, pp. 427–439, 2021.
https://doi.org/10.1007/978-3-030-59509-8_39

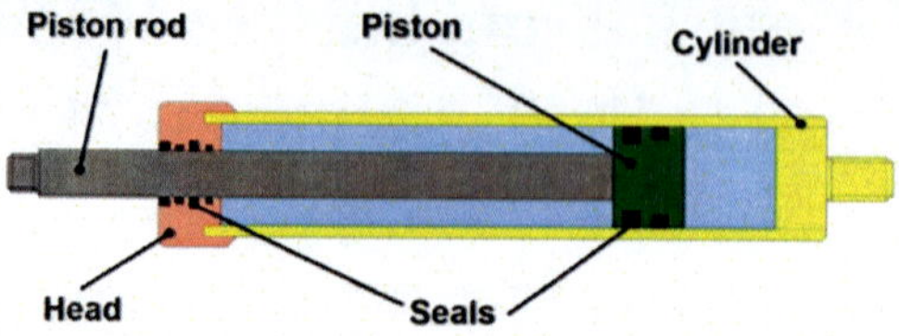

Fig. 1. The design of the exemplary hydraulic actuator.

to approximately 80 °C, however, for actuators for special applications, the operating temperature can be as high as 200 °C.

The cylinder is a pipe ended on the one side with a bottom [2] and on the other attached to the so-called head. It forms the housing of a moving piston-piston rod assembly. It is important that the cylinder has adequate tightness. Any leaks and pressure drops of the working fluid could lead to a decrease of actuator performance.

The piston is usually in the form of a cylinder with grooves for sealing and bearing elements cut out on its lateral surface. It is an element separating two working spaces in the cylinder from each other. More precisely, it prevents the flow of working fluid between the chambers of the cylinder.

The piston is guided by the piston rod. On the one side it is attached to the piston, on the other connected to the driven element [3].

Each of these three actuator components must meet different requirements for operating conditions. Damage to any component could result in a serious failure of the entire hydraulic system, and thus even create a threat to human life.

2 Basics of the Material Selection

Material selection is one of the most important aspects of machine elements design. It complements the structural issues and strictly depends on the purpose of the actuator and its overall dimensions. In the paper [4], the relationship between function, material, shape and method of producing a given element is presented. These four issues should complement each other and the omission of any can be considered a mistake.

It is generally believed that the use of a given material should not generate too high costs. However, the price-benefit ratio should be more important. The use of expensive materials with the best parameters often cannot be avoided, for example, in the case of responsible components of aircraft engines, as a failure in such cases could cause catastrophic consequences.

During choosing the material, it is necessary to draw up a list of the operating conditions of the element, and hence, a set of requirements that the material must meet. Not only mechanical, thermal, corrosive or economic properties, but also its environmental impact should be taken into account here. During material selection, it may be helpful to prepare so-called material properties graph with the help of computer software. It allows to graphically present materials that simultaneously meet two design assumptions. Thanks to this solution, it is possible to find material with optimal parameters. The graph can display data only for one material group (e.g. iron alloys) or simultaneously for several groups.

Figure 2 shows an example of the property graph obtained in CES EduPack. The graph illustrates the values of Young's modulus and density for ferrous materials used in the manufacture of cylinders. Each dot or oval on the chart indicates one particular material. The oval shape of the marker means a wider range of values for a given parameter. In addition, each material group is marked with a separate colour, what improves the readability of the chart.

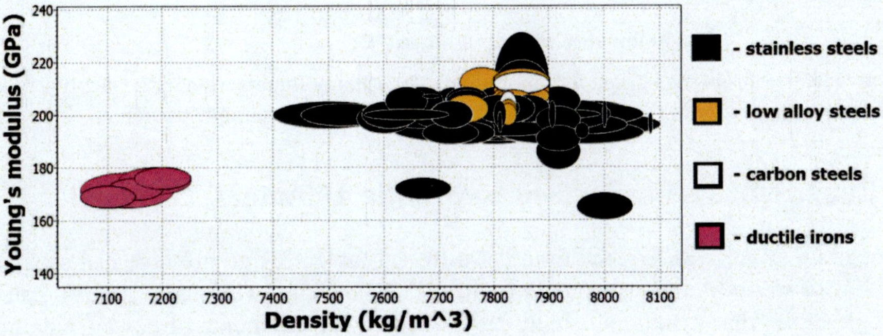

Fig. 2. Material properties graph for ferrous materials used in the manufacture of cylinders - comparison of Young's modulus and density.

Another example of graph can be a presentation of the price and fatigue strength of a given material (Fig. 3). Figure 4 shows the other materials discussed in this article: 7000 series aluminium alloys (Al-Zn alloys) and thermoplastic polymers. Ecological aspect such as the carbon footprint associated with primary production is exhibited here.

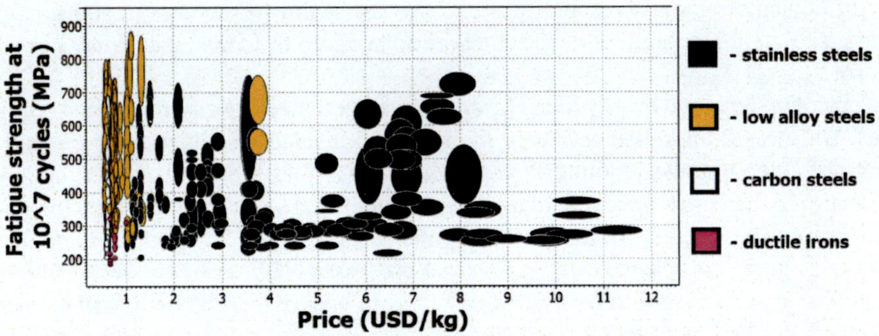

Fig. 3. Material properties graph for ferrous materials used in the manufacture of cylinders - comparison of fatigue strength at a number of cycles of 10^7 and price.

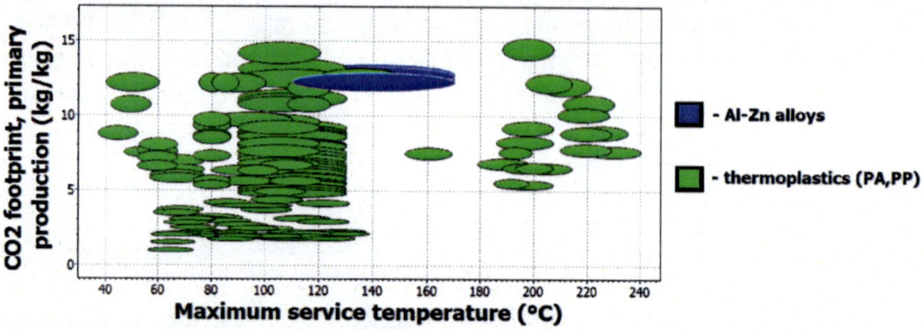

Fig. 4. Material properties graph for aluminium alloys and polymers used in the manufacture of cylinders - comparison of the carbon footprint and maximum operating temperature.

3 Requirements for Parts of Hydraulic Actuators

Actuators may fail due to a multitude of different factors. High pressure can cause the material to strain. Another problem is the risk of buckling (loss of stability), caused by high forces that cylinders usually have to carry [5]. Elements of power hydraulics systems are exposed to corrosion, especially the cylinder and piston rod [3].

In article [6], tribological properties between carbon steel and a seal were examined. Unfortunately, iron oxide from steel may form on the sealing surface - this causes accelerated wear of the elements or increases noise during operation. Therefore, it is necessary to use such materials and modifications in the cylinders that meet the requirements of tribology.

An important aspect that affects the durability of hydraulic cylinders is also the cleanliness of the working fluid. This is related to the increasing requirements for the geometrical accuracy of the actuator built [7]. Fluid particles can accelerate the abrasive wear of cooperating surfaces, and thus damage the sealing elements. In addition, due to the insufficient smoothness of the surfaces adjacent to the seals, these may not fulfil their role.

Currently, more and more producers are focusing on meeting environmental requirements. Businesses approach seriously the issues connected to reduction of the carbon footprint, defined as the amount of greenhouse gases that are emitted into the atmosphere during the entire life cycle of a given product [8]. The European Union has taken measures to reduce gas emissions in its member countries [9]. A significant amount of greenhouse gases are released from industrial processes [10], therefore the selection of production processes should be carried out so as to ensure the highest possible energy efficiency. Another way to meet ecological requirements may be to make structural modifications. There is also a tendency to reduce the weight of machine components.

3.1 Requirements for Cylinders

Cylinders are actuator components that are in contact with the external environment. Actuators often work in the air or even under water, which is why the outer surfaces of the pipes must be characterized by high corrosion resistance and adverse weather conditions. From the inside, the cylinder pipes work with both piston seals and the working fluid [2]. Therefore, in addition to anti-corrosive properties, the quality and geometric accuracy of the internal surfaces of pipes are important [11].

3.2 Requirements for Pistons

The piston surface must be as smooth as possible. Too high roughness will not allow the rings to adhere to the piston surface, which will make it difficult to ensure a proper seal. Due to the fact that the piston is in contact with oil - it must be made of corrosion resistant material.

3.3 Requirements for Piston Rods

The piston rod is a force transferring element, therefore, it should be made of a material with adequate strength. Its surface must also be smooth because it comes into contact with the seal. Low roughness can prevent leakage outside the cylinder.

Piston rods break fatigue, they are often subjected to variable cyclic loads [12]. The material here must be resistant to fatigue and abrasion. It should also be well weldable, because welding is a common way of mounting the ear, much cheaper than threading [13]. The piston rod may be compressed or stretched axially - side loads should be avoided [14]. In single-acting cylinders, the only problem is the risk of buckling. On the other hand, in double-acting cylinders there is also stretching, and thus there is a possibility of material fatigue [15]. Piston rods, as elements sliding out of the cylinder, are often exposed to precipitation [16], as well as to sand and other impurities that can scratch the surface of the piston rod - this issue should also be considered during material choosing.

4 Materials Used for Hydraulic Actuators

Materials used for hydraulic cylinders can be divided into several groups: iron alloys, light metal alloys, composite materials and polymeric materials. Iron alloys can be considered the most common material used for cylinder components. However, over time, steels and cast iron begin to give way to materials from the other three groups.

4.1 Iron Alloys

Among the iron alloys used for actuators, we can distinguish steels and cast iron. The carbon content closely affects the steel structure, which, in turn, affects the mechanical properties. It is well known that the more carbon in steel, the less ductile it is, its impact strength is decreases and weldability deteriorates. However, as carbon concentration increases, hardness and tensile strength increase too. In alloy steels there are additional elements that are added to achieve better material parameters. The degree of change in properties depends on the amount of alloying element.

For cast irons, the low melting point is an advantage [17]. Cast irons are cheap materials, they are also characterized by good mechanical strength and machinability [18]. Their advantage is also the ability to damp vibrations [19].

Hydraulic cylinders are made of both carbon and alloy steels. One of the most popular metals for the production of pipes is unalloyed steel with minimum yield strength of 355 MPa in three different varieties. These materials are structural steel S355 JR (St52) [20, 21] and the same steel in standardized condition S355 J2G3 [11]. Other sources report cylinders made of machine steel E355 [22, 23]. In some cases [23–25] steel S275 JR is used for cylinders. [24, 25] mention the use of this steel also for piston rods of hydraulic actuators. Another popular carbon material of which piston rods are made is C45 steel [15, 26, 48]. This matter can also be used for pistons [27]. It is one of the most popular structural materials on the market due to its good mechanical properties and general availability. It is difficult to weld, but its parameters can be easily changed by heat treatment.

Cylinders are also made of similar steel intended for pipes marked as R35 or R45 [28]. On the other hand, in work [29] it was proposed to use BS 970 070M20 steel for cylinders, which was selected, because it combines the required lightness with a very good yield point, tensile strength, corrosion resistance and wear resistance. In the production process of actuators, in which slightly higher tensile strength is required, steel E470 is used [30].

In addition to carbon steel, alloy steels are also used. A common material for making piston rods is 42CrMo4 steel [23, 31], containing two main alloying elements: chromium and molybdenum. Both additions increase the hardenability and strength of steel [18]. The article [31] analyzes the failure of the piston rod of a cable fatigue testing machine. In this case, part of the element was made of a hollow rod, and a crack formed at the point where the hollow and non-hollow part were connected. As one of the ways to avoid such phenomena in the future, it was proposed to use steel with better properties: 30CrNiMo8. This steel additionally contains an admixture of nickel, which has a positive effect on the brittle fracture strength of steel [18].

The piston rods are made of 20MnV6 steel [13]. The addition of vanadium increases the hardenability and hardness. The actuators, for more demanding applications, are also made of austenitic stainless steels such as AISI 304 [25] or martensitic: AISI 420 [32] and AISI 410 [33]. Generally, martensitic stainless steels are resistant to stress corrosion provided that they are properly heat treated. The failure of the piston rod analyzed in [33] resulted from, among others, improper heat treatment.

In the production of hydraulic cylinders chromium steel 40Cr [27] or its equivalent - 40X [34] are also used. This steel has a very favorable ratio of yield strength to tensile strength. In [35], it was proposed to use 30KhGSA chromium-manganese-silicon alloy steel for piston rods. This steel has very high strength and anti-wear properties. It can be heat treated, it is difficult to weld.

Detailed procedures for selecting the material for the cylinder and piston rod are presented in [36]. The author approached each of these elements separately, first setting the requirements, and then selecting the material using the previously mentioned CES EduPack program. Of the three materials proposed for the piston rod, alloy steel EN24T was chosen due to its good hardenability, high strength and resistance to abrasive wear and corrosion.

In the process of designing hydraulic cylinders, ductile irons are also of great importance. They are characterized by high strength with good plasticity [18]. In the aforementioned article [36], for the cylinder the choice fell on ductile iron ASTM A536 60-40-18 due to its lightness and strength, as well as low cost. Disadvantages of using this material, such as the material's poor resistance to salty and acidic environments, have also been recognized. The use of ductile iron for pistons is described in [37]. ASTM A536 65-45-12 cast iron was selected here. It has a slightly higher tensile strength than 60-40-18. Generally, there is a tendency to use cast iron in the case of not very high pressures in the cylinders (not exceeding 10 MPa). For larger ones, steel materials are usually used [27].

4.2 Light Metal Alloys

Light alloys are characterized by a favourable ratio of strength to weight - in this respect they perform much better than steel [17]. In works [23, 24] and [38] cylinders made of aluminum were tested. The cylinder, piston and piston rod are made of AL 7075 T6 alloy. These alloys have very good mechanical strength, however their corrosion resistance is quite limited [18]. The density of AL 7075 T6 is only 2.8 g/cm^3.

4.3 Composite Materials

Composites are materials in which (usually cheap or light) material, constituting the matrix, is reinforced with a material of higher strength - due to which the new material acquires completely new properties.

A good example of reducing cylinder weight is the use of composite materials in combination with light alloys. In [24] and [38], the construction of a hydraulic cylinder is described, in which a combination of aluminum alloy and composite is used. The cylinder pipe is made of two layers: internal of the already mentioned aluminum alloy AL 7075 T6 and external of a composite material: epoxy resin reinforced with carbon fiber. In [24], the masses of cylinders were compared: only steel made of structural steel, aluminum and aluminum with the hybrid cylinder described above. The last one turned out to be almost six times lighter than a steel cylinder and almost twice as lighter than the aluminum one. Unfortunately, the use of composite materials would also increase manufacturing costs [38].

In [25], the mass of a telescopic actuator cylinder made of carbon fiber reinforced epoxy resin composite was checked and economic issues were analyzed. The results were compared with common structural steel, aluminum and stainless steel. In this case, the use of the composite allowed for a mass reduction of up to 96% and increased costs by more than 150%. The work [39] presents a double cylinder pipe - the inner one is made of steel, and a pipe made of a carbon fiber reinforced polymer (CFRP) has been mounted on it. CFRP is corrosion resistant and lightweight. It is also characterized by high mechanical and fatigue strength [40]. The article [41] describes the possibility of using AL_2O_3 particles in polymer matrix for cylinders, which allows increasing strength and abrasion resistance in a multilayer cylinder made of CFRP.

It should be remembered that composite materials exhibit strong anisotropy of properties, which is why it is important to properly manufacture that material.

4.4 Polymeric Materials

Plastic technology is developing very quickly. These are low density materials, often with self-sealing properties. They reduce noise during operation, their production is easy and relatively cheap. Plastics as a material for the manufacture of hydraulic actuators are mentioned in [42]. Article [43] describes a piston rod made of plastic. Tests carried out for low pressures (at the level of several MPa) have shown that the components manufactured in this way can operate correctly. Their strength, manufacturing accuracy and resistance to various factors provide the basis for research into the use of plastics in power hydraulics in a wider range. In [44] polyamide (PA) and polypropylene (PP) were specified as elements from which prototype hydraulic cylinders were made.

The use of polymeric materials, due to their low stiffness and lower strength than steel, often requires major changes in the structure of particular element [17].

5 Surface Modification of Cylinders, Pistons and Piston Rods

There is a whole range of methods for surface modification of hydraulic system components, including both coatings and surface layer modifications of the materials.

When it comes to surface layer modifications, the most commonly used surface treatments include quenching and tempering - combination of quenching and tempering at high temperatures [30, 45]. This is a treatment that is usually used for piston rods. The hardening itself increases the hardness of the steel and its strength, however, it decreases the plastic properties and causes brittleness. Tempering, in turn, allows reducing stress and increasing the ductility of steel (its impact strength and yield point). As a result, a material with favorable strength and plastic properties is obtained [46].

Work [47] describes the surface hardening applied in the manufacture of cylinders. In [48], authors mention piston rods with a surface modified in this way. This treatment is used in cases where the change of parameters only on the steel surface is needed. Such a procedure prevents excessive stress and thermal deformations [18]. Surface hardening of piston rods is often carried out using induction. Studies [49, 50] describe the results of research that allow us to state that induction hardening significantly improves compressive strength and buckling resistance.

Among the thermochemical treatment methods, nitriding, mainly used for piston rods [30, 51], cannot be omitted. Nitrided piston rods, in addition to high corrosion resistance, are also resistant to abrasive wear [51].

Currently the most popular technique for coating piston rods is chromium plating [35, 45, 50, 52, 53]. These coatings are applied to improve tribological, anti-wear and anti-corrosive properties [54]. However, chrome coatings have disadvantages. Microcracks may appear on the chromed surface, which accelerate the corrosion process. Therefore, a nickel layer is often applied under the chromium layer, which provides additional corrosion protection [50].

Article [55] describes several reasons to replace hard chrome as a galvanic coating for hydraulic system components. As the most important argument, the author indicates the harmful influence of chromium on the health of employees working in the manufacture of the coating - they are more likely to suffer from lung cancer, asthma or damage to the epithelium of the nose and skin. In addition, commonly used hexavalent chromium is harmful to the environment.

An alternative to electrolytic chroming can be high velocity oxygen fuel coatings (HVOF) [55]. This way allows applying, among others, layers with tungsten carbide and cobalt [55, 56]. Such coatings are several times more resistant to wear. They provide better corrosion protection than chrome and increase the impact resistance of elements. Unfortunately, their application is even twice as expensive as standard galvanizing. In addition, it is difficult to maintain such smooth surface as in the case of chromium [55]. Atmospheric plasma spraying (APS) can be an alternative to HVOF. This method can be used to produce metal, ceramic or polymer layers, as well as combine different groups of materials in one coating. An example of APS coating in power hydraulics is the Cr_2O_3 layer sprayed on piston rods.

WC/CoCr and Cr_3C_2/NiCr [52] can also be used as coatings for actuator components. Chromium and nickel significantly increase corrosion resistance, while their abrasive properties are similar to hard chromium. The disadvantage is the high price. An example of a cheaper solution may be the application of thermal sprayed TiC coatings [52]. This process for the production of coatings on piston rods is also indicated in the article [45].

In [35], it is proposed to deposit protective layers of melted cored wires (CW) on the piston rod surfaces. This solution is much cheaper than galvanic chromium plating, unfortunately it has disadvantages. If the chromium content in the molten layer exceeds 13%, there is a decrease in local corrosion resistance. In order to solve this problem, in the aforementioned work research on the anti-corrosion properties of several coatings with different chemical composition and structure was discussed. They all contained over 14% chromium. The results showed that the layers with a ferritic structure had the least tendency to form chromium carbides, which negatively affect the anti-corrosion properties.

Diamond-like carbon (DLC) amorphous carbon coatings are often used on piston rods [30]. In [57], an actuator was tested, in which the mentioned material was also used on the cylinder and piston. DLC coatings have very good tribological and anti-wear properties. They are resistant to high temperatures and aggressive environments [58].

In article [32], two coatings that can be applied in hydraulic systems were compared: TiN and DLC. Both are chemically inert and have good mechanical strength. TiN has a slightly better abrasion resistance, however, DLC has better tribological parameters. The research carried out in the mentioned work showed that DLC has more favorable corrosion potential and hence better corrosion resistance. Coatings can also be used to regenerate actuators components. In [59] a nanocomposite n-Al_2O_3/Ni-Co coating was used to repair local cracks on the piston rod. Another example of recreating the surface of a piston rod is presented in [26]. The possibility of powder plasma transferred arc welding (PPTAW) of Fe-C-Cr-V coatings was discussed there. Unfortunately, for small diameters this technology could create a high risk of deformation. As an alternative, plasma powder coating for rods using corrosion-resistant 315Kh19F3 chrome vanadium cast iron was mentioned.

In [39], a solution was proposed in the way of applying a nanocomposite coating to the inner surface of the composite cylinder, in which the matrix is epoxy resin and the reinforcement is Al_2O_3 and SiO2 nanoparticles. These are materials with very high hardness, and thus abrasion resistance.

The article [60] describes the use of a nanostructured tungsten carbide coating applied by means of a chemical vapor deposition (CVD) process. Such a coating has better abrasion and corrosion resistance than hard chrome. In addition, its properties are better than those of thermal sprayed WC/Co coatings.

6 Conclusion

Parts of hydraulic actuators, such as pistons, piston rods and cylinders, often have to work in difficult conditions, and any failures can cause serious consequences. That is why choosing the appropriate material, that meets all the requirements and goals, is so important.

Currently, actuators made of materials other than steel are being tested more and more often - this is due to the tendency to reduce the impact of production processes on the natural environment and reduce manufacturing costs.

Composite and polymeric materials are often still in the testing stage, but they can be a good alternative to heavy ferrous materials.

Important issues in the design process also include the modification of the component surface, if it is reasonable taking into account operating conditions and costs. Due to the need to replace harmful galvanic chromium plating, completely new coating methods are being developed and implemented to achieve similar or better material properties for surface layers with simultaneous care for the application ecology.

Acknowledgement. The first and second authors would like to thank for the financial support from the project co – financed by the European Union under the European Social Found as a part of the Operational Program: Knowledge, Education, Development, project No. POWR.03.02.00 – 00 – I042/16 – 00.

References

1. Stryczek, S.: Napęd hydrostatyczny, 4th edn, vol. 1. Wydawnictwa Naukowo-Techniczne, Warszawa (2003)
2. Bauer, W.: Hydraulic components design. In: Hydropneumatic Suspension Systems, 4th edn, pp. 95–140. Springer, Heidelberg (2011). https://doi.org/10.1007/978-3-642-15147-7_4
3. Uzny, S., Kutrowski, Ł.: Obciążalność rozsuniętego teleskopowego siłownika hydraulicznego przy uwzględnieniu wyboczenia oraz wytężenia materiału. Modelowanie inżynierskie **37**(68), 125–131 (2018)
4. Ashby, M.F.: The design process. In: Materials Selection in Mechanical Design, 4th edn, pp. 15–29. Butterworth-Heinemann, Oxford (2011). https://doi.org/10.1016/b978-1-85617-663-7.00002-3
5. Uzny, S., Kutrowski, Ł.: Strength analysis of a telescopic hydraulic cylinder elastically mounted on both ends. J. Appl. Math. Comput. Mech. **18**(1), 89–96 (2019). https://doi.org/10.17512/jamcm.2019.1.08

6. Muraki, M., Kinbara, E., Konishi, T.: A laboratory simulation for stick-slip phenomena on the hydraulic cylinder of a construction machine. Tribol. Int. **36**(10), 739–744 (2003). https://doi.org/10.1016/s0301-679x(03)00054-9
7. Osiecki, A.: Hydrostatyczny napęd maszyn, 2nd edn. Wydawnictwa Naukowo-Techniczne, Warszawa (2004)
8. Kijewska, A., Bluszcz, A.: Analiza poziomów śladu węglowego dla świata i krajów UE. Systemy Wspomagania w Inżynierii Produkcji **6**(2), 169–177 (2017)
9. 2030 climate and energy goals for a competitive, secure and low-carbon EU economy. European Commission Press Release, Brussels, 22 January 2014. http://ec.europa.eu/commission/presscorner/detail/en/IP_14_54. Accessed 18 Feb 2020
10. Harvey, F.: New technology could slash carbon emissions from aluminium production. The Guardian article, 10 May 2018. http://www.theguardian.com/environment/2018/may/10/new-technology-slash-aluminium-production-carbon-emissions. Accessed 18 Feb 2020
11. Dašić, P., Manđuka, A., Pantić, R.: Research of optimal parameters of machining big hydraulic cylinders from the aspect of quality. Ann. Oradea Univ. Fascicle Manag. Technol. Eng. **VII**(XVII), 1563–1571 (2008)
12. Jakubczak, H., Rojek, J.: Zmęczeniowe pękanie siłowników hydraulicznych. Diagnostyka **36**, 61–66 (2005)
13. Kawiak, M.: Spawanie tłoczyska siłowników hydraulicznych. Przegląd Spawalniczy **85**(2), 6–10 (2013)
14. Chalamoński, M.: Równomierność ruchu tłoka siłownika hydraulicznego. Diagnostyka **30**, 97–100 (2004)
15. Bohman, E.: Understanding buckling strength of hydraulic cylinders. The Hydraulics & Pneumatics article, 9 March 2017. http://www.hydraulicspneumatics.com/technologies/cylinders-actuators/article/21887243/understanding-buckling-strength-of-hydraulic-cylinders. Accessed 18 Feb 2020
16. Kowalski, K., Złoto, T.: Exploitation and repair of hydraulic cylinders used in mobile machinery. TEKA Comm. Motorization Energetics Agric. **14**(4), 53–58 (2014)
17. Ashby, M.F., Jones, D.R.H.: Light alloys. In: Engineering Materials 2: An Introduction to Microstructures and Processing, 4nd edn, pp. 205–220. Butterworth-Heinemann, Oxford (2012). https://doi.org/10.1016/B978-0-08-096668-7.00011-5
18. Dobrzański, L.A.: Stale i inne stopy żelaza. In: Podstawy nauki o materiałach i metaloznawstwo, 2nd edn, pp. 653–675. Wydawnictwa Naukowo-Techniczne, Warszawa (2006)
19. Pawłowski, W., Kępczak, N.: Teoretyczne badania właściwości dynamicznych łóż obrabiarki wykonanych z żeliwa i hybrydowego połączenia żeliwa z odlewem mineralnym. Mechanik **8**(9), 199–203 (2015)
20. Marczewska, I., Bednarek, T., Marczewski, A., Sosnowski, W., Jakubczak, H., Rojek, J.: Practical fatigue analysis of hydraulic cylinders and some design recommendations. Int. J. Fatigue **28**(12), 1739–1751 (2006). https://doi.org/10.1016/j.ijfatigue.2006.01.003
21. Bednarek, T., Sosnowski, W.: Practical fatigue analysis of hydraulic cylinders – part II, damage mechanics approach. Int. J. Fatigue **32**(10), 1591–1599 (2010). https://doi.org/10.1016/j.ijfatigue.2010.02.013
22. Nicoletto, G., Marin, T.: Failure of a heavy-duty hydraulic cylinder and its fatigue re-design. Eng. Fail. Anal. **18**(3), 1030–1036 (2011). https://doi.org/10.1016/j.engfailanal.2010.12.019
23. Mantovani, S.: Feasibility analysis of a double-acting composite cylinder in high-pressure loading conditions for fluid power applications. Appl. Sci. **10**(3), 826 (2020). https://doi.org/10.3390/app10030826
24. Solazzi, L.: Feasibility study of hydraulic cylinder subject to high pressure made of aluminum alloy and composite material. Compos. Struct. **209**, 739–746 (2019). https://doi.org/10.1016/j.compstruct.2018.11.021

25. Solazzi, L., Buffoli, A.: Telescopic hydraulic cylinder made of composite material. Appl. Compos. Mater. **26**(4), 1189–1206 (2019). https://doi.org/10.1007/s10443-019-09772-8
26. Nefed'ev, S.P., Dema, R.R., Kharchenko, M.V., Pelymskaya, I.S., Romanenko, D.N., Zhuravlev, G.M.: Experience in restoring hydraulic cylinder rods by plasma powder surfacing. Chem. Pet. Eng. **52**(11–12), 785–789 (2017). https://doi.org/10.1007/s10556-017-0271-4
27. Luo, P., Hu, J., Tan, S.: Design and realization of hydraulic cylinder. Region Water Conservancy **1**(1), 27–34 (2018)
28. Przybylski, W.: Zastosowanie obróbki nagniataniem w technologii siłowników hydraulicznych. Postępy Nauki i Techniki **6**, 196–201 (2011)
29. Boye, T., Adeyemi, O., Emagbetere, E.: Design and finite element analysis of double - acting, double - ends hydraulic cylinder for industrial automation application. Am. J. Eng. Res. (AJER) **6**, 131–138 (2017)
30. Tonelli, L., Martini, C., Ceschini, L.: Improvement of wear resistance of components for hydraulic actuators: dry sliding tests for coating selection and bench tests for final assessment. Tribol. Int. **115**, 154–164 (2017). https://doi.org/10.1016/j.triboint.2017.05.021
31. Tavares, S.M.O., Viriato, N., Vaz, M., de Castro, P.M.S.T.: Failure analysis of the rod of a hydraulic cylinder. Procedia Struct. Integrity **1**, 173–180 (2016). https://doi.org/10.1016/j.prostr.2016.02.024
32. Dalibón, E.L., Pecina, J.N., Moscatelli, M.N., Ramírez Ramos, M.A., Trava-Airoldi, V.J., Brühl, S.P.: Mechanical and corrosion behaviour of DLC and TiN coatings deposited on martensitic stainless steel. J. Bio Tribo-Corros. **5**(2), 1–9 (2019). https://doi.org/10.1007/s40735-019-0228-6
33. Moreira, D.C., Furtado, H.C., Buarque, J.S., Cardoso, B.R., Merlin, B., Moreira, D.D.C.: Failure analysis of AISI 410 stainless-steel piston rod in spillway floodgate. Eng. Fail. Anal. **97**, 506–517 (2019). https://doi.org/10.1016/j.engfailanal.2019.01.035
34. Sevagin, S.V., Mnatsakanyan, V.U.: Ensuring the required manufacturing quality of hydraulic-cylinder rods in mining machines. In: IOP Conference Series: Materials Science and Engineering, vol. 709, p. 044095 (2020). https://doi.org/10.1088/1757-899x/709/4/044095
35. Holyakevych, A.A., Orlov, L.M., Pokhmurs'ka, H.V., Student, M.M., Chervins'ka, N.R., Khyl'ko, O.V.: Influence of the phase composition of the layers deposited on the rods of hydraulic cylinders on their local corrosion. Mater. Sci. **50**(5), 740–747 (2015). https://doi.org/10.1007/s11003-015-9780-5
36. Otsima, M.: Material selection process for hydraulic cylinder. In: EAT309 Mechanical Design – Part 5 Materials Review And Selection (2015)
37. O'Rourke, B.: Pressure ratings and design guidelines for ductile iron manifolds. In: 2014 IFPE Technical Conference: Where all the Solutions Come Together and Connections are Made. NFPA, Las Vegas (2014)
38. Solazzi, L.: Design and experimental tests on hydraulic actuator made of composite material. Compos. Struct. **232**, 111544 (2020). https://doi.org/10.1016/j.compstruct.2019.111544
39. Lubecki, M.: Selected design issues in hydraulic cylinder made of composite materials. In: Badania i Rozwój Młodych Naukowców w Polsce: Nauki techniczne i inżynieryjne: Materiały, Polimery, Kompozyty, Młodzi Naukowcy, Poznań (2019)
40. Liu, Y., Zwingmann, B., Schlaich, M.: Carbon fiber reinforced polymer for cable structures—a review. Polymers **7**(10), 2078–2099 (2015). https://doi.org/10.3390/polym7101501
41. Scholz, S., Kroll, L.: Nanocomposite glide surfaces for FRP hydraulic cylinders – evaluation and test. Compos. Part B: Eng. **61**, 207–213 (2014). https://doi.org/10.1016/j.compositesb.2014.01.044
42. Stryczek, P., Przystupa, F., Banaś, M.: Research on series of hydraulic cylinders made of plastics. In: Global Fluid Power Society PhD Symposium (GFPS). IEEE, Samara (2018). https://doi.org/10.1109/gfps.2018.8472385

43. Tran, X.B., Hafizah, N., Yanada, H.: Modeling of dynamic friction behaviors of hydraulic cylinders. Mechatronics **22**(1), 65–75 (2012). https://doi.org/10.1016/j.mechatronics.2011.11.009
44. Harnisch, M.: Kunststoffe n fluidtechnischen Antrieben Oelhydraulik und Pneumatik, pp. 14–16 (2013)
45. Tuominen, J., Näkki, J., Pajukoski, H., Miettinen, J., Peltola, T., Vuoristo, P.: Wear and corrosion resistant laser coatings for hydraulic piston rods. J. Laser Appl. **27**(2), 022009 (2015). https://doi.org/10.2351/1.4914503
46. Deng, X., Ju, D.: Modeling and simulation of quenching and tempering process in steels. Phys. Procedia **50**, 368–374 (2013). https://doi.org/10.1016/j.phpro.2013.11.057
47. Denisov, L.V., Boitsov, A.G., Siluyanova, M.V.: Surface hardening in hydraulic cylinders for airplane engines. Russ. Eng. Res. **38**(12), 1080–1083 (2018). https://doi.org/10.3103/s1068798x18120237
48. Dai, L.-Y., Lin, S.-F., Yang, S.-Z., Pan, G.-F., Guo, N., Dai, L.-L.: Cracking cause analysis of 45 steel piston rod. Heat Treat. Metals **36**(2), 119–121 (2011)
49. Awad, M., Hultgren, J., Roberts, W.: Increased resistance to buckling of piston rods through induction hardening, OVAKO Article, Stockholm (2018). http://www.ovako.com/globalassets/products/hard-chromed/ih_buckling_wp.pdf. Accessed 18 Feb 2020
50. Israelsson, P.: Better steels make better cylinders. The Hydraulics & Pneumatics article, 2 September 2016. https://www.hydraulicspneumatics.com/technologies/cylinders-actuators/article/21885260/better-steels-make-better-cylinders. Accessed 18 Feb 2020
51. Wach, P., Michalski, J., Tacikowski, J., Kowalski, S., Betiuk, M.: Gazowe azotowanie i jego odmiany w przemysłowych zastosowaniach. Inżynieria Materiałowa **29**(6), 808–811 (2008)
52. Bobzin, K., Öte, M., Linke, T.F., Malik, K.M.: Wear and corrosion resistance of fe-based coatings reinforced by TiC particles for application in hydraulic systems. J. Therm. Spray Technol. **25**(1–2), 365–374 (2015). https://doi.org/10.1007/s11666-015-0316-1
53. Gayathri, N., Karthick, N., Shanmuganathan, V.K., Adhithyan, T.R., Madhan Kumar, T., Gopalakrishnan, J.: Productivity improvement and cost reduction in hydraulic cylinders. Int. J. Eng. Technol. **7**(3.34), 382 (2018). https://doi.org/10.14419/ijet.v7i3.34.19232
54. Aizhambaeva, S.Z., Maximova, A.V.: Development of control system of coating of rod hydraulic cylinders. In: IOP Conference Series: Materials Science and Engineering, vol. 289, p. 012020 (2018). https://doi.org/10.1088/1757-899x/289/1/012020
55. Flitney, B.: Alternatives to chrome for hydraulic actuators. Seal. Technol. **2007**(10), 8–12 (2007). https://doi.org/10.1016/s1350-4789(07)70460-9
56. Kuzmin, V., Gulyaev, I., Sergachev, D., Vashchenko, S., Kovalev, O., Kornienko, E., Palagushkin, B.: Supersonic DC plasma torch for deposition of high-density wear-resistant coatings. Mater. Today Proc. **19**, 2152–2156 (2019). https://doi.org/10.1016/j.matpr.2019.07.230
57. Walczak, P.: Analiza modelu matematycznego układu sterowania kierownicą turbiny wodnej małej mocy. Logistyka **6**, 10823–10831 (2014)
58. Madej, M., Ozimina, D., Pająk, M.: Właściwości powłok węglowych uzyskiwanych w procesach fizycznego osadzania z fazy gazowej. Mechanik **88**(4), 151–156 (2015). https://doi.org/10.17814/mechanik.2015.4.190
59. Wang, H., Ma, G., Xu, B., Yong, Q., He, P.: Design and application of friction pair surface modification coating for remanufacturing. Friction **5**(3), 351–360 (2017). https://doi.org/10.1007/s40544-017-0185-3
60. Zhuk, Y.: Nanostructured CVD tungsten carbide coating on aircraft actuators and gearbox shafts reduces oil leakage and improves durability. J. Mater. Eng. Perform. **28**(4), 1914–1923 (2019). https://doi.org/10.1007/s11665-019-03936-0

Plastics a New Trend in Design of Fluid Power Elements and Systems

Jarosław Stryczek(✉)

Wrocław University of Science and Technology, Wybrzeże, Wyspiańskiego 27, 53-370 Wrocław, Poland
jaroslaw.stryczek@pwr.edu.pl

Abstract. The application of plastics instead of traditional metal materials is a new development trend in the field of fluid power. The Fluid Power Research Group (FPRG) from Wroclaw University of Science and Technology conduct research in this field. In previous works [6, 7, 11, 16–18], which were published on that topic, design solutions where shown, as well as research of models of typical hydraulic elements like gerotors, valves and cylinders, which were made of plastics. on the problem of designing, manufacturing and testing of fluid power elements made of plastics. The aim of this paper is to present the development of the works on hydraulic elements made of plastics. The main novelties are: a gerotor pump with a plastic body, research on global deformation of plastic cylinder, visualization and computer simulations of the valve poppet/seal assembly in plastic hydraulic valves, as well as innovative hydraulic systems.

Keywords: Fluid power · Elements and systems from plastic

1 Introduction

The application of plastics instead of traditional metal materials is a new trend in the field of fluid power. In the scientific literature, this issue is relatively rarely discussed and, generally, in a rather fragmentary manner. Typically, only selected hydraulic components and the problems related to their design and research are described. For example, paper [1] presents the design and research into a gerotor mini-pump featuring gears made of POM. Work [2], however, describes an oil pump for a motorcycle engine, made of PPA and PTFE. The materials were used to make the shaft, gears and the housing of the pump. In work [3], a design solution of a gear mini-pump with a plastic housing is presented, in which bearing seats were directly formed, thus eliminating metal bearing sleeves. A plastic was also used to build valve blocks [4]. Thanks to the application of plastics and injection molding technology, it was possible to make a complex arrangement of internal channels.

Paper [5] presents a possibility of building hydraulic cylinders made of plastics and composites for various industries such as machine industry, automation and robotics or chemical industry. Technical information on the subject of hydraulic elements made of

© The Editor(s) (if applicable) and The Author(s), under exclusive license
to Springer Nature Switzerland AG 2021
J. Stryczek and U. Warzyńska (Eds.): NSHP 2020, LNME, pp. 440–452, 2021.
https://doi.org/10.1007/978-3-030-59509-8_40

plastics is also presented in catalogues of products of numerous companies, where both the designs and their utility characteristics are described. For instance, [22] presents a plastic ball valve and a quick-hitch and [23], plastic cylinders.

In this context, it is noted that the literature lacks a systematic approach to the issue of theoretical foundations, design, manufacture and testing of hydraulic components made of plastics. The Fluid Power Research Group from Wrocław University of Science and Technology [6, 7, 11, 17, 18] have been conducting such a systemic research for the last several years. This activity is to effect in the obtaining of a set of basic hydraulic components made of plastics such as a gerotor pump, a relief valve, an on/off valve and a hydraulic cylinder. It is assumed that with the use of those elements, it will be possible to build simple hydraulic systems working in a hydraulic circuit within the pressure range of $p = 6.3$ MPa.

The purpose of this paper is to present the state of work on the issue of the design and production of models and tests of hydraulic elements and systems made of plastics conducted by the FPRG group, especially in relation to the previous published paper [6].

2 Plastics Selection

The selection of plastics is the basic stage in the process of designing hydraulic components. The following material selection criteria should be taken into account:

- high strength characterized by yield strength Re and Young's modulus;
- resistance to high temperature T;
- dimensional stability characterized by shrinkage S and low linear elongation W resulting from the injection molding process;
- low water absorption A occurring as a result of the material reacting with the working medium;
- low price and market availability;
- the workability by machining and injection molding.

According to [6] three basic types of materials were selected for the construction of the hydraulic elements, namely POM, PPS and PEEK, whose technical parameters are presented in Table 1.

The table also shows the comparison of values of selected technical parameters for typical structural steel. The table shows that the strength of plastics is clearly lower than that of steel. This suggests that hydraulic components made of plastics cannot be mechanically and hydraulically loaded as much as metal components. To verify the R_e yield stress values given in the material catalogues, the stress-strain analysis was carried out in [6, 11] for the basic POM material (Fig. 1). The figure shows that there is no clear yield point for POM so it can be assumed in the range of $Re = 60 \div 80$ MPa.

Thus, the catalogue values in Table 1 are confirmed. Therefore, POM was treated as a basic material characterized by sufficient strength values enabling machining, as well as cheap and commercially available.

It was also assumed that PPS and PEEK, which feature higher strength, will also be applied. PEEK, however, is much more expensive, which will increase production costs of hydraulic elements.

Table 1. Comparison of technical parameters of plastic and structural steel [6].

L.p.	Technical parameter	Symbol	Unit	POM	PPS	PEEK	Steel
1	Yield strength	*Re*	MPa	60	180	230	300
2	Young's modulus	*E*	MPa	3000	16000	12700	$2{,}1 \times 10^6$
3	High temperature	*T*	°C	100	160	240	300
4	Shrinkage	*S*	cm/cm	0,0285	0,005	0,005	–
5	Linear elongation	*W*	cm/K	0,0001	0,00005	0,00004	–
6	Water absorption	*A*	%	0,7	0,02	0,4	–

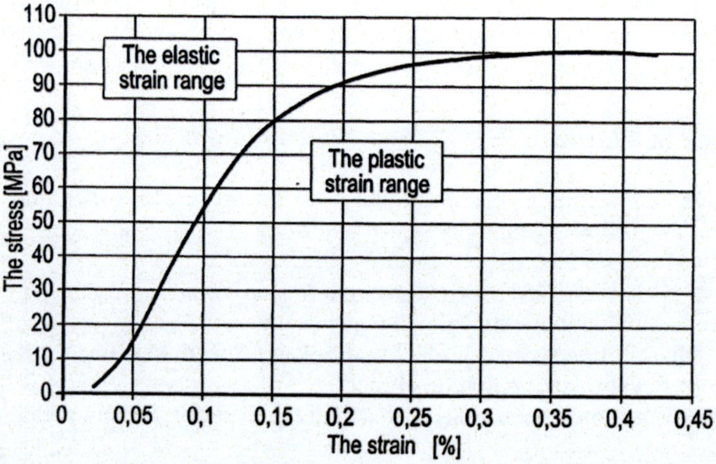

Fig. 1. Stress – strain diagram of POM created on the basis of own research at the Laboratory of Mechanical Engineering Department of Wrocław University of Science and Technology [6, 11]

3 Plastic Rotors, Plastic Gerotor Pump

Research into the possibility of applying plastics in the field of fluid power began with the application of plastics for a gerotor assembly constituting a critical system of gerotor pumps. It was decided to use rotors and a gerotor pump because it is a type of pump which is both simple in its design and efficient in its operation. A pump with an aluminum alloy housing and a steel shaft illustrated in Fig. 2a, b, was selected according to [7]. The pump features, successively, gear assemblies made of POM, PPS, PEEK (see Table 1).

The geometry and kinematics of the gear assembly was developed on the basis of the author's design method given in [8, 9]. All the gear systems featured the same type of internal epicycloidal gearing and the number of teeth $z_1/z_2 = 6/7$. An example gear system is shown in Fig. 2b. The teeth of both rotors are closely fitted so that the intertooth radial clearance is close to zero, $h_r \approx 0$. At the same time, the gear assemblies are fitted to the housing of the pump so as to ensure a minimum axial (side) clearance of $h_a = 0.05$

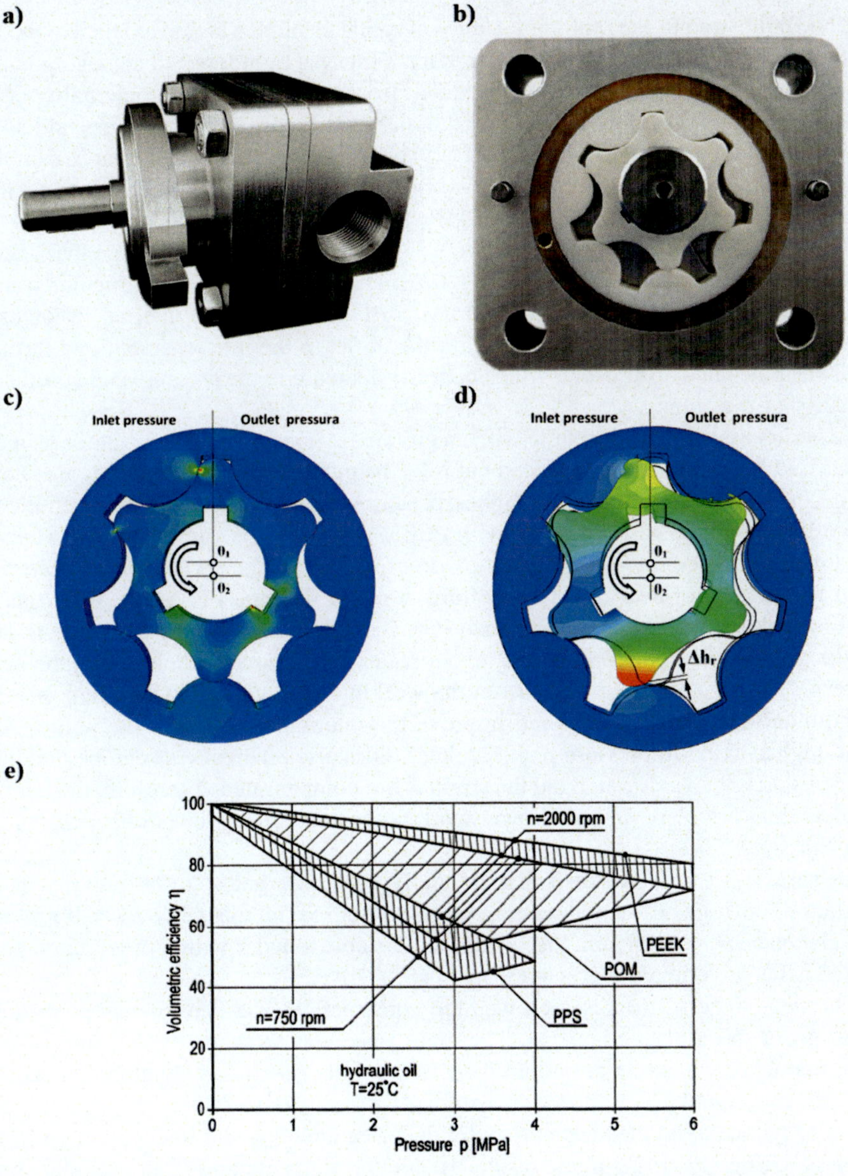

Fig. 2. Aluminium gerotor pump featuring POM gears: a) general view, b) gerotor assembly made of POM, c) stresses in the gear system determined by means of FEM, d) strains in the gear system determined by means of FEM, e) volumetric efficiency of the pump with gears made of various materials [6, 7, 10, 11].

allowing the gears to work with the housing, and simultaneously ensuring the necessary tightness inside the pump.

The application of the minimum values of radial clearance hr and axial clearance ha is a fundamental condition ensuring the pump's internal tightness and achieving higher operating pressures p and efficiency η_r and η. In order to verify the functionality of the rotors made of POM in predictable operational conditions, the FEM stress and strain analysis was conducted on them [10, 11]. The stresses in the tested gear assemblies shown in Fig. 2c did not exceed the allowable values equal to the yield point specified in Table 1, i.e. $\sigma = 21$ MPa < $Re = 60$ MPa.

The strains of the gear system shown in Fig. 2d, however, were so large that uncontrolled intertooth radial clearances $\Delta h_r \approx 0.2$ mm appeared. They are particularly unfavorable for the position of the gears shown in Fig. 2d, because internal leakages in the pump and uncontrolled backflow from the outlet to the inlet chamber and through radial clearance hr + Δhr occur, which causes a decrease in the pressure and volumetric efficiency of the pump.

Experimental tests of the pump with an aluminum housing and other rotor sets made of POM, PPS and PEEK were carried out [6]. The pump performance curves are shown in Fig. 2e. The figure shows that the highest pressure $p = 6$ MPa and volumetric efficiency $\eta_r = 80\%$ was obtained for a pump with PEEK gears. This is due to the high strength and dimensional stability of the gears made of PEEK. In the second place, there is a pump with the POM gears, and in the third, a pump with the PPS gears, even though POM is a material of lower strength than PPS (see: Table 1). The reason why it is so is that the gears made of POM feature a higher water absorption coefficient than the gears made of PPS. By absorbing water from the working medium, the POM gears increase their dimensions (they 'swell') and improve the internal tightness of the pump. As a result, higher working pressure p and higher volumetric efficiency η_r can be obtained. PPS gears do not absorb water and the strain is not compensated. Therefore, the internal tightness does not increase and a lower working pressure p and lower efficiency η_r are obtained.

To sum up, it can be stated that by applying PEEK gears in the pump with an aluminum housing and by maintaining the intertooth radial clearance $h_r \approx 0$ mm and axial clearance $h_a \leq 0.05$ mm, the system will be able to work within pressure range of $p = 6.3$ MPa, at volumetric efficiency $\eta_r = 60 \div 70\%$.

The next stage of the research was the construction of a gerotor pump featuring plastic gears and a plastic housing. The main issue was to shape the housing in such a way that its elastic strain are minimized both axially (a–a), namely along the axis of the shaft, and radially (r–r), that is, perpendicular to the axis of the shaft (see: Fig. 3c). This is to reduce axial clearance ha and radial clearance h_r, and to ensure high pump tightness. Using [12, 13], a pump of $q = 10$ cm^3/rev capacity and of an irregular shape was designed (see: Fig. 3a, b).

Considering the axial direction (a–a), the front and rear housings were strengthened and the number of screws connecting the housings and clamping them together was increased to six. Considering the radial direction (r–r), the central and rear housings were strengthened by thickening them from the outlet (high pressure) side. The FEM stress and strain analysis of the pump's POM housing [14] was conducted. The stresses in the housing did not exceed the allowable values specified for POM in Table 1, i.e.$\sigma = 53$ MPa < $Re = 60$ MPa. Example strains of the pump's housing are presented in

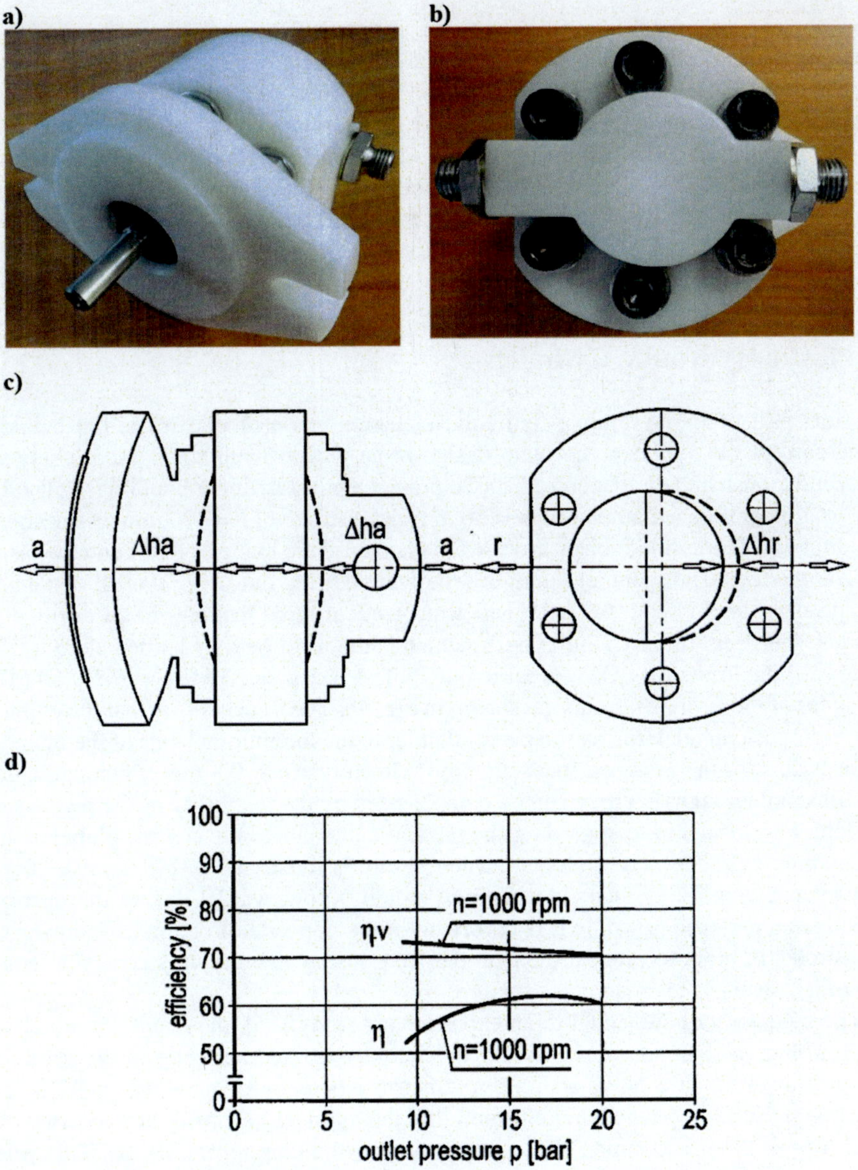

Fig. 3. Gerotor pump featuring plastic gears and housing: a) general view, b) rear housing view, c) strains diagram of the pump housing, d) pump volume and total efficiency.

Fig. 3c. The figure shows that the largest axial strains Δh_a occurred on the plane of contact between the central and front as well as the central and rear housings and amounted to about $\Delta h_a = 0.12$ mm. However, the largest radial strains Δh_r are observed in the central housing on the output side O and were $\Delta h_r = 0.02$ mm. The pump with the gears made of PEEK and the housing made of POM was subjected to experimental tests, the results of which are presented in Fig. 3d. The figure shows that the pump worked correctly at a rotational speed of $n = 1000$ rpm in the pressure range up to 2 MPa at volumetric efficiency $\eta_r \approx 70\%$ and total efficiency $\eta \approx 50 \div 60\%$. Higher rotational speeds and pressures could not be achieved because during the prolonged operation under load the temperature of the housing increases, which causes uncontrolled strain of the housing, disturbance of the pump tightness and the inability to achieve higher operating pressures and efficiency. The study shows that it is necessary to continue searching for a material of a higher strength and dimensional stability and to work on the improvement of its shape.

4 Plastic Hydraulic Cylinder

The next part of the system of hydraulic elements made of plastics is the cylinder. In the case of the cylinder, the main design issue was to maintain tightness between the cylinder bush and the piston rod and to protect against cylinder buckling. Following [15], a cylinder with a diameter of $\phi = 50$ mm and a stroke of $l = 400$ mm was designed (see: Fig. 4a). The cylinder bush, covers and piston rod are made of POM and the piston is made of steel. The cylinder strength tests were carried out using the FEM method. At a pressure load of $p = 6.3$ MPa and a mechanical force loading the piston of $F = 12.5$ kN, the stress in the cylinder bush, which is the most heavily loaded element, did not exceed the limits for POM, specified in Table 1, i.e. $\sigma = 23$ MPa $< Re = 60$ MPa. Examples of the cylinder strains are shown in Fig. 4b. The figure shows that the cylinder tube undergoes radial deformation perpendicular to the longitudinal axis of the cylinder. These strains for the assumed loads are approximately $\Delta r \approx 0.3$ mm. Practically, they are still small enough that they can be compensated by the flexibility of the piston seal. In addition to local strains regarding the bush, the piston and the covers, global strains of the entire cylinder structure occur. Their nature is illustrated in Fig. 4c. The figure shows that the entire cylinder is deformed radially (downwards) during the upstroke of the piston under load, while it is deformed in the opposite direction (up) during the downstroke. In order to prevent the cylinder from losing stability, it is critical to attach both of its covers to the machine structure.

The cylinder was subjected to an experimental study [16], the results of which are shown in Fig. 4d. The figure shows that the cylinder worked correctly in the operating pressure range up to 6 MPa, at high volumetric efficiency of $\eta_r = 96 \div 97\%$. The experimental studies have also confirmed that the cylinder can work at various piston speeds of $v = 20 \div 100$ mm/s. The cylinder also passed long-term tests of 1000 cycles. To sum up, it is stated that the POM plastic cylinder can work in the pressure range of $p = 6.3$ MPa.

5 Valves Made of Plastics

In order to complete the hydraulic system combining elements made of plastics, two types of valves are necessary: the maximum (relief) valve protecting the system against overload and the on/off valve being in fact a two-way two-position valve (2/2), which functions as a directional control valve. A simple valve design solution according to [17]

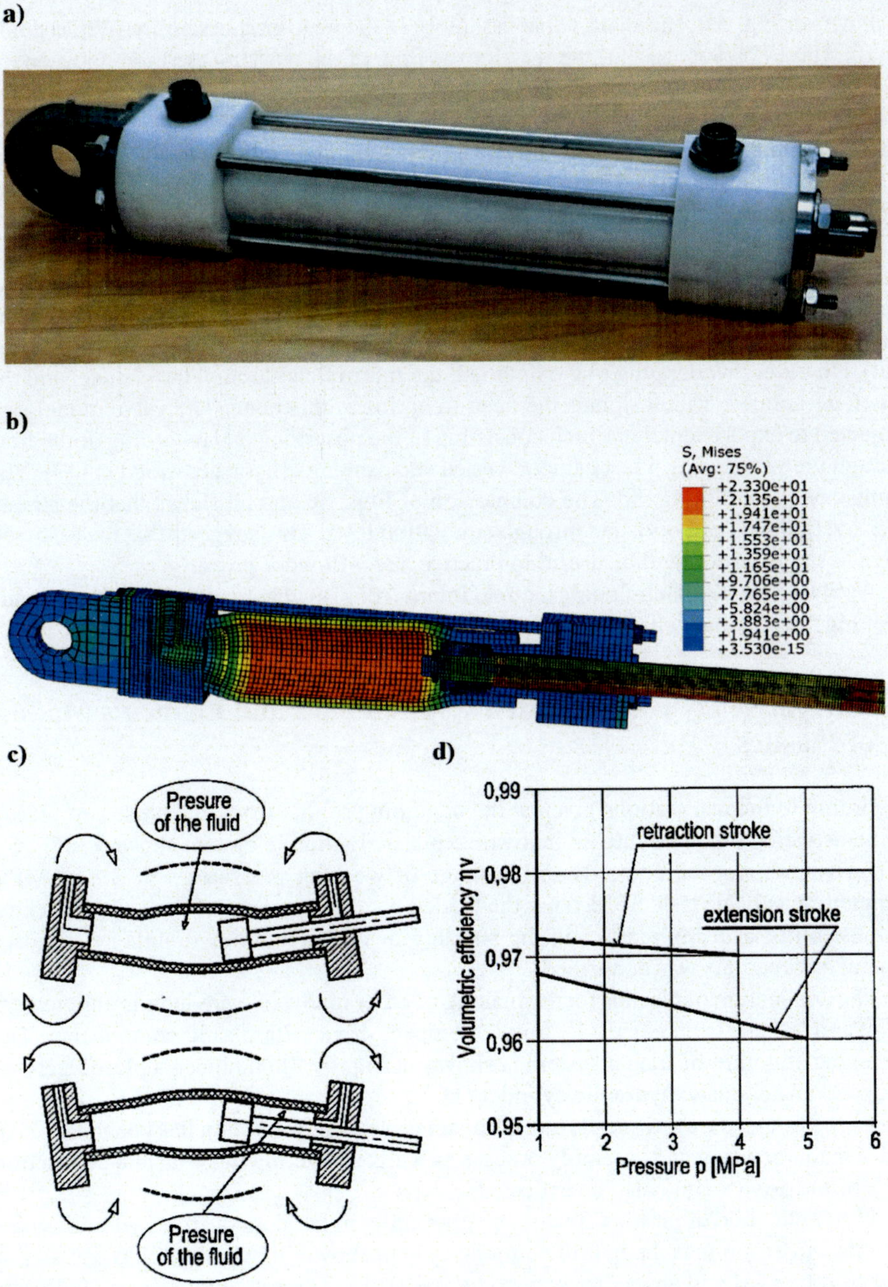

Fig. 4. Hydraulic cylinder made of plastics: a) general view, b) strains of the cylinder bush, c) global strains of the cylinder, d) volumetric efficiency of the cylinder.

is shown in Fig. 5a. The main valve assembly is the head-seat assembly. While going up, the valve head opens (on) the way for the flow of the working medium between the inlet (in) and the outlet (out), while going down (off), the way is closed. The main design issue is to properly shape this assembly so as to ensure tightness during closing (off) and uninterrupted flow, i.e. continuous flow, without cavitation between the inlet (in) and the outlet (out) during opening (on) at varying flow rate and pressure. A model valve featuring the housing made of technical glass (Plexiglas) and a cone-shaped head/cylindrical seat unit made of POM was designed and manufactured.

Simulation tests of the flow through the valve by CFD method were carried out and the results in a form of stream lines are shown in Fig. 5c. The figure shows that the stream lines run continuously and smoothly from the valve inlet (in) to the valve outlet (out). However, overlapping of the lines and the return flows around the channel and the outlet are noticed, which signals the need to optimize this node. The valve model was subjected to experimental research consisting in the observing and recording of the flow through the valve using a fast camera, which was, among others, described in [19]. The results are shown in Fig. 5d. The comparison of Figs. 5c and 5d shows that the stream lines overlap the lines obtained through simulation tests. This suggests that the assumed valve design solution will be useful in practice and will work properly.

Applying the research findings, a maximum (relief) valve made of POM was made (see: Fig. 5e). Its flow characteristics are shown in Fig. 5f.

6 Innovative Hydraulic Systems, Assemblies and Elements Made of Plastics

According to the reservations given at the beginning of this work, the presented plastic elements were combined into an innovative plastic hydraulic system shown in Fig. 6a, b. The system worked correctly in the range of working pressures $p = 2 \div 6$ MPa, completing its basic task, i.e. driving the loaded piston in and out of the cylinder. Thus, the possibility of using plastics in the building of low-pressure hydraulic systems for general applications was confirmed.

The application of plastics for the making of other innovative products is anticipated. An example here is an electro-hydraulic compact drive with plastic components. The schematic structure of such a system is shown in Fig. 6c. It combines a plastic gerotor pump (3) and a plastic hydraulic cylinder (1).

As the figure shows, a simple and compact drive design solution has been made. An easier manufacturing and assembly process as well as lowering material and production costs of the drive should also be expected.

The application of plastics creates an opportunity to build innovative hydraulic components. An example is the hybrid pumping system shown in Fig. 6d [20]. It consists of a set of rotors (3) and an electromagnetic drive (1, 2). The rotating magnetic field sets the gerotor system in the rotary motion, and the gerotor system performs the pumping. Thus, the electric motor and the clutch driving the pump are eliminated, which simplifies the entire system. The hybrid pumping system is powered and controlled remotely. It can also operate as a new type of electronically and remotely controlled pressure energy generator which can be implemented in various complex hydraulic systems [21].

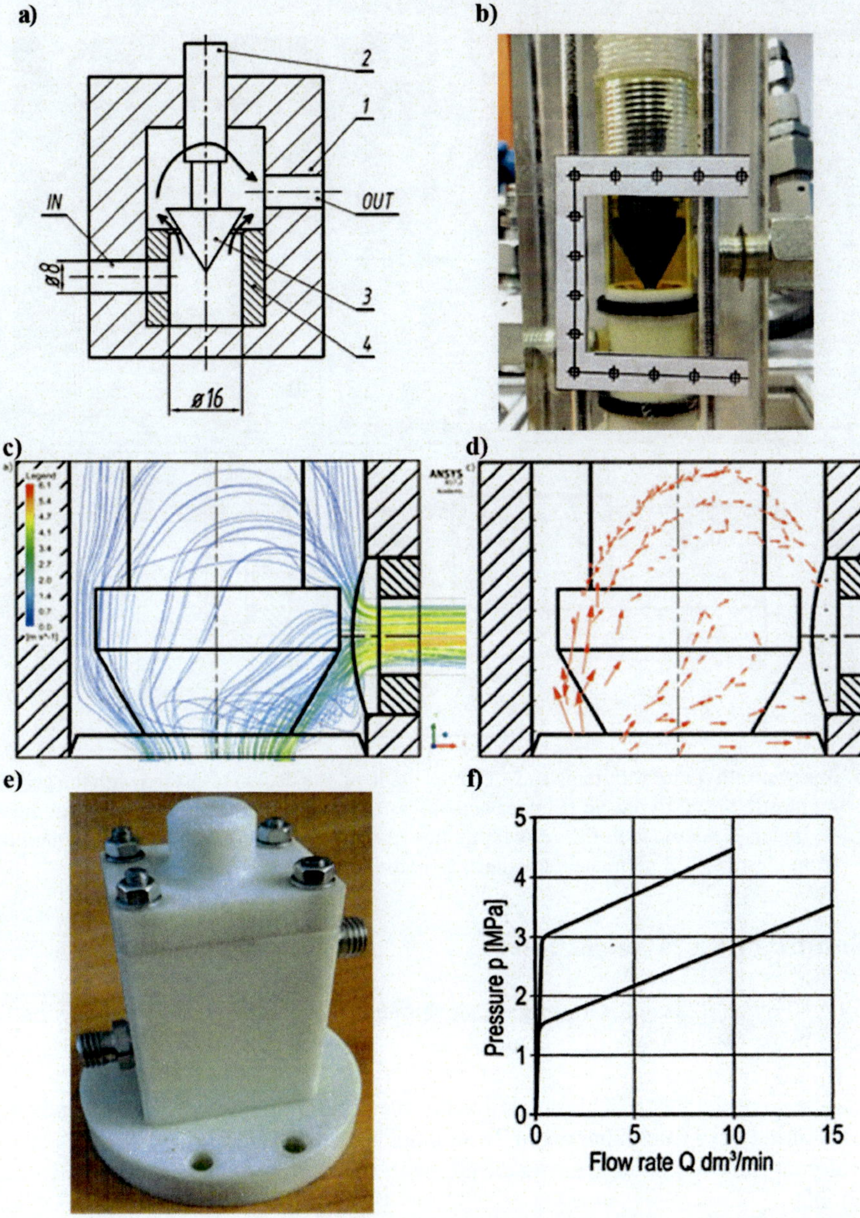

Fig. 5. 2/2 valve made of plastics: a) valve diagram, b) view of the valve made of Plexiglas c) streamlines determined by the CFD method, d) streamlines determined experimentally, e) view of the valve made of POM, f) valve flow characteristics.

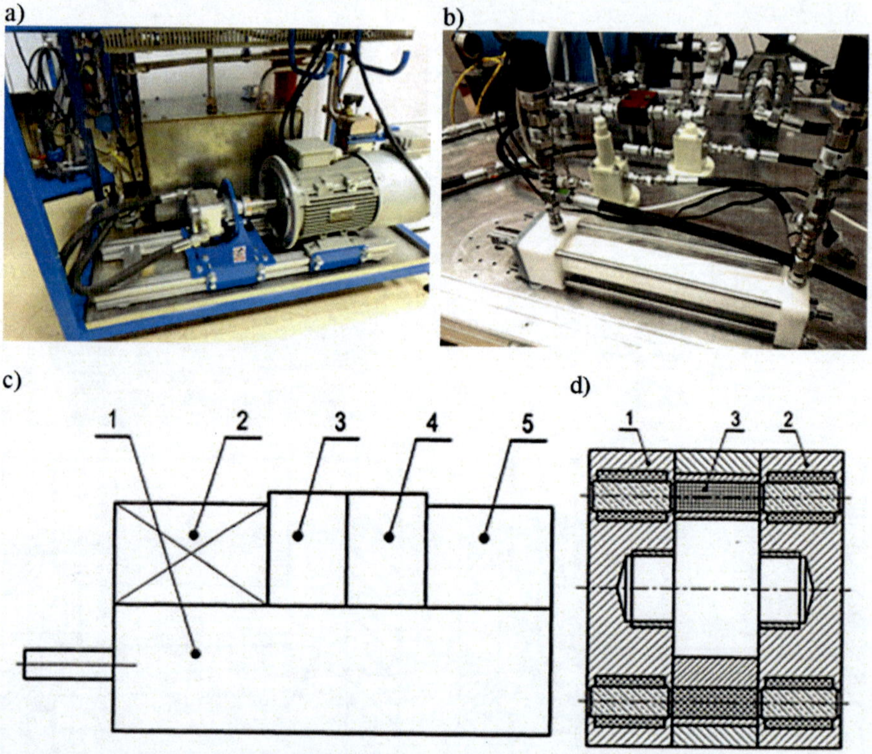

Fig. 6. Innovative hydraulic systems and elements made of plastics: a) test stand with the gerotor pump and plastic gears, b) plastic cylinder and valves of the stand, c) electrohydraulic compact drive: 1-cylinder, 2-electric motor, 3-gerotor pump, 4-manifold, 5-accumulator, d) hybrid pumping system from plastics: 1, 2-permanent magnets, 3-rotating magnet.

7 Summary and Conclusions

The Fluid Power Research Group have developed a method for designing hydraulic elements made of plastics, which involves:

- preparation of the theoretical foundations for designing the elements, which includes the selection of plastics, shaping and strain analysis using FEM, CFD flow tests;
- developing ways of manufacturing the elements by means of machining and injection methods;
- development of test stands and methods for experimental testing of the elements and systems.

That method was applied to design and manufacture a set of plastic hydraulic elements: a gerotor pump, a maximum (relief) valve, an on/off valve, a hydraulic cylinder. A general-purpose hydraulic system was built from those elements, which worked correctly in the pressure range of $p = 2 \div 6$ MPa.

The application of plastics in fluid power creates opportunities for the building of new innovative elements and systems such as the compact drive with plastic components or the hybrid pumping system.

References

1. Gamez-Montero, P.J., Antoniak, P., Castilla, R., Freire, J., Krawczyk, J., Stryczek, J., Codina, E.: Magnet-sleeve-sealed mini trochoidal-gear pomp prototype with polymer compozite gear. Energies (10) (2017)
2. Mancini, D.S., Hilario Cioffi, M.O., Bianchi, E.C.: Replecement of metallieparts for polimer composite materials in motorcucle oil pumps. J. Reinnforced Plast. Compos. (10) (2016)
3. Radionov, L., Rekadze, P., Stryczek, J.: A gear micropump without bearings production. In: Applied Mechanies and Materials, vol. 775 (2015)
4. Bonnano, A.: Materialli polimeriei applicaziani. Oleodynamica Pneumatica, no. 4 (2008)
5. Harnisch, M.: Kunsttoffe in fluidtechnischen Antrieben. Oelhydraulik und Pneumatik, No. 11–12 (2013)
6. Stryczek, J., Banaś, M., Krawczyk, J., Marciniak, L., Stryczek, P.: The fluid power elements and systems made of plastics. Procedia Eng. **176**, 600–609 (2017)
7. Stryczek, J., Bednarczyk, S., Biernacki, K.: Gerotor pump with POM gears: design, production, technology, research. Arch. Civ. Mech. Eng. **13** (2014)
8. Stryczek, J.: Projektieren der Zykloiden Verzahnungen hydraulischer Verdraengermaschinen. Mech. Mach. Theory J. **25**(6), 597–610 (1990)
9. Stryczek, J.: Koła zębate maszyn hydraulicznych (monography in polish) – Gears of hydraulic machines. Publishing House of Wrocław University od Science and Technology, Wrocław (2017)
10. Stryczek, J., Bednarczyk, S., Biernacki, K.: Strength analysis of the polyoxymethylene cycloidal gears of the gerotor pump. Arch. Civ. Mech. Eng. **14**, 647–660 (2014)
11. Biernacki, K., Stryczek, J.: Analysis of stress and deformation in plastic gears used in gerotor pumps. J. Strain Anal. (10) (2010)
12. Stryczek, J., Biernacki, K., Krawczyk, J.: Designing gear pump bodies using FEM. In: Proceedings of the 14th International Scientific Conference: Computer Aided Engineering. Springer (2019)
13. Biernacki, K., Stryczek, J.: Hydraulic gear machine. Utility model. Patent no 68305/2016, Poland (2016)
14. Krawczyk, J., Stryczek, J.: Design and experimental research of plastic gerotor pump. In: Procedings of Global Fluid Power Society Ph.D. Symposium, Samara (2018)
15. Stryczek, P.: Design and research on hydraulic cylinder with olastic components. In: Procedings ASME 9th Fluid Power Net International Ph.D. Symposium on Fluid Power, Florianopolis (2016)
16. Stryczek, P., Przystupa, F., Banaś, M.: Research on series of hydraulic cylinders made of plastics. In: Proceedings of Global Fluid Power Society Ph.D. Symposium, Samara (2018)
17. Marciniak, L., Banaś, M., Stryczek, J.: The design and theoretical and experimental study of plastic hydraulic valves. In: Proceedings of ASME, 9th Fluid Power Net International Ph.D. Symposium on Fluid Power, Florianopolis (2016)
18. Banaś, M., Antoniak, P., Marciniak, L., Stryczek, J.: Visualisation of flow phenomena in hydraulic throttle valves of plastic. In: Proceedings of the 14th International Conference on Vibration Engineering and Technology of Machinery (VETOMAC XIV), Lisbon (2018)
19. Antoniak, P., Stryczek, J.: Visualisation study of the flow processes and phenomena in the external gear pump. Arch. Civ. Mech. Eng. **18**, 1103–1115 (2018)

20. Antoniak, P., Stryczek, J.: Hybrydowy zespół pompujący. Zgłoszenie patentowe, p. 428702
21. Stryczek, J., Antoniak, P., Banaś, M.: An idea of the electronic control system for the multifunctional hydraulic machine. J. Vibroengineering (12) (2010)
22. http://www.saga.info.pl/armaturaztworzywsztucznych
23. Katalog firmy Parker Hannifin

Author Index

© The Editor(s) (if applicable) and The Author(s), under exclusive license
to Springer Nature Switzerland AG 2021

J. Stryczek and U. Warzyńska (Eds.): NSHP 2020, LNME, pp. 453–454, 2021.
https://doi.org/10.1007/978-3-030-59509-8